Introduction to Financial Models for Management and Planning

CHAPMAN & HALL/CRC FINANCE SERIES

Series Editor

Michael K. Ong

Stuart School of Business
Illinois State of Technology
Chicago, Illinois, U. S. A.

Aims and Scopes

As the vast field of finance continues to rapidly expand, it becomes increasingly important to present the latest research and applications to academics, practitioners, and students in the field.

An active and timely forum for both traditional and modern developments in the financial sector, this finance series aims to promote the whole spectrum of traditional and classic disciplines in banking and money, general finance and investments (economics, econometrics, corporate finance and valuation, treasury management, and asset and liability management), mergers and acquisitions, insurance, tax and accounting, and compliance and regulatory issues. The series also captures new and modern developments in risk management (market risk, credit risk, operational risk, capital attribution, and liquidity risk), behavioral finance, trading and financial markets innovations, financial engineering, alternative investments and the hedge funds industry, and financial crisis management.

The series will consider a broad range of textbooks, reference works, and handbooks that appeal to academics, practitioners, and students. The inclusion of numerical code and concrete real-world case studies is highly encouraged.

Published Titles

Introduction to Financial Models for Management and Planning, **James R. Morris and John P. Daley**

Stock Market Volatility, **Greg N. Gregoriou**

Forthcoming Titles

Decision Options: The Art and Science of Making Decisions, **Gill Eapen**

Emerging Markets: Performance, Analysis, and Innovation, **Greg N. Gregoriou**

Portfolio Optimization, **Michael J. Best**

Proposals for the series should be submitted to the series editor above or directly to:
CRC Press, Taylor & Francis Group
4th, Floor, Albert House
1-4 Singer Street
London EC2A 4BQ
UK

CHAPMAN & HALL/CRC FINANCE SERIES

Introduction to Financial Models for Management and Planning

James R. Morris

University of Colorado, Denver
U. S. A.

John P. Daley

University of Colorado, Denver
U. S. A.

CRC Press
Taylor & Francis Group
Boca Raton   London   New York

CRC Press is an imprint of the
Taylor & Francis Group, an **informa** business

A CHAPMAN & HALL BOOK

Chapman & Hall/CRC
Taylor & Francis Group
6000 Broken Sound Parkway NW, Suite 300
Boca Raton, FL 33487-2742

Library of Congress Cataloging-in-Publication Data

Morris, James R. (James Russell), 1944-
    Introduction to financial models for management and planning / James R. Morris and John P. Daley.
    p. cm. -- (Chapman & hall/crc finance series)
  Includes bibliographical references and index.
  ISBN 978-1-4200-9054-3 (alk. paper)
    1. Corporations--Finance--Computer simulation. 2. Corporations--Finance--Mathematical models. 3. Finance--Computer simulation. 4. Finance--Mathematical models. I. Daley, John P. II. Title. III. Series.

HG4012.5.M673 2009
658.1501'5195--dc22
                                    2008045946

Visit the Taylor & Francis Web site at
http://www.taylorandfrancis.com

and the CRC Press Web site at
http://www.crcpress.com

To my family: JAM, RSM, and KDM
Thanks for your love and support.

—Jim

I dedicate this book to my wife and
partner for life, Cynthia. Her untiring support
made my contribution possible.

—John Daley

Contents

Preface

This book is an introduction to the fascinating world of corporate financial modeling. It offers an overview of the power and flexibility of financial models and their role in financial planning. This book is intended for financial managers who wish to improve their financial decision making and for students who wish to become financial managers. In our increasingly complex world, the relationships between the choices faced by a manager and their subsequent outcomes have been obscured. A properly structured financial model can take into account much of this increased complexity and provide decision makers with a powerful planning tool, one that helps them identify the consequences of their decisions before they are put into practice.

The purpose of this book is to help the reader learn how to develop and use computer-based models for financial planning. By financial planning we mean the planning of the investment and financing decisions for a firm. Our approach is to provide the reader with tools for his or her financial "toolbox," and then show how to use those tools to build models. Throughout, we emphasize the structure of the models. By emphasizing structure, we mean the development of models that are consistent with the theory of finance and, at the same time, are practical and usable. It is not the intent of this book to instruct the reader about the various software programs that we use. It is assumed that the reader either knows how to use most of the software or can find software instruction from other sources.

The main focus of the book is upon modeling the problems of financial management. That is, the problems associated with the planning and decisions faced by a firm's financial manager. Even though our primary emphasis is upon models related to corporate financial management, we also introduce the reader to a variety of models related to security markets, stock and bond investments, portfolio management, and options.

This book introduces tools that relate to the financial management of the operating business. The tools include interactive cash budgets and *pro forma* financial statements that balance even under the most extreme assumptions, valuation techniques, forecasting techniques that range from simple averages to time series methods, Monte Carlo simulation, linear programming, and optimization. The toolbox is used to solve the problems of planning the firm's investment and financing decisions that lend themselves to financial modeling. These include evaluating capital projects, planning the financing mix for new investments, capital budgeting under capital constraints, optimal capital structure, cash budgeting, working capital management, mergers and acquisitions, and constructing efficient security portfolios.

One particularly valuable aspect of this book is that we continually focus the models on firm valuation. It makes little sense to build a financial model unless it provides a link between the decision being analyzed and the desired objective. Maximizing firm value is our objective; financial theory supplies the links. Therefore, firm value is the nexus of model structure and the theory of finance.

An ancillary benefit of our focus on firm value is that by developing models that include valuation sectors, the reader learns about the theory of valuation and how to apply it to practical problems—how to build models that are consistent with the theory of valuation and simultaneously enhance one's understanding about how numerous financial decisions influence firm value.

One of the challenges of writing a book such as this is deciding what to include and what to exclude. As we wrote each chapter, we would say to ourselves, "But we did not cover this topic, and we need to elaborate on that." However, if we covered all the great topics in finance that were amenable to modeling, we would have to write several volumes. Unfortunately, we had to leave more out than we could include. What drove our decision about what to include was what volume of material could be covered in a one-semester university course. In addition, the topics chosen were primarily the corporate financial management topics that the student is likely to face in the environment of financial management. In each case, our coverage and models can only scratch the surface. For the material we have included in the text, the intent is to teach the student about building models—to get them started, and later they can provide the necessary details and elaborations as their professional tasks require.

ORGANIZATION

The book is organized into six major sections. In Part I (Chapters 2 and 3), we introduce the *tools* for financial planning. These tools include basic financial analysis (Chapter 2) and the different concepts of cash flow and growth (Chapter 3).

In Part II, the structure of a financial simulation model (Chapter 4) and Monte Carlo simulation (Chapter 5) are explained. Chapter 4, Financial Statement Simulation, is the core of the book in which we explain the structure of a financial statement simulation model. The student learns how to develop a simple model that works correctly and flexibly under all scenarios with a structure that is consistent with the theory of finance. Chapter 5 introduces Monte Carlo simulation and shows how it can be used to explore the consequences of our decisions. While we have noted that this text is not intended to give software instruction, this chapter is an exception. Because most students have not been exposed to Monte Carlo simulation or the software necessary to perform it, we introduce the Monte Carlo software @Risk. With this diverse set of tools in hand, the student is equipped to proceed with the remaining chapters of the book. Access to @Risk software for Monte Carlo simulation is provided with the book.

Most financial planning models are driven by a forecast of a basic input variable such as sales. So an important part of modeling is an ability to develop a forecast. Part III (Chapters 6 through 8) is a practical introduction to methods for forecasting a firm's sales and costs. Because forecasting is a discipline unto itself, the most that we can hope to accomplish in a few chapters in a book devoted to modeling is to introduce the reader to a few basic forecasting methods. Forecasting is introduced in Chapter 6 with simple time trend extrapolation methods. This chapter also discusses how to assess and compare the results of different forecasts. Chapter 7 shows how to use using linear regression to develop structural econometric forecasts.

We show how to relate the firm's sales to measures of economic activity such as GDP and interest rates.

Chapter 8 covers smoothing methods such as moving averages, exponential smoothing, and seasonal adjustment, and provides a brief overview of the time series method of ARIMA modeling. At the end of each of these chapters there are forecasting problems for student practice.

Part IV (Chapters 9 through 12) elaborates on the basic financial planning model of Chapter 4, adding detail and explaining how we can expand the model to deal with more specialize questions. Valuation is a focus for most of our models, and it is important to link decisions to value. Chapter 9 shows how to model the value of equity and deals with more specialized questions such as how to model value per share for a firm that is issuing new equity. Chapter 10 considers the investment in long-term assets. It explains how fixed assets should be handled in the planning model, how to model the capital budgeting decision, and finally how to model the decision to acquire and merge with another firm. It presents an example and models for analyzing the capital budgeting decision and the merger decision.

Part of the problem of analyzing capital investment revolves around the financing of the project, so Chapter 11 deals with the financing decision and the firm's capital structure. The first part of the chapter continues with the investment project modeled in Chapter 10, and shows how we can model its financial structure. Then, we use Monte Carlo simulation to show how to model a firm's optimal capital structure. The last section of the chapter provides models dealing with the more specialized topic of duration and debt swaps.

The last topic of Part IV is working capital in Chapter 12. Chapter 12 delves into the modeling of working capital accounts. We develop different models for managing cash, marketable securities, and receivables.

Part V is devoted to modeling investment securities and investment portfolios. Chapter 13 shows how to model security prices as a binomial process and as a random walk Weiner process. Armed with the models of security prices, Chapter 14 models the portfolio decision and shows how to construct a mean-variance efficient portfolio, and how to model the efficient frontier. Chapter 15 explains basic option models using both the binomial model and the continuous time Black–Scholes model.

The last section, Part VI (Chapters 16 and 17), is about optimization—how to use linear programming to find the best investment and financing decisions. We introduce linear programming in Chapter 16 with a

Weingartner-type capital rationing model and explain the transition from an investment model to a complete planning model. Chapter 17 is devoted to the application of linear programming to working capital planning. We use an Orgler-type model to show how we can plan our short-term investment and financing decisions. These optimization chapters provide the student with detailed help in using the Solver optimization add-in to Excel®.

HOW TO USE THIS BOOK IN THE UNIVERSITY SETTING

At the university level, this book can be used for a graduate level or advanced undergraduate level course in finance. The course where we use this material is part of our MS-Finance program, where the students typically already have taken courses in financial management, investments, statistics, and operations management.

Because the methods of financial modeling are learned by practice and experience, we view a course in financial modeling as a learning-by-doing course. We seldom give examinations. We structure our financial modeling course around a set of problems that require the student to construct models that help with planning and decision making. The imperative is that the models should be consistent with the theory of finance. To fulfill this imperative, it is necessary for the student to combine financial theory with modeling. To do this, the student needs to review the theory and figure out how to apply it at a practical level in a model. The result is that the student learns the theory, and more importantly, learns how that theory is applied in the real world.

The problems in the text provide the opportunity to apply the text material to a comprehensive set of fairly realistic situations. The problems posed to the students require them to set up the models and solve them. By the end of the course the students will have enhanced their skills and knowledge of spreadsheet software, statistical software and methods, Monte Carlo simulation, and optimization. These are valuable skills that are in demand by the businesses that employ our students.

The solutions provide insight into business problems, of course, but it is the model structuring process and the linking of financial theory to real-world scenarios that provide the important lessons. Students learn how to make the links between the business problem and the structure of a planning model. The ability to develop the structure of the planning model is the most valuable feature of our course. At the end of the course, we consistently get very positive feedback from the students, with comments such as, "This has been the most valuable course in my finance program," "I've learned more from this class than any other class in my MBA program," and "This class really helped to bring the material together from my various other courses." The students' struggle to develop functioning models brings together the disparate ideas and theories they have learned in their business education programs. It is in the integration of the areas of study that we also get very favorable feedback from our students. We have had comments such as "This course forced me to review my other courses such as statistics, operations management, and marketing, and really helped me bring them together."

It would be very difficult to cover all the material in this book in a semester. It takes a lot of time to explain the concepts to the students, and the students spend a great deal of time outside of class building their models. In our course, we usually require the students to complete 7–12 of the larger end of chapter problems during a term. This is a heavy workload. Consequently, there is not really sufficient time in a typical semester to cover all the chapters and have the students do the extended problems that accompany each chapter.

The instructor is encouraged to pick and choose which topics will be covered during the term. It is not necessary to cover all of the chapters, nor is it necessary to cover them in sequence. Most of the chapters stand on their own. However, the key chapters are Chapter 4, Financial Statement Simulation; Chapter 5, Monte Carlo Simulation; and Chapter 9, Modeling Value. These are the chapters that feed into material in the chapters that follow, and constitute the heart of the topic of financial modeling. If you cover these chapters, you can generally pick and choose the other chapters you cover without much loss of background or continuity.

The financial analysis Chapters 2 and 3 show the student how to set up models to analyze the firm's condition. In addition, Chapter 3 sets the stage for subsequent modeling chapters by explaining the details of cash flow and growth. Our students indicate that they think this material is valuable. However, if the students are already well versed in financial analysis and cash flow, these chapters can be skipped without losing much for understanding later chapters.

If students have already had substantial exposure to statistical forecasting methods, the forecasting Chapters 6, 7, and 8 can be skipped. On the other hand, while most students have had a statistics course, the typical statistics course spends very little time on forecasting. Consequently, the students gain a lot from working the very practical problems in Chapters 6 through 8.

Most of Chapter 10 is devoted to capital budgeting. The chapter and the problems give student the opportunity to practice modeling the capital budgeting decision. However, most finance students have been exposed to these concepts in other courses. The last section of the chapter explains how to model a merger decision, and this is something students may not have covered elsewhere. Nevertheless, with scarce time, this chapter is not necessary for understanding subsequent material.

On the other hand, Chapter 11 presents material many students have not been exposed to before. Of particular interest is the use of Monte Carlo simulation to analyze the debt financing decision. While this chapter is not required for continuity, it is unique and is an important part of the book.

The security investment chapters in Part V are not necessary for continuity with other chapters, and this material can be covered independently of the earlier chapters. However, these topics are current, and students learn a lot by trying to model the concepts and methods they only see as theories in other courses. They may have learned about a random walk, but it sinks in when the student builds a model. The same is true for the efficient frontier in Chapter 14 or options in Chapter 15.

Finally, are the optimization Chapters 16 and 17. As with any of the topics, if you have time, students should be exposed to these methods and get the opportunity to build the optimization models.

In a nutshell, this book offers you a wide variety of topics that are amenable to modeling. All are practical. In general, it is not necessary to cover the book from start to finish. You have great flexibility to freely choose from the smorgasbord of topics according to the interests of the instructor and the students. Whichever topics you choose, the students will learn a lot. The exposure to modeling different decision problems and to different methods gives the student a foundation to approach the wide variety of modeling tasks that it would be impossible to cover in a single volume.

A WORD ABOUT SOFTWARE

The software that we use for our modeling includes spreadsheets, statistics programs, the Solver add-in to Excel, and the Monte Carlo add-in, @Risk. We find that most of our students are already proficient in the basics of using spreadsheets such as Excel, and they typically have been exposed to statistical packages such as Minitab, or perhaps EViews or SPSS in their statistics courses. In addition, they have usually had some exposure to linear programming software such as the Solver add-in to Excel.

However, most students have not been exposed to @Risk, the software used in the Monte Carlo section of the course. It is a useful tool that can be used in a wide variety of modeling applications. The @Risk add-in for Excel is one of the most valuable parts of this book. As noted at the back of the book, the purchaser of the book has access to a one-year license for a fully functional version of @Risk, along with the range of other companion softwares by Palisade.

Because most students have been exposed to much of the software we use, we do not focus the book on teaching the software. We try to give some guidance and instruction in the software as we go along, but we do not provide elementary software instruction. If the student needs an elementary introduction to the standard software such as Excel, it is assumed that they can get this from another source. On the other hand, for the more specialized software such as Solver and @Risk, we provide basic instruction in the chapters where these tools are used.

Our financial modeling course is one of the most popular courses in our graduate finance program. Students complain about the workload, but at the end, they consistently have high praise for the experience and the fact that they have learned to build a wide variety of very applicable models. We hope you will find this material equally valuable.

FOR THE INSTRUCTOR

At the end of each chapter we have some exercises for the students. Most of these are substantial problems that ask the student to build the models discussed in the chapter. All of the data and the solutions to the problems are included in the instructor's disk. While most of the data for the problems are shown with the problem in the back of the chapter, these data are included in the instructor's disk so it can be given to the students without them having to copy all the data from the text. Many instructors will find ways to improve upon our problems. We would invite you to share your problem sets with us so they can be used widely. Finally, we have tried to eliminate mistakes in our text. However, inevitably there is much that we did not find and correct. We would appreciate corrections and suggestions for improvement.

Thank you for using our book. We hope that you find it instructive and useful.

James Morris and John Daley

Authors

James Morris is a professor of finance at the University of Colorado, Denver. He received his BS, MBA, and PhD degrees from the University of California, Berkeley. He has served previously on the faculties of the University of Houston and the Wharton School of the University of Pennsylvania. He has published research papers dealing with capital structure, cost of capital, working capital management, financial modeling, and firm valuation in top academic journals such as *Journal of Finance, Journal of Financial and Quantitative Analysis*, and *Management Science*. In addition, he is accredited in business valuation by the American Society of Appraisers and has published in its practitioner journal, *Business Valuation Review*.

John Daley is an instructor at the University of Colorado, Denver, where he has been teaching finance since earning his PhD from the University of Washington in 1999. He has earned the degrees of BA in psychology from Stanford University, MM in Trombone from the University of Southern

California, MBA from the University of Colorado at Denver, and PhD in finance and business economics from the University of Washington. Concurrently with his position at the University of Colorado at Denver, he is the Principal Trombone of the Colorado Symphony Orchestra, a position he has held since 1978.

CHAPTER 1

An Overview of Financial Planning and Modeling

Planning
Always plan ahead. It wasn't raining when Noah built the ark.

—RICHARD CUSHING

If you don't know where you are going, any road will get you there.

—LEWIS CARROLL

If you have accomplished all that you have planned for yourself, you have not planned enough.

—EDWARD EVERETT HALE

In preparing for battle, I have always found that plans are useless, but planning is indispensable.

—DWIGHT EISENHOWER

T HE OBJECTIVE OF THIS introductory chapter is to provide perspective on the purpose of financial modeling and planning and to explain the ingredients of financial planning.

1.1 WHAT IS PLANNING?

Most managers engage in some aspect of planning. After all, the first step in business is the establishment of the enterprise—a step that is the result of the entrepreneur's dream about the wealth that could accrue from providing some good or service to others. That initial dream is part of the process of business planning. From its establishment throughout the firm's existence the manager may be thinking about the future, wondering what will happen tomorrow, asking what will be the results of current decisions, and perhaps even dreaming of new enterprises.

However, not every effort to know the future is considered planning. For example, forecasting is not active planning but a passive process. It involves the projection of future events, but it does not involve the consideration of the actions that can be taken to affect the future. While forecasting is a necessary input to business planning, planning is much more than forecasting. Planning combines elements of forecasting with the specifications of decisions, actions, and intended results. A broad definition of planning is the specification of future decisions and actions to reach a stated goal. Very simply put, planning is deciding what you are going to do to get where you want to go.

It is useful to think of a traveler planning a trip. The traveler first sets the primary objective by deciding where to go. He may also have sub-goals or constraints regarding when he wants to reach his goal and how much he wants to spend. The travel plan specifies the route, the signposts that will guide him on the route, where to turn, and where to stop.

If there is no uncertainty about future events on the trip, planning is relatively simple. The best route is chosen in advance with the assurance that it is the correct one: no detours will be necessary and the traveler will not get lost. No alternative routes or actions need to be chosen because under conditions of certainty no random events can interfere with the completion of the trip.

However, any trip involves some uncertainty. Random events may interfere with the plan. A landslide may block the route, a storm can close the airport, and the plane may crash. The smart traveler takes the uncertainty into account by planning alternative routes to the goal should obstructions be encountered. For example, in considering whether the plane may crash, the traveler assesses whether the benefits of flying are worth the risks. In the presence of uncertainty, it is especially important for the traveler to make plans that anticipate possible difficulties so as to reduce the resulting inconvenience and delay should the plan not be met.

Similarly, the greater the uncertainty in the business environment, the greater the necessity for and benefits from planning. One might think that with greater uncertainty planning becomes less useful. After all, "the best laid plans of mice and men often go awry," so why plan if even the best laid plans are destined to go awry? However, if the plans are truly best laid, they make allowance for things going awry. As Murphy's Law says, "Everything that can go wrong, will." The best laid plans should take this unfortunate truth into account.

Whether complex and detailed or broad and simple, a plan should specify the objectives and the actions that lead to these objectives. The concept and process of planning are pointless unless one decides first what is to be achieved. Deciding on a destination seems so obvious that it hardly needs to be discussed. Nonetheless, the setting of goals is so fundamental to business planning that it needs extra emphasis. We plan in order to achieve a goal. Without first stating the goal there is no way to decide along the way which actions should be taken and no way to judge which of the actions were successful. Thus, the first step in planning a trip or a business venture is deciding on the goal or objective.

Once the primary objective is set it is necessary to specify some criteria for judging progress toward that objective. The traveler uses signposts along the road to tell whether he is on the correct route and how far it is to his destination. The business planner also must develop some signposts, standards by which he can measure the firm's progress. For example, if the primary objective is the maximization of the market value of the firm, a set of measures of financial performance such as return on assets, profit margin, and asset turnover can help measure progress toward the objective. If these standards are properly specified, they can be used as signposts (sub-goals or intermediate objectives) toward which management's decisions can be directed. Thus, day-to-day management decisions need not necessarily be considered with the objective of maximizing the firm's value. Instead, operational decisions are made based on more mundane standards, such as attaining a target profit margin. Of course, it is crucial that reaching these sub-goals be consistent with attaining the primary objective.

Often, defining the link between the decision and the objective is the primary source of complexity and difficulty in business planning. If you know that a particular decision will lead unambiguously to the objective, there is very little problem in either planning or decision making. However, the effect a decision will have on progress toward the objective is usually unclear. There are often complex and poorly understood relationships between management's actions and the objective. For example, suppose the goal is to maximize the value of the firm's stock, and management is considering issuing bonds to finance a capital investment. To consider these actions, the planner needs to know how issuing bonds and adopting the investment will affect the value of the stock. While the planner will not know all the effects, he can utilize the theory of finance to shed some light on the links between the investment and financing decisions and the value of the stock, and then trace through these complex relations to determine the best decision. A primary goal of this book is to show how to model the links between financial decisions and firm value so that the planner can analyze the impact of his decisions on the firm's objectives.

1.2 WHAT IS FINANCIAL PLANNING?

Finance is concerned with the problems and decisions relating to assets and liabilities, the two sides of a firm's balance sheet. Almost all management decisions that we regard as financial fall under one of these categories. Investment decisions identify which assets should be purchased. Financing decisions concern the sources of the funds (liabilities) necessary to purchase these assets. Financial planning is concerned with setting the objectives for future investment and financing decisions, judging the effects of various decisions on progress toward the objectives, and then deciding, based on these judgments, which investment or financing alternatives should be undertaken.

The range of problems addressed under the heading of financial planning is enormous. The problems range from very detailed specifications of working capital decisions over a planning horizon as short as a day or a week, to very broad plans about a business strategy over the next decade. For example, a commercial bank must decide how to satisfy the weekly reserve requirements set by the Federal Reserve Bank. The bank has a range of choices about which days of the week will show reserve surpluses and which will show deficits, about how it can invest any excess funds, and about how it can obtain funds to cover any deficits. All these decisions involve borrowing and lending on a daily or even hourly basis. In this case the financial plan has a very short horizon and may involve very detailed analyses of a small number of specialized asset and liability accounts.

An example at the other end of the spectrum is the development of a long-range plan for strategic investment and expansion. A broad strategic analysis might consider the lines of business in which the firm should be involved. Such an analysis tends to have a very long planning horizon, so it does not develop fine details in terms of specific assets and liabilities. Rather, it emphasizes business strategy in the broadest sense. This type of strategic planning fits the description by Brealey, Myers, and Allen (2008) of strategic investments that may not directly have attractive values. Rather, they constitute the purchase of options that will enable the firm to undertake attractive investments in the future because the firm is positioning itself strategically now.

In the quintessential financial planning problem, the firm's sales and production forecasts are taken as given, and the investment and financing plan necessary to support the sales must be developed. An example of such a problem will be presented in Chapter 4.

1.3 THE INPUT TO FINANCIAL PLANNING

Obviously, there are many different needs and applications for financial planning. Nonetheless, there are a number of ingredients that all financial plans should have. To give us some insight into these ingredients, let us return to the travel plan analogy discussed earlier.

Some of the necessary ingredients for a thorough plan have already been discussed; still, it is informative to elaborate on them. First, the destination must be specified. Second, the various possible alternative routes and methods of transportation should be enumerated. Without a careful and complete listing of the possibilities, some of the most attractive routes might be overlooked. Third, each route should be evaluated as to whether it actually leads to the desired destination. In addition, the costs, advantages, and disadvantages of each route should be determined. Fourth, the resources available for making the trip must be considered. The amount of money and time available, suitable clothing and luggage, and the availability and condition of a car all are resources that can affect decisions about the trip. Because each resource is limited, it imposes a constraint on what can be accomplished during the trip. There are other constraints, too. It may be desirable to see particular sights along the way, constraining the route selected. Visiting certain countries might be prohibited by legal restrictions. Finally, the traveler must decide what kind of detail is wanted in the plan, and how specific the time schedule should be. The plan could be made very broad, specifying only the countries to be visited and the approximate dates. Or, the plan could be made extremely detailed, scheduling the activity for each hour of each day. The advantages of more detailed planning must be weighed against the extra time and effort involved in making the detailed plan. Furthermore, the traveler may not really want to be tied to the strict requirements of a tightly scheduled plan.

The following list of ingredients for a travel plan is not exhaustive; there may be other ingredients. Still, it is a good start. The ingredients are

a. The destination or goal

b. The alternative routes

c. The relation between the route and destination

d. The resources and constraints

e. The level of detail

Obviously, every travel plan would not explicitly consider each item on the list. The trip may be so routine and simple that an elaborate plan is unnecessary. Some travelers may not even require an explicit listing; they may develop the plan in their head without bothering to write it down. Still, most of the list's ingredients will be considered in any plan, even if only perfunctorily.

A financial plan is similar to a travel plan. Whatever the financial planning problem, whether it involves detailed analysis or a very broad consideration of strategy, the financial planning process is intended to aid in making future decisions to reach the firm's goal: a financial trip from the present into the future. The ingredients of the plan form the input to the planning process. Indeed, most financial plans should begin with an enumeration of the necessary input. The factors or ingredients that form the major input categories of a financial plan look much like the ingredients to the travel plan:

a. The goal

b. The decision alternatives

c. The link between the decisions and the goal

d. The resources for implementing the decisions and constraints inhibiting goal attainment

e. The planning horizon and the desired amount of planning detail

The planner may specify these elements explicitly or only have an intuitive idea of some or all of them. In either case, these elements are needed to develop a complete financial plan. There are considerable benefits to be derived from going through the exercise of carefully and explicitly specifying the elements in this list.

1.3.1 The Goals of Financial Decisions

With respect to investment and financing decisions, there is general agreement among financial economists that the proper objective is to maximize the value of the firm's equity shares. That is, management should choose the decision alternatives that make the market price of the stock as high as possible. A number of arguments support this position. First, management serves as the fiduciary agent of the stockholders and has the responsibility to act in the best interests of the stockholders. The market price of the stock serves as a barometer of stockholder preferences. If stockholders and investors like management's decisions, they will purchase the stock and thereby drive up its price. Conversely, if they do not like management's decisions, investors will sell their stock and thereby drive down its price. Second, maximization of share prices makes all stockholders wealthier, which is one of the reasons they buy shares in the company. Third, from a broader point of view, decisions that maximize share value are decisions that use the firm's resources efficiently and result in a more efficient allocation of resources for society as a whole.

Of course, management and stockholders may have other objectives, ranging from offering stable employment to the workforce to protecting the environment from pollution. In some instances the additional goal is consistent with share value maximization; in others it is not. In the latter case, management chooses decisions that maximize share price subject to the constraint that the additional goal also be met.

1.3.2 The Decision Alternatives

The purpose of financial planning is to specify the future actions and decisions that will lead to the goal of making the value of the firm's equity shares as large as possible. So the second necessary ingredient is a complete list of the possible decisions and actions. While such a list may seem rather obvious, this step is frequently overlooked. The mistake many managers make in planning is to ask, "What will happen if I make this decision?," when, instead, they should ask, "What decision can I make that will maximize the wealth of the stockholders?" By failing to consider the many

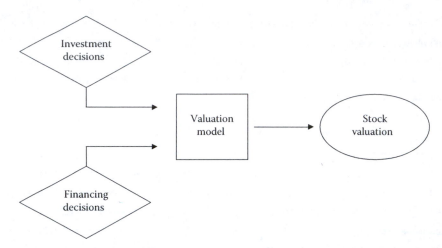

EXHIBIT 1.1 Valuation model provides a link between decisions and objective.

different alternative decisions and asking which one is best, they overlook many potentially superior decision possibilities.

The planner should develop a list of decision alternatives that is as comprehensive as possible. The list should include all possible actions, ranging from doing nothing to actions that may at first seem infeasible or even, perhaps, counterproductive. Certainly, many possibilities will be quickly eliminated. Still, it is useful to engage in the brainstorming exercise of listing them prior to eliminating them.

1.3.3 The Links between Decisions and Goals

To decide which actions will maximize value, one must know how the action will affect share value. To do this, the link between the action and value must be identified. As shown in Exhibit 1.1, the necessary link is a model of valuation that tells us what the hypothetical value of the firm is, given the prospective decision. This is where the theory of finance makes its contribution—the theory of finance provides us with a framework, or way of thinking about, financial problems that explains how various decisions will affect the market value of the firm and its stock. This book will not present a complete survey of the theories of valuation.* However, we discuss the main ideas and develop valuation models that are always in

---

* Valuation models are explained in greater detail in Brealey, Myers, and Allen (2008); Brigham and Ehrhardt (2008); Damodaran (2002); and Jacob and Pettit (1988); a very practical reference is Pratt, Reilly, and Schweihs (2000).

the background of our discussions on planning. Throughout the book, we relate the decision problem to the objective of value maximization by way of a valuation model.

In terms of the problems of financial planning, one must relate the decision problem at hand to the value of the firm and the stock. However, you do not necessarily have to determine the value of the stock in all cases. Rather, the objective for the problem at hand can be stated in operational terms that are consistent with the objective of stock value maximization. For example, in capital budgeting the investment project is chosen so as to maximize the net present value (NPV); and in deciding on the best mix of debt and equity, the mix that minimizes the weighted average cost of capital (WACC) is chosen. In each case the objective is stated in narrower, more specific terms that are consistent with share price maximization.

1.3.4 Resources and Constraints

Because the purpose of financial planning is to decide on future actions to reach a goal, it is essential that the plan take into account the factors that will be available to assist in attaining the goal. In addition, there are inevitably factors that limit the firm's actions or its ability to attain the goal. We call the positive factors resources and the negative ones constraints. The dichotomy is not always clear, however. Because nearly all resources are limited, they also act as constraints.

Resources and constraints occur in both the present and the future. Those in the present describe the company's current situation; those in the future are anticipated to affect the company in the future. Present and future resources and constraints can be classified as

a. Company-specific

b. Industry-specific

c. Related to the economy as a whole

d. Influenced by the legal and political environment

Resources and constraints are important inputs to the planning process and should be considered explicitly in evaluating the desirability and feasibility of a plan of action. The financial planning effort should begin with a careful assessment of present resources and constraints and then develop the plan based on the assumptions and forecasts for future resources and constraints.

1.3.4.1 Company-Specific Resources and Constraints

The firm's present condition is the starting point from which it will develop and the base on which it will build in the future. The firm's present condition is described by its balance sheet, its production technology, and the sources and nature of the supply of its productive inputs—raw materials, products, labor, and management.

1.3.4.1.1 Balance Sheet. The balance sheet describes a firm's assets and liabilities. The firm's assets are simultaneously resources and constraints. They are part of the resources available to implement a plan, but, because they are limited, they also are constraints on efforts to attain the goals. For example, liquid assets are potentially available to finance planned expenditures, but because they are limited, they limit or constrain expenditures financed from this source. A firm's fixed assets are a resource representing investment in capacity to produce output in the future. However, future potential production is limited by existing capacity unless more capital investment is undertaken. The firm's liabilities represent obligations to make promised payments in the future and are therefore constraints on future actions.

1.3.4.1.2 Production Technology. The firm's ability to generate output and profits is determined by the capacity of its plant and equipment. Hence, production capacity is a resource for implementing its plans. Conversely, the firm's future opportunities are constrained by the nature of its technology and its current capacity. The firm's production technology determines what inputs must be used, their relative amounts, and the level of output attainable. The firm's production equipment, plant, and so on, determine the costs and profit margins. Production technology also influences the variability of the firm's income, and thus the firm's business risk. For example, the mix of fixed and variable costs of production determines the firm's operating leverage, its break-even point, and the variability of its operating income.

1.3.4.1.3 Sources of Supply. Established sources of raw materials and good relationships with suppliers are resources for the future. Interruption of these supplies may be a major source of risk for future plans, and limitations on these supplies constrain the rate of production. The costs of materials influence the mix of inputs in the production process and are an important factor in the firm's current and future profitability.

1.3.4.1.4 Labor and Management. People in an organization are an important resource. They can also be a source of problems and difficulties for the firm. As an economy becomes more oriented toward service and information, personnel increasingly become the major resource for reaching the firm's goals. This makes the availability of qualified personnel a primary constraint. In evaluating any financial plan, the reliability and productivity of the firm's labor supply must be considered. Similarly, the strength, ambition, and creativity of management are major factors in determining whether the firm can carry out its plans, as well as whether the firm can respond effectively to a changing environment and the subsequent surprises that can undo their plans.

1.3.4.1.5 Products. The nature of the firm's products, the uniqueness of its products, and any special product advantages, such as patents, are important factors in the plans for the future. The financial planning effort must evaluate the firm's product mix and the future prospects for its products and relate both the product mix and the products themselves to the firm's industry and market.

1.3.4.2 Industry-Specific Constraints

The ability of the company to attain its goals is highly dependent on the environment of and the future prospects for the industry within which it operates. In most cases the industry context presents constraints on the attainment of goals rather than resources that assist the firm, although the opportunities for growth in the industry may be considered a resource.

1.3.4.2.1 Demand. Probably the most important input to any planning effort is the forecast of future demand for the product. In fact, for most financial planning efforts the sales forecast is the factor on which almost all other forecasts depend—the variable that drives the model. One of the first steps in generating a forecast of company sales is preparing a forecast of industry sales. Industry sales constrain a firm because they represent the maximum size of the financial pie available to all firms in the industry. In addition to a basic forecast of industry sales, there are numerous other more specific factors under the heading of demand that may be important inputs to the financial planning effort. These other factors include pricing, advertising by the company and its competitors, growth of market, and product life cycle.

Demand for the product depends on the price charged for it. Therefore, a major input to the analysis of the industry's and company's future sales is an estimate of the demand function for the product, that is, an estimate of how demand changes as the price is changed. Pricing policies of other firms in the industry also serve as important constraints on the company's decisions and affect its financial plans. The future demand for the product and the growth in the demand will depend on prices, the economy, and sociological and demographic trends. All of these factors may have to be considered in developing a sales forecast.

1.3.4.2.2 Industry Structure.   The structure of the industry will influence sales by the company and constrain many of the company's actions. Company sales may be expressed as the product of the total industry sales and the company's share of the market. In turn, the share of the market depends on the number of firms in the industry, their competitiveness, and the degree of concentration in the industry. In competitive industries with many firms, the sales of a particular company will constitute only a small share of the total market, and the firm will be constrained in its ability to change prices and to increase its share of the market. Just the opposite is true for more concentrated, less competitive industries. Competition and industry structure may be influenced by whether patents and licenses protect the various products and production methods. In addition, the nature of the production process, which may exhibit decreasing, constant, or increasing returns to scale, will affect industry concentration and competition. Finally, marketing and distribution practices in the industry will be important factors that constrain the ability of the company to attain its goals.

1.3.4.3 Constraints Imposed by the Economy

Every company is to some degree dependent on the vicissitudes of local, national, and even international economic conditions. Indeed, it is well accepted that one of the best measures of a firm's risk is the extent to which its fortunes are systematically related to the economy as a whole. Thus, we cannot ignore the dependence of the company's fortunes on what happens in the economy as a whole. This makes a forecast of general economic conditions (and perhaps, more specific, regional forecasts) one of the most important inputs to any financial planning effort. Of course, different companies and industries vary in terms of how much they are influenced by economic conditions. Some industries, such as automobiles, housing, and machine tools, are particularly vulnerable to the economic cycle. Others, such as utilities and grocery chains, are not as sensitive. In still other sectors, such as banking, a single, specialized economic variable, such as interest rates, may be of prime importance in developing a company forecast.

1.3.4.4 Constraints Imposed by the Legal and Political Environment

The rules and regulations promulgated by government and the demands of political reality typically impose constraints and limitations on the actions the firm can undertake and may influence the results of the actions taken by the firm. In developing a financial plan, it is important to recognize and take explicit account of these constraints. Because one purpose of financial planning is to plan for the unexpected and to be prepared to respond to surprises, it may also be important to attempt to forecast future legal and political conditions—one of the most difficult forecasting tasks. The company should be prepared to respond to society's demands for such things as pollution control, equal opportunity, and product liability. In addition, it should consider the effects on the company of major political upheavals such as wars, foreign revolutions, expropriations, and devaluations. If a company depends on international trade, it should try to anticipate the effects of currency exchange rate fluctuations.

1.3.5 The Planning Horizon and the Amount of Detail

The last item on the list of ingredients in a financial plan is the length of the planning horizon (the length of time over which the plan will apply) and the amount of detail to be generated for the financial plan. Before developing the financial plan, the analyst must decide how many months or years should be considered in the plan. In addition, the analyst must decide how detailed to make the plan. The appropriate length of the planning horizon and the desired amount of detail are important decisions that are part of the initial input to the planning process.

1.3.5.1 The Length of the Planning Horizon

A financial plan can have a planning horizon of almost any length—one day, one month, one year, five years, or longer. The choice of the planning horizon is important because it affects the results and usefulness of the plan. The planning horizon influences the amount of effort and cost necessary to develop the plan as well as the amount of detail incorporated in the plan. Therefore, it is important to view the planning horizon as a crucial ingredient of the process. Factors that should be considered in deciding on the planning horizon are the nature of the planning problem, the amount of output detail desired from the plan, the horizon of reasonable forecasts, and the period over which the planning horizon makes a difference.

The nature of the planning problem obviously influences the planning horizon. As a general rule, you should make the planning horizon sufficiently long to show the full effects of the decision over a full cycle. If you are concerned with the long-term effects of a broad strategic decision, then a long planning horizon is appropriate. For example, analyzing the effects of a research and development (R&D) program involving new product development over the next 10 years calls for a long planning horizon. It makes little sense to limit the analysis to, say, one year. However, if the problem is determining the financing requirements over the firm's seasonal production cycle, a one-year planning horizon would usually be appropriate because the cycle would tend to repeat itself each year and seasonal financing would not normally be arranged for more than one year. In this case, a six-month planning horizon would be too short to show the effects of the cycle and the results of the decisions being considered.

The nature of the problem also influences the division of the planning horizon into sub-periods, the time intervals that make up the planning horizon. For example, a one-year planning horizon could be divided into 12 monthly sub-periods or four quarterly sub-periods. The length of the sub-period determines how frequently prospective decisions and forecasted results can be changed in the model. If a broad, long-term strategic policy is under consideration, in which the basic decisions would not be varied more often than annually, then you probably would not be interested in developing a plan that generates quarterly or even more frequent results. Conversely, if the problem is one of managing the marketable securities portfolio, where daily decisions may be important, it would make sense to use daily, or perhaps weekly, sub-periods, and it would be pointless to consider a five-year planning horizon composed of annual sub-periods.

1.3.5.2 The Amount of Detail

The length of the planning horizon and the number of sub-periods influence the amount of detail considered in the financial plan. In determining the length of the planning horizon and the sub-periods, you need to consider how much detail is useful. It is too easy, particularly with computers, to generate more detail than can be evaluated or that is meaningful. If the planning horizon is comprised of many sub-periods and the decisions and results are considered for each sub-period, an excessive amount of detail can be produced. One frequently finds that the additional detail is a disadvantage. Too much detail tends to overwhelm the user so that she loses track of how the results relate to the decisions and objectives. In addition, generating and evaluating detail is costly in terms of effort and time. Therefore, the information and detail must be balanced against the cost and effort involved.

It is useful to distinguish between cross-sectional detail and time series detail. Cross-sectional detail refers to the amount of data, variables, and decisions considered for a given period. Financial statements broken down into many account categories for a given period, say June 2010, generate a large amount of cross-sectional detail. Time series detail refers to data, variables, and decisions considered at different points in time. A forecast of net income for each month of a 10-year planning horizon would include much time series detail, but because there is only one variable (net income) for each sub-period (one month), there would be little cross-sectional detail.

Exhibit 1.2 demonstrates the idea. Both charts have 10 pieces of data. In the first chart, greater time series detail is shown at the expense of less cross-sectional detail. In the second chart, more cross-sectional detail is shown with less time series detail.

The amount of detail should be minimized, kept to a level sufficient to help the analyst understand the problem. If the problem involves a long

Time series detail

	Period 1	Period 2	Period 3	Period 4	Period 5
Sales					
Income					

Cross-sectional detail

	Period 1	Period 2
Sales		
Operating cost		
Operating income		
Interest expense		
Income		

EXHIBIT 1.2 Time series detail versus cross-sectional detail.

planning horizon, you can control the size of the problem by decreasing the amount of cross-sectional detail (fewer accounts for each sub-period) and limiting the amount of time series detail [fewer (longer) sub-periods across the planning horizon]. If the planning problem involves a short planning horizon and requires a great deal of cross-sectional detail (e.g., many different investment and financing accounts), you can control the amount of detail by shortening the planning horizon and lengthening the sub-periods.

The length of the planning horizon should be limited to the horizon of reasonable forecasts. If the financial plan depends on forecasts (of sales, interest rates, gross national product, or the like), it does not make much sense to extend the analysis beyond the time over which you can make reasonable forecasts. For example, sales forecasts for specific months make sense for one year into the future, but it is nonsense to try to distinguish between June and July sales five years hence. Consequently, developing a forecast of month-by-month financing needs is appropriate for the next year or so, but inappropriate over a five-year planning horizon. However, estimates of long-term growth trends may be quite reasonable and are useful for broad, general financial planning with a very long planning horizon. For example, an electric utility may need to plan its construction to meet growth in energy demand over the next 25 years. While it cannot forecast month-by-month demand 25 years hence, it can make useful forecasts of long-term growth based on demographic trends. In this case, a 25-year planning horizon divided into annual sub-periods may be appropriate.

1.4 INGREDIENTS OF A FINANCIAL MODEL

In the first part of this chapter we introduced concepts of financial planning using the analogy of a travel plan. We then presented and explained the ingredients of a financial plan. This section extends the concepts of financial planning to financial modeling. Just as a road map connects the traveler's current location with her destination, a financial model connects the firm's current financial state with its future financial objective. In this section we explain the concept of a financial model and the role played by it in financial planning; we then show how the ingredients of a financial model correspond to the ingredients of a financial plan.

1.4.1 What Is a Model?

A model can be defined as "a small copy or imitation of an existing object; or, a preliminary representation of something, serving as a plan from which the final object is to be constructed." Everyone is familiar with model airplanes and recognizes that they are intended to imitate real airplanes. However, the manner in which they imitate the real thing depends on their intended use. For a child, the model only needs to have the shape of an airplane. A simply shaped block of wood is sufficient to excite the child's imagination. On the other hand, for an aircraft manufacturer, the model must be more detailed. It must resemble the full-sized prototype in enough important respects that the performance potential or other aspects of the real plane can be tested. Yet even in this case the model need not include every detail of the real plane to be useful. Only the features necessary to test specific aspects of the plane are required. Features of the real plane irrelevant to the particular aspect being tested can be omitted. For example, to test airworthiness, only the shapes of the fuselage and wings must be precise, whereas to test cockpit ergonomics, only the interior dimensions must be precise.

It is easy to understand what a model airplane is and how it imitates a real airplane. It is more difficult to understand what a financial model is shown as Equations 1.1 and 1.2, and how it is supposed to imitate the real thing. In business finance, the object to be modeled is the set of interrelationships and linkages between the firm's environment, prospective decisions and actions, and the firm's objectives. For example, the question for the financial model may be, What is the effect of an increase in interest rates on the earnings and value of the firm? Interest rates, earnings, and value are considerably more abstract than the shape of an airplane's fuselage, wings, or interior. Nonetheless, these relationships often can be expressed verbally or mathematically with enough essential detail to trace out their potential effects. This, then, is the essence of a financial model—a set of verbal or mathematical statements that express the system of relations and links between the firm's environment, its decisions and actions, and its objectives.

Consider the following example. Suppose we wish to analyze the effects of the firm's pricing decision on its sales. The following simple model shown as Equations 1.1 and 1.2, shows the links between the price, units sold, and total revenue. This model consists of two equations and three variables: Total Revenue$_t$, Units Sold$_t$, and Price per Unit$_t$, where the subscript t denotes the time period. Management decides what price to charge for the product; the model returns the total revenue to be expected. Price per Unit is called the decision variable. Units Sold is a variable that management cannot set. But management can influence Units Sold through the value they set for the decision variable, Price per Unit. The question to be answered is, What price should be set so that the firm can best attain its objective, perhaps maximizing Total Revenue?

A model of Total Revenue is

$$\text{Total Revenue}_t = \text{Units Sold}_t \times \text{Price per Unit}_t \qquad (1.1)$$

$$\text{Units Sold}_t = 500 - 30 \times \text{Price per Unit}_t \qquad (1.2)$$

Equation 1.2, what economists call a demand function, links the firm (its pricing decision) to its competitive environment (the number of units sold). It shows the quantity of units demanded at different prices. Over the range of prices to be considered, each $1 increase in the price of a unit causes the quantity of the product demanded to decrease by 30 units. Thus, if Price per Unit is $5, Units Sold will be 350; but if Price per Unit is increased to $6, Units Sold decreases to 320. The competitive structure of the industry, the nature of consumer demand for the product, and the general condition of the economy all influence the way product demand responds to the price of the product. The demand function summarizes these effects. Through the demand function, these factors constrain management's ability to set an arbitrarily high price to attain its goal.

Equation 1.1 represents the link between the firm's decisions (price) and its objective (maximize revenue). It is essentially a definition that shows how price and demand combine to determine Total Revenue. If we assume that the goal of management is to maximize Total Revenue, then this expression is what is called the objective function—the variable that is to be optimized. In this case, the decision variable, Price per Unit, is supposed to be chosen so as to make the objective, Total Revenue, as large as possible.

Exhibit 1.3 shows how Total Revenue varies as Price per Unit changes. When Price per Unit is less than $8.33, an increase in Price per Unit increases Total Revenue. When the Price per Unit is greater than $8.33, an increase in Price per Unit decreases Total Revenue. Therefore, Total Revenue attains its maximum value when Price per Unit equals $8.33. Hence, the optimal decision—the one that maximizes Total Revenue—is to set the Price per Unit at $8.33. At this price, 250 units would be sold for total revenue of $2083.

This model contains parameters in addition to equations and variables. Parameters are the constants (numbers) in the equations that operate (in this case, multiply or add) on the variables to yield the value of the

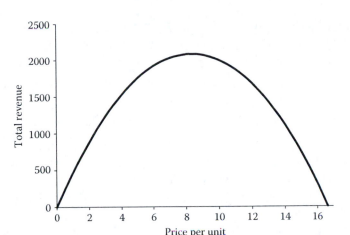

EXHIBIT 1.3   Revenue as a function of price.

equations. In Equation 1.2 the parameters are 500 and 30. These parameters represent the way the quantity demanded changes with price. Different parameters imply a different relationship between price and units sold. The parameters in any model are usually estimates of how one variable responds to another. They represent economic behavior or technological relationships.

The general form of an equation shows the relationships between variables without specifying specific values for the parameters. For example, the general form of the demand function Equation 1.2 is

$$\text{Units Sold} = a + b \times \text{Price per Unit}, \qquad (1.3)$$

where *a* and *b* are generalized parameters. The general form is based on the modeler's estimate of the general relationships between variables. To make Equation 1.3 usable, the values of parameters *a* and *b* must be estimated.

This means that the builder of our simple model faces two tasks: (1) determining the general form of the relationships, and (2) estimating the parameter values. Though in some cases the modeler may just guess structural forms and the parameter values, the preferred procedure is to identify the best structural form and estimate the parameters using historical data. Hence, forecasting and econometrics are an important part of the modeling process. These methods are introduced in the forecasting chapters.

Time is another explicit element in Equations 1.1 and 1.2. The variable Units Sold$_t$ refers to sales during period t. Because Units Sold must be defined in terms of sales per period, the length of period t is an important part of the model. The parameters in Equation 1.2 would be very different for periods of one month versus one year.

Equations 1.1 and 1.2 do not specify the length of or the number of periods in the planning horizon. The model could apply over a number of periods, with an equation assigned to each period in the planning horizon. If the same equations with the same parameters apply identically to each of, say, 10 periods, the model could be written in the general form as

$$\text{Total Revenue}_t = \text{Units Sold}_t \times \text{Price per Unit}_t \qquad (1.4)$$

$$\text{Units Sold}_t = a - \text{b} \times \text{Price per Unit}_t \quad \text{for } t = 1, \ldots, 10, \qquad (1.5)$$

where $t = 1, \ldots, 10$ means that the expression is repeated for each period t from 1 through 10. Parameters can also vary across time. When this is the case, a subscript is added to the parameter, for example, a_t, b_t for $t = 1, \ldots, 10$. This indicates that there are different parameters a and b in the demand function for each of the 10 periods in the planning horizon.

Though simple, the structure of this two-equation model contains most of the key elements of more complex financial models. It is made up of variables, equations, and parameters. Some of the variables represent prospective decisions, others the effects of those decisions. The equation for Total Revenue shows the relation between these variables and the goal or objective. We refer to this expression as the objective function. Finally, our simple model has parameters whose values must be estimated using historical data. More complex models will contain other expressions representing more complex relations between variables, and perhaps expressions that portray constraints and limitations on the firm's decisions. Whether simple or complex, the model links the decision to the objective. The equations in the model portray the influence of resources and constraints on the firm's ability to attain the goals set by the financial planning process. The relationships between variables, expressed as a set of equations, comprise the financial model.

1.4.2 What Does a Model Do?

As a small copy of the firm, the financial model serves as a testing platform. Using the model, the manager can assess the effect of various decisions before actually implementing them and thereby avoid subjecting the firm itself to experimentation.

A financial model of the firm traces through the complex financial relationships in the firm. This is helpful and sometimes essential because of the complexity of the relationships between the firm's financial variables. For some problems it may be possible to discern the general direction of the result of a decision simply by inspection, but it may be difficult to figure out the magnitude of the effect without a model. For other problems, it may not be possible even to discern the direction of an effect without a model. For either type of problem, the intent of the model is to aid and improve decision making by providing more information to the decision maker—the model's output.

1.5 HOW TO DEVELOP THE MODEL?

There are as many different ways to build a financial model as there are potential model builders. Because any given problem can be approached from many directions, there seldom is just one correct model suited to a given problem. Given such a variety of possible approaches, the modeler will find it helpful to begin by considering a set of questions that help organize and structure the modeling effort. These questions relate to the problem, the output, and the input. The answers should be found in the financial plan.

1.5.1 The Decision Problem
1.5.1.1 What Is the Problem?
This is the most fundamental question in approaching the modeling effort. The model must be suited to the decision problem. This requires a thorough assessment of the nature of the financial planning problem. This assessment should be as broad as possible, exploring all facets of the problem. That is, the model must be at least as broad as the problem and be suited to the analysis of all important facets of the problem and all reasonable decisions. A model that is too narrow and limited may not allow the planner to consider some important decision alternatives, and thereby cause her to miss possibly useful solutions. Thus, the initial definition of the problem should be broad enough so that the subsequently developed model is sufficiently general to allow a full analysis of the problem.

1.5.1.2 What Questions Must Be Answered?
For most problems there are questions, that, when answered, suggest the solution. You may say to yourself, "If I just knew the answer to *this question,* then the decision would be simple." The model builder should enumerate these questions as completely as possible. Such a list of questions

helps define the model's structure and influences the kind of output data and answers one wants from the model. For example, suppose we are evaluating a new marketing program. We pose some basic questions: Will the program increase sales? Will earnings and cash flow increase? Will it benefit the stockholders with an increase in the value of the stock? These questions indicate which variables should be the focus for our model: sales, earnings, cash flow, and stock value.

1.5.1.3 What Is the Time Horizon of the Problem?

Financial planning problems are almost always concerned with time. There are various aspects of time in modeling, and these are important determinants of the structure of the model. The length of the planning horizon and the sub-periods should be consistent with the timing of the decision. If the problem being considered is of a long-term nature, or has long-term consequences, then the planning horizon and sub-periods should be long. If the problem involves frequently made decisions over short time intervals, then the planning horizon and sub-periods should be short.

1.5.1.4 How Important Is the Problem?

The importance of the problem for the goals and survival of the firm should be considered when deciding how much time, effort, and expense to commit to model development. If a complex and expensive modeling effort is necessary to provide solutions to an unimportant problem, the effort is probably not justified. If a problem can be handled with a scaled-down model, then, by all means, scale down. In any case, these issues should be considered before embarking on a costly model development effort that promises little benefit.

1.5.2 The Output

1.5.2.1 What Kind of Information Is Needed?

After considering the planning/modeling problem facing the firm and broadly defining what questions need answering, the planner should decide what specific information must be generated by the model. Typically, this information will answer the specific questions identified in the first phase of the modeling effort (What is the problem?). Continuing with the previous example, this would mean a model that outputs revenue, cash flow, and so on.

All too often, a model is developed that generates mountains of information, little of which actually helps the planner make a better decision.

This is particularly true when models are designed before deciding what output is needed. For example, if the problem is to evaluate an investment project, a model that merely generates future *pro forma* financial statements likely will be insufficient for a valid analysis. If the investment analysis requires calculating the expected NPV and assessing the risk of the investment, the model should be developed to do that.

1.5.2.2 How Will the Information Be Evaluated?

You should decide ahead of time how the output is to be evaluated. For example, suppose the problem involves comparing alternative financing decisions. By deciding in advance how you are going to rank the alternatives, you can develop the model so that its output makes such an evaluation feasible. Many models generate standard financial statements with so much detail that the user is unable to evaluate the alternative decisions correctly. This mistake can be avoided by designing the model to produce output that is both relevant to the problem and useful to the decision process. This means relating the information (output) to the firm's objectives. Because the correct decision is the one that will lead to the objective, you cannot evaluate the decision unless you have information that shows the effect of the decision on the objective.

1.5.2.3 What Kind of Detail Is Necessary?

In deciding what output is necessary to make a planning decision, it is important to specify the level of detail needed to evaluate progress toward the planning objective. The planner/modeler should ask two questions. First, what is the smallest amount of detail needed to make the decision? This is the lower limit on the amount of detail required. Second, at what point does more detail lead to the same decision as less detail? This is the upper limit on the amount of detail required. Too much detail is more often a problem than is too little. Structure the model so that it generates no less detail than the lower limit and no more detail than the upper limit.

1.5.2.4 Who Will Use the Information?

The type and format of the output should be tailored to the audience who will use it. It may be appropriate to generate a large volume of finely detailed data for use by the planning analyst. However, as one moves up through the levels of the organization, the users of the data may be less concerned with fine detail and more concerned with a broad picture of the firm. For the executive who takes a broad view of the firm's future, the output should be consolidated and summarized so that it can be quickly and easily grasped.

1.5.3 The Structural Input

After the planner has broadly defined the problem and decided what kind of output is necessary, the model itself is developed and solved with two kinds of input: structural input that determines the structure of the model, and data input that determines the solution to the model. The structural input is the set of equations and relationships that make up the model. The data input are the values for the unknown variables in the structural input. For example, Equations 1.4 and 1.5,

$$\text{Total Revenue}_t = \text{Units Sold}_t \times \text{Price per Unit}_t$$

$$\text{Units Sold}_t = a - b \times \text{Price per Unit}_t$$

are part of the structure of the sales model. The data input (from forecasts or decisions) is fed into the model by specifying

$$\text{Price per Unit}_t = \$5.00$$

$$a = \$500$$

$$b = -30$$

In a spreadsheet model, the structure is the set of formulas in the cells of the spreadsheet and the data input are the numbers that we feed into the model.

Generally, the structural input is based on the set of ideas and questions that motivate the modeling process. The structural input quantifies the plan. It determines the number and form of the equations, the number and types of variables in the equations, and the form of the objective function. The ideas and questions that underlie the structural input are the factors discussed previously as the ingredients of the financial planning process: the goal, decisions, linkages, constraints, and planning horizon.

1.5.3.1 What Is the Goal?

Most models should have an objective function that yields summary measures of the results of the decisions put into the model. In some models the objective function will be a specific equation that generates a measure of the goal, such as the value of the stock. In others, there may not be a specific equation; instead, the model provides output that summarizes progress toward the goal of the decision process. Equation 1.4 is the objective function in our simple model because it provides as output the value of the thing we are concerned with, total revenue in this case.

1.5.3.2 What Are the Decision Alternatives?

One purpose of the financial model is to help identify the set of decisions out of all possible alternatives sets that best fulfill our goal. Hence, a fundamental input to the model is a list of the decision variables. The decision variables should be clearly specified and the model structured so that they can be easily manipulated and changed to explore alternative decisions. For example, in structuring a spreadsheet model, the decision variables should be placed in an input section of the model where the user can change their values easily to see the effect on the objective. In our sales model, the decision variable to be manipulated is Price per Unit$_t$.

1.5.3.3 What Are the Linkages between the Decisions and the Goal?

One reason to build a model is to determine how various decisions affect the firm's goals. An objective function summarizes these linkages. The model should be structured so that the decision variables feed through the model and into the objective function that represents the firm's goals. In the sales model, we feed the pricing decision through the model and we get sales as the output. The two equations of the model provide the linkage from Price per Unit$_t$ to Total Revenue$_t$. Note that we need the whole model (both Equations 1.4 and 1.5) to provide the linkage.

1.5.3.4 What Are the Constraints?

The environment in which the decision maker operates constrains his choices over the decision variables. Constraints can arise from limitations on the amount of resources available to implement the financial plan; from technological relationships that limit choices; and from the industry, economy, or legal environment. There also may be constraints that are true by definition.

1.5.3.4.1 Definitional Constraints. A major part of most financial models, particularly financial statement models, is a set of equations that are definitions. Typically, these are based on accounting rules. We treat these definitional equations as constraints because they constrain the model to conform to these definitions. Definitional equations maintain consistency within the model, aggregate groups of accounts into broader categories, and link account balances from one period to the next. An example of a definition that aggregates accounts is

$$\text{Total Assets}_t = \text{Current Assets}_t + \text{Fixed Assets}_t$$

An example of a definitional equation that links accounts from one period to the next is

$$\text{Earned Surplus}_t = \text{Earned Surplus}_{t-1} + \text{Retained Earnings}_t$$

Equation 1.4 in the sales model is a definitional constraint that simply holds the system together with the definition that "sales is equal to price times quantity."

1.5.3.4.2 *Constraints from the Industry and Economy.* In the sales model, the demand Equation 1.5 is a constraint imposed by the economy and the industry that sets limits on the pricing decision. There also is an implicit constraint in that it is understood that the price cannot be negative. However, in a computer model it may be necessary to make the non-negativity constraint explicit (Price per Unit$_t \geq 0$) because the computer would not otherwise understand that price must be positive.

If we expanded our model into something more detailed, we might encounter various other constraints such as limitations imposed by technological relationships, resources, the economy, and regulations.

1.5.3.4.3 *Technological Constraints.* Technology often places constraints on the actions available to management. For example, management might like to produce more product, but the technology of its manufacturing process limits how much can be produced. Suppose a production function describes the maximum number of units that can be produced with a specified amount of labor and capital equipment, according to

$$\text{Units Produced}_t = 5 \times \text{Labor}_t^{0.33} \, \text{Capital}_t^{0.67} \tag{1.6}$$

This production function limits how much can be produced for a given endowment of resources. So we also need to consider the constraints imposed by our limited resources.

1.5.3.4.4 Resource Constraints.   Assume the maximum number of units of Capital and Labor available are 40 and 60, respectively. Our constraints are stated as

$$\text{Capital}_t \leq 40 \tag{1.7}$$

$$\text{Labor}_t \leq 60 \tag{1.8}$$

which allows us to use less than the amounts available but prevents us from using more.

In both the production function and the resource constraints, the general form of the expressions are part of the structure of the model, and the parameters (5, 0.33, and 0.67 in Equation 1.6 and 40 and 60 in Equations 1.7 and 1.8) are the data input. Throughout this book, as we build models you will be encouraged to separate the data input from the structural model. Doing so allows the structural model to remain general, while the input section is variable. This makes it easy to change the parameters in the model without having to change the model itself.

1.5.3.5 *What Is the Planning Horizon?*

The length of the planning horizon and the number of sub-periods contained therein are important structural inputs to the model. Together they influence the number of equations and variables in the model, and the volume of output from the model. Many of the considerations related to the planning horizon have been discussed previously, with respect to the length of the sub-periods and the amount of detail in the plan. Suffice it to say that as the number of sub-periods increases, the number of variables, the number of equations, and the complexity of the model also increase. Evaluate these factors before actually specifying the model.

Carefully considering the questions relating to the problem, the output, and the structural input provides the modeler with a rudimentary structure with which to begin developing the model in its general form. This consists of a set of equations (and/or inequalities), a set of decision variables that are the arguments in the equations, generalized parameters that enter the equations as constants or coefficients of the decision variables, and possibly an objective function. At this stage in the development, most of the parameters are shown only in their general form (such as *a*, *b*, and *c*). Data must be input to give specific values to the parameters before the model can be solved.

1.5.4 The Data Input

Up to this point, only the general form of the model has been developed. No specific numerical values for the parameters of the model have been supplied. To make the model usable, these parameters must be specified. This is the role of the input data—to fill in the blanks (e.g., the values for a and b) in the general model so that all values except those for the decision variables are specified.

Data input can be classified into three broad categories: data that describe the current state of the system, data that describe the relationships between variables, and data that forecast future conditions.

1.5.4.1 The Current State of the System

Financial planning plots the trajectory of the firm through time—where it was, where it is, and where it hopes to be. It can be thought of as specifying decisions and anticipating their consequences over a number of periods of time in order to reach a stated objective. Obviously, where the firm goes in the future depends on where it starts. The current conditions of the firm and its environment are important determinants of where it goes in the future. Therefore, the financial plan and the financial model must take these current conditions into account. This is what is meant by the current state of the system—the condition or situation from which the firm begins its projected trip into the future. The firm's resources and constraints describe the current state of the system. These were discussed as input to the financial planning process. Internal data describe the firm's current state; external data its environment.

1.5.4.1.1 Internal Data: The Firm. The firm's current condition consists of its endowment of resources, its productive capacity, and its obligations and commitments. The firm's financial statements describe most of these factors. They establish the initial condition from which the future will evolve, particularly for accounting models that generate future *pro forma* financial statements. In addition to the financial statements, the model may also require input data that describe the firm's production technology, sources of supply, labor and management, and products.

Whatever the input data required, there must be congruence between the model and the data. The level of input detail demanded by the model must be matched to the amount of detail available in the accounting statements used as input data. If the required detail exceeds what is available from the firm's accounting system, then the model should be revised so that the data required match the data available.

1.5.4.1.2 External Data: The Environment.   The current conditions in the firm's operating environment also influence the actions the firm can take and where it can go in the future. Most of the factors that describe current conditions relate to economic conditions, such as interest rates, stock prices, and prices of goods and labor.

1.5.4.2 Relations between Variables

Variables in the financial model may be related to each other in a way that is not completely known to the modeler. Nonetheless, the modeler can estimate these relationships by assuming a mathematical structure (an equation) for the relationship and using historical data to estimate its parameters. For example, suppose accounts receivable is assumed to be related to past sales according to the distributed lag relationship in

$$\text{Accounts Receivable}_t = a + b_1 \times \text{Sales}_{t-1} + b_2 \times \text{Sales}_{t-2} + b_3 \times \text{Sales}_{t-3}. \tag{1.9}$$

The parameters a, b_1, b_2, and b_3 in Equation 1.9 are unknown. Using past observations of accounts receivable and sales, these parameters can be estimated by least squares regression techniques. The estimates for a, b_1, b_2, and b_3 become data input for the model.

1.5.4.3 Forecasts of Future Conditions

If we knew the future, we would not need to do much modeling and planning. Unfortunately, we cannot know what the future will bring. Indeed, a major purpose of planning and modeling is to help the firm deal with the uncertainty of the future. This requires forecasting future conditions and relationships. Hence, forecasts are an important input to the modeling process. Because forecasts are basic to solving the model and developing a financial plan, an important issue is the sensitivity of the plan and its results to forecast errors. The modeler should maintain a healthy skepticism of the accuracy of the forecasts and consider how forecast errors affect the financial plan and goal attainment.* Forecasting techniques will be discussed in later chapters to introduce the modeler to the tools necessary to consider and evaluate alternative forecasting methods.

If one has asked and answered the questions relating to the nature of the planning problem, the output required, and the type and availability of input data, much of the modeling effort will have been completed.

---

* An anonymous wag once said that, "Forecasting is the substitution of error for chaos."

1.6 TYPES OF MODELS

There are two general approaches to financial modeling: simulation and optimization. Simulation models show the results of a decision or action that is specified by the user; optimization models solve for the best set of decisions. Though simulation models are more widely used and understood, optimization models also can be very useful. To aide the prospective user in choosing the approach that best suits his needs, both approaches are discussed in this book.

1.6.1 Simulation

Simulation is best understood as a technique that allows the user to answer the question, "What will happen if ...?" Simulation is the process of imitating the firm or organization so that the consequences of alternative decisions, actions, and strategies can be analyzed. Simulation allows management to view the hypothetical results of undertaking a particular decision, making it possible for them to evaluate the decision's consequences prior to implementing it.

The simulation model links the prospective decision to the hypothetical results. The model consists of a set of mathematical expressions that reflect the relationships and links between the firm's environment and the organization, between units of the organization, and between decision and result. Thus, the simulation model provides the necessary bridge between the input (the firm's current state) and the output (the firm's projected future state) of the simulation process.

As shown in Exhibit 1.4, the input to the model normally consists of the current state of the system (the initial condition of the firm and its environment), forecasts of future conditions, and decisions or actions to be considered. A simulation model takes the set of decisions as input, and then combines this input with current and future economic conditions to generate the output—the consequences of those decisions. These consequences, in turn, partly describe the state of the system in a subsequent period.

The simulation approach does not tell the user which decision is best. Rather, the user specifies a decision, the model traces through the relationships, and then the user evaluates the results. The consequences of alternative decisions and strategies can be examined, but the assessment of which strategy is best is left to the user. For example, suppose the decision being considered is whether to finance a new investment project by issuing new common stock or by borrowing from a bank. The typical simulation would trace through the consequences of each of these alternatives and present

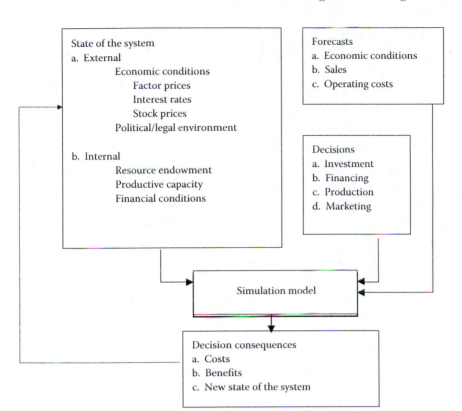

EXHIBIT 1.4   The flow of information for a simulation model.

the user with sets of financial statements and measurements of the objective such as value of the stock. The user then uses the simulated output to evaluate the results of the decisions.

Simulation models offer several advantages over optimization models: they are usually easier to develop and solve; managers usually find the simulation process easier to understand; and managers may find the output, in the form of standard financial statements, easier to understand and accept. In addition, managers may have more confidence in, and provide more support for, a decision that they make themselves based on the prospective consequences generated by the simulation model.

One problem with simulation models is that they can easily generate enough output to overwhelm the user, leaving her unable to evaluate the results and determine the best decision. For example, suppose that in the debt-versus-equity financing problem mentioned above, the output consists of an income statement and balance sheet for each of many periods in the planning horizon and for each decision alternative. In this case, the volume of total output could be so overwhelming and confusing that it is of little help to the decision maker.

Another problem is that users of simulation models develop models capable of generating detailed financial statements, but fail to specify the criteria necessary for determining the best decision. If many different decision variables are to be considered and many different financial factors are to be compared, it is difficult to compare financial statements and judge one as superior. For example, one decision alternative may result in a more liquid current position but greater long-term debt, while another results in the opposite scenario. This leaves the user to decide how to evaluate the trade-off between liquidity and leverage. In such a situation, it is useful to add an objective function that measures the relative desirability of the results of the various decisions under consideration.

An objective function facilitates the ranking of alternative decisions, making it possible to analyze different decision scenarios and select the best one. However, there still may be too many decision alternatives to inspect all of the results, even when they are summarized by an objective function. For example, suppose a planning problem involves just 10 different decision variables—the amount of cash balance to carry, level of inventory, investment in fixed plant, and so forth—and suppose each of these decision variables can take on any of 15 different values. The number of different decision combinations would be $15^{10} = 5.77 \times 10^{11}$. It would certainly take a long time for the planner to examine the result of each possible decision combination. What is needed is an efficient way to search the myriad decision combinations to find the best one. This is the task performed by optimization models. An optimization model can quickly determine which of the many possible combinations of decisions best meets the specified objective.

1.6.2 Optimization

An optimization model determines which decision alternative leads to a desired objective. Whereas a simulation model simply shows the consequences of a decision specified by the user, the optimization model solves for the best decision.

An optimization model is like a simulation model to which has been added an expression that represents the goal of the decision and a set of equations and inequalities that constrain the choices. The constraints are usually closely related to the sets of equations that constitute the simulation model. The expression representing the goal of the decisions is the objective function.

The objective function is an equation that portrays the influence of the decision variables on the objective. The objective is the thing that the user wants to make as large as possible (maximize) or as small as possible (minimize). For example, if the user wants to identify the decisions that maximize the market value of the firm's common stock, the objective function would be a mathematical expression that shows the effect of the decision variables on the market value of the stock.

The decision variables represent the choices available to the decision maker. They might include the level of investment in inventory and how it is to be financed, with the financing choices being the amount to be borrowed and the number of new shares to be issued. If so, the objective function would show the influences of the level of inventory, the amount borrowed, and the number of new shares issued on the market value of the firm's stock. The purpose of the optimization model would be to find the values of these three variables that make the value of the stock as large as possible.

The constraints in an optimization model are sets of equations (or inequalities) that express the limitations that prevent the model from choosing the decision variables that make the objective arbitrarily large or small. Some of the constraints may express sub-goals and other objectives that the user wants to attain in addition to the primary goal specified by the objective function. Other constraint equations might be identical to those in a simulation model.

The optimization approach has two advantages over the simulation approach. First, it forces the user to clarify and clearly state the objectives. Second, it relieves the user of the confusing task of searching through a mountain of simulated results to evaluate numerous decision alternatives. However, because optimization models remove the decision of managers, they may receive less managerial support. Chapters 16 and 17 explain optimization models in greater detail.

1.7 WHAT DO WE GET OUT OF THE MODELING PROCESS?

The financial planning and modeling process can be expensive in terms of time, effort, and money. What benefits do we derive from this process? Are they worth more than the costs? To evaluate the cost/benefit trade-off we need to consider what we get out of the process and compare that to the costs. There are explicit and implicit payoffs from the modeling process. We need to consider both.

1.7.1 Explicit Benefits

The primary benefit of the modeling process is the opportunity for management to test their ideas and decisions before actually implementing them. Management can experiment with the model rather than the firm, exploring how their decisions relate to their overall objectives. With the ability to better anticipate the results of their decisions, managers should be able to make better decisions. In addition, a well-documented plan allows the planner to review and analyze the plan after it has been implemented in order to determine where triumphs and mistakes occurred. Moreover, the ease of exploring decision alternatives gives managers more time to consider broader issues, such as setting, defining, and evaluating policy alternatives. Thus, models help managers use their talents more effectively.

1.7.2 Implicit Benefits

Many benefits from the planning process are intangible and implicit in the process itself. The process forces the planner to systematize her thoughts about objectives, decisions, assumptions, and the interactions among parts of the firm and between variables. As pointed out by Ackoff:

> The value of planning to managers lies more in their participation in the process than in their consumption of its product …. Such participation stimulates the development of a deeper understanding of the business and its environment, and it forces the systematic formulation and evaluation of alternatives that would not otherwise be considered. (Ackoff, 1970, p. 137)

The process of building a model forces the planner to think carefully about the complex relationships between her decisions and the financial variables that constitute the firm. Managers must wrestle with the most basic questions about the firm's objectives and how to measure progress toward those objectives. In addition, a financial plan helps to build consistency between different parts of the firm in terms of assumptions, forecasts, decisions, and results. Finally, involving management in the planning process increases the likelihood they will trust and support the implementation of the planning output.

The costs of modeling and planning are also difficult to measure. Some of the costs are direct and explicit—the costs of the software and time of the planning staff. However, even when the costs are fully counted, it will be difficult to objectively evaluate the net payoff because it is so difficult to know what decisions would have been made if the process had never occurred. So, while we encourage you in your modeling efforts, unfortunately we do not yet have a model that will allow you to objectively and accurately evaluate the net benefits from planning and modeling.

1.8 SUMMARY

The purpose of this text is to help the student learn to construct and use models for corporate financial planning. To do this, it is important to provide some perspective, and the purpose of this introductory chapter has been to set the stage so that the models are not developed and used in a vacuum. We have tried to provide a framework for thinking about how we make plans, how models relate to those plans, and how we want to structure the models so that they can help us to realize the plans.

There are many important inputs to a financial plan, and these inputs correspond to the parts of the planning model. At the broadest level, the ingredients are the goals we are trying to attain, the decisions we have to make, and the links between the decisions and the goals. In the context of financial planning, the most important goal is to maximize the value of the firm, and one of the primary functions of the model is to show how the decisions relate to the goals.

In the future chapters we will explain different aspects of corporate financial planning models, and present different kinds of models. Part I of the text is the starting point, where we learn how the basic tools of financial analysis can provide the foundation for planning and developing planning models. In Chapter 2, we review some basic financial ratio analysis and get started in our model building with a model that will facilitate the analysis of a company. In Chapter 3, we continue to review tools and concepts of financial analysis that will be used subsequently in other financial planning models.

Equipped with this foundation of financial analysis, Section II shows how to build very useful financial planning models. In Chapter 4, we develop a simplified, but complete financial model of a firm. That model will provide the basis for many of the models and methods discussed in the remainder of the book. In Chapter 5, we show how to use Monte Carlo simulation to understand the uncertainty that the firm faces.

Part III of the text is about forecasting. Forecasts are one of the most basic inputs to financial planning models, so we need to have some understanding of forecasting and some ability to forecast to generate this basic input. Chapters 6, 7, and 8 are a primer on forecasting methods, so the model builder has some tools for developing the forecasts that go into the models.

Part IV adds details to our model. As explained in this introductory chapter, setting the goal and our decisions to reach the goal are crucial to planning and modeling. In finance, the goal is usually stated in terms of the value of the firm, so we need to understand and model firm value. Chapter 9 explains the details of modeling firm value. In Chapter 10 we model the firm's capital investment decisions, and in Chapter 11 we build models that allow us to examine how we are going to finance the firm and its investment plans.

The chapters in Part V are concerned with modeling security investments. Chapters 12 and 13 show how to model stock prices and how to construct efficient security portfolios, respectively. Options on securities are one of the places where modeling has been particularly useful, and Chapter 14 shows how to construct basic models of options.

The last section of the book, Part VI, covers financial optimization models. You are introduced to optimization methods and learn about some of the interesting and useful models for optimizing the firm's capital investment and working capital decisions.

If you have the time and are so diligent as to work your way through all these chapters, you will have been exposed to a wide variety of financial models. All of them are practical and useful. In addition, along the way, you will have learned not just about models, but about finance. At the end of our course in financial modeling, our students consistently comment about how much they learned about the theory of finance, how building models helped them to understand how the ideas of finance fit together, how the parts of the firm fit together, and how financial theory relates to the practical world. We certainly hope you will gain this same perspective from the following chapters.

PART I

Tools for Financial Planning and Modeling: Financial Analysis

CHAPTER 2

Tools for Financial Planning I: Financial Analysis

2.1 INTRODUCTION

If you are planning your trip and laying out your route, you need to know your starting point, you need to know your strengths and weaknesses, and you need to know the resources you have to support your trip. In addition, once you start the trip, you will need some criteria for judging your progress toward your destination. That is, you need signposts along the way to tell whether you are on the right road and whether you are making acceptable progress. Financial analysis provides methods and tools for assessing your current condition and strengths and weaknesses, understanding how you got to that condition, and provides data that describe the starting point for the trip into the future. This chapter and the next explain some of the basic tools of financial analysis that are useful for financial planning and modeling.

The financial planning tools that we will discuss in this chapter are financial analysis:

- Analysis of financial ratios

- Break-even analysis

- Analysis of operating and financial leverage

Chapter 3 adds to our planning toolbox with explanations of budgeting and planning:

- Cash flow

- Cash budgeting

- Sustainable growth

- Generation of *pro forma* financial statements

We assume you have some familiarity with the tools of financial analysis that we will discuss, and so we present the methods here as a refresher. More detailed discussions can be found in Brigham and Ehrhardt (2008), Foster (1986), Helfert (1991), and Higgins (1996).

2.2 FINANCIAL RATIO ANALYSIS

In this section we engage in our first modeling effort. The modeling objective is to construct a set of template models into which we can insert data from any target company so as to generate output spreadsheets that enable us to analyze the financial ratios of the target company and compare its performance to its industry. Financial ratios are important in financial modeling for two reasons. First, they provide guidelines for judging performance and identifying sources of strength and weakness in the firm's performance. Second, we use the firm's ratios as input parameters in the planning model of the firm that we will develop in Chapter 4.

The template system consists of several sheets in an Excel® workbook as shown in Exhibit 2.1. This workbook is structured so that we enter the data for the target company and related industry data in the input spreadsheets. The output spreadsheets then process the data so that we can

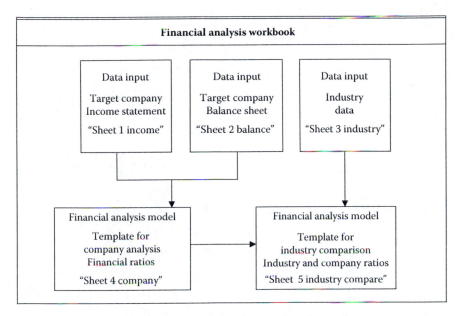

EXHIBIT 2.1   Spreadsheet workbook for financial analysis of a target company.

analyze the company, compare it to its industry, and obtain data for our planning model. In the following example, we enter the data from the target company's financial statements into sheets 1, 2, and 3. Sheets 4 and 5 process the data into a format suitable for our needs. To analyze another company, we simply replace the first company's data with data for the second company in the input sheets, and the results for the second company are shown in the output sheets.

2.2.1   Example: The Odd & Rich Corporation

The target company for financial analysis and subsequent modeling will be the Odd & Rich Corporation (O&R Corp.). The O&R Corp. is an industrial firm that manufactures a range of plastic components that are used as input by other manufacturers of consumer products, automobiles, aircraft, and computers. The manufacturing process involves plastic injection molding and automated assembly. The company uses its own sales force in addition to independent representatives and distributors, who sell the product as independent contractors. The company is somewhat cyclical, but it has been growing at a respectable rate in recent years, and it is expected that its sales will continue to grow in the foreseeable future.

	A	B	C	D	E	F
1	DATA INPUT					
2	User enters Income Statement data for					
3	target company					
4						
5				ODD & RICH CORP		
6						
7				Income Statement		
8				($ Millions)		
9						
10		2006	2007	2008	2009	2010
11		-4	-3	-2	-1	0
12						
13	Sales	748.5	792.6	800.2	944.0	1,000.0
14	Cost of Goods Sold	470.1	496.7	504.8	566.7	607.7
15	Gross Profit	278.4	295.9	295.5	377.3	392.3
16	Selling, General & Administrative	123.4	130.1	131.1	169.3	122.3
17	Operating Income Before Deprec.	155.1	165.8	164.4	208.0	270.0
18	Depreciation	57.8	53.6	54.3	49.9	53.8
19	Operating Profit	97.3	112.2	110.1	158.1	216.2
20	Interest Expense	12.8	3.6	18.7	3.1	3.4
21	Non-Operating Income/Expense & Special Items	(2.2)	2.7	6.8	0.0	0.0
22	Pretax Income	82.3	111.3	98.2	155.0	212.8
23	Income Tax	29.7	29.8	26.4	62.0	85.0
24	Net Income Before Extraordinary Items	52.7	81.5	71.8	93.0	127.8
25	Extraordinary Items	0.0	0.0	0.0	0.0	0.0
26	Net Income	52.7	81.5	71.8	93.0	127.8
27						
28	Dividends	21.1	32.6	43.1	55.8	76.7
29	Retained Earnings	31.6	48.9	28.7	37.2	51.1

EXHIBIT 2.2 Input data "Sheet 1 Income." O&R Corp. income statement.

Assume the current date is year-end 2010, and we have gathered income statement and balance sheet data for the O&R Corp. for the last five years (2006–2010) and entered it into a spreadsheet as shown in Exhibits 2.2 and 2.3, ranging from the earliest year (2006), designated as year −4 shown on the left, to the year just ended (2010), designated as year 0 shown on the far right. These reports correspond to sheets 1 and 2 referenced in the diagram of Exhibit 2.1. We have also assembled average ratios for other companies in the same industry that we entered into the spreadsheet as shown in Exhibit 2.4, which corresponds to the data input sheet 3 in Exhibit 2.1.

We distinguish between data input and the financial analysis model itself. Sheets 1, 2, and 3 are the data that will feed into the financial analysis model that is comprised of sheets 4 and 5 in Exhibit 2.1. The *model* consists of the formulas for calculating financial ratios and the formatting of output, so we have the information in a usable form. That is, sheets 4 and 5 have no numbers in them—only formulas that operate on the input data from sheets 1, 2, and 3. For example, as shown in Exhibit 2.5, Sales in cell B10 are picked up from the input sheet with the formula ="Sheet 1 Income"!B13, which means that the content of

	A	B	C	D	E	F
1	**DATA INPUT**					
2	User enters Balance Sheet data for					
3	target company					
4						
5				**ODD & RICH CORPORATION**		
6				Balance Sheet		
7				($ Millions)		
8						
9		2006	2007	2008	2009	2010
10		-4	-3	-2	-1	0
11						
12	**Assets**					
13	Cash & Equivalents	129.1	149.8	143.1	160.4	170.0
14	Net Receivables	169.1	189.1	166.1	177.5	208.1
15	Inventories	127.3	132.0	128.0	142.8	148.2
16	Other Current Assets	119.0	117.6	119.1	189.5	183.7
17	Total Current Assets	544.5	588.6	556.2	670.2	710.0
18	Gross Plant, Property & Equipment	1,142.6	1,200.6	1,238.3	1,356.7	1,517.8
19	Accumulated Depreciation	288.4	342.0	395.6	449.2	502.8
20	Net Plant, Property & Equipment	854.2	858.6	842.7	907.5	1,015.0
21	Other Assets	237.7	280.0	342.9	435.9	409.0
22	Total Assets	1,636.4	1,727.3	1,741.8	2,013.7	2,134.0
23						
24	**Liabilities**					
25	Long Term Debt Due In One Year	8.7	9.3	11.8	6.9	7.0
26	Notes Payable	0.0	21.5	35.6	0.0	0.0
27	Accounts Payable	152.9	173.5	144.3	175.5	279.1
28	Taxes Payable	18.7	14.0	15.8	41.9	45.3
29	Accrued Expenses & Other Liabilities	117.0	135.0	128.2	342.0	268.6
30	Total Current Liabilities	297.3	353.2	335.7	566.4	600.0
31	Long Term Debt	31.4	26.7	21.8	37.7	73.3
32	Deferred Taxes & Other Liabilities	42.6	33.5	21.5	0.0	0.0
33	Common Stock	20.0	20.0	20.0	20.0	20.0
34	Capital Surplus	80.0	80.0	80.0	80.0	80.0
35	Retained Earnings	1,165.1	1,214.0	1,262.9	1,311.8	1,360.7
36	Less: Treasury Stock	0.0	0.2	0.0	2.1	0.0
37	Common Equity	1,265.1	1,313.8	1,362.9	1,409.6	1,460.7
38	Total Liabilities & Equity	1,636.4	1,727.3	1,741.8	2,013.7	2,134.0

EXHIBIT 2.3   Input data "Sheet 2 Balance." O&R Corp. balance sheet.

cell B10 is equal to the content of cell B13 on the sheet named "Sheet 1 Income."

In this case, we want the model to generate financial ratios and other measures that will enable us to evaluate the financial performance of the target company. Exhibit 2.5 shows the template equations for the income statement section from "Sheet 4 Company" showing that each cell equation points to "Sheet 1 Income," which is the sheet in which the user entered the basic data. With just the formulas entered in the analysis sheet, when a different company is the target, the user needs to change only the input sheet data. The balance sheet (in cells A29:F59 of "Sheet 4 Company") is set up in the same manner, with each cell formula pointing to the corresponding cell in "Sheet 2 Balance."

	A	B	C	D	E	F
1	**DATA INPUT**					
2	User enters Industry Average Ratios					
3	for target company					
4						
5				**Industry Average Ratios**		
6				**Plastic Products**		
7				**SIC**	**3089**	
8						
9		2006	2007	2008	2009	2010
10		-4	-3	-2	-1	0
11	**Liquidity**					
12	Current Ratio	2.3	2.1	2.1	2.2	2.1
13	Quick Ratio	1.4	1.2	1.2	1.2	1.2
14						
15	**Activity: Operating Efficiency**					
16	Receivables Turnover	6.6	6.7	6.8	8.8	7.9
17	Average Collection Period (Days)	57.9	58.4	54.2	48.4	49.4
18	Inventory Turnover	5.3	5.2	5.8	6.2	6.3
19	Days to Sell Inventory	85.0	90.8	81.5	76.2	75.6
20	Operating Cycle (Days)	142.9	149.1	135.7	124.6	125.1
21	Sales/Net Fixed Assets	3.9	3.7	4.0	4.5	5.3
22	Total Asset Turnover	1.1	1.1	1.1	1.3	1.3
23						
24	**Profitability %**					
25	EBITDA/Sales	13.8	13.6	12.6	13.8	13.5
26	EBIT/Sales	7.8	7.6	6.8	8.0	8.3
27	Pretax Profit Margin	5.0	0.9	3.4	4.1	5.8
28	Net Profit Margin (NI / Sales)	2.6	(0.7)	2.3	3.0	4.7
29						
30	**Return on Investment %**					
31	Return on Assets	3.3	(2.7)	0.7	2.9	5.4
32	Return on Equity	9.4	(0.1)	(16.4)	8.6	16.1
33	Return on Invested Capital	4.3	(5.2)	(0.9)	3.5	8.3
34						
35	**Leverage**					
36	Interest Coverage Before Tax	261.8	129.2	36.2	145.5	67.0
37	Interest Coverage After Tax	231.3	119.3	32.6	98.1	58.5
38	Long-Term Debt/Common Equity	91.8	175.6	456.1	346.1	168.0
39	Total Debt/Invested Capital	64.8	65.9	67.7	60.3	60.3
40	Total Debt/Total Assets	36.6	42.8	41.4	40.5	38.3
41	Total Assets/Common Equity	4.3	4.8	8.6	7.0	4.3
42						
43	**Dividends**					
44	Dividend Payout (%)	9.4	14.4	11.4	13.9	9.6
45	Dividend Yield (%)	1.0	1.2	1.0	1.2	0.9

EXHIBIT 2.4   Input data "Sheet 3 Industry." Industry average financial ratios.

The income statement and balance sheet are repeated on the model sheet to facilitate the computation of the financial ratios that are the main output of the company analysis model. There are three ratios sections in the model that are shown in Exhibits 2.6 through 2.9, with "A" being the numerical output, and "B" showing the formulas.

The output of Sheet 4 is shown in four parts as Exhibits 2.6A, 2.7A, 2.8A, and 2.9A. Exhibit 2.6A summarizes some of the standard financial ratios

	A	B
1		
2		
3		='Sheet 1 Income'!C5
4		='Sheet 1 Income'!C7
5		='Sheet 1 Income'!D8
6		
7		='Sheet 1 Income'!B10
8		='Sheet 1 Income'!B11
9		
10	Sales	='Sheet 1 Income'!B13
11	Cost of Goods Sold	='Sheet 1 Income'!B14
12	Selling, General & Admin	='Sheet 1 Income'!B16
13	Operating Income Before Deprec	=Sales-CGS-SGA
14	Other Inc/Exp	='Sheet 1 Income'!B21
15	EBITDA	=B13+B14
16	Depreciation	='Sheet 1 Income'!B18
17	EBIT	=EBITDA-Depr
18	Interest	='Sheet 1 Income'!B20
19	EBT	=EBIT-Intr
20	Tax	='Sheet 1 Income'!B23
21	Income Before Extraordinary Items	=EBT-Tax
22	Extraordinary Items	='Sheet 1 Income'!B25
23	Net Income	=IBEI+ExI
24		
25	Dividends	='Sheet 1 Income'!B28
26	Retained Earnings	=NI-Div

EXHIBIT 2.5 Income statement model template data drawn from input "Sheet 1 Income."

that help us to evaluate the company's performance. An extra column on the right side computes the average over five years for the various ratios. Exhibits 2.7 and 2.8 shows the financial statement in "common-size" format, where the income data are stated as a percent of sales and the balance sheet is stated as a percent of total assets. That is, the income statement ratios are calculated for each line and each year as the income or cost amount divided by that year's sales. Similarly, the common-size balance sheet section divides each balance sheet item by the total assets for that year, and Exhibit 2.9 shows the balance sheet accounts as a percent of sales.

Two ways to gain perspective from the financial ratios are (1) to look at the trend of ratios over time, and (2) to compare them to other companies in the same industry and with the average ratios for the industry. The trend of the ratios over time, is a good starting point for our analysis because significant changes or trends in ratios frequently signal potential problems or promising improvements. While scanning across the line of ratios may

(A)

ODD & RICH CORP — Financial Ratios

#		2006 (-4)	2007 (-3)	2008 (-2)	2009 (-1)	2010 (0)	Average
65	**Profitability Ratios**						
66	**Return on Investment**						
67	EBITDA / TA	9.3%	9.8%	9.8%	9.7%	11.7%	10.1%
68	EBIT / TA (BEP)	5.8%	6.7%	6.7%	7.4%	9.3%	7.2%
69	NI / TA (ROA)	3.2%	4.7%	4.1%	4.3%	5.5%	4.4%
70	NI / Equity (ROE)	4.2%	6.2%	5.3%	6.6%	8.7%	6.2%
71							
72	**Profit Margin**						
73	Gross Profit Margin	37.2%	37.3%	36.9%	40.0%	39.2%	38.1%
74	CGS / Sales	62.8%	62.7%	63.1%	60.0%	60.8%	61.9%
75	SGA / Sales	16.5%	16.4%	16.4%	17.9%	12.2%	15.9%
76	Op Cost / Sales	79.3%	79.1%	79.5%	78.0%	73.0%	77.8%
77	EBITDA / Sales	20.4%	21.3%	21.4%	22.0%	27.0%	22.4%
78	Deprec. / Sales	7.7%	6.8%	6.8%	5.3%	5.4%	6.4%
79	EBIT / Sales	12.7%	14.5%	14.6%	16.7%	21.6%	16.0%
80	Interest / Debt (-1)	9.0%	9.0%	32.5%	7.6%	7.6%	13.4%
81	Tax Rate	36.0%	26.8%	26.9%	40.0%	39.9%	33.9%
82	NI bEI / Sales	7.04%	10.28%	8.97%	9.85%	12.78%	9.8%
83	NI / Sales	7.04%	10.28%	8.97%	9.85%	12.78%	9.8%
84							
85	**Operating Efficiency**						
86	Sales / TA	0.46	0.46	0.46	0.44	0.43	44.9%
87	Sales / NFA	0.88	0.92	0.95	1.04	0.99	95.5%
88	Sales / AR	4.43	4.19	4.82	5.32	4.81	471.2%
89	AR / Sales	22.59%	23.86%	20.75%	18.81%	20.81%	21.4%
90	Receivables Period (days)	81.3	85.9	74.7	67.7	74.9	7691.4%
91	Inventory Turnover	3.7	3.8	3.9	4.0	4.1	389.3%
92	Inventory Period (days)	97.5	95.7	91.3	90.7	87.8	9259.5%
93	Payable Period (days)	117.1	125.7	102.9	111.5	165.3	12451.4%
94	Payable / Sales	20.4%	21.9%	18.6%	18.6%	27.9%	21.4%
95	Operating Period (days)	178.8	181.6	166.0	158.4	162.7	16950.9%
96	Cash Cycle (days)	61.7	55.8	63.1	46.9	-2.6	4499.5%
97							
98	**Leverage**						
99	Total Debt / TA	22.7%	23.9%	21.8%	28.1%	29.1%	25.1%
100	IB Debt / TA	2.5%	3.3%	4.0%	2.1%	3.5%	3.1%
101	Debt / Equity	29.4%	31.5%	27.8%	42.8%	46.1%	35.5%
102	IB Debt / Equity	3.2%	4.4%	5.1%	3.2%	5.5%	4.3%
103	EBIT / Interest	7.44	31.88	6.25	51.00	63.59	3203.2%
104	Total Assets / Equity	1.29	1.31	1.28	1.52	1.59	139.9%
105							
106	**Liquidity**						
107	CA / CL	1.83	1.67	1.66	1.18	1.18	150.4%
108	(Cash + AR) / CL	1.00	0.96	0.92	0.60	0.63	82.2%
109	Cash / (Op Cst/360)	78.3	86.1	81.0	78.5	83.8	8153.4%
110	Cash / Sales	17.2%	18.9%	17.9%	17.0%	17.0%	17.6%
111							
112	**Other Relationships**						
113	Dividend Payout	40.0%	40.0%	60.0%	60.0%	60.0%	52.0%

(B)

Financial Ratios

#	A	B	C	G (Average)
61		=D29		
63		=B33	=C33	
64		=B34	=C34	
65	**Profitability Ratios**			
66	**Return on Investment**			
67	EBITDA / TA	=EBITDA/TA	=C15/C46	=AVERAGE(B67:F67)
68	EBIT / TA (BEP)	=EBIT/TA	=C17/C46	=AVERAGE(B68:F68)
69	NI / TA (ROA)	=NI/TA	=C23/C46	=AVERAGE(B69:F69)
70	NI / Equity (ROE)	=NI/Equity	=C23/C58	=AVERAGE(B70:F70)
71				
72	**Profit Margin**			
73	Gross Profit Margin	=(Sales-CGS)/Sales	=(C10-C11)/C10	=AVERAGE(B73:F73)
74	CGS / Sales	=CGS/Sales	=C11/C10	=AVERAGE(B74:F74)
75	SGA / Sales	=SGA/Sales	=C12/C10	=AVERAGE(B75:F75)
76	Op Cost / Sales	=B74+B75	=C74+C75	=AVERAGE(B76:F76)
77	EBITDA / Sales	=EBITDA/Sales	=C15/C10	=AVERAGE(B77:F77)
78	Deprec. / Sales	=Depr/Sales	=C16/C10	=AVERAGE(B78:F78)
79	EBIT / Sales	=EBIT/Sales	=C17/C10	=AVERAGE(B79:F79)
80	Interest / Debt (-1)	=C18(STD+LTD)	=C18/(STD+LTD)	=AVERAGE(B80:F80)
81	Tax Rate	=Tax/EBT	=C20/C19	=AVERAGE(B81:F81)
82	NI bEI / Sales	=IBEI/Sales	=C21/C10	=AVERAGE(B82:F82)
83	NI / Sales	=NI/Sales	=C23/C10	=AVERAGE(B83:F83)
84				
85	**Operating Efficiency**			
86	Sales / TA	=Sales/TA	=C10/C46	=AVERAGE(B86:F86)
87	Sales / NFA	=Sales/NFA	=C10/C44	=AVERAGE(B87:F87)
88	Sales / AR	=Sales/AR	=C10/C37	=AVERAGE(B88:F88)
89	AR / Sales	=AR/Sales	=C37/C10	=AVERAGE(B89:F89)
90	Receivables Period (days)	=360/B88	=360/C88	=AVERAGE(B90:F90)
91	Inventory Turnover	=CGS/Invent	=C11/C38	=AVERAGE(B91:F91)
92	Inventory Period (days)	=360/B91	=360/C91	=AVERAGE(B92:F92)
93	Payable Period (days)	=AP/(CGS/360)	=C49/(C11/360)	=AVERAGE(B93:F93)
94	Payable / Sales	=AP/Sales	=C49/C10	=AVERAGE(B94:F94)
95	Operating Period (days)	=B90+B92	=C90+C92	=AVERAGE(B95:F95)
96	Cash Cycle (days)	=B95-B93	=C95-C93	=AVERAGE(B96:F96)
97				
98	**Leverage**			
99	Total Debt / TA	=(CL+LTD+OLTLiab)/TA	=(C52+C54+C55)/C46	=AVERAGE(B99:F99)
100	IB Debt / TA	=(STD+LTD)/TA	=(C50+C54)/C46	=AVERAGE(B100:F100)
101	Debt / Equity	=(CL+LTD+OLTLiab)/Equity	=(C52+C54+C55)/C58	=AVERAGE(B101:F101)
102	IB Debt / Equity	=(STD+LTD)/Equity	=(C50+C54)/C58	=AVERAGE(B102:F102)
103	EBIT / Interest	=EBIT/Intr	=C17/C18	=AVERAGE(B103:F103)
104	Total Assets / Equity	=TA/Equity	=C46/C58	=AVERAGE(B104:F104)
105				
106	**Liquidity**			
107	CA / CL	=CA/CL	=C40/C52	=AVERAGE(B107:F107)
108	(Cash + AR) / CL	=(Csh+AR)/CL	=(C36+C37)/C52	=AVERAGE(B108:F108)
109	Cash / (Op Cst/360)	=Csh/((CGS+SGA)/360)	=C36/((C11+C12)/360)	=AVERAGE(B109:F109)
110	Cash / Sales	=Csh/Sales	=C36/C10	=AVERAGE(B110:F110)
111				
112	**Other Relationships**			
113	Dividend Payout	=Div/NI	=C25/C23	=AVERAGE(B113:F113)

EXHIBIT 2.6 (A) Financial ratio section of financial analysis model template "Sheet 4 Company." (B) Formulas for first two columns and averages of financial ratio section "Sheet 4 Company."

(A)

	J	K	L	M	N	O
1						
2						
3		**ODD & RICH CORP**				
4		Income Statement				
5		Common Size: % of Sales				
6						
7	2006	2007	2008	2009	2010	
8	-4	-3	-2	-1	0	
9						Average
10 Sales	100.0%	100.0%	100.0%	100.0%	100.0%	100.0%
11 Cost of Goods Sold	62.8%	62.7%	63.1%	60.0%	60.8%	61.87%
12 Selling, General & Admin	16.5%	16.4%	16.4%	17.9%	12.2%	15.89%
13 Operating Income Before Deprec	20.7%	20.9%	20.5%	22.0%	27.0%	22.24%
14 Other Inc/Exp	-0.3%	0.3%	0.9%	0.0%	0.0%	0.18%
15 EBITDA	20.4%	21.3%	21.4%	22.0%	27.0%	22.42%
16 Depreciation	7.7%	6.8%	6.8%	5.3%	5.4%	6.39%
17 EBIT	12.7%	14.5%	14.6%	16.7%	21.6%	16.04%
18 Interest	1.7%	0.5%	2.3%	0.3%	0.3%	1.03%
19 EBT	11.0%	14.0%	12.3%	16.4%	21.3%	15.00%
20 Tax	4.0%	3.8%	3.3%	6.6%	8.5%	5.22%
21 Income Before Extraordinary Items	7.0%	10.3%	9.0%	9.9%	12.8%	9.78%
22 Extraordinary Items	0.0%	0.0%	0.0%	0.0%	0.0%	0.00%
23 Net Income	7.0%	10.3%	9.0%	9.9%	12.8%	9.78%
24						
25 Dividends	2.8%	4.1%	5.4%	5.9%	7.7%	5.18%
26 Retained Earnings	4.2%	6.2%	3.6%	3.9%	5.1%	4.61%

(B)

	I	J	K
1			
2			
3			=B3
4			=B4
5		Common Size: % of Sales	
6			
7		=B7	=C7
8		=B8	=C8
9			
10	=A10	=Sales/Sales	=C10/C10
11	=A11	=CGS/Sales	=C11/C$10
12	=A12	=SGA/Sales	=C12/C$10
13	=A13	=B13/Sales	=C13/C$10
14	=A14	=B14/Sales	=C14/C$10
15	=A15	=EBITDA/Sales	=C15/C$10
16	=A16	=Depr/Sales	=C16/C$10
17	=A17	=EBIT/Sales	=C17/C$10
18	=A18	=Intr/Sales	=C18/C$10
19	=A19	=EBT/Sales	=C19/C$10
20	=A20	=Tax/Sales	=C20/C$10
21	=A21	=IBEI/Sales	=C21/C$10
22	=A22	=ExI/Sales	=C22/C$10
23	=A23	=NI/Sales	=C23/C$10
24			
25	=A25	=Div/Sales	=C25/C$10
26	=A26	=RE/Sales	=C26/C$10

EXHIBIT 2.7 (A) Target company income statement common-size percent of sales "Sheet 4 Company." (B) Formulas for first two columns and averages of common-size income statement section "Sheet 4 Company."

(A)

	I	J	K	L	M	N	O
29			ODD & RICH CORP				
30			Balance Sheet				
31			Common Size: % of Total Assets				
32							
33		2006	2007	2008	2009	2010	
34		-4	-3	-2	-1	0	Average
35	Assets						
36	Cash	7.9%	8.7%	8.2%	7.5%	7.3%	7.9%
37	Accounts Receivable	10.3%	10.9%	9.5%	8.3%	9.0%	9.6%
38	Inventory	7.8%	7.6%	7.3%	6.6%	6.4%	7.2%
39	Other CA	7.3%	6.8%	6.8%	8.8%	7.9%	7.5%
40	Current Assets	33.3%	34.1%	31.9%	31.2%	30.6%	32.2%
41							
42	Gross Fixed Assets	69.8%	69.5%	71.1%	63.2%	65.5%	67.8%
43	Accum Deprec	17.6%	19.8%	22.7%	20.9%	21.7%	20.6%
44	Net Fixed Assets	52.2%	49.7%	48.4%	42.3%	43.8%	47.3%
45	Other Long Term Assets	14.5%	16.2%	19.7%	26.5%	25.6%	20.5%
46	Total Assets	100.0%	100.0%	100.0%	100.0%	100.0%	
47							
48	Liabilities						
49	Accounts Payable	9.3%	10.0%	8.3%	8.2%	12.0%	9.6%
50	Short Term Debt	0.5%	1.8%	2.7%	0.3%	0.3%	1.1%
51	Other CL	8.3%	8.6%	8.3%	17.9%	13.5%	11.3%
52	Current Liabilities	18.2%	20.5%	19.3%	26.4%	25.9%	22.0%
53							
54	Long Term Debt	1.9%	1.5%	1.2%	1.8%	3.2%	1.9%
55	Other LT Liabilities	2.6%	1.9%	1.2%	0.0%	0.0%	1.2%
56	Common Stock	6.1%	5.8%	5.7%	4.6%	4.3%	5.3%
57	Earned Surplus	71.2%	70.3%	72.5%	61.1%	58.7%	66.8%
58	Common Equity	77.3%	76.1%	78.2%	65.6%	63.0%	72.1%
59	Total Liabilities & Equity	100.0%	100.0%	100.0%	93.8%	92.1%	

(B)

	I	J	K
29			=D29
30			=D30
31		Common Size: % of Total Assets	
32			
33		=J7	=K7
34		=B34	=C34
35	=A35		
36	=A36	=Csh/TA	=C36/C$46
37	=A37	=AR/TA	=C37/C$46
38	=A38	=Invent/TA	=C38/C$46
39	=A39	=OCA/TA	=C39/C$46
40	=A40	=CA/TA	=C40/C$46
41			
42	=A42	=GFA/TA	=C42/C$46
43	=A43	=AccumDep/TA	=C43/C$46
44	=A44	=NFA/TA	=C44/C$46
45	=A45	=OAssets/TA	=C45/C$46
46	=A46	=TA/TA	=C46/C$46
47			
48	=A48		
49	=A49	=AP/TA	=C49/C$46
50	=A50	=STD/TA	=C50/C$46
51	=A51	=OCL/TA	=C51/C$46
52	=A52	=CL/TA	=C52/C$46
53			
54	=A54	=LTD/TA	=C54/C$46
55	=A55	=OLTLiab/TA	=C55/C$46
56	=A56	=CS/TA	=C56/C$46
57	=A57	=ErnSrpls/TA	=C57/C$46
58	=A58	=Equity/TA	=C58/C$46
59	=A59	=TLiab/TA	=C59/C$46

EXHIBIT 2.8 (A) Target company balance sheet common-size percent of total assets "Sheet 4 Company." (B) Formulas for first two columns and averages of common-size balance sheet section "Sheet 4 Company."

(A)

	J	K	L	M	N	O
		Assets & Liabilities as a Percent of Sales				
62						
63						
64	2006	2007	2008	2009	2010	
65	-4	-3	-2	-1	0	Average
66 **Assets**						
67 Cash / Sales	17.2%	18.91%	17.88%	16.99%	17.00%	17.6%
68 Accounts Receivable / Sales	22.6%	23.86%	20.75%	18.81%	20.81%	21.4%
69 Inventory / Sales	17.0%	16.65%	16.00%	15.13%	14.82%	15.9%
70 Other CA / Sales	15.9%	14.84%	14.88%	20.07%	18.37%	16.8%
71 Current Assets / Sales	72.7%	74.26%	69.51%	71.00%	71.00%	71.7%
72						
73 Gross Fixed Assets / Sales	152.7%	151.48%	154.74%	143.72%	151.78%	150.9%
74 Accum Deprec / Sales	38.5%	43.15%	49.44%	47.58%	50.28%	45.8%
75 Net Fixed Assets / Sales	114.1%	108.33%	105.30%	96.14%	101.50%	105.1%
76 Other Long Term Assets / Sales	31.8%	35.33%	42.85%	60.40%	59.19%	45.9%
77 Total Assets / Sales	218.6%	217.92%	217.66%	227.53%	231.69%	222.7%
78						
79 **Liabilities**						
80 Accounts Payable / Sales	20.4%	21.89%	18.03%	18.59%	27.91%	21.4%
81 Short Term Debt / Sales	1.2%	3.89%	5.92%	0.73%	0.70%	2.5%
82 Other CL / Sales	18.1%	18.80%	18.00%	40.67%	31.39%	25.4%
83 Current Liabilities / Sales	39.7%	44.57%	41.95%	59.99%	60.00%	49.2%
84						
85 Long Term Debt / Sales	4.2%	3.37%	2.72%	3.99%	7.33%	4.3%
86 Other LT Liabilities / Sales	5.7%	4.22%	2.68%	0.00%	0.00%	2.5%
87 Common Stock / Sales	13.4%	12.59%	12.50%	10.37%	10.00%	11.8%
88 Earned Surplus / Sales	155.7%	153.17%	157.82%	138.96%	136.07%	148.3%
89 Common Equity / Sales	169.0%	165.76%	170.32%	149.33%	146.07%	160.1%
90 Total Liabilities & Equity / Sales	218.6%	217.93%	217.66%	213.32%	213.40%	216.2%
91						
92 **Other Relationships**						
93 Deprec / GFA-1		4.7%	4.5%	4.0%	4.0%	4.3%
94 Deprec / NFA-1		6.3%	6.3%	5.9%	5.9%	6.1%

(B)

	I	J	K
			Assets & Liabilities as a Percent of Sales
62			
63			
64		=B64	=C64
65		=B65	=C65
66	=A35		
67	Cash / Sales		=C36/C$10
68	Accounts Receivable / Sales		=C37/C$10
69	Inventory / Sales		=C38/C$10
70	Other CA / Sales		=C39/C$10
71	Current Assets / Sales		=C40/C$10
72			
73	Gross Fixed Assets / Sales		=C42/C$10
74	Accum Deprec / Sales		=C43/C$10
75	Net Fixed Assets / Sales		=C44/C$10
76	Other Long Term Assets / Sales		=C45/C$10
77	Total Assets / Sales		=C46/C$10
78			
79	Liabilities		
80	Accounts Payable / Sales		=C49/C$10
81	Short Term Debt / Sales		=C50/C$10
82	Other CL / Sales		=C51/C$10
83	Current Liabilities / Sales		=C52/C$10
84			
85	Long Term Debt / Sales		=C54/C$10
86	Other LT Liabilities / Sales		=C55/C$10
87	Common Stock / Sales		=C56/C$10
88	Earned Surplus / Sales		=C57/C$10
89	Common Equity / Sales		=C58/C$10
90	Total Liabilities & Equity / Sales		=C59/C$10
91			
92	**Other Relationships**		
93	Deprec / GFA-1		=C16/GFA
94	Deprec / NFA-1		=C16/NFA

EXHIBIT 2.9 (A) Target company balance sheet assets and liabilities as a percent of sales "Sheet 4 Company." (B) Formulas for first two columns of balance sheet percent of sales "Sheet 4 Company."

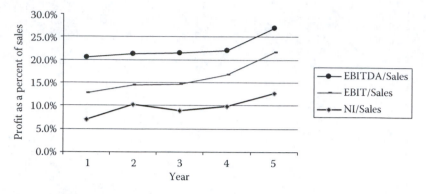

EXHIBIT 2.10   Excel graph of O&R Corp.'s profit margins.

quickly alert you to changes that may be important, graphing the trend of ratios is an even more effective means to highlight the trends. For example, Exhibit 2.10 shows an Excel graph of the O&R Corp.'s profit margins clearly showing an upward trend. The picture of this trend helps us quickly see one of the important aspects of the O&R Corp.'s operations. Similarly, pictures of the trends of other ratios can help us visualize what is happening to the firm. The usefulness of your financial analysis model can be enhanced by including graphs of the more important ratios.

Exhibit 2.11 compares our target company with average ratios for the industry taken from the input data sheet 3. For comparing our company with the industry, it is important that both the industry and company ratios be computed in the same manner. The average ratio for the industry simply gives us something with which we can compare our company's ratio. However, it is important to remember that the industry average is only an indication of what other companies are doing. It does not mean that the other companies are doing it right. Whole industries can be unhealthy, and their average may be nothing to aspire to. Nevertheless, using these averages provides a starting point to gain perspective about how our target company is doing compared to others.

In comparing with the industry, we need to look for indications that the company is performing either better or worse than the industry average, and look at trends and significant changes. Scanning for significant differences between the company and industry and watching for trends can alert us to potential problems or sources of strengths for our company. For example, the industry average net profit margin ranged from a low

	Industry	Plastic Products						
	Company	Odd & Rich Corp						
	SIC	3089						

Comparison of Company and Industry Performance Ratios

		Year -4	Year -3	Year -2	Year -1	Most Recent Year
PROFITABILITY						
Operating Margin Before Depr (%)	Company	20.72%	20.92%	20.54%	22.03%	27.00%
	Industry	13.79	13.57	12.64	13.80	13.51
Operating Margin After Depr (%)	Company	13.00%	14.16%	13.75%	16.75%	21.62%
	Industry	7.83	7.64	6.77	8.01	8.29
Pretax Profit Margin (%)	Company	11.00%	14.05%	12.27%	16.42%	21.28%
	Industry	4.97	0.90	3.39	4.08	5.79
Net Profit Margin (%)	Company	7.04%	10.28%	8.97%	9.85%	12.78%
	Industry	2.55	-0.74	2.29	3.00	4.68
LEVERAGE						
Long-Term Debt/Common Equity (%)	Company	0.02	0.02	0.02	0.03	0.05
	Industry	91.75	175.60	456.08	346.12	167.99
Total Debt/Invested Capital (%)	Company	3.1%	4.2%	4.8%	3.1%	5.2%
	Industry	64.84	65.93	67.70	60.27	60.35
Total Debt/Total Assets (%)	Company	2.4%	3.3%	4.0%	2.2%	3.8%
	Industry	36.59	42.79	41.36	40.47	38.35
ACTIVITY						
Inventory Turnover	Company	3.69	3.76	3.94	3.97	4.10
	Industry	5.28	5.16	5.81	6.22	6.31
Receivables Turnover	Company	4.43	4.19	4.82	5.32	4.81
	Industry	6.64	6.66	6.83	8.84	7.92
Total Asset Turnover	Company	0.46	0.46	0.46	0.47	0.47
	Industry	1.13	1.12	1.14	1.30	1.30
DIVIDENDS						
Dividend Payout (%)	Company	40.0%	40.0%	60.0%	60.0%	60.0%
	Industry	9.35	14.41	11.40	13.86	9.55

Comparison of Company and Industry Performance Ratios

		Year -4	Year -3	Year -2	Year -1	Most Recent Year
PROFITABILITY						
Return on Assets (%)	Company	3.22%	4.72%	4.12%	4.62%	5.99%
	Industry	3.25	-2.71	0.73	2.93	5.38
Return on Equity (%)	Company	4.16%	6.20%	5.27%	6.60%	8.75%
	Industry	9.36	-0.14	-16.41	8.56	16.11
Return on Investment (%)	Company	2.44%	3.65%	2.07%	2.57%	3.33%
	Industry	4.29	-5.20	-0.88	3.53	8.34
LEVERAGE						
Interest Coverage Before Tax	Company	7.44	31.88	6.25	51.00	63.59
	Industry	261.83	129.23	36.22	145.51	67.00
Interest Coverage After Tax	Company	5.1	23.6	4.8	31.0	38.6
	Industry	231.29	119.26	32.62	98.05	58.49
Total Assets/Common Equity	Company	1.29	1.31	1.28	1.43	1.46
	Industry	4.35	4.77	8.61	6.95	4.30
ACTIVITY						
Average Collection Period (Days)	Company	81.34	85.89	74.72	67.70	74.92
	Industry	57.88	58.35	54.20	48.35	49.45
Days to Sell Inventory	Company	97.49	95.68	91.29	90.71	87.80
	Industry	84.97	90.78	81.54	76.23	75.62
Operating Cycle (Days)	Company	178.83	181.57	166.01	158.41	162.72
	Industry	142.86	149.14	135.75	124.58	125.07

EXHIBIT 2.11 Sheet 5 output. Comparison of company and industry ratios.

of −0.74% to a high of 4.68%, with a recent upward trend. O&R Corp.'s margin has been well above the industry, and is trending upward. This is clearly a very positive indicator. However, its turnover of total assets has averaged about 0.45 with little trend, whereas the industry has enjoyed a turnover almost three times as high at an average of about 1.2, with clear improvement over the last five years. This indicates a major weakness in O&R Corp.'s performance. After this preliminary examination of the company and industry ratio and trends, we can do a more systematic decomposition of our company's performance with a DuPont analysis.

2.2.2 DuPont Analysis

The main purpose of financial analysis is to find and understand the strengths and weaknesses of the company. The "DuPont analysis" is a very useful structure that provides a systematic way of pinpointing the sources of strengths and weaknesses in a firm's operations. It decomposes the return on equity (ROE) into three components: the net profit margin, asset turnover, and the equity multiplier. The DuPont format is

$$\text{Return on Equity} = (\text{Net Profit Margin}) \times (\text{Total Asset Turnover})$$
$$\times (\text{Equity Multiplier})$$

$$\left(\frac{\text{Net Income}}{\text{Equity}} \right) = \left(\frac{\text{Net Income}}{\text{Sales}} \right) \times \left(\frac{\text{Sales}}{\text{Total Assets}} \right) \times \left(\frac{\text{Total Assets}}{\text{Equity}} \right) \quad (2.1)$$

For the O&R Corp., in the most recent year (year 5) this is calculated as

$$\left(\frac{127.8}{1460.7} \right) = \left(\frac{127.8}{1000} \right) \left(\frac{1000}{2134} \right) \left(\frac{2134}{1460.7} \right)$$
$$8.75\% = 12.78\% \times 0.469\% \times 1.46\%.$$

These ratios are not particularly meaningful without some benchmark for comparison. Industry ratios provide such benchmarks. For the industry in which the O&R Corp. competes, these ratios are*

Return on Equity	Net Profit Margin	Total Asset Turnover	Equity Multiplier
16.1%	4.7%	1.3	4.3

---

* The product of the three industry average component ratios does not equal the industry average ROE given in Exhibit 2.4 because the industry average data may exclude some companies because of missing data or negative numbers.

The O&R Corp.'s ROE of 8.7% is well below the industry average of about 16%, so we would like to know why the company is not doing nearly as well as the average company in its industry.

The first component we consider is the net profit margin. With a net profit margin of 12.8%, the O&R Corp. is earning 12.8¢ for each $1.00 of sales, almost three times higher than the industry average of 4.7 cents. The second component is asset turnover. At 0.469 times per year, its asset turnover is very low compared to the industry average of about 1.3. The third component is the equity multiplier. An equity multiplier of 1.46 is about a third of the industry average of 4.3. We see immediately that in spite of generating a much higher profit margin than the industry, the O&R Corp. earned a lower ROE because both its asset turnover and equity multiplier were considerably lower than the industry.

Now we want to dig a little deeper to understand what is going on with O&R Corp. To do that, we need to consider what affects these three major performance factors. Exhibit 2.12 shows selected performance measures that link to these three factors and to ROE and suggests the things we need to consider further to understand O&R Corp.'s performance.

Because O&R Corp.'s profit margin is very good, we don't need to be too concerned with that. However, had it been too low, we would need to dig into its components as suggested by the diagram. If data are available, we would examine the firm's pricing structure to see if it is in line with the industry,

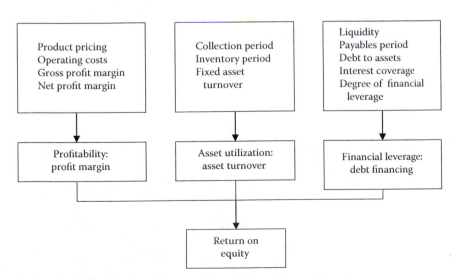

EXHIBIT 2.12   DuPont format. Three factors of performance.

and what factors, such as product demand and the competitive structure of the industry, would prevent the firm from increasing its prices. Then, we look at the cost structure using the common-size income statement that shows cost as a percent of sales. Are the various cost categories such as costs of goods sold (CGS); selling, general and administrative expenses (SGA); depreciation; and interest in line with industry cost ratios? What has been their trend? If any one of these general cost categories seems to be out of line, we should dig deeper to see what component of that cost is the culprit.

2.2.2.1 Asset Utilization

The second component in the DuPont format is asset turnover, which we already know is much lower than the industry average. The turnover of total assets measures the efficiency with which assets are being used. Is the firm generating too few sales given its investment in assets, or, to put it another way, has it invested too much in assets given its level of sales? Whichever way we state it, with its turnover well below the industry average, the company is not getting as much out of its investment in assets as other companies in the industry. Total asset turnover only gives us a clue to the problem. We need to dig deeper to see which type of asset is being used inefficiently. This entails looking at the turnover and other measures of efficiency and activity for each major asset category. In O&R Corp.'s case, these main categories are accounts receivable, inventories, and fixed assets. Exhibit 2.13 shows the turnover rates for each category for both the company and its industry.

In each category, O&R Corp.'s turnover rates are much lower than the industry. It has too much money tied up in each of these categories. The lower turnover of inventories and receivables indicates that the company is carrying items on its shelves longer than the industry, and then taking too long to collect the receivables from its customers. The

	Receivables	Inventory	Net Fixed Assets
Formula	$\dfrac{\text{Sales}}{\text{Accounts receivable}}$	$\dfrac{\text{Cost of goods sold}}{\text{Inventory}}$	$\dfrac{\text{Sales}}{\text{Net fixed assets}}$
O&R	$\dfrac{1000}{208.1} = 4.8$	$\dfrac{607.7}{148.2} = 4.1$	$\dfrac{1000}{1015} = 0.99$
Industry	7.9	6.3	2.3

EXHIBIT 2.13   O&R Corp. rates of turnover of assets.

lower turnover of fixed assets shows that the company invests proportionally more in plant and equipment (fixtures, equipment, and real estate) than others in the industry. Thus, all three categories are problems for this company.

2.2.2.2 Operating Cycle

You can gain a better perspective of the company's management of its inventory and receivables by examining the operating cycle. The operating cycle is the average length of time between inventory being purchased from a supplier and cash being collected from a customer. This length of time consists of the inventory period and the receivables period. Both are directly related to asset turnover.

Inventory period is the average time that an item is held in inventory from the date of purchase to the date of sale. It is calculated as

$$\text{Inventory Period} = \frac{\text{Inventory}}{\text{Cost per Day}}$$

$$= \frac{\text{Inventory}}{(\text{CGS}/360 \text{ days})}$$

$$= \frac{148.2}{(607.7/360 \text{ days})} = \frac{148.2}{1.69} = 7.8 \text{ days} \qquad (2.2)$$

or

$$= \frac{360 \text{ days}}{\text{Inventory Turnover}} = \frac{360 \text{ days}}{4.1} = 87.8 \text{ days}.$$

On the average, O&R Corp. is carrying items on its shelves for 87.8 days, as compared to the industry average of about 76 days.

After selling an item, the company must collect the cash from the customer. If no credit is granted, the cash is collected immediately. But if the item is sold on credit, it takes some time to collect from the customer. The average period of time it takes to collect from the customer is called the receivables period (or collection period). This is calculated as

$$\text{Receivables Period} = \frac{\text{Receivables}}{\text{Sales per Day}} = \frac{\text{Receivables}}{(\text{Sales}/360 \text{ days})} \qquad (2.3)$$

$$= \frac{208.1}{(1000/360 \text{ days})} = \frac{208.1}{2.78} = 74.9 \text{ days}$$

or

$$= \frac{360 \text{ days}}{\text{Receivables Turnover}} = \frac{360 \text{ days}}{4.8} = 74.9 \text{ days}.$$

It takes the company an average of 74.9 days to collect from its customers as compared to about 49 days for the industry.

The length of the operating cycle is the time it takes from purchasing the inventory to selling it plus the time between selling an item to a customer and collecting the cash from the customer. That is, the time between buying the inventory and collecting the cash from selling it is the sum of the inventory period and the receivables period:

Operating Cycle = Inventory Period + Receivables Period,

which for O&R Corp. is

Operating Cycle = 87.8 days + 74.9 days = 162.7 days.

O&R Corp. takes an average of 162.7 days from the time it purchases the raw materials until the date that it collects cash from the customer, as compared to about 125 days for the industry.

Looking at the turnover of various asset categories shows us that O&R Corp. has too much investment tied up in receivables, inventories, and fixed assets. If they could bring their turnover in line with the industry, they would reduce the capital tied up in asset investment and thereby increase their ROA and ROE. There are two additional aspects of O&R Corp.'s operating cycle and turnover that could be sources of inefficiency. These are cash and accounts payable (AP). We leave it to the reader to dig into these to see if O&R Corp. is carrying too much or too little liquidity, or perhaps not managing their trade credit properly.

2.2.2.3 The Equity Multiplier and Financial Leverage

The last component in the DuPont decomposition of the ROE is the equity multiplier, a measure of financial leverage. The equity multiplier is the ratio of total assets to stockholder's equity. It is the inverse of the equity-to-total assets ratio, so is related to the use of debt in a firm's capital structure. If a lot of debt is used, total assets will greatly exceed equity and the multiplier and its effect on ROE will be large. If very little debt is used, the multiplier will be close to one and its effect on ROE will be small. Thus, the equity multiplier measures the use of leverage from debt financing. In O&R Corp.'s case, the equity-to-asset ratio is

$$\frac{\text{Stockholders' Equity}}{\text{Total Assets}} = \frac{1460.7}{2134}$$
$$= 0.684$$
$$= 68.4\%,$$

so 68.4% of the funds used to purchase the firm's assets were supplied by stockholders through purchases of shares of stock and through retained earnings. The remaining 32% of O&R Corp.'s capital came from sources other than equity.

The equity multiplier tells us the total amount of assets supported by each one dollar of equity and is calculated as

$$\text{Equity Multiplier} = \frac{\text{Total Assets}}{\text{Stockholders' Equity}}$$
$$= \frac{2134}{1460.7}$$
$$= 1.46. \qquad (2.4)$$

This tells us that each dollar of equity supports the purchase of $1.46 worth of assets. In the case of O&R, the firm's equity multiplier of 1.46 is considerably lower than the industry's 4.3. If the firm used more debt, it could potentially increase its ROE. However, the very high multiplier for the industry might suggest that other companies in the industry are too highly levered. If so, we would not want to follow their lead if it imposed too much risk.

The problem is that debt is a two-edged sword. Just as more leverage magnifies gains when the firm prospers, it magnifies losses when the firm does poorly. In addition, increasing the equity multiplier may decrease another component in the DuPont format, the profit margin. If the firm uses more debt it will have to pay more interest. Its net profit margin may decrease as its financial risk increases. Whether the firm is better off or not will depend on the trade-off between the positive effects of leverage on ROE and the negative effects of leverage on profit margin and financial risk. Analysis of leverage, liquidity ratios, and debt coverage ratios can help us to determine the firm's ability to service the debt.

We start our analysis of O&R Corp.'s debt with various leverage ratios. These include the total debt ratios and the debt-to-equity ratio. The broadest measure of debt is the total debt ratio that takes into account all the firm's liabilities, and is measured as

$$\text{Debt Ratio} = \frac{\text{Total Liabilites}}{\text{Total Assets}}$$
$$= \frac{673.3}{2134}$$
$$= 31.6\%, \tag{2.5}$$

where

$$\text{Total Liabilities} = \text{Current Liabilities} + \text{Long-term Debt.}$$

O&R Corp.'s debt ratio means that about 32% of the investment in the firm's assets is financed by debt sources, including liabilities such as trade credit and accruals, as well as interest-bearing debt such as notes payable and long-term debt. O&R Corp.'s total debt ratio is much lower than the industry average of about 77%.

Since total liabilities includes many items that are not debt that must be repaid and are unlikely to lead to financial difficulty, we modify our measure of leverage to reflect the type of debt that is of greater concern for our analysis of leverage: the interest-bearing debt. Interest-bearing debt includes short-term debt, such as notes payable, and long-term debt. Focusing on the interest-bearing debt (denoted by ib), O&R Corp.'s debt ratio is

$$\text{Debt Ratio}_{ib} = \frac{\text{Short-term Debt} + \text{Long-term Debt}}{\text{Total Assets}}$$
$$= \frac{7.1 + 73.3}{2134}$$
$$= 3.8\%. \tag{2.6}$$

An alternative leverage measure emphasizes invested capital. Invested capital refers to permanent capital that was contributed by both lenders and equity investors, and consists of interest-bearing debt and equity. O&R Corp.'s invested capital is

$$\text{Invested Capital} = \text{Debt}_{ib} + \text{Equity}$$
$$= \text{Short-term Debt} + \text{Long-term Debt} + \text{Equity}$$
$$= 7.0 + 73.3 + 1460.7$$
$$= \$1541.0 \tag{2.7}$$

making its debt-to-invested capital ratio 80.3/1541 = 5.2%, as compared to 60.3% for the industry.

Comparing the debt ratios with the industry may not tell us much more than that our company uses more or less debt than the average company in the industry. Except in the extreme cases, it does not really tell us whether the company's situation is good or bad. A couple of problems of evaluating a debt ratio and comparing it with the industry are that (1) the absolute level of the ratio, either for the company or the industry, does not really tell us how much risk is involved; and (2) if we do not know that the industry average is a sensible standard, knowing that the company has more or less debt than the industry average will not necessarily tell us whether the company's position is good or bad.

Interest coverage ratios are more useful for evaluating the risk of the debt position because they measure the cushion of earnings that is available to service the debt. The most common interest coverage ratio is calculated as

$$\text{Interest Coverage} = \frac{\text{Earnings Before Interest and Taxes (EBIT)}}{\text{Interest}}$$
$$= \frac{216.2}{3.4}$$
$$= 63.6, \tag{2.8}$$

which means that income available to pay interest is almost 64 times greater than the interest obligation. With an industry average of about 67, both the company and the industry have very safe coverage levels, in spite of the fact that the industry appears to be highly levered when measured with other ratios.

This brief analysis of O&R Corp. has pointed out some of its strengths and weaknesses. We know that the company has a very attractive profit margin, but still earns a lower ROE than the average of its industry. The reason for this lower return is that the company has very poor asset utilization and it may be under-leveraged. But we have only touched the surface with our analysis. There are many aspects of financial analysis such as the firm's liquidity that we did not discuss simply because our objective is to get you started on modeling rather than teach all the details of financial analysis. We leave it to you, the reader, to dig deeper into the company with the methods that are discussed in greater depth in the other texts we have referenced. In the next two sections we will look at additional tools for assessing the risk and performance of the firm.

2.3 BREAK-EVEN ANALYSIS

One significant issue in financial planning is the division of costs between fixed and variable components. This dichotomy has important implications for the firm's ability to operate profitably, and for the variability and riskiness of the firm's income. A useful set of tools for financial planning relate to the analysis of fixed and variable costs. These tools are break-even analysis and the analysis of operating and financial leverage. In this section we discuss break-even analysis.

2.3.1 Fixed and Variable Costs

Before discussing break-even and the measures of operating and financial leverage, we need to recast the income statement in terms of fixed and variable costs. Consider Exhibit 2.14, which displays the income statement for the O&R Corp. in two formats. The first format is the typical one as might be seen in an annual report. Costs are categorized as to the general purpose of the expense, but not as fixed or variable. CGS and SGA include both fixed and variable components. For example, depreciation expense is frequently included in CGS. Furthermore, all costs and expenses are shown with no intermediate calculation of income. In the revised format, expenses are classified as fixed or variable and income is shown at intermediate

Standard Income Statement	
Sales	$1000.0
Costs of goods sold (CGS)	661.5
Selling, general, and administrative expenses (SGA)	122.3
Interest expense	3.4
Income tax	85.0
Net income (NI)	$127.8
Revised Format Fixed and Variable Cost	
Sales	$1000.0
Less: Variable operating expenses	563.0
Fixed operating expenses	167.0
Earnings before interest, taxes, depreciation and amortization (EBITDA)	270.0
Fixed cost: Depreciation	53.8
Earnings before interest and taxes (EBIT)	216.2
Interest expense	3.4
Earnings before tax (EBT)	212.8
Income tax	85.0
Net income (NI)	$127.8

EXHIBIT 2.14 O&R Corp. Income statements in two formats.

levels, such as earnings before interest, taxes, depreciation, and amortization (EBITDA) and EBIT, on its way to the bottom line, net income.

Two questions that arise in financial analysis and financial planning are: What is the break-even point, and how likely is it that the firm will be able to operate above that point? Break-even analysis is the tool that helps answer these questions. We define the break-even point as the level of output for which revenue just covers all costs. For production processes that involve variable costs only, the firm will be profitable at all output levels so long as the sale price exceeds the variable cost per unit. However, when some costs are fixed and do not vary with the level of output, some minimum level of output is necessary before revenue covers both fixed and variable costs.

In break-even analysis, fixed costs are costs that do not vary (in the short run) with the number of units of product produced or sold; variable costs are costs that change with the volume of production or sales. Examples of fixed costs are salary expenses for personnel that are employed regardless of the level of production; rental and lease expenses that depend on time but not on sales volume; and, of course, non-cash expenses such as depreciation and amortization. Examples of variable costs include sales commissions, labor costs related to the manufacturing process that depend on the volume of output, and the costs of materials used in production. The classification of costs as fixed or variable depends on the time horizon of the analysis. While all costs are variable in the long run, many costs are independent of sales volume in the short run, and are therefore fixed in the short run. The shorter the horizon, the more costs are fixed.

We use the example of the O&R Corp. in our discussion of break-even analysis, and consider EBITDA break-even—that is, the level of sales where EBITDA will be zero. The first step is to separate fixed from variable costs. O&R Corp. has fixed cost components both in the CGS and in selling, general and administrative costs. These include the costs of management salaries, rent, insurance, property taxes, and so on. Assume that these are about $167 million per year, excluding depreciation.* The remaining costs

---

* Fixed and variable costs were estimated from O&R Corp.'s five-year income data with the regression

$$\text{Total Operating Cost} = F + v \,(\text{Sales}),$$

where the intercept F is the estimate of fixed cost, and the coefficient v the estimate of the variable cost as a percent of sales. With five observations, the estimated equation was

$$\text{Total Operating Cost} = \underset{(t\,=\,2.53)}{167.3} + \underset{(t\,=\,7.56)}{0.58\,(\text{Sales})} \qquad R^2 = 0.93 \ (\text{adjusted})$$

For ease of exposition of the O&R Corp. example, the variable cost rate was assumed to be 0.563 to fit with the fifth year's income statement figures in Exhibit 2.14.

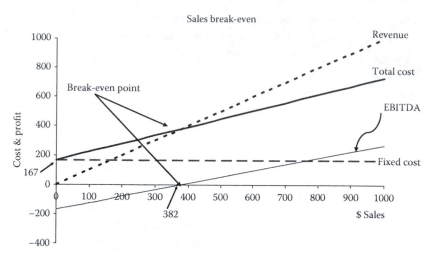

EXHIBIT 2.15   O&R Corp. Break-even chart.

are variable. They include items such as the cost of merchandise, direct labor, and sales commissions. The variable costs are estimated to be 56.3% of sales. In terms of units sold, the average item sells for $10, with an average variable cost of $5.63. The $4.37 difference between sale price and variable cost is the contribution margin per unit.

Exhibit 2.15 shows a graph of revenue and costs.* Sales are shown on the horizontal axis; revenue, costs, and profit are shown on the vertical axis. Sales revenue is shown by the dashed line that starts at the origin and slopes upward at a 45° angle. Fixed costs are shown by the flat, horizontal dashed line that intersects the vertical axis at $167 million. The horizontal line indicates that these costs remain at $167 million regardless of the level of sales or the number of units sold. Total costs are the sum of fixed and variable costs and are shown as the solid line slopes upward from $167 on the vertical axis. The slope of this line is 0.563, the rate at which variable and total costs increase as the level of sales increases.

The point of intersection between the total cost and revenue lines is the break-even point, which occurs at a sales level of $382 million. At levels of sales below $382 million, total operating costs exceed revenue and the firm suffers a loss; at sales levels higher than $382 million, revenue exceeds operating costs and the firm has positive EBITDA. At sales of $382 million,

---

* The appendix to this chapter provides a brief instruction on how to use a data table to help make Exhibit 2.15.

revenue equals total cost and the firm breaks even (EBITDA is zero). Hence, we call sales of $382 million the break-even point.

The break-even point is imposed by the presence of fixed costs. In the absence of fixed costs, the operation would be profitable at all sales levels so long as the sales price exceeds the variable cost per unit. In the presence of fixed costs, the sales level must be high enough so that the contribution margin (sales price minus variable cost) covers the fixed costs. With a percentage variable cost of 56.3%, O&R Corp.'s percentage contribution margin is $(1 - 0.563) = 0.437$, and the break-even sales level (denoted as Sales*) is calculated as[†]

$$\text{Sales}^* = \frac{\text{Fixed Cost}}{(1 - \text{Variable Cost Rate})}$$

$$\text{Sales}^* = \frac{\$167 \text{ million}}{(1 - 0.563)}$$

$$= \$382.2 \text{ million.} \qquad (2.9)$$

This example defined break-even as the sales level where EBITDA would be zero. Break-even can also be defined in terms of cash flow, EBIT, or net income. In each case, the method is the same: to determine the sales level where the incremental profit covers fixed costs, we divide fixed costs by the variable profit rate (percent or dollar contribution margin). For example, stepping down one level on the income statement to EBIT, O&R Corp.'s fixed costs amount to $220.8 million, including depreciation of $53.8 million, in addition to the fixed operating costs of $167 million we have already discussed. The variable cost rate is still 56.3%, so the sales level for EBIT break-even is calculated as

$$\text{Sales}^* = \frac{\text{Fixed Cost}}{(1 - \text{Variable Cost Rate})}$$

$$\text{Sales}^* = \frac{\$220.8 \text{ million}}{(1 - 0.563)}$$

$$= \$505.3 \text{ million.}$$

[†] For an analysis in terms of physical units, the break-even number of units, Q^*, is calculated as

$$Q^* = \frac{\text{Fixed Cost}}{(\text{Price} - \text{Variable Cost per Unit})} = \frac{\$167 \text{ million}}{(\$10 - \$5.63)}$$

$$Q^* = \frac{\$167 \text{ million}}{(\$4.37)} = 38.21 \text{ million units,}$$

where the sale price is $10 per unit, and the variable cost is $5.63 per unit.

Thus, at a sales level of $382 million, EBITDA is zero, and because of the fixed expense of depreciation, EBIT will be negative. However, at the higher sales level of $505.3 million, this additional fixed costs will be covered, and EBIT will be zero instead of negative. If sales exceed $505.3 million, EBIT will be positive.

2.4 ANALYSIS OF OPERATING AND FINANCIAL LEVERAGE

Fixed costs not only impose break-even points, but they also make the firm's earnings more variable and uncertain than they would be if there were no fixed costs and all costs were variable. In fact, using fixed cost methods of production and financing tend to cause earning to fluctuate more than sales. This magnification effect is called leverage, of which there are two types: operating leverage and financial leverage. Since operating and financial leverage can increase the risk and uncertainty for the firm and its stakeholders, it is important to understand and measure it. In this section we introduce three leverage measures: degree of operating leverage (DOL) and degree of financial leverage (DFL), which together, give us the degree of combined leverage (DCL).

The magnification effect of leverage is shown in Exhibit 2.16. It shows the effects of changes in sales on EBITDA, EBIT, EBT, and NI. The base

			Degrees of Leverage Sensitivity Analysis				
					Base Level		
	Variation From Base		-10%	-1%		1%	10%
Sales			900.0	990.0	1,000.0	1,010.0	1,100.0
Variable Operating Cost			506.7	557.4	563.0	568.6	619.3
Fixed Operating Cost			167.0	167.0	167.0	167.0	167.0
EBITDA			226.3	265.6	270.0	274.4	313.7
Depreciation			53.8	53.8	53.8	53.8	53.8
EBIT			172.5	211.8	216.2	220.6	259.9
Interest			3.4	3.4	3.4	3.4	3.4
Earnings Before Tax			169.1	208.4	212.8	217.2	256.5
Income Tax @	40%		67.6	83.4	85.1	86.9	102.6
Net Income			101.5	125.1	127.7	130.3	153.9
% Change in Sales			-10%	-1%		1%	10%
% Change in EBITDA			-16.185%	-1.619%		1.619%	16.185%
% Change in EBIT			-20.213%	-2.021%		2.021%	20.213%
% Change in EBT			-20.536%	-2.054%		2.054%	20.536%
% Change in NI			-20.536%	-2.054%		2.054%	20.536%

EXHIBIT 2.16   O&R Corp. Sensitivity of earnings to changes in sales.

level of sales of $1000 is shown in the middle, the columns on the left of base show the impact of sales declining 1% and 10%, respectively, from the base, and the columns on the right show the effects of sales increases of +1% and +10%, respectively. The bottom five rows show the percentage change from the base level for each of the four earnings measures. We see that the percentage change in earnings is a multiple of the percentage change in sales. For example, a 1% change in sales up or down results in a 2.02% change in the same direction in EBIT and a 2.054% change in Net Income. Similarly, if sales changes by 10%, EBIT changes by 20.21%, and NI change by 20.54%. In this case, EBIT changes 2.02 times the percent change in sales, and NI changes by 2.054 times the change in sales.

The sensitivity of earnings to changes in sales is like elasticity in economics. In economics the price elasticity of demand is measured by the percentage change in demand resulting from a 1% change in the price: (% change demand)/(% change in price). In the case of leverage we have an elasticity of earnings with respect to changes in sales. The elasticity of EBIT with respect to changes in sales is

$$\frac{(\% \text{ Change in EBIT})}{(\% \text{ Change in Sales})} = \frac{2.02\%}{1\%} = 2.02,$$

and the elasticity of NI with respect to change in sales is

$$\frac{(\% \text{ Change in NI})}{(\% \text{ Change in Sales})} = \frac{2.054\%}{1\%} = 2.054.$$

This elasticity or sensitivity is what is measured by the degrees of leverage: DOL, DFL, and DCL.

2.4.1 Degree of Operating Leverage

Earnings are elastic relative to sales (elasticity greater than 1) because of the presence of fixed costs. If all costs were directly proportional to sales, then earnings would vary directly with sales on a one-for-one basis. However, when some costs are fixed, independent of sales, increases in sales can occur without concomitant increases in all the costs, so that earnings can increase proportionally more than sales. The use of fixed cost methods of operation (e.g., the use of machinery rather than labor) is referred to as using operating leverage, and we measure the DOL, with the calculation of the sales elasticity of EBIT. This elasticity or DOL can be calculated (at a given level of sales) as

$$\text{DOL} = \frac{\text{Sales} - \text{Variable Costs}}{\text{Sales} - \text{Variable Cost} - \text{Fixed Costs}}$$
$$= \frac{\text{Sales}(1-v)}{\text{Sales}(1-v) - F} = \frac{\text{Sales}(1-v)}{\text{EBIT}}, \tag{2.10}$$

where v denotes variable cost as a proportion of sales and F the fixed costs. At a sales level of \$1000 million, O&R Corp. has

$$\text{DOL} = \frac{1000(1-0.563)}{1000(1-0.563) - 220.8} = \frac{437}{216.2} = 2.02.$$

At this sales level, a 1% change in sales will result in a 2.02% change in EBIT, as shown in Exhibit 2.16. However, it is important to note that the DOL depends on the sales level. At a different sales level, the DOL will be different. In fact, the closer the firm is operating to its break-even point, the greater the DOL.

2.4.2 Degree of Financial Leverage

Another source of fixed cost is the interest expense resulting from debt financing, because debt financing provides leverage to the equity owners. Thus, the ideas about break-even points, leverage, and elasticity that apply to operating leverage also apply to the use of the fixed cost source of financing; that is, debt. In the absence of debt financing, any variation in EBIT is translated directly into an equal proportional variation in net income. However, with debt financing, because of the necessity to make a fixed interest payment, any variation in EBIT will be leveraged (or magnified) into a more than proportional change in net income. This effect is referred to as financial leverage and is measured by the elasticity of net income with respect to EBIT. This elasticity is the DFL, and is calculated with the formula

$$\text{DFL} = \frac{\text{EBIT}}{\text{EBIT} - I}, \tag{2.11}$$

where I denotes the interest payment. For example, at a sales level of $1000, the DFL is

$$DFL = \frac{216.2}{216.2 - 3.4} = 1.016,$$

meaning that a 1% change in EBIT would be translated into a 1.016% change in after-tax net income.

2.4.3 Degree of Combined Leverage

The equity owners of the firm are subject to both operating leverage and financial leverage. Variations in sales are translated into magnified variations in net income by the combined effects of fixed cost production and fixed cost financing methods. This effect is summarized and measured as the DCL, which is calculated as

$$
\begin{aligned}
DCL &= DOL \cdot DFL \\
&= \left[\frac{\text{Sales} - \text{Variable Costs}}{\text{EBIT}} \right] \left[\frac{\text{EBIT}}{\text{EBIT} - I} \right] \\
&= \frac{\text{Sales}(1 - v)}{\text{EBIT} - I}
\end{aligned}
\tag{2.12}
$$

at a given level of sales. For our example, at a sales level of $1000, the DCL is

$$DCL = \frac{1000(1 - 0.563)}{216.2 - 3.4} = \frac{437}{212.8} = 2.05,$$

or expressed as the product of DOL and DFL

$$
\begin{aligned}
DCL &= DOL \cdot DFL \\
&= 2.021 \cdot 1.016 = 2.05.
\end{aligned}
$$

This means that at a sales level of $1000, a 1% change in sales would be translated into a 2.05% change in net income and earnings per share. With relatively little financial leverage, most of the combined leverage is due to O&R Corp.'s fixed production costs.

The ideas of elasticity and DOL are of use to the planner in contemplating the effects of the structure of the firm on the firm's future performance. It should be remembered, however, that both these measures depend on the level of sales. Generally, as the level of sales increases (above the break-even point), the degree of leverage declines.

This elementary analysis assumed that costs and revenues are linearly related to the number of units sold. That is, the graphs of revenues and costs as shown in Exhibit 2.15 are straight lines, as opposed to curved lines. This assumption is a useful approximation for many situations and is well suited to introducing the concepts. In real life, the relationships may be more complex, and indeed, the concept of break-even point itself may be more complicated than is apparent from this standard one-period example.

2.5 CONCLUSION

This chapter introduced some of the tools that we can use for financial analysis. The reason this is important is that to plan for the future we need to know our starting point. The tools of financial analysis help us to evaluate the financial condition of the firm. They tell us about the firm's present condition, and provide insight into the reasons how and why the firm reached its present condition, which is the starting point for our trip into the future. Knowing more about our starting point will help us to develop better plans for the future. In addition, the financial relationships within the firm that are revealed by our financial analysis provide the input for developing our financial models that will help us fine tune our financial plans. The next chapter will add to our stock of tools for financial analysis and planning by honing our understanding of cash flows, cash budgets, and sustainable growth.

PROBLEMS

1 O&R Corporation

Financial statements for O&R Corp. were shown in Exhibits 2.2 and 2.3. Using that financial data as the basic input, develop a spreadsheet model that will show the various ratios that were discussed in this chapter. Your spreadsheet for the O&R Corp. should be set up in a way that it separates the data input from the calculations of the financial ratios. That is, the data input should be in a separate section or separate sheet of the model, and the formulas that calculate the performance ratios should be in their own section or own sheet. This formula section should include only formulas and labels, but none of the input data. Thus, the ratios section will be a template for calculating ratios that can be used for evaluating performance for any company whose data you input.

2 K.D. Nickel Corporation

K.D. Nickel Corporation and Willards, Inc. are both department store retailers with stores located throughout the country. Six years of past financial statements for each of these companies, in addition to average ratios for their industry, are shown below. Your task is to set up a spreadsheet model so that you can perform a financial analysis of these two companies, comparing them with each other and with the industry averages. The objective of the financial analysis is to determine each company's strengths and weaknesses by digging as deeply into their performance as the data allow.

Modeling task: The spreadsheet you construct should be a template for doing financial analysis. By template, we mean the spreadsheet with formulas that will compute the various ratios, etc., that you use. The template should not contain any of the data input. The data input should come from a separate spreadsheet. In this way, you can use the template for any company, and simply input that company's data from a separate data sheet onto the analysis template.

Should you not want to enter the data yourself, the financial data are to be found on the class disk included with this book.

INSERT:
Financial Statement Data for K.D. Nickel

K.D. Nickel
Income Statement
K.D. Nickel
Balance Sheet

Financial Statement Data for Willards, Inc.

Willards, Inc.
Income Statement
Willards, Inc.
Balance Sheet

Average Ratios for Department Stores

K.D. NICKEL

Ticker Symbol	KDNC					
SIC Code	5311					

INCOME STATEMENT

($ MILLIONS)

	2010	2009	2008	2007	2006	2005
Sales	17,786.0	32,347.0	32,004.0	31,846.0	32,510.0	31,380.0
Cost of Goods Sold	10,772.0	21,948.0	22,193.0	22,458.0	23,665.0	22,466.0
Gross Profit	7,014.0	10,399.0	9,811.0	9,388.0	8,845.0	8,914.0
Selling, General, & Admin	5,809.0	8,658.0	8,459.0	8,637.0	7,164.0	6,759.0
Operating Income Before Deprec.	1,205.0	1,741.0	1,352.0	751.0	1,681.0	2,155.0
Depreciation & Amort	394.0	667.0	717.0	695.0	710.0	637.0
Operating Profit	811.0	1,074.0	635.0	56.0	971.0	1,518.0
Interest Expense	261.0	407.0	426.0	477.0	673.0	663.0
Non-Operating Income & Spec Iter	-4.0	-83.0	-6.0	-465.0	233.0	100.0
Pretax Income	546.0	584.0	203.0	-886.0	531.0	955.0
Income Taxes	182.0	213.0	89.0	-318.0	195.0	361.0
Income Before Extraord						
Items & Discontin Op	364.0	371.0	114.0	-568.0	336.0	594.0
Preferred Dividends	25.0	27.0	29.0	33.0	36.0	38.0
Available for Common	339.0	344.0	85.0	-601.0	300.0	556.0
Extraord Items & Discontin Op	-1,292.0	34.0	-16.0	-137.0	0.0	0.0
Net Income	-953.00	378.00	69.00	-738.00	300.00	556.00
EPS -Basic Before Extra Items	1.25	1.28	0.32	-2.29	1.16	2.20
Dividends Per Share	0.259	0.743	0.754	0.806	0.780	0.848

K.D. NICKEL

Ticker Symbol	KDNC					
SIC Code	5311					

($ MILLIONS)

BALANCE SHEET

	2010	2009	2008	2007	2006	2005
ASSETS						
Cash & Equivalents	2,994.0	2,474.0	2,840.0	944.0	1,233.0	511.0
Net Receivables	233.0	705.0	698.0	893.0	1,138.0	4,415.0
Inventories	3,156.0	4,945.0	4,930.0	5,269.0	5,947.0	6,031.0
Other Current Assets	130.0	229.0	209.0	151.0	154.0	168.0
Total Current Assets	6,513.0	8,353.0	8,677.0	7,257.0	8,472.0	11,125.0
Gross Plant, Property & Equipme	5,637.0	8,154.0	8,317.0	8,062.0	8,195.0	8,333.0
Accumulated Depreciation	2,122.0	3,253.0	3,328.0	2,948.0	2,883.0	2,875.0
Net Plant, Property & Equipment	3,515.0	4,901.0	4,989.0	5,114.0	5,312.0	5,458.0
Other Assets	8,272.0	4,613.0	4,382.0	7,371.0	7,104.0	7,055.0
TOTAL ASSETS	18,300.0	17,867.0	18,048.0	19,742.0	20,888.0	23,638.0
LIABILITIES						
Notes Payable	260.0	288.0	935.0	250.0	955.0	2,362.0
Accounts Payable	1,167.0	1,792.0	1,551.0	1,948.0	1,480.0	1,496.0
Taxes Payable	119.0	123.0	158.0	195.0	179.0	232.0
Accrued Expenses	1,051.0	1,373.0	1,077.0	656.0	1,613.0	1,597.0
Other Current Liabilities	1,157.0	583.0	778.0	1,186.0	238.0	283.0
Total Current Liabilities	3,754.0	4,159.0	4,499.0	4,235.0	4,465.0	5,970.0
Long Term Debt	5,114.0	4,940.0	5,179.0	5,448.0	5,844.0	7,143.0
Other Liabilities	4,007.0	2,398.0	2,241.0	3,800.0	3,351.0	3,356.0
EQUITY						
Preferred Stock	304.0	333.0	363.0	399.0	446.0	475.0
Common Stock	137.0	134.5	132.0	131.5	130.5	125.0
Capital Surplus	3,394.0	3,288.5	3,198.0	3,162.5	3,135.5	2,725.0
Retained Earnings	1,590.0	2,614.0	2,436.0	2,566.0	3,516.0	3,844.0
Common Equity	5,121.0	6,037.0	5,766.0	5,860.0	6,782.0	6,694.0
TOTAL EQUITY	5,425.0	6,370.0	6,129.0	6,259.0	7,228.0	7,169.0
TOTAL LIABILITIES & EQUITY	18,300.0	17,867.0	18,048.0	19,742.0	20,888.0	23,638.0

	I	J	K	L	M	N	O
47	**WILLARDS INC**						
48							
49	Ticker Symbol	WLLS					
50	SIC Code	5311					
51							
52	INCOME STATEMENT						
53		($ MILLIONS)					
54		2010	2009	2008	2007	2006	2005
55	Sales	7,848.1	8,169.6	8,388.3	8,817.8	8,921.2	8,011.7
56	Cost of Goods Sold	5,170.2	5,254.1	5,507.7	5,802.1	5,762.4	5,286.1
57	Gross Profit	2,677.9	2,915.5	2,880.6	3,015.6	3,158.8	2,725.6
58	Selling, General, & Admin	2,174.3	2,232.1	2,264.2	2,295.9	2,275.9	1,979.2
59	Operating Income Before Deprec.	503.5	683.4	616.5	719.8	882.9	746.4
60							
61	Depreciation & Amort	290.7	301.4	310.8	303.2	292.7	239.7
62							
63	Operating Profit	212.9	382.0	305.7	416.6	590.2	506.8
64	Interest Expense	187.8	185.4	195.8	229.0	241.8	199.8
65	Non-Operating Income & Spec Ite	-9.1	14.6	1.6	-46.7	-64.5	-87.9
66	Pretax Income	16.0	211.1	111.6	140.9	283.9	219.1
67	Income Taxes	6.7	74.8	45.8	44.0	120.2	83.8
68							
69	Income Before Extraord						
70	Items & Discontin Op	9.3	136.3	65.8	96.8	163.7	135.3
71	Preferred Dividends	0.0	0.0	0.0	0.0	0.0	0.0
72	Available for Common	9.3	136.3	65.8	96.8	163.7	135.2
73	Extraordinary Items	0.0	-534.7	6.0	-102.7	0.0	0.0
74	Net Income	9.344	-398.405	71.798	-5.85	163.721	135.237
75							
76	EPS -Basic Before Extra Items	0.11	1.61	0.78	1.06	1.55	1.26
77	Dividends Per Share	0.16	0.16	0.16	0.16	0.16	0.16

	I	J	K	L	M	N	O
1							
2	**WILLARDS INC**						
3							
4	Ticker Symbol	WLLS					
5	SIC Code	5311					
6							
7		($ MILLIONS)					
8	BALANCE SHEET						
9		12/31/10	12/31/09	12/31/08	12/31/07	12/31/06	12/31/05
10	ASSETS						
11	Cash & Equivalents	160.9	142.4	153.0	194.0	198.7	72.4
12	Net Receivables	1,191.5	1,338.1	1,074.9	979.2	1,104.9	1,192.6
13	Inventories	1,632.4	1,594.3	1,561.9	1,616.2	2,047.8	2,157.0
14	Other Current Assets	39.0	55.5	24.7	53.5	72.2	15.7
15	Total Current Assets	3,023.7	3,130.3	2,814.5	2,842.9	3,423.7	3,437.7
16							
17	Gross Plant, Property & Equipm	5,070.5	5,134.6	5,120.4	5,015.9	4,961.0	5,407.9
18	Accumulated Depreciation	1,873.0	1,764.1	1,664.7	1,507.6	1,341.8	1,723.2
19	Net Plant, Property & Equipment	3,197.5	3,370.5	3,455.7	3,508.3	3,619.2	3,684.6
20	Other Assets	189.9	175.2	804.3	848.0	875.3	1,055.2
21	TOTAL ASSETS	6,411.1	6,675.9	7,074.6	7,199.3	7,918.2	8,177.6
22							
23	LIABILITIES						
24	Long Term Debt Due In One Yea	499.7	140.7	100.5	211.3	110.6	166.7
25	Notes Payable	50.0	0.0	0.0	0.0	0.0	0.0
26	Accounts Payable	457.5	429.1	562.5	399.0	351.7	524.1
27	Taxes Payable	106.5	69.8	19.4	17.6	32.4	5.9
28	Accrued Expenses	222.4	246.8	245.7	248.9	315.9	397.1
29	Other Current Liabilities	0.0	0.0	0.0	0.0	0.0	0.0
30							
31	Total Current Liabilities	1,336.1	886.5	928.1	876.7	810.6	1,093.8
32	Long Term Debt	2,072.8	2,743.2	2,676.6	2,928.2	3,450.9	3,561.2
33	Deferred Taxes	617.2	645.0	644.0	638.6	702.5	681.1
34	Other Liabilities	147.9	137.1	157.5	126.0	121.5	0.0
35	EQUITY						
36	Preferred Stock	0.0	0.0	0.0	0.0	0.0	0.4
37	Common Stock	1.2	1.2	1.2	1.2	1.2	1.2
38	Capital Surplus	45.6	61.9	49.6	69.7	252.1	407.1
39	Retained Earnings	2,190.3	2,201.2	2,617.6	2,558.9	2,579.6	2,432.8
40	Common Equity	2,237.1	2,264.2	2,668.4	2,629.8	2,832.8	2,841.1
41	TOTAL EQUITY	2,237.1	2,264.2	2,668.4	2,629.8	2,832.8	2,841.5
42							
43	TOTAL LIABILITIES & EQUITY	6,411.1	6,675.9	7,074.6	7,199.3	7,918.2	8,177.6
44							
45	Shares Outstanding - Common	83.5	84.8	83.9	85.0	98.8	106.9

APPENDIX A: USING NAMES IN THE EXCEL SPREADSHEET

Reading and understanding the formulas in the spreadsheet are easier when you use variable names instead of cell addresses. For example, a formula shown as Income = Sales − Cost is easier to read and decipher than C8 = C6 − C7. This appendix provides a brief explanation for using variable names in your spreadsheet. For more detailed instructions, you should access Excel's Help or one of the many Excel instruction books that are widely available.

To demonstrate the use of names, Exhibit A.1 shows a simple income model in Excel. Cell B8 shows the standard Excel cell reference method for a formula, with Income being calculated as =B6 − B7. Cell C8 expresses the same income formula in terms of variable names "Sales" and "Cost." In this case, the cell C6 was named "Sales" and cell C7 was named "Cost," so the formula =Sales − Cost is the same as =C6 − C7, but it is easier to read.

Certain cells in the model are named, and the names appear in the formulas. In addition to Sales referring to C6, C7 is named "Cost," each of the cells D6:F6 (meaning cells D6 through F6) is named "Revenue," and cells E7:F7 are named "CGS." In addition, columns D:F are named for their years, 2008:2010. This way, the variable can be designated by its year, as in line 10 with "Revenue 2009" and "Revenue 2010" indicating the revenue in each of those years.

There are a couple of ways to name a variable. The first, and most direct is to select the cell you want to name. The cell address will show in the Name box on the Formula Bar in the upper left corner of the sheet. In place of the cell address, type the name of the variable. For example, in Exhibit A.2, the Name box is on the upper left above cell A1 showing the word "Sales," which refers to highlighted cell C6. Cell C6 was named Sales by clicking on the cell to select or highlight it, and then typing the name

B6		ƒx 748			
	A	B	C	D	E	F
1					ODD & RICH CORP	
2					Income Statement	
3						
4		2006	2007	2008	2009	2010
5						
6	Sales	748	793	800	944	1000
7	Cost	550	573	627	632	789
8	Income	=B6-B7	=Sales-Cost	=D6-D7	=Revenue-CGS	=Revenue-CGS
9						
10	Sales Growth		=(Sales/B6)-1	=(D6/Sales)-1	=(Revenue 2009'/Revenue 2008')-1	=(Revenue 2010'/Revenue 2009')-1
11						

EXHIBIT A.1   Named cells in model.

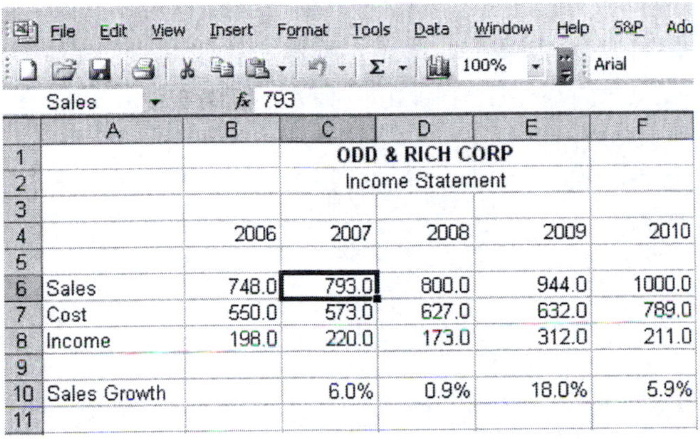

EXHIBIT A.2   Variable names in Excel models.

Sales into the name box as shown. Once the cell is named, formulas can be written with the name instead of the cell address, C6.

The second way to name cells is to use the Name option from the Insert menu. Starting with Insert on the menu bar, the sequence of steps is: Insert > Name > Define. Define allows you to specify names for a selected cell or selected range of cells. Exhibit A.3 indicates that cells D6:F6 have

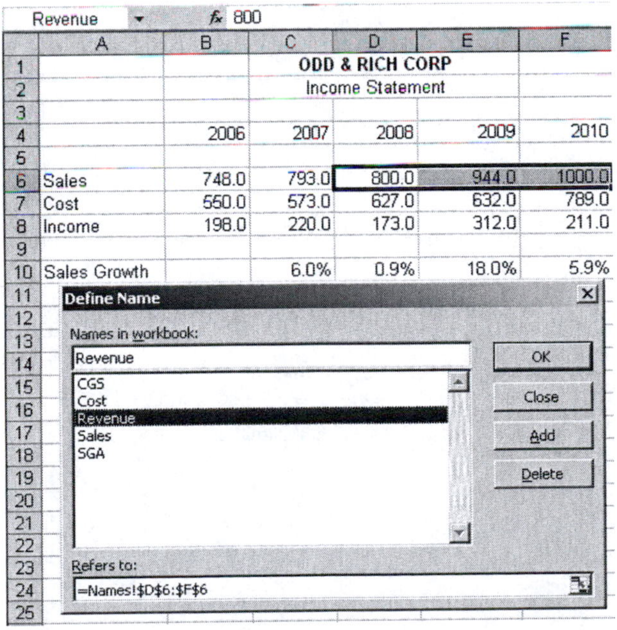

EXHIBIT A.3   Define Name dialog box.

been defined with the name "Revenue." Formulas in columns D:F can then use the name Revenue. For example, the formula for Income in column E might be expressed as Income = Revenue – E7, where, having not yet been named, E7 is Cost in 2009.

Once the variables are defined in the model, the variable name can be included in the formulas by applying the names. Two steps to apply the name are

$$Insert > Name > Apply$$

The Apply command brings up the dialog box to Apply Names as shown in Exhibit A.4. Click Revenue and OK will insert Revenue into all formulas that use the cells that have been named Revenue. Selecting all the names in the list will apply all of them in their respective formulas.

Note in Exhibit A.1 that line 10 has formulas with the years. Dates or other column headings allow the variables to be designated as the cell at the intersection of the row and column. For example, cell E10 calculates sales growth with the formula

$$=(Revenue\ '2009'/Revenue\ '2008') – 1,$$

where Revenue is the name of the sales variable, and the year "2008" refers to Revenue in the previous column D. The space between Revenue and "2008" indicates the intersection with the column named "2008." Columns can be named with the Create dialog so long as the name begins with a letter. The name should not be a number or begin with a number such as 2000 because Excel will not recognize it as a name. Preceding the number with the apostrophe as with 2000 will avoid that problem, or formatting the line with date format will help Excel to recognize the name.

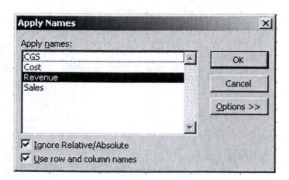

EXHIBIT A.4   Apply Names dialog box.

APPENDIX B: CONSTRUCTING A DATA TABLE

To do a break-even analysis of O&R Corp., it was helpful to construct a spreadsheet that will show earnings at different levels of sales. That is, we want to do a sensitivity analysis of O&R Corp.'s earnings by exploring how earnings change as the sales level changes. We can do this simply by changing the sales that are input into the income statement on the spreadsheet. However, as we change the sales input and see the different earnings output we will probably want to record and remember the results, so we know what the earnings level will be associated with each different sales level that we input. An easy way to do this is to use Excel's Data Table capability.

A Data Table is simply a table showing the results when we change some of the inputs to the spreadsheet. Exhibit B.1 includes a Data Table that shows the different levels of EBITDA that would result when the number of units sold is varied. Assumptions about prices, variable cost, and fixed cost are in section A3:D17 of the sheet. Section F3:M12 is a partial income statement with each column showing the earnings calculation as the number of units sold is varied from 0 to 25 units.

Formulas for column I of the income statement section are shown below, with some of the formulas referring to the assumptions in column D. Column I calculates EBITDA for the assumed number of Units Sold in Cell I5, which is zero. If we want to see how EBIT increases as Units Sold is increased, all we would have to do would be either (a) change the Units Sold in I5, or (b) copy the formulas in column I across the sheet and input a different number of Units Sold in each column of row 5, or (c) allow a Data Table to do it for us. In this case, we use a Data Table (Exhibit B.2).

	A	B	C	D	E	F	G	H	I	J	K	L	M
1	Odd & Rich Corp												
2									Data Table				
3	Assumptions					Break-Even Analysis							
4													
5	Sales Price Per Unit			10		Units Sold			0	10	15	20	25
6	Variable Cost Per Unit			5.63		Revenue			0	100	150	200	250
7	Contribution Margin Per Unit			4.37									
8						Sales			0	100	150	200	250
9	Variable CGS/Sales			50.0%		Variable Operating Costs			0	56.3	84.45	112.6	140.75
10	Variable SGA/Sales			6.3%		Fixed Operating Costs			167	167	167	167	167
11	Variable Operating Cost			56.3%		Total Costs			167	223.3	251.45	279.6	307.75
12	Fixed CGS			86		EBITDA			-167	-123.3	-101.45	-79.6	-57.75
13	Fixed SGA			81									
14	Total Fixed Operating Cost			167									
15	Depreciation	(fixed)		53.8									
16	Interest	(fixed)		3.4									
17	Total Fixed Cost			224.2									

EXHIBIT B.1   Data table for sensitivity analysis.

	F	I
1		
2		Data Table
3	Break-Even Analysis	
4		
5	Units Sold	0
6	Revenue	=I5*$D$5
7		
8	Sales	=I6
9	Variable Operating Costs	=($D$9+$D$10)*I6
10	Fixed Operating Costs	=$D$12+$D$13
11	Total Costs	=I9+I10
12	EBITDA	=I6-I9-I10

EXHIBIT B.2   Formulas in the Data Table.

In this example, the Data Table is actually just cells I5:M6, with formulas in rows 8:12 drawing from the Data Table output. To implement the Data Table the steps are as follows:

1. Construct the base model with formulas as shown in column I. Note that the Units Sold in I5 is 0 and the Revenue in I6 is entered as the formula = I5 × $D$5, where D5 is input data. In this model, Units Sold is the Independent Variable that will be changed in the Data Table for our sensitivity analysis. Revenue in I6 is the dependent variable.

2. Fill in the input values of the Independent Variable across row 5 in columns J:M with the values you want to test as the Independent Variables (e.g., 10, 20, 30, 40). Prior to completing the data table, the cells are as follows:

	F	G	H	I	J	K	L	M
1								
2				Data Table				
3	Break-Even Analysis							
4								
5	Units Sold			0	10	20	30	40
6	Revenue			0				
7								

Note that cells in the Revenue row J6:M6 are empty. There are no formulas or numbers in these cells. However, I6 does have the formula = I5 × $D$5.

3. Select the rectangle of cells, I5:M6, that will be the Data Table with the Independent Variables in the first row and the dependent variable in the second row.

4. On the Menu Bar, click Data ⇒ Table. This brings up the Table dialog box. For the Row input cell, click in cell I5 that represents the base level of the Independent Variable.

F	G	H	I	J	K	L	M
			Data Table				
Break-Even Analysis							
Units Sold			0	10	20	30	40
Revenue			0				
Sales			0	**Table**			✕
Variable Operating Costs			0				
Fixed Operating Costs			167	Row input cell:	$I$5		
Total Costs			167	Column input cell:			
EBITDA			-167				
				OK		Cancel	

5. Click OK. Excel will fill in the Dependent Variable row (row 6) with the amount of revenue corresponding to each of the Independent Variable values in row 5. We immediately get the filled in Data Table as follows:

	F	G	H	I	J	K	L	M
1								
2				Data Table				
3	Break-Even Analysis							
4								
5	Units Sold			0	10	20	30	40
6	Revenue			0	100	200	300	400
7								
8	Sales			0	100	200	300	400
9	Variable Operating Costs			0	56.3	112.6	168.9	225.2
10	Fixed Operating Costs			167	167	167	167	167
11	Total Costs			167	223.3	279.6	335.9	392.2
12	EBITDA			-167	-123.3	-79.6	-35.9	7.8

What the Data Table did was fill in the values for line 6 as if you had simply repeated the formula of I6 across the line. This provides you with an easy way to do a sensitivity analysis where you want to construct the sensitivity table showing the values of a dependent variable corresponding to a range of values of an independent variable. In this case, the Data Table includes cells I5:M6, and lines 8–12 were not really specified as part of the Data Table. However, these model lines take the Data Table output of line 6 and complete the remainder of the income statement so that you can see the sensitivity impact of various Units Sold on EBITDA.

This Data Table was for the case where you have just one independent variable. Excel also has the capacity to build a Data Table with two independent variables.

Two ways to gain perspective from the financial ratios are (1) to look at the trend of ratios over time; and (2) to compare them to other companies in the same industry and with the average ratios for the industry. The trend of the ratios over time is a good starting point for our analysis because significant changes or trends in ratios frequently signal potential problems or promising improvements. Scanning across the line of ratios may quickly alert you to changes that may be important, and graphing the trend of ratios is even more effective in highlighting the trends. For example, Exhibit 2.10 shows an Excel graph of O&R Corp.'s profit margins clearly showing an upward trend. The picture of this positive trend might not be worth a thousand words, but it surely helps us quickly see one of the important aspects of O&R Corp.'s operations. Similar pictures of the trends can help us visualize what is happening with the firm, and the usefulness of your financial analysis model can be enhanced by including graphs of the more important ratios.

The Tools for Financial Planning II: Growth and Cash Flows

IN THE PREVIOUS CHAPTER we looked at some methods for financial analysis. In this chapter we will continue to study some useful tools and methods for financial analysis and planning. The tools we will discuss include methods for projecting *pro forma* financial statements, the concept of sustainable growth, concepts of cash flow, and the development of cash budgets. Even though it may not seem like these topics are helping us build financial models, they are, in fact, very important for understanding the structures of the models we will build in the following chapters. Building correct models will be dependent on having a clear understanding of cash flows, cash budgets, *pro forma* financial statements, and growth. Reviewing the definitions and interrelationships between these factors will make it easier to understand subsequent modeling, and help you to develop your own models that work correctly.

3.1 PROJECTING *PRO FORMA* FINANCIAL STATEMENTS

In this section we introduce a simple method for forecasting the firm's future financial statements. We use a percent of sales format for developing *pro forma* financial statements. This framework provides the basis for much of our discussion of financial planning models in subsequent chapters. Many of the examples in this book use a sales-driven model. This means that the primary input to our model is a forecast of sales, and that forecast drives the rest of the model. As shown in Exhibit 3.1, with the sales-driven model the sales forecast determines the required asset levels based on the idea that there is a minimum level of assets necessary to support sales; in turn, the asset requirements determine the level of financing necessary to purchase the assets.

To explain the percent-of-sales approach for developing financial projections, we build on the example of the O&R Corp. that we used in the last chapter. We would like to use our knowledge of O&R to determine what its financial condition will be in the future as it grows. The past financial statements and the ratios we computed provide the starting point for our analysis. Exhibit 3.2 shows the financial statements for O&R Corp. for the two years, 2009 and 2010, and the far right column of each statement is the account balance as a percent of sales.

With the percent of sales approach, our forecast of sales drives the forecast of income, expenses, assets, and liabilities. We assume that there is a link between the firm's sales level and the expenses, assets and liabilities. In the simplest percent of sales model, this link is represented by the assumption that most of these categories maintain a constant ratio to sales. This is often a reasonable approximation, because the structure of the firm's operations determines its expenses, assets, and liabilities. For example, O&R manufactures plastic components. In order to be able to generate $1.0 billion in sales it will need to have a certain minimum number of machines, amount of warehouse space, delivery trucks, and administrative offices. To service its customers it will need an inventory, and each plant will have to have an appropriate number of employees. To attract the number of customers necessary to buy $1.0 billion worth of merchandise, it will need some minimum level of marketing. In addition,

EXHIBIT 3.1   The flow of a sales-driven financial model.

ODD & RICH CORP
Income Statement
($ Millions)

	Year		Percent of Sales
	2009	2010	2010
Sales	944.0	1,000.0	
Cost of Goods Sold	566.7	607.7	60.8%
Selling, General & Administrative	169.3	122.3	12.2%
EBDIT	208.0	270.0	27.0%
Depreciation	49.9	53.8	5.4%
EBIT	158.1	216.2	21.6%
Interest	3.1	3.4	0.3%
EBT	155.0	212.8	21.3%
Tax	62.0	85.0	8.5%
Net Income	93.0	127.8	12.8%
Dividends	55.8	76.7	7.67%
Retained Earnings	37.2	51.1	5.11%

ODD & RICH CORP
Balance Sheet
($ Millions)

	Year		Percent of Sales
	2009	2010	2010
Assets			
Cash	160.4	170.0	17.0%
Accounts Receivable	177.5	208.1	20.8%
Inventory	142.8	148.2	14.8%
Other CA	189.5	183.7	18.4%
Current Assets	670.2	710.0	71.0%
Gross Fixed Assets	1,792.7	1,927.0	192.7%
Accumulated Depreciation	449.2	503.0	50.3%
Net Fixed Assets	1,343.5	1,424.0	142.4%
Total Assets	2,013.7	2,134.0	213.4%
Liabilities			
Accounts Payable	175.5	279.1	27.9%
Short-Term Debt	6.9	7.0	0.7%
Other CL	383.9	313.9	31.4%
Current Liabilities	566.3	600.0	60.0%
Long-Term Debt	37.8	73.3	7.3%
Common Stock	100.0	100.0	10.0%
Earned Surplus	1,309.6	1,360.7	136.1%
Common Equity	1,409.6	1,460.7	146.1%
Total Liabilities & Equity	2,013.7	2,134.0	213.4%

EXHIBIT 3.2   Financial statements for Odd & Rich Corp.

to finance its inventories and facilities, it will probably need to borrow some funds or obtain equity financing. This tells us that there is a link between the firm's sales and the expenses, assets and liabilities that support the firm's sales. Assume that with its present mode of operation, it is unlikely that the firm can generate a much higher level of sales without increasing its expenses, assets, and liabilities in pace with sales.

Naturally, the assumption of a constant percent of sales will not be appropriate in all situations. Even for a firm where it fits in general, there will be exceptions. Some expenses will be fixed, and will not vary with sales, or the firm may be able to accommodate moderate increases in sales without having to increase its asset base. For other firms, there may be only the weakest link between sales and assets. For example, a service firm such as an accounting firm may be able to serve many more clients without investing in more assets, or a software company can sell many more disks of software without expanding its plant or even without hiring additional employees. So, like any method, we need to be careful how we apply the percent of sales approach and make sure that our assumptions match the reality of the situation. In those cases where it fits, it can be a very useful way to begin our planning and modeling.

In the case of O&R, we will assume that the constant percent of sales model is appropriate.[*] 'We have forecasted O&R's sales to increase by 15% from their level of $1000 in 2010 to $1150 in 2011. We want to see how this growth will affect the firm's needs for funds for 2011. To implement this, we will assume that each of the expense and asset categories in Exhibit 3.2 continues to maintain the proportional relationship with sales that it has had in the past. Most of the account percentages are summarized in Exhibit 3.3. We will use these percentages to estimate each of the income statement and balance sheet accounts by multiplying the forecast sales with the percentage from the exhibit.

With costs proportional to sales, most of the income and expense items are straightforward. For example, for 2011, the cost of goods sold (CGS) and $EBITDA_t$ would be estimated as

$$\text{Cost of Goods Sold}_{2011} = \left(\frac{\text{CGS}}{\text{Sales}} \right) \text{Sales}_{2011}$$
$$= (0.608) \times 1150$$
$$= \$699.2 \text{ million}$$

---

[*] For the remainder of this chapter we will assume that O&R's costs are directly proportional to sales, and that there are no fixed costs.

Expenses and Earnings		
Cost of goods sold/sales	(CGS/sales)	60.8%
Selling, general, and administrative/sales	(SGA/sales)	12.2%
Earnings before interest, taxes, depreciation, and amortization/sales	(EBITDA/sales)	27.0%
Earnings before interest and taxes/sales	(EBIT/sales)	21.6%
Net income/sales	(NI/sales)	12.8%
Dividend payout ratio	(Div/NI)	60.0%
Assets and Liabilities		
Cash/sales	(Cash/sales)	17.0%
Accounts receivable/sales	(AR/sales)	20.8%
Inventories/sales	(Invent/sales)	14.8%
Other current assets/sales	(OCA/sales)	18.4%
Current assets/sales	(CA/sales)	71.0%
Net fixed assets/sales	(NFA/sales)	142.4%
Total assets/sales	(TA/sales)	213.4%
Account payable/sales	(AP/sales)	27.9%
Other current liabilities/sales	(OCL/sales)	31.4%

EXHIBIT 3.3 O&R Corp. expenses, earnings, assets, and liabilities as a percent of sales.

and

$$\text{EBITDA}_{2011} = \left(\frac{\text{EBITDA}}{\text{Sales}}\right)\text{Sales}_{2011}$$
$$= (0.27) \times 1150$$
$$= \$310.5 \text{ million.}$$

The calculation of selected items on the income statement and balance sheet is shown in Exhibit 3.4.

For each type of asset, the ratio (Asset/Sales) is assumed to be the same each year, so we determine the level of a given asset required to support the sales by calculating

$$\text{Asset}_t = \left(\frac{\text{Assets}}{\text{Sales}}\right)\text{Sales}_t. \tag{3.1}$$

Pro Forma Income

Sales_1	$= \text{Sales}_0 (1+g)$	$= \$1000 (1.15)$	$= \$1150$
$\text{Cost of Goods Sold}_1$	$= \text{Sales}_1 (\text{CGS/Sales})$	$= 1150 (0.608)$	$= 699.2$
$\text{Selling, Gen \& Admin}_1$	$= \text{Sales}_1 (\text{SGA/Sales})$	$= 1150 (0.122)$	$= 140.3$
EBITDA_1	$= \text{Sales}_1 (\text{EBITDA/Sales})$	$= 1150 (0.27)$	$= 310.5$
EBIT_1	$= \text{Sales}_1 (\text{EBIT/Sales})$	$= 1150 (0.216)$	$= 248.4$
Net Income_1	$= \text{Sales}_1 (\text{NI/Sales})$	$= 1150 (0.128)$	$= 147.2$
Dividends_1	$= \text{NI}_1 \times \text{Payout Ratio}$	$= 147.2 (0.60)$	$= 88.3$
$\text{Retained Earnings}_1$	$= \text{NI}_1 - \text{Dividends}_1$	$= 147.2 - 88.3$	$= 58.9$

Pro Forma **Balance Sheet: Assets and Liabilities**

Cash_1	$= \text{Sales}_1 (\text{Cash/Sales})$	$= 1150 (0.17)$	$= \$195.5$
$\text{Accounts Receivable}_1$	$= \text{Sales}_1 (\text{AR/Sales})$	$= 1150 (0.208)$	$= 239.2$
Inventories_1	$= \text{Sales}_1 (\text{Invent/Sales})$	$= 1150 (0.148)$	$= 170.2$
$\text{Other Current Assets}$	$= \text{Sales}_1 (\text{OCA/Sales})$	$= 1150 (0.184)$	$= 211.6$
Current Assets_1	$= \text{Sales}_1 (\text{CA/Sales})$	$= 1150 (0.71)$	$= 816.5$
$\text{Net Fixed Assets}_1$	$= \text{Sales}_1 (\text{NFA/Sales})$	$= 1150 (1.424)$	$= 1637.6$
Total Assets_1	$= \text{Sales}_1 (\text{TA/Sales})$	$= 1150 (2.134)$	$= 2454.1$
$\text{Accounts Payable}_1$	$= \text{Sales}_1 (\text{AP/Sales})$	$= 1150 (0.279)$	$= 320.9$
$\text{Other Current Liabilities}$	$= \text{Sales}_1 (\text{OCL/Sales})$	$= 1150 (0.314)$	$= 361.1$
Earned Surplus_1	$= \text{Earned Surplus}_0 + \text{Ret.Earn}_1$	$= 1360.7 + 58.9$	$= 1419.6$

EXHIBIT 3.4   Calculation of expense, earnings, assets, and liabilities using percent of sales approach.

For example, we assume that the total asset-to-sales ratio will be maintained at 2.134. Therefore, with sales growing at 15% from $1000 million to $1150 million in 2011, the total assets required in 2011 will be

$$\text{Total Assets}_{2011} = (2.134) \times 1150$$
$$= \$2454.1 \text{ million.}$$

A similar calculation would apply for each of the individual asset accounts as shown in Exhibit 3.4.

Although we have discussed the link between sales and assets, we need to provide some additional explanation for the sources of funds and the liability accounts. The liabilities are the sources of funds for purchasing the assets needed to support sales. We split these sources into three categories: internal equity financing, spontaneous liabilities, and discretionary liabilities.

Internal equity financing refers to retained earnings. Retained earnings for the period is calculated as

$$\text{Retained Earnings}_t = \text{Net Income}_t - \text{Dividends}_t.$$

Assuming that dividends are based on a dividend payout ratio that is maintained at a constant level each period, retained earnings are

$$\text{Retained Earnings}_t = (1 - \text{Payout Ratio})\text{Net Income}_t,$$

where $(1 - \text{Payout Ratio})$ is called the retention ratio. With net income being a constant percentage of sales based on the net profit margin, we have

$$\text{Retained Earnings}_t = (1 - \text{Payout Ratio})\left(\frac{\text{NI}}{\text{Sales}}\right)\text{Sales}_t. \qquad (3.2)$$

With a net profit margin of 12.8%, and a 60% payout ratio, O&R's retained earnings for 2011 are expected to be

$$\text{Retained Earnings}_t = (1 - 0.60)(0.128) \times 1150$$
$$= \$58.9 \text{ million.}$$

The retained earnings for 2011 will be added to the earned surplus (retained earnings accumulated from past periods), so O&R's earned surplus at the end of 2011 will be

$$\text{Earned Surplus}_{2011} = \text{Earned Surplus}_{2010} + \text{Retained Earnings}_{2011}$$
$$= 1360.7 + 58.9$$
$$= \$1419.6 \text{ million.}$$

Spontaneous liabilities refer to liability accounts that are related directly to the production and sales cycle. These include accounts such as accounts payable, wages payable, and various liability accrual accounts that normally increase with production and sales. Since these accounts are closely linked to production and sales, their increase is considered as spontaneous with respect to sales. This spontaneous increase in liabilities helps finance part of the required increase in assets. To the extent that the liabilities are maintained at a constant proportion of sales, the financing provided will be a constant proportion of the increase in sales. In our example, accounts payable is such a spontaneous liability account. The accounts payable for 2011 are expected to be

$$\text{Accounts Payable}_{2011} = \left(\frac{\text{AP}}{\text{Sales}}\right)\text{Sales}_{2011}$$
$$= (0.279) \times 1150$$
$$= \$320.9 \text{ million.} \tag{3.3}$$

In this example, we will assume that Other Current Liabilities also increase spontaneously with sales, so the level of Other Current Liabilities in 2011 will be

$$\text{Other Current Liabilities}_{2011} = \left(\frac{\text{OCL}}{\text{Sales}}\right)\text{Sales}_{2011}$$
$$= (0.314) \times 1150$$
$$= \$361.1 \text{ million.}$$

Discretionary liabilities are those sources of funds that do not necessarily accrue with the firm's production and sales. An explicit action has to be taken to obtain these funds. The most common discretionary sources are borrowing and the issuance of new equity shares. These are sources of funds external to the firm, and the decision to use these sources is discretionary as opposed to spontaneous. In the case of our percent of sales approach, the discretionary sources of funds are the "plug" variables that make the balance sheet balance. These discretionary sources of financing were omitted from Exhibits 3.3 and 3.4 because they will be calculated as residuals.

To demonstrate the process of estimating these plug variables, Exhibit 3.5 shows the provisional balance sheet with the assets, internal equity financing, and spontaneous liabilities resulting from the projected growth in sales. Financing from the discretionary sources is not yet shown, so there is an imbalance. If we take no additional action to obtain financing, total assets exceed total liabilities by $172.2 million. We call this deficiency the "required external financing (REF)," because it is the amount of external, discretionary financing that we will need to make things balance. In this case, assume that long-term debt is our discretionary source and

	2010	+	Change in 2011	=	Year End 2011
Total assets	$2134.0	+	320.1	=	2454.1
Accounts payable	279.1	+	41.8	=	320.9
Other current liabilities	313.9	+	47.2	=	361.1
Notes payable	7.0		(No change)		7.0
Long-term debt	73.3		(No change)		73.3
Common stock	100.0		(No change)		100.0
Earned surplus	1360.7	+	58.9	=	1419.6
Total liabilities and equity	2134.0	+	147.9	=	2281.9

Required external financing = Excess of assets over liabilities = $172.2

EXHIBIT 3.5 Provisional asset and liability balances prior to obtaining additional discretionary financing.

we borrow enough to meet our needs. Long-term debt will increase by $172.2 million from $73.3 million to $245.5 million. With this additional debt everything will balance—sources will equal uses, and liabilities will equal assets as shown in Exhibit 3.6.

Exhibit 3.6 shows the sources of funds that finance the purchase of assests, and at the bottom, the completed balance sheet based on our

Uses		Sources	
——————————————— Flow of funds———————— — — — — —			
Purchase of assets	320.1	Increase in accounts payable	$41.8
		Increase in other current liabilities	47.2
		Retained earnings	58.9
		Additional borrowing	172.2
		Total financing	$320.1
—————————————— Assets and liabilities —————————————			
Cash	195.5	Accounts payable	$320.9
Accounts receivable	239.2	Other current liabilities	361.1
Inventories	170.2	Notes payable	7.0
Other current assets	211.6	Long-term debt	245.5
Current assets	816.5	Common stock	100.0
Net fixed assets	1637.6	Earned surplus	1419.6
Total assets	$2454.1	Liabilities and equity	$2454.1

EXHIBIT 3.6 Equality of sources and uses with discretionary financing.

percent of sales projection; Exhibit 3.4 showed the income statement. These *pro forma* financial statements were the immediate objective of our percent of sales approach. The percent of sales approach is a simple but powerful method that we will use throughout this book. However, our example glossed over a lot of detail in order to demonstrate the concept of REF. For example, by assuming that the net profit margin was constant over time, we implicitly assumed that all expenses were strictly proportional to sales. This could pose a difficulty when we issue debt to fund our external financing needs. With the debt changing, interest expense may not be strictly proportional to sales, and this will cause our net profit margin not to be strictly proportional to sales as sales grow. Problems like this are why we will spend a considerable amount of time learning how to structure more detailed planning models. In our future chapters we will dig into these problems and learn how to handle them. In the next two sections we will extend this simple percent of sales example to discuss two useful concepts for financial planning: REF and sustainable growth.

3.2 GROWTH AND THE NEED FOR FINANCING

One of the more difficult phases in the life cycle of a firm is the period of rapid growth. It would seem that growth would be a blessing, yet there are countless stories of firms that failed because they were unable to plan and manage their growth. The difficulty is that as a firm grows, its need for funds to finance expansion rapidly outruns its ability to generate funds from its operations. Faced with the opportunity to expand, but lacking the internal financing, these firms often turn to debt as the source of financing, and frequently they find themselves with debt servicing obligations that exceed their ability to pay. Failure is eventually brought about because the needs of expansion overwhelm the firm's resources for expansion. The difficulties and dangers of too rapid expansion can be mitigated by anticipating the firm's financing requirements, and developing a management policy that controls growth at a pace consistent with the firm's ability to finance it. Tools for growth planning include the concepts of required external financing (REF) and sustainable growth. These two closely related concepts are important because they help us understand the financing implications of the firm's growth and plan the appropriate financing strategies. The concepts of REF and sustainable growth help answer questions about

1. How much financing will be necessary to support a given growth rate in sales?

2. How rapidly can the firm's sales increase without having to obtain external financing?

3. How rapidly can sales grow if we constrain our borrowing?

3.2.1 Required External Financing

REF [also referred to as external funds needed (EFN)] is the additional funds that must be obtained from sources external to the firm in order to meet its financing needs, such as would be imposed by growth. The concept of REF is based on the assumption that the firm requires a certain level of assets and liabilities to support its sales. As sales grow, it will need to expand its asset base to be able to generate the sales. The required growth in assets will need to be financed. Part of the financing will be internal and/or spontaneous; part may have to be external. This logic is general, and does not necessarily depend on a stable percent of sales situation. However, we will discuss REF in the context of a percent of sales model, and continue to use O&R Corp. as our example.

In the last section our objective was to project *pro forma* financial statements. In the process we calculated the amount of external funds that O&R required to meet the needs imposed by its sales growth. This was referred to as REF in Exhibit 3.5, and amounted to $172.2 million. Now we will elaborate on the external financing and develop a short cut to calculate REF.

Working with the forecast that O&R's sales will grow at 15% during 2011, we want to determine the amount of external financing necessary to support that growth. Proceeding like we did in the previous section, we determine the total amount of asset investment that will be required. Then, we estimate the amount of funds that will be generated from internal and spontaneous sources. The remainder is the amount of funds that we need to obtain from external sources.

Assume that the amount of assets required to support sales is more or less a constant proportion of sales [i.e., the ratio (Total Assets/Sales) is the same each year], then the assets required to support forecasted future sales in period t are (Equation 3.1)

$$\text{Asset}_t = \left(\frac{\text{Assets}}{\text{Sales}}\right)\text{Sales}_t$$

and, if sales increase from one year to the next, the investment in assets required to support the increase in sales from years $t-1$ to t is

$$\text{Required Investment}_t = \left(\frac{\text{Total Assets}}{\text{Sales}}\right) \times (\text{Sales}_t - \text{Sales}_{t-1}) \quad (3.4)$$

With O&R's total asset-to-sales ratio at 2.134, and sales growing at 15% from $1000 million to $1150 million in 2011, the investment required during 2011 will be

$$\begin{aligned}
\text{Required Investment}_{2011} &= 2.134 \times (\$1150 - \$1000) \\
&= (2.134) \times 150 \\
&= \$320.1 \text{ million.}
\end{aligned}$$

This investment in new assets must be financed, and the sources include internal equity, spontaneous financing, and external financing. Our question is how much external financing will we need?

We have already discussed internal equity financing and seen that it is the product of sales, net profit margin, and the retention ratio:

$$\text{Retained Earnings}_t = (1 - \text{Payout Ratio})\left(\frac{\text{NI}}{\text{Sales}}\right)\text{Sales}_t \quad (3.5)$$

With a net profit margin of 12.8%, and a 60% payout ratio, O&R's retained earnings for 2011 are expected to be

$$\begin{aligned}
\text{Retained Earnings}_t &= (1 - 0.60)(0.128) \times 1150 \\
&= \$58.9 \text{ million.}
\end{aligned}$$

The spontaneous financing will be the increase in spontaneous current liabilities, calculated as

$$\text{Spontaneous Financing}_t = \left(\frac{\text{Spontaneous Liabilities}}{\text{Sales}}\right) \times (\text{Sales}_t - \text{Sales}_{t-1})$$

$$(3.6)$$

The spontaneous liabilities for O&R include both Accounts Payable and Other Current Liabilities. Combined, these are equal to about 59.3% of sales, and O&R, spontaneous financing in 2011 will amount to

$$\text{Spontaneous Financing}_{2011} = (0.593) \times (\$1150 - \$1000)$$
$$= \$89.0 \text{ million.}$$

This means that if the relation between sales and spontaneous current liabilities remains unchanged, there will be an increase in these accounts of $89 million due to the increase in sales. Note that the ratio of accounts payable to sales is related to the length of the payables period. If the firm takes longer to pay its accounts payable, the ratio (Spontaneous Current Liabilities/Sales) will increase, so more financing would be available from this source.

After considering internal and spontaneous financing, the remainder of the funds for expansion must come from external, discretionary sources—borrowing or issuing equity securities. The amount of REF is

Required Investment	$320.1 million
− Internal Equity Financing	− 58.9
− Spontaneous Financing	− 89.0
= REF	$172.2 million

To have sufficient assets to support its projected sales in 2011, O&R must invest $320.1 million in additional assets. It needs to find the funds to purchase these assets. $147.9 million of the funds will be provided by a spontaneous increase in current liabilities and retained earnings. The remainder must be financed by going outside the firm to obtain $172.2 million from either borrowing or by issuing new equity shares. Of course if sales grow slower than 15%, O&R will not need as much external financing.

Given our percent of sales assumptions, we have a shortcut for figuring the REF. Combining Equations 3.4, 3.5, and 3.6, REF can be expressed as

$$\text{REF}_t = \left[\left(\frac{\text{Total Assets}}{\text{Sales}}\right) - \left(\frac{\text{Spontaneous Liabilities}}{\text{Sales}}\right)\right] \times (\text{Sales}_t - \text{Sales}_{t-1})$$

$$- \text{Sales}_t \times \left(\frac{\text{Net Income}}{\text{Sales}}\right) \times (1 - \text{Payout Ratio}), \quad (3.7)$$

which, for 2011, would be

$$
\begin{aligned}
\text{REF}_{2011} &= (2.134 - 0.593) \times (1150 - 1000) - (1150) \times (0.128) \times (1 - 0.60) \\
&= (1.541) \times (150) - (0.0512) \times (1150) \\
&= 231.1 - 58.9 \\
&= \$172.2 \text{ million.}
\end{aligned}
$$

Note the kinds of trade-offs that can be considered with this analysis. An increase in the turnover of total assets (a decrease in Total Assets/Sales) can make a difference in the amount of external financing that is required. For example, if total asset turnover can be increased to the industry average of 1.3 from the company's present level of 0.469 (Total Assets/Sales decreasing from 2.134 to 0.769), the external financing requirement would be eliminated. Alternatively, with the other ratios unchanged, if the payout ratio is reduced to zero, the REF can be reduced to $84 million. Another trade-off is the rate of growth itself. The need for external financing is imposed by growth, and if sales grow more slowly, the firm will not have to seek as much external financing. In the next section we discuss the link between growth and the need for external financing.

3.2.2 Sustainable Growth

In the last section we took the rate of growth in sales as given and asked how much external financing would be required to support that growth. In this section we ask how rapidly we could grow without having to seek external financing.

We saw that if O&R's sales grow at 15% next year it will need to obtain about $172 million of external financing. However, the amount of external financing depends on the rate of growth, if O&R grows at a slower rate it will not require as much new investment and external financing. Exhibit 3.7 shows this relationship for O&R, with REF on the vertical axis and growth in sales on the horizontal axis. At very low growth, REF is negative. This means that the firm not only does not need external funds, but it generates funds that can be returned to its investors. As the growth rate increases, the need for funds increases. So at very high growth rates a great deal of external financing is needed. Note the point where the REF line

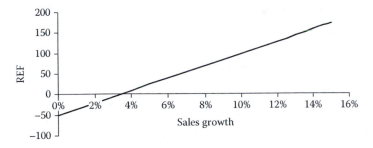

EXHIBIT 3.7 Required external financing.

crosses the horizontal axis, which is at a growth rate of about 3.44%. At this point, the need for external funds is zero; all of the firm's funding needs are met from internal sources comprised of spontaneous current liabilities and retained earnings. Given O&R's operating characteristics as represented by the various ratios, its sales could grow at this rate and it would not have to resort to any financing other than the funds provided by trade credit and retained earnings. This is a rate of growth that is sustainable without the need to seek external financing. Faster growth is certainly possible, but it will have to be financed with external sources.

This leads us to the concept of sustainable growth. Sustainable growth is a rate of growth in sales that can be sustained without having to resort to external sources of financing. We consider two different versions of sustainable growth. The first, denoted by g^*, is the fastest rate that sales can grow without forcing the firm to use any external financing except those funds provided by spontaneous credit sources and internal equity financing. The second version, denoted by g^{**}, is the rate of sales growth the firm can sustain while keeping its debt ratios constant and meeting its equity needs with internal equity financing. We will now explain the two versions of sustainable growth.

In Exhibit 3.7, the point where the REF line crosses the horizontal axis is the growth rate where no external financing is required. The firm would not have to resort to any discretionary borrowing or to issue any new equity, all of its financing needs are supplied from spontaneous liabilities and internal equity (retained earnings). In this example, the growth associated with zero REF is 3.44%. We can use Equation 3.7 to solve for the zero REF growth rate by setting REF to zero, and solving for growth. The solution is the firm's sustainable growth, g^*, that is given by the formula

$$g^* = \frac{\left(\dfrac{\text{Net Income}}{\text{Sales}}\right) \times (1 - \text{Payout Ratio})}{\left[\left(\dfrac{\text{Total Assets}}{\text{Sales}}\right) - \left(\dfrac{\text{Spontaneous CL}}{\text{Sales}}\right)\right] - \left(\dfrac{\text{Net Income}}{\text{Sales}}\right) \times (1 - \text{Payout Ratio})}$$

$$= \frac{(0.128) \times (1 - 0.60)}{(2.134 - 0.593) - 0.128 \times (1 - 0.60)}$$

$$= 0.0344 = 3.44\%. \tag{3.8}$$

If the firm's annual growth in sales exceeds 3.44%, outside financing (either debt or equity) will be required; and the higher the growth rate, the more external funds will be required. If the firm's annual growth rate in sales is less than 3.44%, it will be able to meet its needs with spontaneous financing and internal equity financing and generate excess cash.

Confining the financing of growth to spontaneous current liabilities and internal equity may constrain the firm's growth unnecessarily. If management is willing to utilize debt financing in addition to these other sources, then greater growth may be possible without exceeding the limits of prudence. We could ask how much growth the firm could finance without increasing the proportional amount of debt in the firm's permanent capital structure. Higgins (1981) provides a formula similar to the one for g^* above that represents the highest sustainable rate of growth, g^{**}, in sales that the firm can generate and finance without changing the financing mix of debt and equity. Higgins' formula is

$$g^{**} = \frac{\left(\dfrac{\text{Net Income}}{\text{Sales}}\right) \times (1 - \text{Payout Ratio}) \times \left[1 + \left(\dfrac{\text{Debt}}{\text{Equity}}\right)\right]}{\left(\dfrac{\text{Total Assets}}{\text{Sales}}\right) - \left(\dfrac{\text{Net Income}}{\text{Sales}}\right) \times (1 - \text{Payout Ratio}) \times \left[1 + \left(\dfrac{\text{Debt}}{\text{Equity}}\right)\right]},$$

$$\tag{3.9}$$

which for O&R, with a debt-to-equity ratio of, yields

$$g^{**} = \frac{(0.128) \times (1 - 0.60) \times (1 + 0.46)}{2.134 - (0.128) \times (1 - 0.60) \times (1 + 0.46)}$$

$$= 0.0363 = 3.63\%.$$

Thus, O&R could sustain sales growth of about 3.63% per year with the assumed profit margin, payout ratio, and asset turnover by financing the expansion in such a way that the firm's total debt (both spontaneous, current liabilities, and discretionary debt) to total equity is maintained at its current level of 0.46 to 1.*

It is important to keep in mind what these sustainable growth concepts mean, and what their limitations are. They are useful first approximations of the limits of the firm's long-run growth rate—its speed limit, if you will. This does not mean that the firm cannot grow more rapidly than the formulas indicate. More rapid growth is attainable. However, the firm can only grow faster by either obtaining outside financing or changing one of the limiting ratios. It can grow faster by increasing its profit margin, by decreasing its dividend payout, by increasing its asset turnover, by increasing its spontaneous borrowing or discretionary borrowing, or, finally, by issuing external equity. In any case, to pursue more rapid growth, something has to give. These formulas help to remind the manager which items have to adjust.

---

* Another widely used formula that relates growth to the firm's financial ratios expresses growth as

$$Growth = (Retention\ Ratio) \times (Return\ on\ Equity)$$

$$= Retention\ Ratio\left[\left(\frac{Net\ Income}{Sales}\right)\left(\frac{Sales}{Total\ Assets}\right)\left(\frac{Total\ Assets}{Equity}\right)\right],$$

where the latter three terms express the firm's return on equity based on the DuPont formula discussed in Chapter 2. However, this formula refers to the growth rate in net income based on the assumption that the

$$Return\ on\ Equity = \left(\frac{Net\ Income}{Sales}\right)\left(\frac{Sales}{Total\ Assets}\right)\left(\frac{Total\ Assets}{Equity}\right)$$

is the return earned on the incremental investment of retained earnings. The formula does not necessarily relate to growth in sales, nor does it represent a limit on the growth rate. It generally is slightly less than g^{**}. Nonetheless, it is the formula that Higgins uses in the recent edition (1996) of his book. He refers to this version of sustainable growth as $PRAT = $ Profit Margin × Retention Ratio × Asset Turns × EquiTy Multiplier. Finally, to be applied precisely, the equity multiplier should be the current period's total assets divided by the previous period's equity.

The use of various ratios to determine required future financing and estimate growth limitations is useful. However, in more realistic applications the formulas are just approximations that tend to gloss over the specifics of the situation. For example, when the various financial ratios are not constant from year to year, the formulas will not be exact. When costs include both fixed and variable components, profit margins will change as sales grow, and the formulas will be inaccurate. Furthermore, when the firm has excess capacity, it may be able to grow without making new investments, so it can grow more rapidly than the sustainable growth formulas indicate. To gain greater accuracy we need more detailed models. In subsequent chapters we will develop models that will allow us to analyze our plans in greater detail.

3.3 CASH FLOW

From the point of view of financial planning, financial management, and firm valuation, cash flow is usually of more importance than earnings as derived with standard accounting methods. This section explains how to calculate cash flow. There are two lessons to be learned in this section regarding cash flows. The first is that there are many different versions of cash flow; the second is that you should know how to calculate cash flow for versions that are appropriate for your application.

Cash flow refers to the amount of cash that has been generated during a specified period of time. But we have to be careful how we define and measure "the amount of cash." For example, one common method for calculating the cash flow is

$$\text{Net Income} + \text{Depreciation Expense.}$$

We will call this "Earnings Cash Flow." This version of cash flow is based on the idea that Net Income understates the amount of cash flowing into the firm because Depreciation is a non-cash expense and therefore should be added back to net income to determine how much cash was earned during the period. However, there are several difficulties with this definition of cash flow. First, this definition assumes that depreciation is the only non-cash expense, when in fact there may be other expenses incurred during the period that did not involve a cash payment. Second, it assumes that our sales revenue actually generated cash, when in fact, sales may not have brought in any cash—for example, they may have been credit sales. Third, there may be other transactions that are not reflected in the income statement that involve cash inflows and outflows. These can include items such as increases in working capital, purchases of fixed assets, repayment of debt, funds provided by borrowing, funds obtained by issuing new stock, and the payment of dividends. None of these are included in the calculation of net income and do not show up on the income statement. The point is, the "Net Income + Depreciation" version of cash flow is not a complete answer to the question, "How much cash did the firm generate?" Cash flow is a lot more complicated than that.

When we ask, "How much cash did the firm generate?," we should be more specific and clarify "Cash for whom?," or "Cash for what purpose?" For example, we could ask, "How much cash was generated by sales for the purpose of paying our expenses?" The answer would involve the determination of cash payments by customers, but would be before any expenses. Or, we could ask, "How much cash was generated to cover our interest expense?" This would be a very different number than the first one, and in neither case would the answer be "Net Income + Depreciation." The point is that we simply have to be careful to state, "Cash flow to whom?," or, "Cash flow for what purpose?"

In this book, we deal with several different definitions of cash flow. These include cash flow from operations, cash flow available for all the firm's investors, and cash flow available for the stockholders. There are other versions we could consider, but these are the most important ones. As noted above, which of these versions we use depends on the question being asked and the intent of the application.

In accounting, questions are usually related to what actually happened. But in financial planning, questions usually relate to what could happen, or perhaps, how much will something (usually an asset) be worth? In these cases, we frequently distinguish between the actual, the necessary, and the forecast. For example, free cash flow is a term often used in planning and valuation contexts. There are numerous definitions of free cash flow, but a commonly accepted definition is "the cash flow that is available to be distributed to investors after the firm has met all its needs for ongoing operations, including investments in working capital and fixed assets." Central to this concept of free cash flow is the idea that for the ongoing, normal operations of the firm, there are investments that are *necessary* to keep the firm going and to support its normal growth. Hence, when planning for or valuing an ongoing business, the free cash flow calculation must include these necessary investment uses. More generally, when we calculate cash flow, we need to keep in mind the intent of our application.

We must specify what the firm needs in order to figure out the cash flow. We will now use the O&R Corp. data to explain these cash flow concepts.

We begin our discussion by using the data from O&R's financial statements in Exhibit 3.2 to construct a Statement of Cash Flows in a standard accounting format. The Statement of Cash Flows for O&R is shown in Exhibit 3.8. This "indirect format" is how cash flow is usually reported in an annual report. It starts with net income and makes adjustments to take account of other transactions that affect cash flow. Transactions are classified into operating, investing, and financing activities. Operating activities refers to transactions that are a normal part of the ongoing

Cash flows from operating activities			
Net income	$127.8		
Adjustment for non-cash items:			
Depreciation expense	+53.8		
Earnings cash flow		$181.6	
Subtract:			
Increase in accounts receivable	30.6		
Increase in inventories	5.4		
Increase in other current assets	−5.8		
Funds used to increase current assets		−30.2	
Add:			
Increase in accounts payable	103.6		
Increase in other current liabilities	−70.0		
Funds provided by increase in current liabilities		+33.6	
Cash flow from operations			185.0
Cash flows from investing activities			
Purchase of fixed assets	134.4		
Cash flow from investing			−134.4
Cash flows from financing activities			
Issuance of short-term debt	0.1		
Issuance of long-term debt	35.6		
Issuance of common stock			
Payment of dividends	−76.7		
Cash flow from financing			−41.0
Net cash flow			+9.6
Plus: Cash balance, 12/31/2009	160.4		
Equals: Cash balance, 12/31/2010			$170.0

EXHIBIT 3.8   O&R Corp. Statement of cash flows, year 2010 ($ millions).

business. Investing activities refers to transactions that are not part of the regular operating activities, such as the purchase or sale of long-term assets. Financing activities refers to (non-operating) transactions that raise new capital or make payments to investors (creditors, stock-holders, etc.).

The purpose of the cash flow statement is to summarize all the firm's transactions in terms of their impact on the cash balance at the end of the year. Thus, the cash flow is the $9.6 shown on the third line from the bottom. In terms of asking "cash flow for what purpose," this statement defines cash flow as that which is available to increase the cash balance. And the answer is that during 2010 the firm generated $9.6 million that was used to increase the cash balance. However, this cash flow is not one of the versions used for most purposes in our financial planning and modeling mentioned earlier: cash flow from operations, cash flow available to the firm's investors, and cash flow available for the stockholders. In the field of finance there is no consistent name for each of these versions, and each version has a slightly different method of computation, depending on the author. We will try to be consistent throughout this book with our terminology and computation. However, along the way we will learn a variety of terms used for each version of cash flow.

3.3.1 Cash Flow from Operations

Cash flow from operations refers to the cash flow generated by the normal operating activities of the firm. Consider a firm producing and selling a product. Its normal operations bring in cash from selling the product, and require the purchase of the raw materials, the payment for production, payments for wages, and so on. In addition, as part of its regular operations, it may invest by increasing its cash balance, accounts receivable and inventories, and it uses trade credit to purchase its materials. The cash flow from operations considers all these items in its calculation. Most of this is shown in the first section of Exhibit 3.8. However, the increase in cash is not shown in that section of Exhibit 3.8 because it is shown at the bottom of the statement as a residual. For our purposes, in other contexts, we will usually include the change in cash balance in the cash flow from operating activities. With this minor variation, the cash flow from operations will be

Net Income + Depreciation and Other Non-cash Expenses
　= Earnings Cash Flow
　　– Increase in Cash and Securities

        – Increase in Accounts Receivable
        – Increase in Inventories
        – Increase in Other Current Assets
        + Increase in Accounts Payable
        + Increase in Other Current Liabilities
      = Cash Flow from Operations

There is an important caveat that we need to make in this definition of cash flow: What we include in each category depends on our intended use for the cash flow calculation. For example, if we are measuring the actual cash flow, we would include all the items just as they are shown on the financial statement. However, if we want to measure the operating cash flow that potentially would be available to the investors, we would consider the *necessary*, or required, changes in these current accounts. Cash & Securities is a case in point. While cash may actually increase by \$9.6 million, the firm may need an increase of only \$3.6 million to meet its normal operating requirements. If so, \$6 million of the actual increase is excess. This excess potentially could be paid out to the firm's investors without impinging on normal operations. So, if the question is what was the actual operating cash flow, the \$9.6 million is included. If the question is "what is the cash flow from operations that is available for the investors?," the answer takes account of only a necessary increase in cash balance of \$3.6 million. An extension of the same idea is that there may be other current accounts that do not relate to the normal operations of the firm. For example, if the firm has large investments in marketable securities that are not necessary for the regular operations, then these investments would not be considered in the calculation of operating cash flow. Similarly, the categories "Other Current Assets" and "Other Current Liabilities" are included above as catchall categories. If these accounts do not relate to normal operations, they would be excluded.

A second caveat relates to the treatment of short-term debt. Short-term debt is a current liability. However, for our purposes it is not included in the cash flow from operations because it is a discretionary source of funds that is part of the cash flow from financing. So, in the following discussion, when we speak of current liabilities or the change in current liabilities, we will not include short-term debt.

In the calculation of cash flow from operations note that increases in current assets and liabilities are included as uses and sources of funds, respectively. We can summarize the impact of these current accounts on operating flow by treating the change in net working capital (NWC) as a cash outflow. NWC is defined as

$$\text{Net Working Capital} = \text{Current Assets} - \text{Current Liabilites}$$

or

$$\text{NWC}_t = \text{CA}_t - \text{CL}_t,$$

where NWC_t denotes net working capital, and CA_t and CL_t denote current assets and liabilities at time t, respectively, with current liabilities excluding short-term debt. The increase in NWC during period t is

$$\begin{aligned}
\text{Increase in NWC}_t &= (\text{CA}_t - \text{CL}_t) - (\text{CA}_{t-1} - \text{CL}_{t-1}) \\
&= (\text{CA}_t - \text{CA}_{t-1}) - (\text{CL}_t - \text{CL}_{t-1}). \quad (3.10)
\end{aligned}$$

For the year 2010 this was

$$\begin{aligned}
\text{Increase in NWC}_{2010} &= (710.0 - 670.2) - (593.0 - 559.4) \\
&= 39.8 - 33.6 \\
&= \$6.2,
\end{aligned}$$

where all current assets are included, and current liabilities excludes short-term debt. Thus, all the current account inflows and outflows can be summarized as the change in NWC, and the calculation of the cash flow from operation is summarized in Exhibit 3.9.

Net income		127.8
+	Depreciation expense	+ 53.8
=	Earnings cash flow	181.6
−	Increase in net working capital	− 6.2
=	Cash flow from operations	175.4

EXHIBIT 3.9 Cash flow from operations.

Note that the cash flow here of \$175.4 differs from the Cash Flow from Operations shown in Exhibit 3.8 (shown as \$185.0) by the amount of \$9.6. This difference stems from the fact that the increase in *Cash* is included in the increase in NWC when calculating Cash Flow from Operations, but it is not included in the Cash Flow from Operations in the Statement of Cash Flows (Exhibit 3.8) where it is the residual.

During the course of the firm's normal operations, there is usually a need to increase its current assets to support the growth in sales. The increase in current assets is an investment that needs to be financed. The funds required to finance the increase have to come from somewhere. Part of these required funds will be supplied by increases in current liabilities. The part of the investment in current assets that is not financed by the increase in current liabilities must be financed from other sources. This is what the increase in NWC represents—the part of the new investment in current assets that must be financed from sources other than current liabilities.

As noted earlier, what we include in the change in NWC depends on the purpose. If we are calculating cash flow available to be distributed to investors in the sense of free cash flow, the increase in NWC should take account of the changes in the current accounts *necessary* for the normal operations of the firm. Later in the book, in the context of building a computer-based model, we also will refer to changes in non-cash working capital. In this latter case, we will mean the necessary changes in all the current accounts except cash, treating cash as a residual as shown in Exhibit 3.8. Another variation is the distinction between spontaneous current liabilities and discretionary current liabilities, as was discussed in the first part of the chapter. In that context, we distinguished between the current liabilities that grow "spontaneously" with the growth in sales, and those that are not spontaneous. In this case, we include the spontaneous current liabilities in the calculation of NWC and exclude liabilities such as short-term debt that do not grow spontaneously but are instead "discretionary."

3.3.2 Cash Flow from Investing

The second section of Exhibit 3.8 shows the Cash Flow from Investing Activities as a cash outflow. This refers to the investment in long-term assets such as plant and equipment or even the purchase of another company. In the format presented in that exhibit, this investment also takes account of the replacement of depreciated assets. In Exhibit 3.8 the cash flow from investing activities is shown as an outflow of $134.4. There are two ways to calculate this from the financial statements:

Method 1: Change in Gross Long-term Assets:

Gross Fixed Assets (GFA), year end 2010:	$1518.0 million	
Other Long-term Assets	409.0	
Gross Long-term Assets, year end 2010:		$1927.0
Less:		
GFA, year end 2009:	1356.7	
Other Long-term Assets	435.9	
Gross Long-term Assets, year end 2009:		1792.6
Equals: Investment in Long-term Assets, year 2010		$134.4

Method 2: Change in Net Long-term Assets Plus Depreciation Expense:

Net Long-term Assets (Including Other Long-term Assets), year end 2010	$1424.0 million	
Less: Net Long-term Assets, Year End 2009	1343.4	
Expenditures for New Fixed Assets		80.6
Plus: Depreciation Expense, Year 2010		53.8
Equals: Investment in Fixed Assets, Year 2010		$134.4

Both of these methods yield the total amount invested which consists of the amount spent to replace the assets that were depreciated during the period, plus the amount invested to purchase new assets beyond what was worn out. In this case, the firm spent $134.4 million to purchase fixed assets, which consisted of $53.8 to buy assets to replace those that were depreciated during the period, and $80.6 million to purchase assets in excess of those that wore out.

This calculation is correct assuming that depreciation expense has been included in the source of funds. That is, if the source of funds is calculated as (Net Income + Depreciation Expense), then because depreciation is treated as a source, you have to include an expenditure for replacement of depreciated assets as part of the fixed asset investment. On the other hand, if depreciation expense is not included in the calculation of the sources of funds, then you do not need to include the replacement as part of the expenditures. In that case, the expenditures for fixed assets would be just the increase in net fixed assets (NFA), $80.6 million.

The balance sheet in Exhibit 3.2 is simplified, and does not show the whole range of accounts that might be shown on a more complex statement. There could be many more types of asset accounts than those shown. In that case, the investment section of the cash flow statement must show the changes in all asset accounts just as was done here for fixed assets and other long-term assets.

3.3.3 Cash Flows from Financing

Exhibit 3.8 shows the Cash Flow from Financing Activities as +$41.0 million. This summarizes the flow of capital to and from the firm's investors, where "investors" includes both the creditors and stockholders. The flows to the creditors include repayment of debt principal. Flows from the creditors to the firm include payments by the creditors in the form of new loans.

In this case, there was a net inflow from the creditors to the firm of $35.7 million in the form of short- and long-term loans. If there had been repayments of debt principal to the lenders, it would have been shown as an outflow from the firm to the lenders. Alternatively, the payments from and to the creditors can be combined as the net flow to the creditors. Interest payments were not included here because they were already taken into account in the calculation of net income. Flows to the stockholders include dividend payments and share repurchases. Flows from the stockholders include purchases of shares by the stockholders. The other investor flow in the exhibit is a dividend payment of $76.7 million that is an outflow from the firm to the stockholders. If the firm repurchased shares from stockholders, the amount they paid to do so would also be shown as an outflow; and if the stockholders purchased additional shares, the amount they invested would be an inflow to the firm.

The definition and calculation of cash flow depend on the purpose of the cash flow statement. The purpose of a cash flow statement such as Exhibit 3.8 is to summarize the cash flow *to the firm* from all sources, with the increase in the cash balance shown as a residual at the bottom because the cash account is the recipient of the cash flows into the firm. How particular flows are treated depends on the purpose of the cash flow statement. Purpose, in this context, refers to the question, "Cash flow to whom?" We will refer to the group or entity *to whom* the cash is flowing as the *object* of the cash flow. In Exhibit 3.8, the object of the cash flow is the firm, or the firm's cash balance, and the cash flow available to increase the cash balance is the residual of the calculation. The firm gets whatever remains after everyone else has been paid. However, the firm and the cash balance are not always the object of the cash flow. There may be other objects such as the stockholders or all the investors. In general, the investor group that is the object of the cash flow is left out of the financing section of the cash flow statement, and the flow available to that group is the residual of the calculation. For example, if our question is, "What is the cash flow that is available to be paid to the stockholders?," then the stockholders are the object of the cash flow statement. The calculation of cash flow should take account of all the other flows and leave as a residual the cash flow that remains for the stockholders. If all the other flows are taken into account, then what is left over is the amount that is available to the stockholders. In that case, the financing section of the cash flow statement would show the flows from and to the other investors such as the creditors, and the flow available to the stockholders would be the remainder at the bottom of the statement.

3.3.3.1 Equity Cash Flow

Equity Cash Flow is the term we will generally use to refer to the cash flow that is available to the stockholders. Several interchangeable terms are commonly applied to the Equity Cash Flow, such as cash flow to equity, cash flow available for the stockholders, and dividend paying ability. We will probably use all of these terms at various times. They all refer to the cash flow that is left for the owners after the firm's other needs for funds have been met. Think of a small business on Main Street. The cash flow to equity is the amount that the owner could take out of the business at the end of the operating period after paying all the bills, paying his creditors whatever was due for them this period, and meeting all the needs for plowing money back into the business. What is left over at the end of the period belongs to the owner and can be withdrawn from the business without impinging on its operations in the future. That is the equity cash flow.

The equity cash flow is calculated in Exhibit 3.10, using data from Exhibit 3.8. Now, the current stockholders are the object of the cash flow calculation. The blank lines represent categories that would be included, but are not present in O&R's case. Note that the equity cash flow equals the dividends that were actually paid to the stockholders. In other cases it is

Cash flow from operating activities			
	Net income	$127.8	
+	Depreciation expense	53.8	
=	Earnings cash flow	181.6	
−	Increase in net working capital	−6.2	
=	Cash flow from operations		175.4
Less:			
Cash flow from investment activities			
	Expenditures for net long-term assets	80.6	
+	Depreciation expense	53.8	
=	Investment in fixed assets		− $134.4
Plus:			
Cash flow from financing activities			
+	Proceeds from short-term borrowing	0.1	
−	Repayment of short-term debt principal	—	
+	Issuance of long-term debt	+35.6	
−	Repayment of long-term debt principal	—	
+	Proceeds from issuance of stock	—	
=	Cash flow from financing		$35.7
Equity cash flow			+$76.7

EXHIBIT 3.10   Equity cash flow.

possible that these would differ because some of the cash that could have been paid to the stockholder was retained as (say) marketable securities, and not paid as dividends.

As with any of our cash flow definitions, this version of equity cash flow has variations. In this case, we calculated the equity cash flow from the point of view of the current stockholders, asking the question, "What is the cash flow available to be distributed to the current stockholders?" We included the proceeds from the issuance of stock as a cash inflow that enhanced the cash flow to equity. That is because the flow of funds into the firm from new stockholders would add to the cash available this period to be paid as dividends to the current stockholders. Offsetting this positive effect of selling new shares is the fact that the ownership position of the existing stockholders would be diluted, and they would get a smaller share of the total dividends that are paid in future periods. An alternative definition of equity cash flow would not treat the proceeds from the issuance of new shares as a cash inflow. Rather, it would exclude all equity transactions and simply ask, "How much cash flow was generated by the firm from all sources other than equity transactions?" For some applications, this latter definition is appropriate. However, for others, the former is best. In this book, we will generally use the former definition where we treat the issuance of stock as a cash inflow that enhances the cash flow available for the current stockholders.

3.3.3.2 Cash Flow to Invested Capital

The cash flow to invested capital means the cash flow that is available to be paid to all the groups that have invested capital in the firm. Other terms are used to refer to this cash flow including "cash flow to investors," "overall cash flow," "cash flow from assets," "enterprise cash flow," "entity cash flow," and "cash flow from the firm." Regardless of the name used, this definition asks, "After taking account of both the cash flow from operations and the need to invest funds in new assets, how much cash flow was generated that would be available for both the creditors and the equity investors?"

In this context, creditors refers to both long-term and short-term lenders, but does not include the trade creditors or other liabilities that are not explicit debt. Generally, any debt that is interest-bearing is included in this category, whether it is long term or short term. We will make a distinction between spontaneous credit and discretionary credit. Spontaneous credit refers to the liabilities that normally increase in step with the growth in sales, such as trade credit. Discretionary credit is credit that does not increase spontaneously and normally requires an explicit action to initiate, such as signing a promissory note. This is usually interest-bearing debt. Even short-term debt such as notes payable are considered discretionary, and we will generally consider the short-term lenders as part of the group of permanent investors. The reason for this is that for most firms short-term debt is constantly renewed and used on an ongoing basis, so there is always some short-term debt on the balance sheet. Hence, we treat it as a permanent source of capital.

Obviously, stockholders also are considered as investors. In addition, there may be other investors who provide permanent capital, but who cannot be clearly defined as creditors or stockholders. Examples include claims that are hybrids of debt and equity such as convertible debt, preferred stock, and other more exotic securities. For our purposes, it does not really matter whether they are considered debt or equity. We put all investors, debt and equity, in the same category, and simply referring to them as "the investors," or the providers of invested capital.

Cash flow from operating activities			
	Net income	$127.8	
+	Depreciation expense	53.8	
+	Interest expense	3.4	
−	Increase in net working capital	−6.2	
=	Cash flow from operations		178.8
Less:			
Cash flow from investment activities			
	Expenditures for net long-term assets	80.6	
+	Depreciation expense	53.8	
=	Investment in fixed assets		−$134.4
Cash flow to invested capital			$44.4

EXHIBIT 3.11 Cash flow to invested capital.

Having specified who is included in the category "investors," now we need to show how to determine the cash flow to investors, or the cash flow to invested capital. The cash flow to investors is the cash flow from operations plus investing activities. It does not yet take account of any of the payments to the creditors (interest-bearing debt) or to the stockholders. For O&R this is shown as Exhibit 3.11.

Note that the cash flow from operations includes a term that adds back interest expense. This is to restate net income without the effect of interest expense. Interest is part of the payment to the creditors, so in asking how much cash flow the firm generated for the creditors, we want to remove interest as an expense and show it as part of the funds available for the creditors. To see how this works out, the cash flow to investors of $44.4 was distributed in the following manner:

Interest Paid to Lenders	+3.4
− Proceeds from Short-term Borrowing	−0.1
+ Repayment of Short-term Debt Principal	−
− Issuance of Long-term Debt	−35.6
+ Repayment of Long-term Debt Principal	−
+ Dividends Paid to Stockholders	+76.7
− Purchases of Stock by Stockholders	−
= Cash Flow Received by Invested Capital	$44.4

This format is from the perspective of the investors, so the flow from the firm *to* the investors is shown as positive and *from* investors to the firm is shown as negative. From the stockholders' point of view, the dividend payment is a positive flow to them. From the lenders' point of view, interest is a flow to the lenders, and the issuance of new debt is a payment of $35.7 from the lenders to the firm. On a net basis, the investors as a group (both lenders and stockholders) have received a total of $44.4 from the firm.

We have to have another caveat about defining cash flow in terms of, "Cash flow for whom?" or "Cash flow for what purpose?" In the above calculation of cash flow to the invested capital, we added back interest expense to net income and showed it as part of the cash flow available to the investors. We did not make any kind of tax adjustment for the interest. Later on, when we are discussing the valuation of the firm and we are discounting the investors' cash flow at the firm's WACC, we will change the definition of cash flow to investors once again. When we get to that point, we will change the way we calculate the cash flow to invested capital by removing the effect of the tax deductibility of interest. That is, in the calculation above, we added back interest of $3.4 to restate cash flow from operation. Subsequently, when we are dealing with the WACC, we will add back "Interest $(1 - T)$," where T denotes the tax rate, instead of "Interest." We won't go into this issue any further now, but this is a warning to be prepared to encounter yet another way to calculate cash flow.

In this section, we have examined the statement of cash flows and have discussed various definitions of cash flow. In the next section, we will consider another type of cash flow statement—the statement of cash receipts and disbursements. This will give us a tool of analyzing the short-run flow of cash in and out of the firm and for determining our potential needs for funds.

3.4 CASH RECEIPTS AND DISBURSEMENTS

A common problem faced by a financial manager is that there is a lag between disbursements for expenses of production and receipts from sales. This is particularly true for a firm that has a seasonal sales cycle. The lag between receipts and disbursements results in a need for seasonal financing. The financial planning problem is to anticipate the financing needs, determine the amount of financing required, and find the best way to meet this requirement. The following example is typical. It demonstrates the seasonal financing needs by developing a statement of cash receipts and disbursements. This statement (cash budget) is applicable to a wide variety of problems, both short- and long-term, but it is especially useful for considering short-term (quarterly, monthly, or even weekly or daily) cash flows.

3.4.1 Example: The Mogul Corporation

The Mogul Corporation manufactures skis that are sold on credit to retail outlets. The sales are seasonal, as retailers stock up on skis in late fall in anticipation of winter sales. The production cycle, the seasonal sales pattern, and the credit terms granted to the retailers all influence the firm's financing requirements. The question is: Given the anticipated sales, what will be Mogul's financing requirements for the coming season.

The starting point for analyzing the firm's financing requirements is to develop a *pro forma* statement of cash receipts and disbursements. This will require a set of assumptions and forecasts about the company's resources and constraints and the future condition of the economy. The major assumptions are shown in Exhibit 3.12; the monthly sales forecasts for the coming year are shown in line 1 of Exhibit 3.13. The fiscal year is assumed to start in July and end the following June, with the sales for the previous year and following year following the same pattern month-to-month.

---

a. Mogul Corporation manufactures skis to meet demand one month ahead of anticipated sales.

b. Purchases of materials (line 7) are 50% of sales and occur two months prior to sales. Purchases are made on trade credit and are paid one month after the purchase. Thus, July purchase disbursements of $4000 are 50% of August sales.

c. Wages and salary payments (line 8) are 20% of sales and are paid in the month the skis are manufactured. Thus, July wages of $1600 are 20% of August sales because the skis sold in August are manufactured in July.

d. Rent (line 9) for the building is $2000 per month.

e. Interest payments on an outstanding loan (line 10) are $600 per month.

f. Quarterly principal payments (line 11) of $1500 are due in March, June, September, and December.

g. Quarterly payment of estimated income taxes (line 12) of $1000 must be made in April, June, September, and December.

h. Miscellaneous other expenses (line 13) will amount to $500 per month.

i. All receipts are from sales. Ten percent of each month's sales are typically on a cash basis (line 2) and 90% are on credit.

j. Forty percent of sales are collected the first month after the sale (line 3), 30% are collected the second month (line 4), 15% are collected three months after the sale (line 5), and 5% are bad debts and are never collected.

EXHIBIT 3.12 Assumptions used for developing cash budget for Mogul Corporation.

Mogul Corporation
Cash Receipts & Disbursements
July - June

	Jul	Aug	Sep	Oct	Nov	Dec	Jan	Feb	Mar	Apr	May	Jun
1 Sales	3000	8000	18000	40000	60000	35000	20000	15000	5000	3000	2000	1000
Receipts												
2 Cash sales	300	800	1800	4000	6000	3500	2000	1500	500	300	200	100
Credit sales												
3 First month	400	1200	3200	7200	16000	24000	14000	8000	6000	2000	1200	800
4 Second month	600	300	900	2400	5400	12000	18000	10500	6000	4500	1500	900
5 Third month	450	300	150	450	1200	2700	6000	9000	5250	3000	2250	750
6 Total Receipts	1750	2600	6050	14050	28600	42200	40000	29000	17750	9800	5150	2550
Disbursements												
7 Purchases	4000	9000	20000	30000	17500	10000	7500	2500	1500	1000	500	1500
8 Wages/salary	1600	3600	8000	12000	7000	4000	3000	1000	600	400	200	600
9 Rent	2000	2000	2000	2000	2000	2000	2000	2000	2000	2000	2000	2000
10 Interest	600	600	600	600	600	600	600	600	600	600	600	600
11 Debt principal	----	----	1500	----	----	1500	----	----	1500	----	----	1500
12 Taxes	----	----	1000	----	----	1000	----	----	----	1000	----	1000
13 Misc.	500	500	500	500	500	500	500	500	500	500	500	500
14 Total Disbursements	8700	15700	33600	45100	27600	19600	13600	6600	6700	5500	3800	7700
15 Net Cash Flow	(6950)	(13100)	(27550)	(31050)	1000	22600	26400	22400	11050	4300	1350	(5150)
16 Beginning Cash	5000	(1950)	(15050)	(42600)	(73650)	(72650)	(50050)	(23650)	(1250)	9800	14100	15450
17 Cumulative Cash	(1950)	(15050)	(42600)	(73650)	(72650)	(50050)	(23650)	(1250)	9800	14100	15450	10300
18 Desired cash	5000	5000	5000	5000	5000	5000	5000	5000	5000	5000	5000	5000
19 Required financing	(6950)	(20050)	(47600)	(78650)	(77650)	(55050)	(28650)	(6250)	0	0	0	0

EXHIBIT 3.13   Cash receipts and disbursements.

Sales are at a seasonal low in June and July, start increasing rapidly in August, reach a peak in November, then decline over the winter and spring months. The sales forecast is the primary input. It "drives" the calculation of the firm's financing needs, the output that is the focus of this exercise. Obviously, any change (or inaccuracy) in the sales forecast would lead to a change (or error) in the forecast of the financing needs. Based on the forecast and assumptions, Exhibit 3.13 is the cash receipts and disbursements statement showing the cash flows and required financing over a one-year planning horizon that is divided into monthly sub-periods. This statement, however, is only a first approximation. It is not yet complete because it does not show interest expenses related to financing the deficit. This is because we need to work through the cash budget first, to know how much must be borrowed prior to calculating the financing costs. We will take these financing costs into account when we explicitly consider the form of financing. Similarly, any earnings on investment of excess cash are not yet considered in this preliminary statement. At this early stage, the purpose is to develop a preliminary estimate of our financing needs without yet taking account of these added details.

Notice that non-cash expenses such as depreciation are not included in the cash budget. In addition, there is no bad debt expense because bad debts are reflected as receivables never collected. Moreover, in order to show the disbursements for June that depend on sales forecasted beyond June and the receipts on credit accounts that depend on sales prior to July, we assumed that the past months' and future months' (not shown) follow the same monthly sales pattern as the current cycle.

In Exhibit 3.13 the net cash flow (line 15) is equal to total receipts (line 6) minus total disbursements (line 14). Net cash flow plus the cash balance at the beginning of the month yields the month-end cash balance. In the exhibit, cumulative cash (line 17) is calculated as if no additional funds are obtained. Thus, it shows what the cash balance would be at the end of the month if no funds were borrowed (or invested). The cumulative cash balance of ($1950) at the end of July indicates that a cash deficit of $1950 must be financed to restore the balance to zero. In addition, because the firm desires a cash balance of $5000, the total required financing (line 19) for July is $6950.

The cumulative cash item carries the deficit (or surplus) from month to month in a cumulative manner. That is, the deficit in July is $1950, in August it is $13,100, and the total of these two amounts is $15,050. This cumulative deficit plus the $5000 desired cash balance means that a total of $20,050 must be financed in August—$6950 left over from July, plus an additional loan of $13,100 to cover the August deficit.

This cumulative required financing is shown as required financing (line 19). The required financing increases from July through October, reaching a peak of $78,650. This $78,650 represents the total amount of financing that will be necessary over the one-year cycle. Note that the required financing declines after October because the firm's net cash flow becomes positive. Since this positive net cash flow allows some of the financing to be repaid, the required financing declines. For example, using November's net cash flow of $1000 to repay the loan reduces required financing by $1000. By March, there will have been sufficient cash inflow to repay any loans and to build up the cash balance. Thus, by the end of March, the required financing is zero and there is excess cash of $4800 (indicating that the cash balance would be $4800 in excess of the $5000 desired cash balance).

Exhibits 3.14 and 3.15 show the funding needs and cash flows from Exhibit 3.13 in a graphical format. Exhibit 3.14 shows the month-by-month funding needs and funds available. The additional required financing represents the marginal deficit for each month that must be funded either through borrowing or through drawing down the cash balance (as in June). Funds Available represents the funds that can be used to repay debt, build up the cash balance, or invest in securities. Exhibit 3.15 shows the cumulative

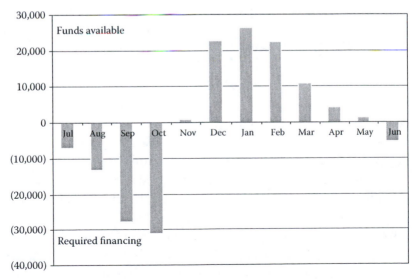

EXHIBIT 3.14   Mogul Corporation—Required financing and excess funds.

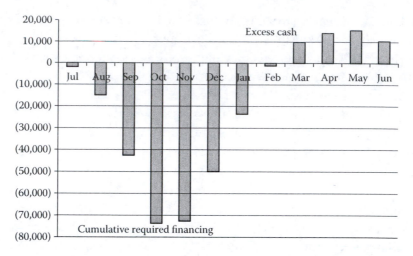

EXHIBIT 3.15　Mogul Corporation—Cumulative required financing.

required financing or excess cash. Required financing builds up to $78,650 in October and then declines over the following months, turning positive by March (assuming net cash inflows are used to repay the debt).

Note that the rapid increase in required financing is the result of expenses leading (occurring in advance of) sales and receipts lagging (occurring after) sales. The lead of expenses and lag of receipts tend to magnify the deficits and make them last longer. One question management should consider is what actions can be taken to narrow the time gap between the disbursements and receipts. That is, in terms of the ingredients of financial planning, the decision alternatives that address the problem must be specified.

In approaching such a planning problem it is important to choose correctly the length of the planning horizon (one year in this example) and the length of the sub-periods (months in this case). If the analyst had looked only at the annual figures, the total annual receipts would be $199,500, the total disbursements would be $194,200, and the annual net cash flow would be $5300. From the annual figures, it is not immediately apparent that O&R will require financing of $78,650 over the cycle. By neglecting to consider the monthly cash flows, the analyst would be completely unprepared to meet the firm's financing needs.

It was noted previously that planning involves more than forecasting; planning is the specification of future decisions and actions to reach a stated goal. In terms of this definition, the cash budgeting problem just analyzed was not really planning, rather it was forecasting. It generated a forecast of future required financing based on a forecast of sales and a set of assumptions about how customers pay their accounts and how the firm would produce and pay for the goods it sold. Planning would extend beyond this simple forecast of financing needs to explore alternative policies and decisions so as to determine the best way to reach the firm's goal. This requires that the goal and the various decisions alternative be specified and analyzed.

It is generally agreed that with respect to financial decisions the proper objective is to make the market value of the firm as large as possible. This very broad objective is made operational by specifying a set of goals that lead to value maximization. For present purposes, assume that the goal is to make decisions that maximize the present value of the net cash flow available to equity over the planning horizon. In the context of this example, the net cash flow available to equity would be the amount of cash flow available that could potentially be paid to the firm's owners after all expenses and after servicing any debt associated with financing the seasonal cash deficits.* Because we are calculating the present value of net cash flows to equity, the appropriate discount rate is the firm's cost of equity. It is estimated to be 20% and is assumed not to vary any meaningful amount with different decision alternatives.

To analyze any decision alternatives one must begin by considering the effect of the present policy on the goal. This requires determining the present value of the net cash flows given the present policy, against which the present value of the net cash flows under the alternative policy will be compared. Exhibit 3.16 is a revised cash budget based on Exhibit 3.13, with total receipts from Exhibit 3.13 shown as the first line of Exhibit 3.16, with the line number 6 on the left indicating that this matches line 6 of Exhibit 3.13; similarly, the second line of Exhibit 3.16 reproduces the total disbursements, line 14, of Exhibit 3.13.

We now explicitly take account of the costs of financing. Exhibit 3.16 assumes the cost of borrowing is 10% per year with interest paid monthly and debt is repaid from any excess cash held above the desired level of $5000. Thus, line "a" shows the interest that would be paid for funds borrowed. The $58 interest paid in August is one month's interest on the $6950 borrowed in July ($6950 × 0.10/12 = $58). Funds Borrowed is shown in line "d", with negative values beginning in November indicating debt repayment. The net increase in cash, shown on line "e", represents the cash

---

* The net cash flows available to equity are on line "e". Net Increase in Cash, in Exhibits 3.16 and 3.17.

Mogul Corporation
Cash Budget
Revised to Include Debt Service

	Jul	Aug	Sep	Oct	Nov	Dec	Jan	Feb	Mar	Apr	May	Jun
6 Total receipts	1,750	2,600	6,050	14,050	28,600	42,200	40,000	29,000	17,750	9,800	5,150	2,550
14 Total disbursement	8,700	15,700	33,600	45,100	27,600	19,600	13,600	6,600	6,700	5,500	3,800	7,700
a. Incremental interest	0	58	168	399	661	658	475	259	74	0	0	0
b. Collection costs	0	0	0	0	0	0	0	0	0	0	0	0
c. Net cash flow	(6,950)	(13,158)	(27,718)	(31,449)	339	21,942	25,925	22,141	10,976	4,300	1,350	(5,150)
d. Funds borrowed	6,950	13,158	27,718	31,449	(339)	(21,942)	(25,925)	(22,141)	(8,926)	0	0	0
e. Net increase in cash	0	0	0	0	0	0	0	0	2,049	4,300	1,350	(5,150)
f. Beginning cash	5,000	5,000	5,000	5,000	5,000	5,000	5,000	5,000	5,000	7,049	11,349	12,699
g. Ending cash	5,000	5,000	5,000	5,000	5,000	5,000	5,000	5,000	7,049	11,349	12,699	7,549
h. Loan balance	6,950	20,108	47,825	79,274	78,935	56,992	31,067	8,926	0	0	0	0
Present value of net increase in cash		$2,313										

EXHIBIT 3.16 Cash budget summary with current policy.

available from operations after paying interest and principal on the funds borrowed, and is the amount that potentially could be paid to the firm's owners after all expenses and debt servicing without decreasing the cash balance or requiring other financing. The present value of the net increase in cash, amounting to $2313 on the last line, represents the contribution of this 12-month period's operations to the value of the equity of the firm. It serves as the value basis for evaluating alternative policies.

There are numerous alternative policies and decisions that could be evaluated to determine the one that is most likely to help the firm reach its objective. These alternatives include changing credit terms to affect the lag between sales and collections, changing the production schedule and purchasing and payment policies so as to affect the lead of disbursements over sales, and considering various ways to fund the financing requirements. Not all decision alternatives will be analyzed here. Rather, just one policy alternative is considered as an example to demonstrate the approach.

The policy change to be evaluated in this case involves an effort to speed up collection of receivables. Assume that increased collection efforts will persuade customers to speed up their payments, thereby narrowing the gap between sales and collections. Ten percent of sales would be cash, and 90% would be credit sales. Additional costs of collection would be 2% of sales per month, and 70% of sales would be collected one month after sales, 15% would be collected in the second month, and there would still be a 5% bad debt rate. As with the base case, the required financing will be funded with a revolving line of credit at a cost of 10% per year with interest paid monthly in arrears with interest paid monthly and debt is assumed to be repaid out of the positive net cash flows. Exhibit 3.17 shows the receipts and disbursements for this alternative. The line numbers correspond to those of Exhibit 3.13, with line 6 showing receipts and line 14 showing the total disbursements from Exhibit 3.13. Line "a" shows the incremental interest that must be paid for the funds borrowed, analogous to line "a" in Exhibit 3.16; and line "b" shows the additional 2% collection costs associated with the increased collection efforts. The net increase in cash (line "e") corresponds to line "e" of Exhibit 3.16 and represents the cash flow available to equity associated with this policy. The present value of the net increase in cash of $207 is to be compared with the $2312 shown on the last line of Exhibit 3.16. This policy will result in a smaller contribution to the value of equity than the present policy evaluated in Exhibit 3.16. Thus, even though the more rigorous collection efforts accelerate the collections, the present value

Mogul Corporation
Cash Budget with Increase Collection Effort
Revised to Include Debt Service

		Jul	Aug	Sep	Oct	Nov	Dec	Jan	Feb	Mar	Apr	May	Jun
6	Total receipts		1,300	3,050	7,850	17,800	36,700	51,500	35,500	20,750	14,000	6,050	3,050	1,950
14	Total disbursement		8,700	15,700	33,600	45,100	27,600	19,600	13,600	6,600	6,700	5,500	3,800	7,700
a.	Incremental interest		0	62	169	388	626	565	310	133	19	0	0	0
b.	Collection costs		60	160	360	800	1,200	700	400	300	100	60	40	20
c.	Net cash flow		(7,460)	(12,872)	(26,279)	(28,488)	7,274	30,635	21,190	13,717	7,181	490	(790)	(5,770)
d.	Funds borrowed		7,460	12,872	26,279	28,488	(7,274)	(30,635)	(21,190)	(13,717)	(2,284)	0	0	1,173
e.	Net increase in cash		0	0	0	0	0	0	0	0	4,897	490	(790)	(4,597)
f.	Beginning cash	5000	5,000	5,000	5,000	5,000	5,000	5,000	5,000	5,000	5,000	9,897	10,387	9,597
g.	Ending cash		5,000	5,000	5,000	5,000	5,000	5,000	5,000	5,000	9,897	10,387	9,597	5,000
h.	Loan balance	0	7,460	20,332	46,612	75,100	67,826	37,191	16,001	2,284	0	0	0	1,173
	Present value of net increase in cash		$207											

EXHIBIT 3.17   Cash budget summary with faster collections.

of receiving these funds sooner is not sufficient to offset the cost of the collection effort, and the old policy should be retained.

This example shows the usefulness of the cash receipts and disbursements statement. It helps to focus attention on the potential financing needs over the planning horizon being considered and shows when financing will be necessary, how much will be necessary, and when it can be repaid. The cash budget can be used to analyze many types of financial planning problems, regardless of the length of the planning horizon. It is particularly suited to the short planning horizon that is broken up into short periods such as days, weeks, or months, as in the foregoing example. In cases where there is a seasonal pattern to the funds flows, the cash budget is an ideal tool.

3.5 CONCLUSION

The purpose of this chapter has been to introduce some of the standard tools and analytical methods of financial planning. These included developing *pro forma* financial statements, estimating REF by relating investment and financing to the growth in sales, understanding the concept of sustainable growth, calculating the numerous versions of cash flow, and developing a cash receipts and disbursement statement.

None of these techniques really requires the technology of computers or financial software, yet each can be used more easily with the assistance of the financial modeling software that is easily accessible on computers. The remainder of the book is concerned with the methods for constructing models for financial planning. The next chapter will build on the concepts of projecting the future financial statements so that we will be able to structure properly a financial planning model.

PROBLEMS

1 Cash Flows: O&R Corp.

Problem 1 of Chapter 2 asked you to develop a template model for financial ratios using the data from the O&R Corp. Now you are asked to expand on that model so it will generate cash flow statements for the company. Show three different cash flow statements for O&R: (a) cash flow in the accounting format of Exhibit 3.8 with the cash balance as the residual; (b) cash flow to the equity investors as was shown in Exhibit 3.10; and (c) cash flow to all investors as was shown in Exhibit 3.11. As with the assignment in Chapter 2, your model should be a template with only labels and formulas in the template, and the data in a separate section of the sheet, or on a different sheet.

2 Cash Budgeting: Taurus Corporation

The Taurus Corporation has a seasonal sales pattern with sales reaching their peak in late summer, and the low point occurs in the winter. The manager plans to arrange a line of credit from the bank to meet its seasonal cash needs. One question is: how big should the line of credit be? The manager has asked you to prepare a *pro forma* cash budget to determine the firm's financing needs. In addition, he wants you to evaluate the effects of a possible change in credit policy.

Facts that need to be considered in this analysis are as follows:

The current date is June 1, 2010.

Taurus manufactures two products: Product A and Product B. The sale price of A is $22 per unit, and the sale price of B is $15 per unit. Past and future expected future unit sales are as follows:

Monthly Unit Sales for Taurus Corp.

	April 2010	May	June	July	August
Product A	1500	1700	2500	3500	4000
Product B	2100	2790	3100	3850	3000
	September	October	November	December	January 2011
Product A	2500	2100	1700	1300	1000
Product B	3000	2700	2200	1500	1200
	February	March	April	May	June
Product A	800	1500	2000	2500	3000
Product B	1600	2100	2300	3100	3900
	July	August	September		
Product A	3800	4200	3000		
Product B	4000	3500	2200		

Each month's sales occur at a constant daily rate. That is, each day's sales are 1/30 of the total monthly sales. Five percent of sales are on a cash basis, and 95% of the sales are to credit customers.

The firm grants a 2% discount for accounts paid within 20 days. Otherwise, the customers are supposed to pay the full invoice amount within 30 days. Of those who buy the firm's product on credit, 15% pay within the 20-day discount period, 50% pay on the 30th day following the sale, 30% ignore the 30-day terms and pay 45 days after the sale, and the remaining 5% never pay and are written off as bad debts 60 days after the sale.

Cost of raw materials purchased for production is 60% of sales, with the purchase occurring two months in advance of sales. Taurus pays for its purchase of raw materials in the month following the date of the purchase and delivery.

Wages and salaries consist of a fixed component that amounts to $20,000 per month, plus a variable component that is 5% of that month's sales. Wages and salaries are paid in the month that they occur. Property rental is $4000 per month. In addition, in June a payment of $50,000 must be made to buy back the stock of a minority stockholder.

The current cash balance is $3000, but management has determined that it wants to maintain a cash balance of at least $20,000 at all times.

If Taurus uses a bank line of credit to fund its cash shortage, it will pay 8% (annual rate) interest each month that there is a positive loan balance. The interest paid in a given month is based on the loan balance outstanding at the end of the previous month. The lender will require that Taurus maintain a compensating balance in its cash account of at least 15% of the loan balance, and loan principal can be repaid only in the months of March, June, September, and December. Assume that if the company has excess cash it will use it to repay debt in the months allowed. If it has excess cash in the other months, it invests it in marketable securities that yield 6% annually. It can liquidate these securities to repay debt in the months allowed.

Assignment

A. Use a spreadsheet program to prepare a cash budget for the next 12 months (June through May) that shows how much will have to be borrowed in each month to maintain a constant cash balance of at least $20,000. Your model should reflect any interest paid on borrowing.

   Your model should consist of two modules: a Formula Module, and a Data Module. The Formula Module should show only the formulas, with none of the data (numbers) shown in the formulas. The Data Module will contain all the numbers that represent parameters, forecasts, and assumptions.

B. The company's collection of accounts receivable is regarded as a little slow. In your spreadsheet evaluate the desirability of the following change in credit policy: The credit customers will be granted a 3% discount if they pay in 10 days, otherwise they are supposed to pay in 30 days. In addition, the company will increase its collection efforts to accelerate the payments of its other customers. The additional collection efforts will cost $2500 per month. The new 3/10 net 30 policy will result in 25%

of the credit customers paying in 10 days and taking the discount. With the additional collection effort, 60% will pay in 30 days, 12% will pay in 45 days, and write-offs will drop to 3%. Is this change beneficial? What will be the effect on the firm's required borrowing?

3 Planning for Growth: Fly-by-Night Airlines, Inc.

The Fly-by-Night Airlines, Inc. is a mid-cap regional airline company that operates on the west coast. It is a point-to-point carrier that competes with Southwest Airlines and United Express, among others. The income statement and balance sheets for 2007–2010 appear below. Assume that the various average financial ratios indicated by these statements will prevail in future years.

Senior management of Fly-by-Night has asked you to analyze their capacity for growth. Over the next five years they would like to expand sales at a rate of 20% per year. However, they are not sure about the implications of this growth. They have posed several questions that they would like you to answer.

A. As a first step, management has requested that you create a common-size income statement and balance sheet (percent of sales for both) in the same format as the given accounting statements. In other words, your common-size statements will look exactly like the given statements except that they will have percentages of sales rather than dollar amounts in the cells. The exceptions (marked by an *) are that Depreciation expense is a percentage of the previous year's Plant, Property, and Equipment; Interest expense is a percentage of last year's total debt; and that Deferred Taxes are a percentage of last year's Taxes. No portion of long-term debt becomes current over the next five years. Verify that the percentages given are indeed the four-year averages (three years for the lagged ratios of depreciation expense, interest expense, and deferred taxes) of the values in your common-size income statement and balance sheet.

Based on the Fly-by-Nights' 2010 accounting statements, the questions that management would like answered are

B. How rapidly could sales grow if they use no discretionary debt financing in future years and meet all their equity financing needs using internal sources. Assume that they do not change their dividend payout ratio (0%) from its past level. Use the g^* formula.

C. Approximately how rapidly could sales grow if they utilize debt financing at a level such that they maintain their most recent total debt–total capital ratio [(Current Liabilities + Debt)/(Total Assets)], and all their equity needs are met from retained earnings. Use the g^{***} (*PRAT*) formula.

D. What would happen to g^* and g^{***} if Fly-by-Night decided to start paying out 30% of their net income as dividends.

E. Assume that they could change their debt-to-total capital and their dividend payout ratios to new levels that would be maintained at that new level in future years, and that the other ratios (net profit margin and asset turnover) do not change. Could they handle the desired 20% growth without changing their capital structure thereafter? (Hint: use g^{***} for your analysis.)

What are the different combinations of dividend payout and total asset-to-equity ratios that could be used to support a 20% annual growth in sales? Graph this relationship. Let the horizontal axis be the payout ratio, and the vertical axis be the asset/equity ratio. Show the combinations of these ratios that will support 20% growth according to the g^{***} formula.

F. Assume that (a) the firm's operating profit margin (EBIT/Sales) is maintained in future years; (b) the interest rate on new debt (both short- and long-term) is 4.8%; (c) no portion of the outstanding debt becomes current for the next five years; (d) the payout ratio is held at 0%; (e) no new external equity financing is used (i.e., any equity financing that is used comes entirely from retained earnings); (f) other ratios (turnover and so on) are held constant, but net profit margin is allowed to change because of the effect of interest expense; (g) intangibles, other assets, and deferred taxes are held constant; and (h) assets necessary to support growth in year 2002 must be in place at the beginning of the year (same for 2003).

Approximately what amount of external financing will be required if the sales grow at 20% per year over the next two years? What will this do to their net profit margins in years 2011, 2012, and 2013?

[Note: Do not use *pro forma* statements to answer these questions. Rather, use the formulas from this chapter along with the average ratios calculated in Part I.]

G. If Fly-by-Night was able to grow its sales more rapidly, would it be more valuable? Give a brief explanation and discussion.

What you turn in: A memo that answers all of the above questions, including the requested graph, in a well-organized, logical manner. Place the common-size statements with average ratios in an appendix.

Financial Date for Fly-By-Night Airlines

Fly-by-Night Airlines
Income Statement
($000)

	2007	2008	2009	2010	Percent of sales 2007-2010
Sales	1,592,200	1,739,400	1,897,700	2,082,000	
COGS	958,700	951,600	961,700	1,103,900	54.7%
SG&A	477,000	580,500	649,900	693,400	32.7%
Gross income	156,500	207,300	286,100	284,700	
Depreciation	65,500	66,300	73,000	83,800	5.6%
Amortization	2,000	2,000	2,100	1,000	
Operating income	89,000	139,000	211,000	199,900	8.6%
Interest expense	38,400	33,600	21,200	16,300	4.8%
Interest income	13,700	18,200	14,600	37,100	
Income before taxes	64,300	123,600	204,400	220,700	
Taxes	26,300	51,200	80,000	86,500	40.2%
Net income	38,000	72,400	124,400	134,200	
Shares outstanding	14,241	14,785	23,388	26,372	

Fly-by-Night Airlines
Balance Sheet
($000)

	2007	2008	2009	2010	Percent of sales 2007-2010
Assets					
Cash & equivalents	101,800	212,900	306,600	329,000	12.6%
Accounts receivable	69,700	72,800	70,600	75,600	4.0%
Inventories	47,800	47,200	44,100	54,300	2.7%
Prepaid expenses	80,900	92,100	107,500	124,000	5.5%
Current assets	300,200	425,000	528,800	582,900	
Plant, property, and equip	1,189,700	1,334,400	1,478,300	1,954,100	80.8%
Accumulated depreciation	326,300	373,800	417,000	486,700	
Net fixed assets	863,400	960,600	1,061,300	1,467,400	
Intangibles	61,600	59,600	57,500	56,500	
Other assets	86,200	88,300	84,200	76,300	4.6%
Total assets	1,311,400	1,533,500	1,731,800	2,183,100	91.7%
Liabilities and shareholder equity					
Accounts payable	65,400	73,900	84,300	104,200	4.5%
Other current liabilities	349,300	370,700	414,400	448,000	21.7%
Short term debt	47,000				
Current portion l.t. debt	24,100	28,700	27,200	66,500	
Total current liabilities	485,800	473,300	525,900	618,700	
Long-term debt	503,600	512,200	317,200	486,700	
Deferred taxes	49,500	72,300	99,200	144,000	
Total liabilities	1,038,900	1,057,800	942,300	1,249,400	
Common stock	17,200	22,000	30,400	31,300	
Paid in capital	166,800	292,800	473,800	482,900	
Retained earnings	88,500	160,900	285,300	419,500	
Total shareholder equity	272,500	475,700	789,500	933,700	
Total liabilities and equity	1,311,400	1,533,500	1,731,800	2,183,100	

4 Required External Financing: K.D. Nickel Company

Problem 2 in Chapter 2 provided data for the K.D. Nickel Company for fiscal years 2006–2010. You were asked to model the financial ratios for Nickel's. Now, assume that selected ratios for each of the next three years (2011–2013) are going to be equal to the *average* of the ratios for the past five years. The ratios that will be equal to the past average will be

Income Statement:
  Operating Costs/Sales (this includes CGS, SGA, and Other Inc./Exp.)
  Depreciation Expense/Sales
  Interest$_t$/Total Debt$_{t-1}$ (interest-bearing debt)
  Tax/EBT
  Dividend/NI
Balance Sheet:
  Total Assets/Sales
  Spontaneous Liabilities/Sales (this includes AP and other CA, and other LT Liabilities)

Interest-bearing debt is assumed to stay at it previous level (year end 2010), as are preferred and common stock.

Develop a spreadsheet model of Nickel's financial statements that will provide simplified income statements and balance sheets for each of the next three years. Show the calculation of REF for each year. External financing for the next three years will be from an undetermined source that will be designated as "Additional Financing." This additional financing will be shown as a liability on the balance sheet. This financing will not involve any explicit interest cost.

Having developed a model, explore the sensitivity of REF to changes in the rate of growth in sales. What happens to REF when sales grow at (a) 0%, (b) 3%, (c) 10%, and (d) 20% annually?

PART II

**Tools for Financial Planning
and Modeling: Simulation**

CHAPTER 4

Financial Statement Simulation

4.1 INTRODUCTION

In Chapter 1 we discussed financial planning and financial models in general terms, and in Chapter 3 we explained how to develop *pro forma* financial statements. Now we will be more specific and discuss financial models in greater detail. This chapter will present a generic simulation model of a firm. The specific equations of the model are presented and discussed, so the reader will have a thorough understanding of the structure of such a model.

There are at least as many different types of financial simulation models as there are uses for them.* The most common financial simulation model is a financial statement model that consists of a set of mathematical expressions

---

* The references at the end of the text include many different discussions of simulation models. Some of the more comprehensive discussions of models and model building include Benninga (2008), Bryant (1982), Francis (1983), Gershefski (1968, 1969), Ho and Lee (2004), Meyer (1977), Naylor (1973, 1979), Sengupta (2004), Schrieber (1970), and Warren and Shelton (1971).

used to generate *pro forma* financial statements so that the effect of alternative decisions on the firm's financial condition can be evaluated. The mathematical equations specify the relationships, or links, between the economic variables, environmental variables, decision variables, and the financial state of the firm. After these relationships are properly specified, the input (environmental and economic forecasts, and decision variables) is fed through the model to generate output consisting of the state of the firm as described by its financial statements.

Our discussion will focus on a financial statement model that is driven by a sales forecast. The sales forecast is the primary input that drives the model, and the model allows us to explore the investment and financing necessary to meet the sales forecast. We can think of the model as being structured in modules, or blocks of equations, with a module for each of the following tasks:

- Accounting

- Income generation

- Investment

- Financing.

Exhibit 4.1 describes the flow between the modules. The income generation module utilizes the economic forecasts, decision specifications, and parameter estimates to generate estimates of revenues and costs that are input to the accounting module, which, in turn, generates trial values for the various accounts of the financial statements. Those trial values are used by the investment and financing modules to calculate the amount of new investment in assets necessary to support the forecast level of sales and the amount and form of new financing necessary to purchase these assets.

The new investment and financing are fed back to the accounting module to generate the *pro forma* financial statements consistent with the sales forecast. These financial statements are part of the output of the model, from which management can evaluate the forecast and their prospective decisions. Each of these modules will be explained and an example model presented. The accounting module will be discussed first because it is of central importance in the model.

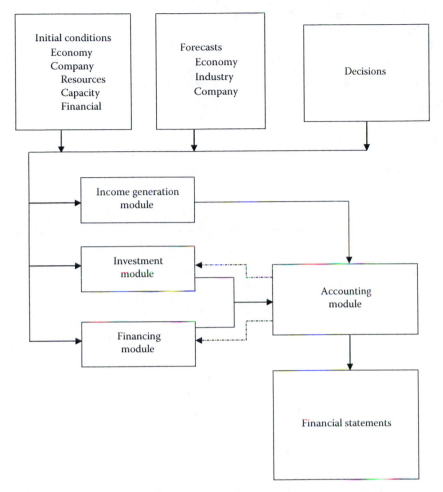

EXHIBIT 4.1 Structure and flow of a financial simulation model.

4.2 THE ACCOUNTING MODULE

4.2.1 Equations of the Module

The skeleton of most financial simulation models is a set of equations that are accounting definitions and relationships. These equations maintain accounting consistency, aggregate the detailed accounts into their broad groups, and link the account balances from one period to the next. These equations generate the financial statements for the firm: the income statement, the balance sheet, and the cash flow statement.

A.1	Sales$_t$	=	Sales$_{t-1}$ (1 + Growth Rate)
A.2	Operating Costs$_t$	=	Sales$_t$ (Operating Cost/Sales)
A.3	EBITDA$_t$	=	Sales$_t$ − Operating Costs$_t$
A.4	Depreciation Expense$_t$	=	Gross Fixed Assets$_{t-1}$ (Depreciation Rate)
A.5	EBIT$_t$	=	EBITDA$_t$ − Depreciation Expense$_t$
A.6	Interest Expense$_t$	=	Debt$_{t-1}$ (Interest Rate)
A.7	Earnings Before Tax$_t$	=	EBIT$_t$ − Interest Expense$_t$
A.8	Income Tax$_t$	=	Earnings Before Tax$_t$ (Tax Rate)
A.9	Net Income$_t$	=	Earnings Before Tax$_t$ − Income Tax$_t$
A.10	Dividend$_t$	=	Cash Available for Dividend$_t$
A.11	Retained Earnings$_t$	=	Net Income$_t$ − Dividend$_t$

EXHIBIT 4.2A Accounting module: Income statement equations.

To demonstrate our modules, we will consider a simplified model for a firm. Exhibits 4.2A, B, and C show, in generalized form, examples of the types of equations necessary to generate the major account categories for the income statement (Exhibit 4.2A), the balance sheet (Exhibit 4.2B), and a statement of cash flow (Exhibit 4.2C). In these exhibits, the subscript t denotes time period t, and variables that must be provided as data or generated elsewhere in the simulation system are enclosed in parentheses. For example, the Growth Rate of sales is a forecast from outside this sector, and is data input in Equation A.1. Similarly, in this sales-driven model, costs are expressed as a percent of sales, and the ratio (Operating Cost/ Sales) is a parameter forecast that is exogenous to this sector. In the income statement sector, Exhibit 4.2A, each equation defines a revenue or expense account in terms of input variables or in terms of other preceding accounts. The balance sheet sector, Exhibit 4.2B, consists of an equation for each account category in the balance sheet, in addition to equations that total the accounts. The cash flow statement, Exhibit 4.2C, consists of expressions that calculate the flow of cash based on the income statement and changes in the balance sheet accounts.

The system of equations in Exhibits 4.2A through C is a simple model, so we can show the principles without being overwhelmed with detail and complexity. For example, we show just two current asset accounts, Cash and Other Current Assets, because this is all we need for an introduction to model construction. We could easily expand the model to show the accounts in as much detail as necessary. For example, we might want to

Assets

B.1	$Cash_t$	=	$Cash_{t-1}$ + Increase in $Cash_t$
B.2	Other Current $Assets_t$	=	Other Current $Assets_{t-1}$ + Other CA $Investment_t$
B.3	Current $Assets_t$	=	$Cash_t$ + Other Current $Assets_t$
B.4	Gross Fixed $Assets_t$	=	(Gross Fixed Assets/Sales) $Sales_t$
B.5	Accumulated $Depreciation_t$	=	Accumulated $Depreciation_{t-1}$ + Depreciation $Expense_t$
B.6	Net Fixed $Assets_t$	=	Gross Fixed $Assets_t$ – Accumulated $Depreciation_t$
B.7	Total $Assets_t$	=	Current $Assets_t$ + Net Fixed $Assets_t$

Liabilities

B.8	Current $Liabilities_t$	=	Current $Liabilities_{t-1}$ + Spontaneous $Financing_t$
B.9	$Debt_t$	=	$Debt_{t-1}$ + Debt $Issued_t$ – Debt $Repaid_t$
B.10	Common $Stock_t$	=	Common $Stock_t$ + Stock $Issued_t$
B.11	Earned $Surplus_t$	=	Earned $Surplus_{t-1}$ + Retained $Earnings_t$
B.12	Stockholders' $Equity_t$	=	Common $Stock_t$ + Earned $Surplus_t$
B.13	Total Liabilities and $Equity_t$	=	Current $Liabilities_t$ + $Debt_t$ + Stockholders' $Equity_t$

EXHIBIT 4.2B   Accounting module: Balance sheet equations.

C.1	Earnings Cash $Flow_t$	=	Net $Income_t$ + Depreciation $Expense_t$
C.2	NWC $Investment_t$	=	Other CA $Investment_t$ – Spontaneous $Financing_t$
C.3	Cash Flow from $Operations_t$	=	Earnings Cash $Flow_t$ – NWC $Investment_t$
C.4	Capital $Expenditures_t$	=	Fixed Asset $Investment_t$
C.5	Capital $Financing_t$	=	$(Debt_t – Debt_{t-1})$ + (Common $Stock_t$ – Common $Stock_{t-1})$
C.6	Net Cash $Inflow_t$	=	Cash Flow from $Operations_t$ – Capital $Expenditures_t$ + Capital $Financing_t$
C.7	Increase in $Cash_t$	=	(Cash/Sales) $Sales_t$ – $Cash_{t-1}$
C.8	Cash Available for $Dividend_t$	=	Net Cash $Inflow_t$ – Increase in $Cash_t$

EXHIBIT 4.2C   Accounting module: Cash flow equations.

add accounts for marketable securities, receivables, and inventories, but it is not necessary now. We will expand our discussion of these other categories later. Of course, every model will differ because of the different type and detail of the input data available and the output data desired. Any

D.1	Increase in Cash$_t$	=	(Cash/Sales) Sales$_t$ − Cash$_{t-1}$
D.2	Other CA Investment$_t$	=	(Current Assets/Sales) Sales$_t$ − Other Current Assets$_{t-1}$
D.3	Current Asset Investment$_t$	=	Increase in Cash$_t$ + Other CA Investment$_t$
D.4	Investment in New Fixed Assets$_t$	=	Net Fixed Assets$_t$ − Net Fixed Assets$_{t-1}$
D.4	Fixed Asset Investment$_t$	=	(Gross Fixed Assets/Sales) Sales$_t$ − Gross Fixed Assets$_{-1}$
D.5	Total Investment$_t$	=	Current Asset Investment$_t$ + Fixed Asset Investment$_t$

EXHIBIT 4.2D   Investment module: Asset investment equations.

E.1	Required Financing$_t$	=	Total Investment$_t$
E.2	Earnings Cash Flow$_t$	=	Net Income$_t$ + Depreciation Expense$_t$
E.3	Minimum Dividend$_t$	=	Net Income$_t$ (Payout Ratio)
E.4	Internal Financing$_t$	=	Earnings Cash Flow$_t$ − Minimum Dividend$_t$
E.5	Spontaneous Financing$_t$	=	(Current Liabilities/Sales) Sales$_t$ − Current Liabilities$_{t-1}$
E.6	Required External Financing$_t$	=	Required Financing$_t$ − Internal Financing$_t$ − Spontaneous Financing$_t$
E.7	Debt Issued$_t$	=	Required External Financing$_t$ (Debt Portion)
E.8	Stock Issued$_t$	=	Required External Financing$_t$ − Debt Issued$_t$
E.9	Total Financing$_t$	=	Spontaneous Financing$_t$ + Internal Financing$_t$ Debt Issued$_t$ + Stock Issued$_t$

EXHIBIT 4.2E   Financing module.

given set of expressions can usually be collapsed into a single, less detailed, more general expression, or expanded by adding more expressions and variables. The detail generated by the model and thus the extent to which sections of the model are expanded by adding more variables and equations or are shrunk by aggregating variables and decreasing the number of equations is dependent on the requirements of the user. The user who is more concerned with the effects of sales or production decisions may elaborate and expand into minute detail the equations simulating revenue (A.1) or operating costs (A.2), and at the same time eliminate much of the unnecessary detail represented by the remainder of the model. On the other hand, the financial officer may prefer to take the revenue and costs as given and emphasize the financial implications of security investments, capital investments, and financing decisions. This could involve expansion of the number of equations and variables relating to these decisions, such as those relating to interest expense (A.6), dividends (A.10), working capital (B.1), (B.2), (B.3), and (B.8), fixed assets (B.4), (B.5), (B.6), and capital accounts (B.9), (B.10), (B.11), and (B.12).

The financial simulation consists of a set of equations that must be fulfilled for each period t over the planning horizon. Thus, each equation in Exhibit 4.2 is subscripted with t to denote period t. In most such models, the solution proceeds sequentially, with the first simulated period being solved based upon the initial conditions of the firm. The solution from the first period then is used as an input for the solution for the second period, and so on, to the end of the planning horizon. Each sector (income, balance sheet, and cash flow) of this set of equations is related to the other sectors. Therefore, they must be solved as a coherent whole. That is not to say that they are necessarily simultaneous equations in a mathematical sense—they may be, but it is not necessary. Certainly, the equations are interrelated, but they can and should be structured so that the total set can be solved sequentially rather than simultaneously. Sequential solution refers not just to the time sequence from the earliest period to the last, but also to the order in which the equations are solved for a given period, for example, from top to bottom. The first input variable, the driver of the model, is usually the sales forecast that typically is generated exogenously. The consequences of this sales forecast are generated by working through the list of equations in descending order. Simultaneity is avoided by letting a variable for period t be a function of either a variable for period $t - 1$ (instead of the same period, t), or of a variable that has already been calculated for the same period, t. For example, Equation A.6 calculates interest expense for period t and is dependent on the amount of debt outstanding at the end of period $t - 1$. The amount of debt outstanding at t is as yet unknown, and the equations would be simultaneous if interest expense for t (A.6) required the debt balance at t instead of $t - 1$.[*]

The first block or sector of equations, as shown in Exhibit 4.2A, is the income sector. It takes the forecast of sales as input and generates a corresponding forecast of expenses and income, simulating the income

---

[*] Spreadsheet software such as Excel will resolve the circularity, but in complex models this may cause difficulty if there are several different competing simultaneous systems in the model. It is better to build the model so as to minimize the amount of simultaneity in the model.

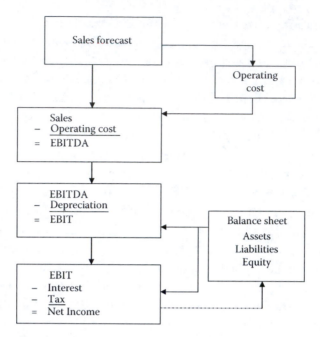

EXHIBIT 4.3   Income statement generation.

statement in steps shown diagrammatically in Exhibit 4.3. The income statement equations consist of both definitions and functional equations. The definitions are those expressions that simply calculate the value of a new variable based on values of previously determined variables, and the expressions are true by definition. For example, EBITDA is defined as Sales minus Operating Costs. This expression, by itself, requires no direct parameter estimation or decision specification. Equations A.3, A.5, A.7, A.9, and A.11 are all definitions. Functional equations are those expressions that require the specification of some exogenous parameter or decision variable as are indicated in parentheses. For example, $Sales_t$ will be related to sales in the previous period, $Sales_{t-1}$, and the forecast of the rate of growth of sales, $Growth\ Rate_t$, where $Growth\ Rate_t$ must be forecasted or specified as a parameter. Operating costs are dependent on the cost ratio (Operating Cost/Sales) that is forecast elsewhere and input as a parameter. How we obtain these parameter forecasts will be discussed later.

Since the essence of financial simulation is to track the financial status of the firm through time, it is important that the results in one period are properly reflected as input for the next period. This function is served, in part, by the balance sheet equations that use balances at the end of the previous period ($t - 1$) as the initial conditions, or beginning balances, for current period (t). Thus, whatever happens in period $t - 1$ affects the results for the next period, t. For example, (B.1) shows Cash$_t$ as dependent on Cash$_{t-1}$, the cash balance at the end of the previous period. In (B.9) the debt balance at the end of the period, Debt$_t$, depends on the previous balance, Debt$_{t-1}$. In addition, expenses and income in period t are related to asset and liability balances at the end of the previous period. The most obvious example is Equation A.4, which calculates Depreciation Expense during period t as a proportion of Gross Fixed Assets (GFA)$_{t-1}$, the balance at the end of the previous period. Similarly, Interest expense (A.6) for period t is proportional to the amount of debt at the end of the previous period.

The cash flow statement, Exhibit 4.2C, provides the link between the income statement and the balance sheet, and is a crucial part of the mechanism for maintaining equilibrium in the model. The starting point is "Earnings Cash Flow" (Net Income + Depreciation) from the income statement, and the subsequent expressions take account of changes in all asset and liability accounts. In Equation C.2, Other CA Investment is the change in current assets, excluding cash. It represents the cash outflow required to increase non-cash current assets. Spontaneous Financing is the increase in current liabilities that occurs in response to increasing sales. NWC Investment is the difference between the increase in non-cash current assets and the spontaneous financing from current liabilities. It represents the investment in (non-cash) current assets that is not financed by the spontaneous increase in current liabilities. Cash Flow from Operations (C.3) is the cash flow from normal operating activities net of the investment in (non-cash) NWC. Capital Expenditures (C.4) is the amount spent to replace and expand plant and equipment and other long-term assets, and is based on the projected ratio of required GFA-to-Sales, and the change in sales (D.4). Capital Financing (C.5) is the amount of funds flowing into the firm from the issuance of new securities (debt and equity). The total of these inflows and outflows is the Net Cash Inflow (C.6), which is the amount of cash from all sources that is available either to increase the cash account or to pay dividends. In this case, the cash balance is increased proportionally to the increase in sales (C.7), and the remaining cash flow is paid as a dividend to the stockholders (C.8).

It is important to make sure that the cash flow goes to a specific spot in the model as opposed to flowing into a financial black hole. In our current model, the Net Cash Inflow is available to be added to the cash account or to be paid as dividends; or it could be invested in marketable securities. However the cash is allocated, one of the categories must receive the residual. For example, in the current format we use the cash to maintain the cash balance at a target percent of sales and payout the remainder (the residual) as dividends. In this way, dividends are a residual, assuring that all the cash flow goes somewhere. We could easily reverse the logic and have the dividend determined first (e.g., based on the dividend payout ratio), and allow the addition to cash be the residual. In that case, because cash would go to the cash account as a residual, cash would not be specified as a percent of sales.

An example of cash flowing into a financial black hole would be if the cash balance was determined as a percent of sales, and dividends were set by a dividend payout ratio, with no specification of where any left over cash would go. For example, consider the following data:

Net Cash Inflow:	$1000	
Addition to Cash Balance	300	[Set so that Cash = X% of Sales]
Dividends	500	[Set so that Dividends = Y% of Net Income]
Excess Cash Flow	$200	

If the model does not specifically put the $200 excess cash in some account, it will simply disappear into a financial black hole, neither being added to an asset account (cash or marketable securities), nor being paid out as dividends. In other words, we should not have both the cash balance driven by sales and dividends driven by income, without otherwise specifying a target for any residual cash flow. For example, an alternative target for the residual cash flow could be security investments. Continuing with the current example, the $200 excess cash flow could be used to buy marketable securities. The point is our model must account for all the cash flow.

4.2.2 The Income Generation Module

The income generation module is the sector that generates variables related to revenues and costs that are, in turn, used in the accounting module. To generate these intermediate inputs to the accounting module, the income sector utilizes forecasts of future economic conditions and forecasts of firm relationships and decisions. These forecasts are input data to the system of equations comprising the income generation module. Of course, every model will approach the task of forecasting revenue and cost differently. The amount of detail in the model will depend on the purpose of the modeling effort. A more detailed income module could go into great detail calculating different categories of revenues and costs. For now, we can concentrate on the basic structure of the model; we use the simple approach of embedding the income generation equations in the Accounting Module. We model revenue using a simple growth model in which revenue grows at the forecast rate, and we model costs proportional to sales. The inputs are the sales growth forecast and the cost parameter forecasts. In Exhibit 4.2A, the driver of the model is the equation

$$\text{Sales}_t = \text{Sales}_{t+1}(1 + \text{Growth Rate}),$$

and the forecast parameters are the items in parenthesis: (Growth Rate), (Operating Cost/Sales), (Depreciation Rate), (Interest Rate), and (Tax Rate).

One of the primary uses of a financial simulation model is to determine the level of future investment and/or financing that may be necessary to meet the firm's objectives. The role of the investment and financing modules is to specify these required levels of investment and financing. We consider how these modules fit into the simulation model.

4.2.3 The Investment Module

The purpose of the investment module is to simulate the firm's investment in assets, and set the investment schedule so that the cash inflows and outflows related to the investment can be used in the other sectors of the simulation model to generate the output necessary to evaluate the firm's financial plans. There are two approaches to modeling asset investment. One is a sales-driven model in which assets are assumed to be functions of sales. In this type of model, investment occurs such that the level of assets is sufficient to support the sales. In the second approach, investment is the driver of the model. Investment plans are input directly for the purpose of evaluating specific investments; sales and production are part of the secondary data describing the investment. In this chapter we will use the first approach, in which investment adjusts to the sales forecast, because its simpler structure is a better fit with our continuing focus on understanding the structure.

The simplest investment model calculates the total stock of assets proportional to the sales. Investment is calculated as the change in required assets from one period to the next. For example, assume that the income generation module specifies Sales$_t$ and the required stock of assets is directly proportional to sales, so that

$$\text{Assets}_t = \left(\frac{\text{Assets}}{\text{Sales}} \right) \times \text{Sales}_t, \tag{4.1}$$

where (Assets/Sales) is the parameter representing the required ratio of assets to sales. This parameter may be obtained by estimating it from historical data regarding sales and asset levels, or it may be based on anticipated technological relationships or policy decisions that are being analyzed with the simulation.

The investment required by the percent-of-sales model (4.1) is

$$\text{Asset Investment}_t = \text{Assets}_t - \text{Assets}_{t-1}$$

$$= \left(\frac{\text{Assets}}{\text{Sales}} \right) \times (\text{Sales}_t - \text{Sales}_{t-1}). \tag{4.2}$$

This model is unrealistically simple, since different asset categories will have different relations to sales. To develop a more detailed model, we would specify different functional links between each asset category and sales. For now, we will distinguish just three asset categories: Cash, Other Current Assets, and Fixed Assets. The expressions that calculate the investment in assets were shown in Exhibit 4.2D.

4.2.4 The Financing Module

The financing module is the set of equations that specifies the sources of funds needed to acquire the assets determined in the Investment Module. The financing in the model is an important part of the equilibrating mechanism that balances assets and liabilities and assures that sources are equal to uses of funds. Exhibit 4.2E shows expressions for a simple financing module.

We can separate the sources into spontaneous and discretionary financing, and distinguish between internal and external financing. Spontaneous financing refers to those sources of funds that are generated more or less automatically by the firm's normal operations and growth. Some liabilities, particularly current liabilities, grow with the firm's operations, and provide financing that grows spontaneously as sales increase. Trade credit and accruals are examples of sources that usually grow with sales, providing a spontaneous source of financing. In our model, we combine these spontaneous sources into one account, Current Liabilities, and call the increase in current liabilities "Spontaneous Financing," as shown in Equation E.5. Discretionary financing refers to sources of funds that require explicit decisions to use them. Discretionary sources include borrowing and the issuance of new securities. These sources are discretionary in the sense that they need to be specified as a decision variable that is entered into the model.

We can also divide sources into internal or external financing. Internal financing consists of the sources internal to the firm's operations. External refers to the funds that come from outside the firm and need to be specified as an explicit decision. We usually think of retained earnings as internal equity financing. However, in our model, we will use a slightly broader definition of internal financing. Internal Financing is shown in Equation E.4 as Earnings Cash Flow less Dividends, which is the same as Retained Earnings + Depreciation. Earnings cash flow (net income + depreciation) is generated by the normal operation of the firm, and provides funds that can be used to purchase new assets, repay debt, or pay dividends. This source of financing is internal to the firm because no application need be made to outsiders to obtain the funds. It is also a spontaneous source of funds, but for our purposes we will use the term "spontaneous financing" to refer to the increase in current liabilities, and the internal financing to refer to internal equity financing.

The financing module takes Total Investment$_t$ from the investment module as its initial input. In the financing module, as shown in Exhibit 4.2E, Required Financing$_t$ is set to equal to Total Investment$_t$ (E.1). Subtracting the sum of Internal Financing$_t$ and Spontaneous Financing$_t$ from Required Financing$_t$ yields Required External Financing (REF$_t$) (E.6). The user of the model decides how much of the External Financing will come from debt by specifying the decision variable (Debt Portion). Any remaining financing will come from the issuance of new stock. By assigning all remaining financing needs to stock, we ensure that all investment will be financed and the model will balance. It is important that investment and financing be consistent with each other and with the overall financial structure of the firm; that is, the investment and financing must be specified in such a way that the firm is in equilibrium.

4.3 EQUILIBRIUM IN THE SIMULATION MODEL

Virtually all financial models must have a mechanism that balances the system. This section discusses the balancing process in the simulation model. A model is balanced or in equilibrium when assets equal liabilities and sources of funds equal uses. We need to build a mechanism into our model that assures this equilibrium.

There are generally two ways to equilibrate the model: flow balancing or stock balancing. A flow balancing model is the one that reaches its equilibrium by making sure that sources of funds equal uses. Balance of the inflows and outflows can be maintained by variations of expressions of the form

$$\text{Total Uses of Funds} = \text{Total Sources}$$

or

$$\text{Net Change in Cash Balance} = \text{Net Cash Inflow.}$$

A stock balancing model equilibrates by making the stock (level) of liabilities equal assets. Stock balancing is accomplished by various expressions of the form

$$\text{Total Assets} = \text{Total Liabilities}$$

or

$$\text{Total Assets}_t = \text{Total Liabilities}_{t-1} + \text{New Financing}_t.$$

Although each of these approaches can be used by itself, it is often useful to include both stock and flow balancing in the same model to serve as check that the model is performing properly. In the example to be discussed shortly, we will include elements of both approaches.

A simulation model cannot make decisions; it can only compute and indicate the results. Consequently, even though provision is made in the accounting equations for sources to equal uses or, for assets to equal liabilities, the model must be told specifically how to reach equilibrium. If new financing is required, the user must tell the model the composition of the new financing: whether it is debt or equity or whatever. In many financial simulation models this is accomplished with a "plug variable" that forces equality between sources and uses or between assets and liabilities. The role of the investment and financing modules is, in part, to specify the amount and composition of these plug variables. For example, suppose that for a given sales forecast the investment module shows that $100 million of additional investment in fixed assets is required. This means that we must obtain $100 million of financing. Suppose $25 million will be generated from spontaneous and internal sources. Without additional adjustment, the model will show assets exceeding liabilities and net worth by $75 million. The financing module makes the necessary adjustment. It calculates a "plug variable" equal to the REF that balances assets with liabilities. It would be up to the user to specify the form of this financing.

The logical steps to reach a stock balancing equilibrium are shown as a flowchart in Exhibit 4.4. Based on a sales forecast, the investment module determines the composition and level of assets necessary to generate the sales. If these required assets differ from the existing level of assets, the

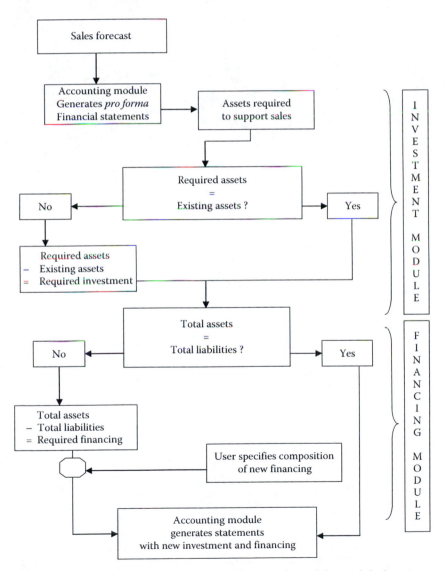

EXHIBIT 4.4   Flowchart for balancing a financial model: stock balancing.

amount of new investment is determined. The financing module takes this number and calculates the level of new financing necessary to acquire the investment. Total new financing is calculated as the difference between total assets (with the new investment) and the level of liabilities and shareholder equity that would exist without the new financing. The amount of the new financing is a plug variable that balances assets and liabilities.

Exhibit 4.5 shows the steps for a flow balancing model. A cash receipts and disbursement model is an example of a flow balancing model where

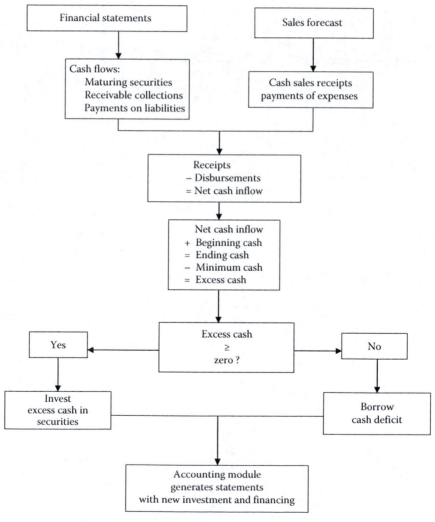

EXHIBIT 4.5   Flow balancing: Flowchart for a cash receipts and disbursement model.

the difference from the stock balancing approach is most apparent. As shown in Exhibit 4.5, this type of model would determine the net cash inflow based on a sales forecast and the schedule of receipts and disbursements. The ending cash balance is compared to the minimum cash balance. The model would check for excess cash or a cash deficit. If ending cash exceeds the minimum balance, the investment module would have the excess invested in marketable securities. If ending cash is below the required minimum, the financing module borrows to finance the deficit. The plug variable in this case is the amount to be invested in marketable securities, or the amount of borrowing necessary to make the ending cash balance equal to the required cash balance.

A flow balancing model is ideally suited to the problems of simulating working capital decisions, but the same task can be accomplished with a stock balancing model. Moreover, the flow balancing model can be used for long-term investment and financing simulations just as easily as can the stock balancing approach. These two approaches are interchangeable because, in a properly specified model, actions that balance the cash flows will also balance the stock of assets and liabilities. In a more aggregative model, especially one that is specified in terms of longer time periods, the stock adjustment approach is often easier to develop. In a more detailed model, especially the one with shorter time periods, the flow adjustment approach is frequently more useful. Either approach can be used. We have found that fewer modeling mistakes occur when we combine elements of both approaches in our models. In the next section our example has elements of both stock balancing and flow balancing.

4.4 BUILDING A LONG-RANGE PLANNING MODEL

4.4.1 Model of O&R Corp.

In this section, we consider an example of a model of a firm's financial statements that can be used to project the firm's future financial condition. Ultimately, the purpose of the model is to be able to evaluate the impact of different policies on the value of the firm's stock. To do this, we need to develop a valuation model for the firm. However, at this stage we just want to develop the first part of the model that will simulate the financial statements. We will add the valuation module later. The starting point for our model will be to construct the modules that were explained in general terms in the earlier section of this chapter. We use the same basic structure of the model as shown in Exhibit 4.2, with some elaborations added as necessary.

As you recall from Chapter 2, the O&R Corp. is an industrial firm that manufactures a range of plastic components that are used as input by other manufacturers of consumer products, automobiles, aircraft, and computers. The company is somewhat cyclical, but it has been growing at a respectable rate in recent years, and it is expected that its sales will continue to grow in the foreseeable future. Our task is to develop a model that will help us to analyze the firm's investment and financing needs over the next five years.

The financial statements for past two years serve as the starting point for a model that will simulate the firm's financial performance for the next five years. Based on an examination of past performance and forecasts for the future, our assumptions about the firm's future growth, costs, and operating ratios are as follows:

Annual Growth in Sales	6%
Operating Costs/Sales	70%
Depreciation/GFA	3%
Tax Rate	40%
Dividend Payout Ratio	60%
Target Ratio of Cash/Sales	17%
Non-cash Current Assets/Sales	54%
GFA/Sales	193%
Current Liabilities/Sales	60%

Our forecast of the economy and financial markets suggests that the rate on U.S. Treasury bills will be about 5%, and the firm will be able to borrow at an interest rate of 9%.

Having collected past financial statements and our assumptions and projections, we develop the data input section of our model, shown as part of our spreadsheet in Exhibit 4.6. The income statements and balance sheets for the past two years are shown in lines 10–43, columns C and D. Other data input, forecasts, and policy parameters are shown in column I.

First, we will cheat a little and show the results of the completed model in Exhibits 4.7A and 4.8A. The completed planning model is constructed in modules following the general format shown earlier in Exhibits 4.2A through E. In Exhibit 4.7A, the Accounting Module and Income Generation modules are combined to model the income statement and balance sheet in lines 51–85, with the past results in columns C and D, and the *pro forma* for the five-year planning horizon in columns E through I. The

	A	B	C	D	E	F	G	H	I
1					ODD & RICH CORPORATION				
2									
3									
4					INPUT SECTOR				
5									
6		Internal Conditions				External Conditions			
7									
8		Initial Condition: Financial Statements				Economic & Environmental Conditions & Forecasts			
9									
10		BALANCE SHEET				Risk Free Interest Rate			5.0%
11			2009	2010		Market Risk Premium			7.0%
12						Interest Rate on Borrowing			9.0%
13		Cash	160.4	170.0		Annual Sales Growth			6.0%
14		Other Current Assets	509.8	540.0		Long Term Growth Rate			6.0%
15		Total Current Assets	670.2	710.0					
16		Gross Fixed Assets	1,792.5	1,926.8					
17		Allowance for Depreciation	449.0	502.8					
18		Net Fixed Assets	1,343.5	1,424.0					
19		Total Assets	2,013.7	2,134.0					
20						Policies & Decisions			
21		Current Liabilities	566.4	600.0					
22		Debt	37.7	73.3		Debt Portion			100%
23		Common Stock	100.0	100.0		Payout Ratio			60%
24		Earned Surplus	1,309.6	1,360.7					
25		Stockholders Equity	1,409.6	1,460.7					
26		Total Liabilities and Equity	2,013.7	2,134.0					
27									
28		INCOME STATEMENT							
29			2009	2010					
30						Parameter Estimates			
31		Sales	944.0	1,000.0					
32		Operating Costs	736.0	730.0		Operating Cost / Sales			70.0%
33		EBITDA	208.0	270.0		Depreciation/GFA			3.0%
34		Depreciation	49.9	53.8		Tax Rate			40.0%
35		EBIT	158.1	216.2		Cash/Sales			17.0%
36		Interest Expense	3.1	3.4		CA (non-cash) / Sales			54.0%
37		Earnings Before Tax	155.0	212.8		GFA/Sales			193.0%
38		Income Tax	62.0	85.0		TA / Sales			2.13
39		Net Income	93.0	127.8		CL / Sales			0.60
40						Beta			0.99
41		Dividends	55.8	76.7		Number of Shares			100.00
42									
43		Retained Earnings	37.2	51.1					
44									
45									

EXHIBIT 4.6   O&R Corp. Data input sector for financial statement model.

Investment, Financing, and Cash Flow modules are shown in Exhibit 4.8A. The Investment Module (lines 89–96) calculates the total amount that needs to be invested in new assets in order to support the sales forecast. The Financing Module (lines 98–114) determines how the investment will be financed, splitting the financing between internal and external financing (line 110).

The cash flow section (lines 116–128) reconciles all the sources and uses of cash by summarizing the changes in balance sheet accounts and the

	A	B	C	D	E	F	G	H	I
47					ODD & RICH CORPORATION				
48					FINANCIAL STATEMENTS				
49					2009 - 2015				
50									
51	INCOME STATEMENT		<<<<<<<<<<	<<<<<<<<Past	Future	>>>>>>>>>	>>>>>>>>>	>>>>>>>>	>>>>>>>>
52			2009	2010	2011	2012	2013	2014	2015
53									
54	Sales		$944.0	$1,000.0	$1,060.0	$1,123.6	$1,191.0	$1,262.5	$1,338.2
55	Operating Costs		$736.0	$730.0	$742.0	$786.5	$833.7	$883.7	$936.8
56	EBITDA		$208.0	$270.0	$318.0	$337.1	$357.3	$378.7	$401.5
57	Depreciation		$49.9	$53.8	$57.8	$61.4	$65.1	$69.0	$73.1
58	EBIT		$158.1	$216.2	$260.2	$275.7	$292.2	$309.8	$328.4
59	Interest Expense		$3.1	$3.4	$6.6	$7.2	$7.6	$7.9	$8.3
60	Earnings Before Tax		$155.0	$212.8	$253.6	$268.5	$284.7	$301.8	$320.0
61	Income Tax		$62.0	$85.0	$101.4	$107.4	$113.9	$120.7	$128.0
62	Net Income		$93.0	$127.8	$152.2	$161.1	$170.8	$181.1	$192.0
63									
64	Dividends		$55.8	$76.7	$91.3	$96.7	$102.5	$108.7	$115.2
65	Retained Earnings		$37.2	$51.1	$60.9	$64.4	$68.3	$72.4	$76.8
66									
67	Check: Div - Min Div				$0.0	$0.0	$0.0	$0.0	$0.0
68									
69	BALANCE SHEET								
70			2009	2010	2011	2012	2013	2014	2015
71									
72	Cash		$160.4	$170.0	$180.2	$191.0	$202.5	$214.6	$227.5
73	Other Current Assets		$509.8	$540.0	$572.4	$606.7	$643.1	$681.7	$722.6
74	Total Current Assets		$670.2	$710.0	$752.6	$797.8	$845.6	$896.4	$950.1
75	Gross Fixed Assets		$1,792.5	$1,926.8	$2,045.8	$2,168.5	$2,298.7	$2,436.6	$2,582.8
76	Allowance for Depreciation		$449.0	$502.8	$560.6	$622.0	$687.0	$756.0	$829.1
77	Net Fixed Assets		$1,343.5	$1,424.0	$1,485.2	$1,546.6	$1,611.6	$1,680.6	$1,753.7
78	Total Assets		$2,013.7	$2,134.0	$2,237.8	$2,344.3	$2,457.2	$2,576.9	$2,703.8
79									
80	Current Liabilities		$566.4	$600.0	$636.0	$674.2	$714.6	$757.5	$802.9
81	Debt		$37.7	$73.3	$80.2	$84.2	$88.3	$92.7	$97.3
82	Common Stock		$100.0	$100.0	$100.0	$100.0	$100.0	$100.0	$100.0
83	Earned Surplus		$1,309.6	$1,360.7	$1,421.6	$1,486.0	$1,554.3	$1,626.8	$1,703.6
84	Stockholders Equity		$1,409.6	$1,460.7	$1,521.6	$1,586.0	$1,654.3	$1,726.8	$1,803.6
85	Total Liabilities and Equity		$2,013.7	$2,134.0	$2,237.8	$2,344.3	$2,457.2	$2,576.9	$2,703.8
86									
87	Check: Assets - Liabilities				$0.0	$0.0	$0.0	$0.0	$0.0

EXHIBIT 4.7A   O&R Corp. Accounting and income modules: *Pro forma* statements generated by planning model.

flows from the income statement. This section provides part of the balancing mechanism by splitting the Net Cash Inflow (line 126) into the part that goes to the Cash balance and the residual that is available to be paid as dividends. By taking into account the income flows and every balance sheet category, we are assured that the cash is not flowing into the financial black hole.

The results shown in Exhibit 4.7A are generated by the formulas of the model. Exhibit 4.7B shows the formulas for the income statement and balance sheet. Two columns of the five-year model are shown: column A shows the variable names, and column E shows the formulas

	A	E
47		**ODD & RICH CORPORATION**
48		FINANCIAL STATEMENTS
49		2003 - 2009
50		
51	INCOME STATEMENT	
52		=D52+1
53		
54	Sales	=D54*(1+Annual_Sales_Growth)
55	Operating Costs	=Sales*Operating_Cost___Sales
56	EBITDA	=Sales-Operating_Costs
57	Depreciation	=IF(Depreciation_GFA*D75>=D77,D77,Depreciation_GFA*D75)
58	EBIT	=EBITDA-Depreciation
59	Interest Expense	=E133
60	Earnings Before Tax	=EBIT-E59
61	Income Tax	=Earnings_Before_Tax*Tax_Rate
62	Net Income	=Earnings_Before_Tax-Income_Tax
63		
64	Dividends	=IF(Cash_Available_for_Dividends>=Net_Income*Payout_Ratio,MAX(E128,0),MAX(Net_Income*Payout_Ratio,0))
65	Retained Earnings	=Net_Income-Dividends
66		
67	Check: Div - Min Div	=Dividends-E106
68		
69	BALANCE SHEET	
70		=E52
71		
72	Cash	=D72+Increase in Cash
73	Other Current Assets	=CA__non_cash___Sales*Sales
74	Total Current Assets	=Cash+E73
75	Gross Fixed Assets	=MAX(GFA_Sales*Sales,0)
76	Allowance for Depreciation	=MIN(D76+Depreciation,E75)
77	Net Fixed Assets	=E75-E76
78	Total Assets	=E74+E77
79		
80	Current Liabilities	=CL___Sales*Sales
81	Debt	=D81+E111
82	Common Stock	=D82+E112
83	Earned Surplus	=D83+E65
84	Stockholders Equity	=E82+E83
85	Total Liabilities and Equity	=E80+Debt+E84
86		
87	Check: Assets - Liabilities	=E78-E85

EXHIBIT 4.7B   O&R planning model. Formulas in accounting and income modules.

for one period of the planning horizon. Many of the variables are named so as to make the formulas easier to decipher. References to column D refer to the year preceding column E. The input data come from Exhibit 4.6.

Some of the expressions need some explanation because they are more complicated than in Exhibit 4.2 so as to take into account various exceptional cases. We need to realize that a computer model is very obedient, but also very stupid. It will do exactly what you tell it to do.

	A	B	C	D	E	F	G	H	I
89	INVESTMENT MODULE								
90					2011	2012	2013	2014	2015
91									
92	Current Asset Investment				42.6	45.2	47.9	50.7	53.8
93	Investment in New Fixed Assets				61.2	61.4	65.1	69.0	73.1
94	Replacement of Dep. Assets				57.8	61.4	65.1	69.0	73.1
95	Fixed Asset Investment				119.0	122.7	130.1	137.9	146.2
96	Total Investment				161.6	167.9	178.0	188.7	200.0
97									
98	FINANCING MODULE								
99					2011	2012	2013	2014	2015
100									
101	Required Financing				161.6	167.9	178.0	188.7	200.0
102	Sources of Funds:								
103	Net Income				152.2	161.1	170.8	181.1	192.0
104	Depreciation				57.8	61.4	65.1	69.0	73.1
105	Earnings Cash Flow				210.0	222.5	235.9	250.1	265.1
106	Minimum Dividend				91.3	96.7	102.5	108.7	115.2
107	Internal Financing				118.7	125.8	133.4	141.4	149.9
108	Spontaneous Financing				36.0	38.2	40.4	42.9	45.4
109	Total Internal & Spontaneous Financing				154.7	164.0	173.8	184.3	195.4
110	Required External Financing				6.9	3.9	4.2	4.4	4.6
111	Debt Issued				6.9	3.9	4.2	4.4	4.6
112	Equity Issued				0.0	0.0	0.0	0.0	0.0
113	Total Discretionary Financing				6.9	3.9	4.2	4.4	4.6
114	Total Financing				161.6	167.9	178.0	188.7	200.0
115									
116	CASH FLOW								
117					2011	2012	2013	2014	2015
118									
119	Earnings Cash Flow				210.0	222.5	235.9	250.1	265.1
120	Other Current Asset Investment				32.4	34.3	36.4	38.6	40.9
121	Spontaneous Financing				36.0	38.2	40.4	42.9	45.4
122	NWC Investment				-3.6	-3.8	-4.0	-4.3	-4.5
123	Cash Flow From Operations				213.6	226.3	239.9	254.3	269.7
124	Capital Expenditures				119.0	122.7	130.1	137.9	146.2
125	Capital Financing				6.9	3.9	4.2	4.4	4.6
126	Net Cash Inflow				101.5	107.5	113.9	120.8	128.1
127	Increase in Cash				10.2	10.8	11.5	12.1	12.9
128	Cash Available for Dividends				91.3	96.7	102.5	108.7	115.2
129									
130	Check: NCF - Chg Cash - Div				0.0	0.0	0.0	0.0	0.0
131									
132	INTEREST AND DEBT SERVICE								
133	Interest Expense		0.0	3.4	6.6	7.2	7.6	7.9	8.3
134	Total Interest Bearing Debt		37.7	73.3	80.2	84.2	88.3	92.7	97.3
135	Total Debt		604.1	673.3	716.2	758.3	802.9	850.2	900.3

EXHIBIT 4.8A   O&R Corp. Investment, financing, and cash flow modules.

Unless you anticipate every eventuality, this obedience can lead to perverse results. For example, Depreciation expense was shown in Exhibit 4.2 (A.4) as

$$\text{Depreciation Expense}_t = \text{Gross Fixed Assets}_{t-1} \times (\text{Depreciation Rate})$$

$$= \text{GFA}_{t-1} \times \left(\frac{\text{Depreciation}}{\text{GFA}}\right).$$

	A	E
89	INVESTMENT MODULE	
90		=E52
91		
92	Current Asset Investment	=E74-D74
93	Investment in New Fixed Assets	=E77-D77
94	Replacement of Dep. Assets	=Depreciation
95	Fixed Asset Investment	=E93+E94
96	Total Investment	=E92+E95
97		
98	FINANCING MODULE	
99		=E52
100		
101	Required Financing	=E96
102	Sources of Funds:	
103	Net Income	=Net_Income
104	Depreciation	=Depreciation
105	Earnings Cash Flow	=E103+E104
106	Minimum Dividend	=MAX(Payout_Ratio*Net_Income,0)
107	Internal Financing	=E105-E106
108	Spontaneous Financing	=E80-D80
109	Total Internal & Spontaneous F	=E107+E108
110	Required External Financing	=E101-E109
111	Debt Issued	=MAX($I$22*E110,0)
112	Equity Issued	=MAX(E110-E111,0)
113	Total Discretionary Financing	=E111+E112
114	Total Financing	=E109+E113
115		
116	CASH FLOW	
117		=E52
118		
119	Earnings Cash Flow	=Net_Income+Depreciation
120	Other Current Asset Investment	=E73-D73
121	Spontaneous Financing	=E80-D80
122	NWC Investment	=E120-E121
123	Cash Flow From Operations	=E119-E122
124	Capital Expenditures	=(E77-D77)+Depreciation
125	Capital Financing	=(Debt-D81)+(E82-D82)
126	Net Cash Inflow	=E123-E124+E125
127	Increase in Cash	=(Cash_Sales*Sales)-D72
128	Cash Available for Dividends	=E126-E127
129		
130	Check. NCF - Chg Cash - Div	=E126-(E72-D72)-Dividends
131		
132	INTEREST AND DEBT SERVIC	
133	Interest Expense	=$I$12*D81
134	Total Interest Bearing Debt	=Debt
135	Total Debt	=E80+Debt

EXHIBIT 4.8B   O&R planning model. Formulas for investment module and financing modules including cash flow and debt service sections.

The problem is that as time passes, if the firm's asset investment has not exceeded its depreciation, its NFA will decline. This can lead to a situation where the Accumulated Depreciation exceeds the GFA, so that NFA become negative. To prevent such a perverse modeling result, we need to modify the Depreciation Expense. We do this with expressions of the following form:

If $GFA_{t-1}(Dep/GFA) \geq NFA_{t-1}$, then Depreciation Expense = NFA_{t-1}, but

If $GFA_{t-1}(Dep/GFA) < NFA_{t-1}$, then Depreciation Expense $= GFA_{t-1}$ (Dep/GFA).

In "Excelese" (the language of Excel®) this is expressed in line 57 of the model as Depreciation Expense$_t = IF[(Dep/GFA) \times GFA_{t-1} > NFA_{t-1}, NFA_{t-1}, (Dep/GFA)GFA_{t-1}]$. In this way, if we reach the point where further depreciation would drive NFA negative, Depreciation Expense will be zero.

Another case where we may need to anticipate perverse modeling is the dividend. If we specify the dividend as a simple percent of Net Income, as

$$Dividend = Net\ Income \times (Payout\ Ratio),$$

we can end up with a negative dividend when we experience a loss. The expression

$$Dividend = MAX[Net\ Income \times (Payout\ Ratio), 0]$$

prevents paying a negative dividend when net income is negative. In addition, we may want to avoid paying a dividend when there is no cash flow to support it. Or, we may need to provide for paying out extra cash flow to the stockholders when the cash flow exceeds the needs of the firm. If Cash Available for Dividend is the amount available after meeting all of the firm's other needs, then the dividend is calculated as If: Cash Available for Dividends > Net Income (Payout Ratio), then Dividend = MAX (Cash Available for Dividends, 0), but If: Cash Available for Dividends ≤ Net Income (Payout Ratio), then Dividend = MAX {MIN [Cash Available for Dividends, Net Income (Payout Ratio)], 0}.

This will payout the smaller of the Cash Available or Net Income (Payout Ratio), but it will not allow the dividend to be negative. In our model, the financing module assures that the Cash Available will be sufficient to pay the dividend determined by the Payout Ratio, so we express this in Excelese in line 64 as

$$Dividend = IF\ [Cash\ Available\ for\ Dividends \geq Net\_Income \\ \times Payout\_Ratio, MAX\ (Cash\ Available\ for \\ Dividends, 0), MAX(Net\ Income \times Payout\ Ratio, 0)].$$

The formulas for the cash flow and the Investment and Financing Modules are shown in Exhibit 4.8B. The Investment Module calculates the required investment based on the increase in assets—both current and fixed. Total Investment (line 96) carries over to the Financing Module where Required Financing (line 101) equals Total Investment. The Financing Module then determines how much of the Required Financing will be provided by Internal Financing (line 107) and Spontaneous Financing (line 108), with the remainder funded by external sources of debt and equity. The REF is split between debt and equity by the user's decision input, Debt Portion, in cell I22 of the Policies and Decisions section.

The expressions in these modules follow Exhibits 4.2 with some minor elaboration. For example, note in lines 111 and 112 the use of the MAX( ) function. The expression

$$\text{Debt Issued} = \text{MAX}(\$I\$22 \times E110, 0)$$

prevents the Debt Issued from being negative in case the REF turns out to be negative, with the same provision being made for Equity Issued.

The equations in the O&R model are simplified so we do not get too bogged down in detail. As you add features to the model you will have to modify the equations. For example, this model does not provide for debt repayment and does not distinguish between short-term and long-term debt. If you add short-term debt to the model, you should provide for its repayment in the Financing Module so that you have enough financing to repay the maturing short-term debt. Similarly, if long-term debt has to be repaid, the Financing Module would have to provide sufficient funds to make the principal payment. Another obvious elaboration would be to vary the growth of sales from one period to the next.

This model sets the target cash balance equal to a percent of sales, and allows the dividend to be the residual that disposes of any cash flow that would otherwise cause the cash balance to exceed the required amount. Line 126 calculates the Net Cash Inflow from all sources, and then the inflow is split between the Increase in Cash (line 127) and Cash Available for Dividends (line 128). In this case, the Increase in Cash is proportional to the increase in sales, and whatever is left over is available for dividends. The dividend payment (line 64) is based on the Cash Available for Dividends. The Minimum Dividend (line 106) in the Financing Module assures that the dividend will be at least the amount required by the Payout Ratio specified by the user (cell I23). However, the dividend is allowed to exceed the payout rate if excess cash flow is available. This assures that Net Cash Inflow goes somewhere—either to build up cash or to be paid out as a dividend. We could have structured the model the other way around. We could let the payout ratio drive the dividends, and use any excess cash flow to build up the cash balance (or marketable securities).

Note how this model reached its equilibrium so that every thing balances. Following the logic of Exhibit 4.4, the sales forecast drives all asset categories. The increase in assets determines Total Investment (line 96). Required Financing (line 101) equals Total Investment, and the Financing Module calculates the internal and spontaneous sources (line 109) that are available to fund the investment. The remainder, REF (line 110), is split between the discretionary sources of debt and external equity. Balance is maintained by making sure that the increase in liabilities and equity (retained earnings, increase in current liabilities, and increase in debt) is equal to the increase in assets. This model is a hybrid of stock and flow balancing. By balancing assets and liabilities, it is stock balancing, but since it does this by focusing on the changes in assets and liabilities, it resembles a flow balancing model. In addition, the Cash Flow section reconciles the equality of sources and uses, and insures that the cash flow goes either to the cash account or is paid out as a dividend.

It is helpful to include features in your model that check whether it is balancing and doing what you want it to do. Lines 67, 87, and 130 are such checks. Line 87 checks to see if assets and liabilities are equal. Obviously, Total Assets–Total Liabilities and Equity should always be zero. However, it is easy to overlook something and have an imbalance, so line 87 provides a quick check. The line can be eliminated after you have tested the model under a variety of assumptions and find that it performs accurately under all scenarios. A problem that frequently occurs for the novice modeler is that the model will perform well when sales are growing, but if sales decline or follow unusual patterns, the model may not balance. So, when testing the model, try a variety of different growth assumptions to see that it balances under all conditions.

The check in line 130 is to make sure that the cash flow is being allocated either to cash or to dividends. This check is: Net Cash Inflow—Increase in Cash—Dividends. If this is non-zero, it would mean the cash is not flowing into one of these uses, and is flowing into a "financial black hole." This error would occur if you set cash equal to a percent of sales, dividends equal to a percent of income, and the cash flow exceeds the amount necessary to meet these two requirements. The check in line 67 is not really required, but allows the user a quick check to see if the firm is paying dividends in excess of the amount required by the payout ratio.

We have developed the basic model and it seems to balance. We now discuss two additional aspects in the following sections. First, we test the model to see that it works as intended. Second, after testing the model, we evaluate the financial results of our plan.

4.4.2 Testing the Model

We have developed our model and we have seen that it seems to balance. However, before concluding that it is a robust model, we need to do some testing to see if it performs well under a variety of circumstances, and to see what its weaknesses might be. For example, as noted earlier, a model may seem to work fine with moderate assumptions about growth, but when the assumption is changed, it may not work well. Alternatively, we may find that the model balances and otherwise performs as designed, but that the results may seem unrealistic, suggesting some changes in the underlying assumptions. For example, if sales decline, would we divest assets proportionally to the decrease in sales? If we think not, and if it is an important issue, we may need to change the model to take account of more realistic actions.

Our first test drive is to try out different growth rates for sales to check that the model always balances. We will try extremes rates of growth for sales: −50%, 0%, and +50% per year. In each instance, we check the asset and liability difference (line 87), the cash flow disposition (line 103), and the dividend payment (line 67). In addition, we will look for other anomalies that might indicate a problem in our modeling. The table below summarizes our main observations for each of the growth rates we tried. What we find is that the model balances in all cases, but at the extremes, we may get some results that may be problematic.

At very high growth rates, the need to invest in new assets requires a very large amount of borrowing because the model says all external funds must come from debt. This is not really a problem for the model but it alerts us to a common problem associated with rapid growth. Growth requires investment and a growing firm can easily outrun its ability to finance the growth. As far as the modeling goes, there is nothing wrong. However, we may decide to include a provision in the financing section that requires equity financing instead of debt if the debt ratios get too high.

Line	Issue	Growth Rate		
		−50%	0%	+50%
87	Liabilities = Assets?	ok	ok	ok
103	Cash flow fully allocated?	ok	ok	ok
67	Dividend = Payout × Income?	Div>NI	Div>NI	ok
Comments		Assets decline Debt constant Debt>TA	Equity declines Debt/Equity increases	Declining dividend Heavy investment Extreme debt

At zero growth, the model balances, but the firm is generating enough cash flow each period that it can pay a dividend that exceeds net income. This causes the equity to decline. There is no provision in the model to repay debt, so the debt balance remains constant, and the debt-to-equity ratio increases.

When the sales are declining rapidly, the model still balances. It continues to pay dividends that exceed income because the firm, in essence, is being liquidated. Equity declines, but as in the zero growth case, debt is not being repaid, so not only does the debt-to-equity ratio increase, eventually debt exceeds total assets. This suggests that we may want to put a provision in our model that repays debt from the cash flow generated by the liquidation.

We also notice a weakness that is common to percent of sales models: the model performs as designed, and assets increase and decrease proportionally to sales. It might be more realistic to assume that some asset categories, particularly fixed assets, increase in discrete increments as sales increase, and do not decrease (in the short run) when sales decrease. For example, if sales increase 10%, fixed assets may not increase at all because the firm has some unused capacity that enables it to increase production without increasing the plant. On the contrary, if sales decrease 10% the firm probably will not immediately dispose of 10% of its productive capacity. It may keep the existing plant for several periods until it can dispose of a whole production unit, especially if the plant is not divisible. In Chapter 10, we will discuss how we can modify our model so that we can take account of our capacity to expand and provide for moderate sales declines without divesting our assets.

These observations do not indicate that the model is not working correctly. Rather, they point to problems that could be encountered by the actual business under conditions of extreme growth or extreme decline. If we feel that extreme scenarios are unlikely to occur, then we may feel that it is unnecessary to take the trouble to modify the model to account for such extreme occurrences. However, it would be worthwhile to try out additional growth scenarios to see what the limits of the model are: that is, at what points does the model start to generate results that we do not regard as realistic?

4.4.3 Tracking Performance

When we build our model, we should include a section that summarizes the results of our projections. Some of the results have already been shown—that is, the *pro forma* financial statements for the planning horizon. However, it is difficult to digest all the data of these statements, so we need to summarize the results in such a way that we can evaluate the firm's prospective performance. Two aspects of that evaluation will be discussed here: analysis of the financial ratios and estimation of the value of the firm.

Exhibit 4.9A shows the financial ratio section of our O&R model, and Exhibit 4.9B shows the formulas for one column. The ratios are shown in order of the DuPont format that emphasizes the interplay of profit margin, asset turnover, and financial leverage in determining the return on equity:

Return on Equity = Profit Margin × Asset Turnover × Equity Multiplier

$$\left(\frac{\text{Net Income}}{\text{Equity}}\right) = \left(\frac{\text{Net Income}}{\text{Equity}}\right) \times \left(\frac{\text{Sales}}{\text{Total Assets}}\right) \times \left(\frac{\text{Total Assets}}{\text{Equity}}\right).$$

	A	B	C	D	E	F	G	H	I
137					ODD & RICH CORPORATION				
138									
139				2009	2010	2011	2012	2013	2014	2015
140	RATIOS								
141	Return On Assets (EBIT/TA)			7.9%	10.1%	11.6%	11.8%	11.9%	12.0%	12.1%
142	Return On Equity (NI/EQUITY)			6.6%	8.7%	10.0%	10.2%	10.3%	10.5%	10.6%
143	Profit Margin (NI/SALES)			9.9%	12.8%	14.4%	14.3%	14.3%	14.3%	14.3%
144	Asset Turnover (SALES/TA)			0.47	0.47	0.47	0.48	0.48	0.49	0.49
145	Asset Intensity (TA / Sales)			2.13	2.13	2.11	2.09	2.06	2.04	2.02
146	Equity Multiplier			1.43	1.46	1.47	1.48	1.49	1.49	1.50
147	Current Ratio (CA/CL)			1.18	1.18	1.18	1.18	1.18	1.18	1.18
148	Debt Ratio (TD/TA)			30.0%	31.6%	32.0%	32.3%	32.7%	33.0%	33.3%
149	Debt to Equity (TD/Equity)			42.9%	46.1%	47.1%	47.8%	48.5%	49.2%	49.9%
150	Debt to Equity (TD ib /Equity)			2.7%	5.0%	5.3%	5.3%	5.3%	5.4%	5.4%
151	Debt to Total Capital			2.6%	4.8%	5.0%	5.0%	5.1%	5.1%	5.1%

EXHIBIT 4.9A   O&R Corp. Measures of financial performance for planning period.

	A	E
140	RATIOS	
141	Return On Assets (EBIT/TA)	=EBIT/E78
142	Return On Equity (NI/EQUITY)	=Net_Income/(E82+E83)
143	Profit Margin (NI/SALES)	=Net_Income/Sales
144	Asset Turnover (SALES/TA)	=Sales/E78
145	Asset Intensity (TA / Sales)	=E78/Sales
146	Equity Multiplier	=E78/E84
147	Current Ratio (CA/CL)	=E74/E80
148	Debt Ratio (TD/TA)	=(E80+Debt)/E78
149	Debt to Equity (TD/Equity)	=(E80+E81)/E84
150	Debt to Equity (TD ib /Equity)	=E81/E84
151	Debt to Total Capital	=E81/(E81+E84)

EXHIBIT 4.9B   O&R Corp. Formulas for financial ratios.

Other standard ratios relating to liquidity (Current Ratio) and financial leverage (Debt Ratio, Debt-to-Equity, etc.) are also shown. In the exhibit TD denotes total debt, including current liabilities such as accounts payable, whereas TD_{ib} denotes interest-bearing debt.

This model does not show many of the wide variety ratios and performance measures that could have been included. The performance measures included in the model should be based on the detail and purpose of the model. In this model we did not break the assets into categories such as accounts receivable, inventories, and various types of fixed assets. If it is important for our analysis to consider these more detailed accounts, then we can easily add more specialized ratios such as the collection period for the receivables, the inventory period, or turnover rates for fixed assets.

What we see from the exhibit is that most of the ratios are forecasted to be quite stable over the planning horizon. This should not be surprising in view of the fact that the assets and most of the expenses are modeled as a constant percent of sales. Our conclusion for O&R is that the firm can handle moderate growth without being burdened by excessive debt.

This brief analysis helps point out one of the most common uses for such a model—the exploration of the consequences of a set of policies. The model enables us to see what might happen under a variety of scenarios if we follow the policies that are specified in the model. If we do not like the projected results, the model allows us to try alternative policies out so that we can find the mix that will best suit our objectives.

Speaking of objectives, remember that in Chapter 1 we emphasized that we should relate the model results to the objectives of our decisions. Generally, our objective is to maximize the value of the equity of the firm. Consequently, we should have a valuation module in our model. We will discuss that in the next section.

4.4.4 Valuation

4.4.4.1 The Valuation Module

The next module of our model provides an estimate of the value of the equity of O&R Corp. While the ratio analysis discussed in the last section can be helpful for showing strengths and weaknesses for our plan, it is not particularly helpful with trying to decide which of many possible plans is best. The solution is a valuation module. Since our objective is to maximize the value of the equity, if we can estimate the value of the equity that results from each policy, we will have a single measure related directly to the objective, allowing us to see which policies have the best chance of reaching our objective.

For the O&R Corp., our valuation will be based on the dividend discount model of the form:

$$\text{Equity Value}_0 = E_0 = \sum_{t=1}^{h} \frac{D_t}{(1+k_E)^t} + \frac{E_h}{(1+k_E)^h}, \qquad (4.3)$$

where D_t denotes the dividend paid at time t, k_E the discount rate, h the number of periods in the forecast horizon, and E_h the terminal value of the equity at the end of the forecast horizon. We will use the capital asset pricing model (CAPM) to estimate the cost of equity, k_E, and the constant growth model to estimate the terminal value, E_h. The horizon is five years. Our valuation model will generate an estimate of the aggregate value of the equity, that is, total equity capitalization, which is equal to the price per share times the number of shares. At this stage we will not be concerned with the value per share. We will deal with the number of shares and the price per share in Chapter 9.

Exhibit 4.10A shows the valuation module, and Exhibit 4.10B shows the formulas for selected columns of the model. The data for calculating the discount rate were input at the top of the model in the section labeled Economic & Environmental Conditions & Forecasts (cells I10:I11), and the beta is in cell I40. The long-run growth rate for the constant growth

	A	B	C	D	E	F	G	H	I
153					ODD & RICH CORPORATION				
154		EQUITY VALUATION MODULE							
155									
156		COST OF EQUITY							
157		Risk Free Interest Rate	5.0%						
158		Market Risk Premium	7.0%						
159		Beta	0.99						
160		Discount Rate	11.921%						
161		PVIF @ cost of equity at t=0			0.893	0.798	0.713	0.637	0.569
162									
163									
164									
165				Period	2010	2011	2012	2013	2014	2015
166		DIVIDEND VALUATION MODEL							
167		Dividends			91.30	96.65	102.48	108.66	115.21
168		PV(dividends)							
169		Present Value of Dividends		366.69	319.11	260.49	189.06	102.94	0.00
170									
171		Terminal Value							
172									
173		Long Term Growth Rate							0.06
174		Dividend t = 6							122.12
175		Terminal Value of Stock at t = 5							2,062.57
176		PV of Terminal Value		1,174.50	1,314.51	1,471.21	1,646.59	1,842.88	2,062.57
177									
178		Aggregate Value							
179		of Stock		1,541.19	1,633.61	1,731.70	1,835.65	1,945.82	2,062.57

EXHIBIT 4.10A   O&R Corp. Valuation module.

	B	C	D	E	I
153			=D137		
154	EQUITY VALUATION MODULE				
155					
156	COST OF EQUITY				
157	Risk Free Interest Rate	=I10			
158	Market Risk Premium	=I11			
159	Beta	=I40			
160	Discount Rate	=C157+C158*C159			
161	PVIF @ cost of equity at t=0			=1/(1+C160)	=H161*(1/(1+$C$160))
162					
163					
164					
165			Period =D52	=E52	=I52
166	DIVIDEND VALUATION MODEL				
167	Dividends			=Dividends	=Dividends
168	PV(dividends)				
169	Present Value of Dividends		=NPV($C$160,E167:I167,0)	=NPV($C$160,F167:I167,0)	0
170					
171	Terminal Value				
172					
173	Long Term Growth Rate				=I14
174	Dividend t = 6				=I167*(1+I173)
175	Terminal Value of Stock at t = 5				=I174/(C160-I173)
176	PV of Terminal Value		=$I175/(1+$C$160)^5	=$I175/(1+$C$160)^4	=$I175/(1+$C$160)^0
177					
178	Aggregate Value				
179	of Stock		=D169+D176	=E169+E176	=I169+I176

EXHIBIT 4.10B   O&R Corp. Formulas for valuation module. Selected columns.

model is input cell I14. This input data are repeated in cells C157:C159. The beta* is estimated to be 0.98, and the cost of equity is

$$k_E = k_f + (k_m - k_f)\beta$$
$$= 0.05 + (0.07)0.98 = 0.119 = 11.9\%. \tag{4.5}$$

The dividends shown in line 167 come from the Dividends shown in line 64 of the income statement. The present value of dividends shown in line 169 is the present value, at each date, of the dividends remaining to be paid until the end of the horizon. For example, the $366.69 shown in D169 is the present value at the end of 2010 ($t = 0$) of the dividends to be received in years 2011–2015; the $319.11 is the present value at the end of 2011 of the dividends paid in years 2013–2015; and so on until $102.94, which is the present value (PV) at the end of 2014 of the dividend of $115.21 to be received at the end of 2015.

The terminal value at the end of period 5 is shown in cell I176 based on the constant growth model

$$E_5 = \frac{D_5(1+g)}{(k_E - g)}$$
$$= \frac{115.21 \times (1.06)}{(0.1192 - 0.06)}$$
$$= \$2062.49 \tag{4.6}$$

The value of the equity in line 179 is the sum of the present value of the dividends and the present value of the terminal value. Each column shows the value at a different date. Cell D179 is the value at the beginning of the horizon, the end of 2010 ($t = 0$); cell E179 is the value at the end of 2011. Thus, with this model, we can follow the projected value of equity through time, so see what might happen with different policies.

* For a more detailed model capable of analyzing the effects of alternative capital structures, the beta should adjust to the amount of financial leverage. In that case, the input beta could be the unleveraged beta, and the leveraged beta would adjust internally in the model using the Hamada (1969) formula

$$\beta_L = \beta_U[1 + (\text{Debt/Equity}) (1 - T)],$$

where L denotes leveraged and U unleveraged. So as to not make the model too complex at this stage, we have not yet dealt with this issue. We discuss the effects of leverage on cost of equity in greater detail in Chapter 9.

4.4.4.2 Evaluation and Sensitivity Analysis

Given the conditions and assumptions listed in the input section, we have estimated the aggregate value of the equity to be $1541.23. Suppose we want to know the impact on the value with different forecasts for variables such as growth rates, or we may want to explore the effects of alternative financing policies.

To determine what happens when the firm experiences different rates of growth of sales for the next five years (holding the post-horizon growth rate constant at 6%), four different growth rates were put into the model, with other variables unchanged from the previous scenario. The results are displayed in the following table.

Sensitivity Analysis: Debt Ratios and Equity Values with Varying Sales Growth Rates

Five-year growth rate (%)	−10	0	6	10
Debt ratio 2011 (%)	33.8	32.4	32.0	34.9
Debt ratio 2015 (%)	52.2	36.4	33.3	46.1
Value of equity ($)	3597.3	2740.8	1541.2	1533.1

The higher debt ratio for the −10% growth scenario is because the firm is reducing its equity over time as it pays out it dividends. The main reason that the equity is more valuable with lower growth is that it is unnecessary to raise new capital. This means that the cash flow to equity is greater with lower growth, accompanied by a lower discount rate for equity due to lower debt.* The fact that the equity is worth more even when sales are declining at 10% seems surprising. Yet, this is also because the firm is throwing off a lot of cash as it declines. In addition, the long-run growth (after five years) was held constant at 6%, so the 10% decline only applies to the first five years.† We conclude that the stockholders would be better off with slower growth for this firm.

This sensitivity analysis focused on varying the growth rate. The financing policy was unchanged from the base model where it was assumed that 100% of external financing came from debt. Now let us explore the

---

* These figures were based on the assumption that the beta and the cost of equity change with the capital structure according to the Hamada formula. The model shown in this chapter does not show that effect.
† There is a perverse aspect to the negative growth scenario. When the growth rate is negative, part of the year 5 dividend is cash derived from liquidated assets. The constant growth model assumes this cash will be available and grow in perpetuity, something that is clearly false. We address this shortcoming in Chapter 9.

implication of different financing policies. We assume 10% growth in sales during the five-year horizon and vary the proportion of debt that is used to fund our external financing requirements. The following table shows the results with the Debt Portion (cell I22) varying from 0% to 100%.

Sensitivity Analysis: Debt Ratios and Equity Values with Varying Debt Portions Sales Growth at 10%

Debt portion (%)	0	25	50	75	100
Debt ratio 2011 (%)	31.3	32.2	33.1	34.0	34.9
Debt ratio 2015 (%)	30.6	34.4	38.2	42.1	46.1
Value of equity ($)	1872.1	1788.3	1703.9	1618.9	1533.1

What we see here is that equity value declines as we use more debt to fund our external financing needs over the next five years. The reason for this is that debt financing and the resulting interest expense eats into the equity cash flow. Less is left for the stockholders, so equity value is smaller if we use more debt. In addition, as we use more debt, the cost of equity increases, further depressing the value of equity. However, note that this equity value is the aggregate value of equity, and not the value per share. Thus far, we have not included share values in our model, so we can say very little about the value per share. The only thing we know is that if we use 100% debt financing and issue no new equity, the number of shares will remain the same over the planning horizon. However, if we issue new equity to finance part of our needs, the number of shares will increase. Thus, even though the aggregate value of equity is higher with 0% debt financing, it could be the case that the value per share is smaller. To address that issue, we would have to model the number of shares. We will leave that for later when we delve more deeply into models of valuation in Chapter 9.

There are many other variables and policies we could explore with sensitivity analysis. In this section we demonstrated the approach for just a couple of variables: growth and debt financing. The variety of ways to manipulate the model is limited only by the needs and imagination of the model builder.

4.5 CONCLUSION

This chapter is the heart of this book about financial modeling. It has discussed methods for structuring financial simulation models. There are many different ways to model a given problem, and given the huge variety of problems that could be addressed, the variety of models are almost endless. Obviously, not all types of models can be presented and explained, so the purpose of this chapter has been to present and discuss the basic form of a "generic" financial simulation model. The types of equations that constitute such a model were explained. This generic model consists of modules that serve different functions in the model—the accounting, income generation, investment module, financing, and cash flow modules. In addition, we introduced valuation concepts in the valuation module. Each module consists of equations that express the relationships within the firm, and between the firm and its environment.

It was noted that a very important component of any such model is a mechanism for balancing the model—that is, helping the model to reach equilibrium. Two major approaches to equilibrating the model were explained: a cash flow balancing system, and an asset–liability balancing system. Each approach represents a way to balance investment and financing and to maintain consistency in the model.

The model that we developed was small, with detail kept to a minimum. The reason for this was to emphasize the structure of the model. Once we have a functioning model that balances and is robust to widely varying assumptions, we can always add additional detail. Later chapters will discuss different ways by which we could model specific accounts to address more specialized decision problem.

PROBLEMS

1 Spreadsheet Model for Growth: Fly-by-Night Airlines, Inc.

The Fly-by-Night Airlines, Inc. company was described in Problem 3, Chapter 3. This is a continuation of that problem, so the income statement and the balance sheets for 2007–2010 that were shown in that problem are assumed to apply to this problem.

Your assignment is to develop a spreadsheet model of the Fly-by-Night Airlines, Inc. Your model should enable you to generate *pro forma* financial statements (income statement, balance sheet, and sources and uses of cash statement) for each of the next five years (2011–2015).

In developing your model, make the following assumptions:

- Future ratios of CGS/Sales, SGA/Sales, Cash/Sales, Receivables/Sales, Inventories/Sales, Prepaid Expenses/Sales, Plant Property and Equipment/Sales, Other Assets/Sales, Accounts Payable/

Sales, and Other Current Liabilities/Sales will be maintained at their average levels for the past four years.

- The ratios of Depreciation Expense to Plant Property and Equipment$_{t-1}$ and the change in Deferred taxes to Taxes$_{t-1}$ will be constant (three-year average).
- The ratio of Taxes to Taxable income will be constant (four-year average).
- Intangibles will be amortized at $1000 per year.
- No short-term financing will be used, and, after the current portion of long-term debt is paid off in 2011, no other portion of the long-term debt will become current.
- Investment and financing are in place at the beginning of each year, that is, the assets required to support sales in 2011 have been purchased and paid for by December 31, 2010.
- Interest is paid "in arrears," meaning that interest paid on December 31, 2011 is based on the debt balance outstanding at the end of the previous period, December 31, 2010.

The goal of the previous two entries is that
Depreciation expense$_t$ is calculated as Depreciation rate $\times$ PPE$_{t-1}$
Interest expense$_t$ is calculated as Interest rate $\times$ Debt$_{t-1}$.

- All external financing will be long-term debt, at an interest rate of 4.8%. Use any excess cash to retire debt. Payout all remaining excess cash as dividends (over and above the planned dividend). Hint: you will need to write a logical statement in the debt repayment cell of your financing module that prevents long-term debt from becoming negative. You will also need to calculate dividends in your financing module, then feed them into the income segment of the accounting module.
- Fly-by-Night expects to earn no interest income in the future.
- The dividend payout is 0%.

Assignment

A. Show the results of your model assuming that sales grow at (1) g^* and (2) g^{***} (based on ratios as of 2010) for the next five years. What are your external financing requirements in each year for each scenario? What is the trend in the profit margin under each scenario?

B. Show the results with sales growing at (1) 20% and (2) 30% per year. What are your external financing requirements in each year for these scenarios? What is the trend in the profit margin under each scenario? What happens to the profit margin as sales growth increases beyond g^{***}?

C. Show the results when sales grow 30% per year and half of all earnings are paid out as dividends. What are your external financing requirements in each year for this scenario? What is the relationship between dividends and external financing requirements?

D. The sustainable growth rate g^{***} is significantly easier to apply and remember than g^{**}. However, it understates somewhat the actual sustainable growth rate. At what rate can Fly-by-Night grow its sales over the next five years and end up with the same Total debt/Total assets ratio that it had at the end of 2010, assuming a zero dividend payout ratio? or a 50% dividend payout ratio? (Hint: use Goal Seek)

Your model should be constructed so that it balances: that is, assets should equal liabilities; sources equal uses. Put checks into your model to verify that it balances. In addition, your model should be constructed in such a way that the sections that contain equations do not have any numbers as parameters. The parameters should be in a separate section, with the two decision variables—sales growth and dividend payout—clearly delineated. Use Excel's naming utility to name your parameters, for example, *cash_sales* for cash divided by sales. This will help you organize your model and make it easier to interpret your formulas.

What you hand in: a brief narrative in memo form explaining your answers to Parts A, B, C, and D. You will probably find it useful to include one or two tables in your memo. In an appendix, attach the *pro forma* statements for each scenario (formatted such that income statement and balance sheet are on one page, sources and uses of cash and the financing module are on another) and a print out of the formulas for your model. (There should be only one formula sheet. You answer the above questions simply by changing the values of your decision parameters and observing the effects.) Moreover, include in your appendix a flow chart/decision tree diagram of the logic used to code the *Debt repaid* line of your financing module.

2 Modeling Growth and Financing: Fly-by-Night Airlines, Inc.

Fly-by-Night Airlines, Inc. was described in Problem 1. This is a continuation of that problem, so the income statement and the balance sheets for 2007–2010 that apply to Problem 1 are assumed to apply to this problem.

Your assignment is to continue to work on the same model, with the same assumptions as were described in Problem 1.

You have developed the basic model of the financial statements. Hopefully, it worked the first time. However it is important to make sure that it balances under a variety of circumstances. For this assignment, assume a normal dividend payout of 50%. Check the balancing by varying the rate of growth of sales. Use high growth rates such as 20% and 50% per year and see how it performs. Then use low growth scenarios such as –10% and –20% per year. Assume that extra cash is used first to repay debt; if any is left over, pay it out as dividends. As before, assets should equal liabilities and all cash flow should be traceable and go to some use.

You will need to modify your model slightly for it to perform correctly. In particular, you will need to add a conditional statement (IF, THEN) to prevent taking a depreciation deduction once NFA reaches zero. (Note: in extremely negative growth, you are selling assets so the balance of NFA could become negative if you depreciate assets after this occurs.) As was the case in Problem 1, you will need a conditional statement to prevent negative values of debt in the balance sheet after total debt is retired, and a separate line to calculate total dividends (regular dividend plus return of free cash flow). After testing the model and making any necessary changes so that it balances, make the following additions and changes. Your model should be set up in such a way that it allows you to specify the proportions of debt and equity as a parameter in its assumptions section. Add several lines to your financing module that allows you to vary the way the firm finances its funds needs—that is, allows you to choose between debt and equity financing. Test it with three different assumptions:

(a) All new financing is done with long-term debt. The interest rate is 5%.
(b) You use a mix of debt and equity such that any new financing is 60% debt and 40% equity.
(c) All new financing is with new equity.

Test your model with varying sales growth rates.

3 K.D. Nickel Company: Financial Statement Simulation

You are attempting to develop a five-year financial planning model for the K.D. Nickel Company. The purpose of your planning effort is to determine Nickel's future requirements for external financing and to estimate the aggregate value of the equity. The date is the beginning of 2011, and from the exercise in Chapter 2 you have past financial data for the company through fiscal year 2010. Your planning model will use annual periods. Decisions and actions occur at the end of each year. The first period in your planning horizon is 2011. Thus, the first year's sales, expenses, cash flows, borrowing, and so on, all occur at the end of 2011, and the second flows and decisions occur at the end of 2012.

For the purpose of making forecasts of Nickel's future financial condition, the following initial assumptions are made.

- Dollar amounts are stated in millions.
- Sales will grow at an annual rate of 3%.
- Costs of Goods Sold (CGS) will be equal to 60% of sales.
- Selling, general and administrative expense will be equal to 30% of sales.
- Depreciation expense will be equal to 8.5% of the previous year's GFA.
- Interest on past debt is 7% of the previous year's debt balance (combine short- and long-term debt). The interest rate on new debt (debt issued after fiscal year 2002) will be 7%.
- The income tax rate is 40%.

Ignore other types of expenses such as non-operating expense, extraordinary items, and discontinued operations. Other Income and Expenses are assumed to be included in the SGA.

After the end of the planning horizon, the net profit margin will be stable and constant, equal to the net profit margin in the last year of the planning horizon.

For future periods the asset and liability ratios are forecasted to be as follows:

- The target cash balance is 15% of sales.
- The average collection period for accounts receivable will be eight days (based on a 360-day year).
- The average inventory period will be 110 days (based on CGS and a 360-day year).
- GFA will be equal to 30% of sales.

Other long-term assets will be equal to a fixed amount of $6500 (million) plus a variable amount equal to 12% of sales. Other long-term assets is a catchall category that includes "other current assets," "other long-term assets," and all assets not specifically mentioned above.

The average period for accounts payable will be 35 days (based on CGS and 360 days).

The firm has other liabilities that are equal to 20% of sales. This figure combines short- and long-term items. Assume they are short term, and stay at 20% of sales.

All external financing needs will be met with debt financing. All new debt financing will be in the form of short-term debt that matures at the end of each year. It is assumed that it will be turned over annually. New short-term debt will have an interest rate of 7%, and previously issued short- and long-term debt will have an interest rate of 7%. The presently outstanding long-term debt matures beyond the end of the planning horizon.

The minimum dividend payout ratio (Dividends/Earnings) is 60% of earnings. After the end of the planning horizon, the payout will be a constant 60% of earnings.

New investment projected for the year is assumed to occur at the end of the year, and new financing for the investment is also assumed to occur at the end of the year.

The book value of preferred stock will be maintained at its current level over the remaining planning horizon. The preferred dividend is 7.4% of the previous year's book value of preferred. The required return on preferred stock is 8%.

There are currently 267 million shares of common stock outstanding at the end of 2010.

The current one-year risk-free interest rate is 2%, the five-year rate is 3.5%, and the 20-year rate is 5%. Assume that these rates will continue to prevail for the remainder of the planning horizon.

The expected risk premium on the market portfolio is assumed to be 7% over the planning horizon. Assume that the cost of equity is based on the CAPM, with the beta assumed to be 0.90 throughout the planning horizon.

Assignment

A. Develop a spreadsheet model for Nickel's that will enable you to generate *pro forma* financial statements for each of the years in the planning horizon, and determine the amounts that are necessary to finance Nickel's operations for the next five years. The required financing should distinguish between spontaneous and external sources.

Assume that new external financing will be short-term debt that matures at the end of each year.

Pay particular attention to how the model reaches its equilibrium of Assets = Liabilities and Sources = Uses.

B. Now change the growth assumption. Assume that the sales growth rate over the five-year planning horizon is equal to: (1) the sustainable growth rate g^*. What is the impact on the financing needs of the firm? (2) Now assume the growth rate is equal to the g^{**}. What happens to the financing needs of the firm? (3) What happens if growth in sales is 10% per year? (4) What happens if sales decline at 10% per year?

In each of these cases, examine impact of the policies and growth variations on the value of equity.

CHAPTER **5**

Monte Carlo Simulation

Murphy's Laws:

- Anything that *can* go wrong, will go wrong.
- If anything simply cannot go wrong, it will anyway.
- If you perceive that there are four possible ways in which things can go wrong, and you circumvent these, then a fifth way, unprepared for, will promptly go wrong.
- Everything goes wrong all at once.

Murphy was an optimist.

5.1 INTRODUCTION

One of the motivations for building financial models and developing plans for the future is to deal with uncertainty, or, to put it another way, to deal with the certainty that something will go wrong. Although we do not know what will happen in the future, we want to be prepared for most eventualities. Building models and developing plans is our way to minimize the pain and costs of encountering our uncertain future in business. However, to this point we have only discussed how to build simulation models. We have not really considered how to use them to deal with the uncertain world of business. That is the subject of this chapter. We introduce a powerful tool for dealing with uncertainty in financial planning: Monte Carlo simulation.

Among the kinds of uncertainty that we might deal with are uncertainties about future economic events: prices, sales levels, market shares, costs, lives of assets and projects, interest rates, and even availability of financing. All these factors, and more, are important to the success of most business ventures, and most of them vary over time—we do not know what they will be in the future. Without knowing what will happen in the future, it is difficult to be sure we are making the right decision now. Yet, we want to make the best decision possible under the circumstances of our limited knowledge and considerable ignorance. In this regard, models can help us. All the modeling and planning in the world cannot assure us that we will have made the best decision when viewed with the benefit of hindsight. All we can hope to do is to arrive at the best decision possible given the information that is available before the fact.

The key word is information. With the best information, properly evaluated, we can improve the likelihood that we are making the best decision. What modeling and planning do is provide a way to process and analyze the available information so that we can get the maximum amount of insight about our uncertain future. The model simply helps us organize what we know about our firm, its relation to its environment, and the possible events that will influence our firm. Planning involves evaluating that information, so we are prepared with multiple plans of action as the future rolls toward us.

No matter how thorough our models or how careful our plan, our decisions will be completely correct only with monumentally good luck. Generally, when we look back on our decisions, they will turn out to have been wrong. Modeling and planning help to minimize the costs of being wrong. If we have prepared for our uncertain future, if we have anticipated that the best laid plans are bound to go awry, then we can be ready to respond to events that turn out differently than expected. Not only do we need to plan, but we need to plan for being wrong.

One way we can plan for being wrong is to consider the various events that can possibly occur and develop contingency plans. We can play out different scenarios ahead of time and consider how we should react in these different scenarios. We can explore the various possible events and assess our situation with these different events. This is where modeling pays off. A model provides a mechanism to explore the future ahead of time. The simulation model gives us an inexpensive way to play out the different events and assess their effects on our firm.

One use for simulation models is scenario simulation—what is known as "what if analysis" or sensitivity analysis. Scenario simulation involves trying out different scenarios and asking, "What if this happens?" or, "What if that happens?" It allows us to trace through the intricate relations that make up our firm by trying different scenarios one at a time. For example, you could ask what would be the value of O&R's stock (a) if sales grow at 3% versus (b) if sales grow at 20%. Having built and tested the model, you only need to input these growth assumptions to view the forecast of the results for each scenario.

But, suppose your colleague, Ms. Grimaldi, says that more than two growth rates are possible. She says that there are many possible growth rates, including zero or even negative growth, and we should also plan for those. To placate her, you decide to run the simulation for various growth rates between −5% and +25%, which you consider the widest range possible. Now, you have the output for many of the growth rate between these extremes. You both are overwhelmed by the output and find it hard to evaluate the results. You say that you ought to look only at the more likely outcomes. She agrees, but says that she is not really sure which rate is most likely to occur. Furthermore, it is not just sales growth that is uncertain— most of the other inputs, such as costs and interest rates, are also uncertain. You should look at the various combinations of growth and expense ratios. In a fit of frustration, you answer that there are so many different scenario combinations that both of you would be overwhelmed and unable to evaluate the results. Besides, most of these combinations are very unlikely. Ms. Grimaldi smiles condescendingly, and says that you should try Monte Carlo. "Monte Carlo?" you say, "I don't have time to travel, so why talk about exotic destinations?" Again she smiles condescendingly, and says, "Monte Carlo simulation, you silly goose. Don't you know about Monte Carlo simulation?" You grimly answer Ms. Grimaldi, "Yeah, smart aleck, I've heard of it, but maybe you could provide additional enlightenment." Ms. Grimaldi starts to tell you about Monte Carlo simulation.

5.2 MONTE CARLO SIMULATION

Monte Carlo simulation extends what we have called scenario simulation by taking explicit account of the probabilities of different scenarios and thereby determining the likelihood of different outcomes. In doing so, Monte Carlo simulation provides a richer information set on which to base our decisions. Not only does it tells us the worst, most likely, and best case outcomes, tempered by a prudent consideration of the pessimistic and optimistic possibilities for each input, but it also tells us the likelihood of these and other possible outcomes.

To see how this works, suppose our model has three uncertain input variables—prices, sales, and interest rates—and that each of these variables can assume one of many possible different values. A simple scenario approach would assign a specific value to each of these variables, what we call an event, and trace out the consequences. Monte Carlo is like running a series of scenario simulations, one for each different possible event, taking into account the probability of that event occurring. It does this by assuming that each uncertain variable is a random variable with a known probability distribution. The Monte Carlo engine then randomly generates or draws many events, tracing out and tracking the consequences of each draw. The proportional number of times we draw a particular value for each of the variables approximates its respective probability. The end result is a frequency distribution of consequences that approximates the true relative likelihoods of the different outcomes.

The real value of making the random draws from probability distributions and then working the results through the model is that we can combine and explore the consequences of a number of different random variables in the system. If there are many different random variables that affect the outcomes of our decisions, it is extremely difficult to discern directly how these variables will interact. But, by simulating their random behavior we can begin to understand the combined effects of the uncertainty of various inputs and take account of the relative likelihood of the consequences.

To see how Monte Carlo simulation works, we will look at a simple example. First, we will set the stage with scenario simulation and then we add Monte Carlo simulation to the mix. Suppose our company currently has sales of $1000 and we expect them to grow 6% in the coming year. Operating costs are expected to be 80% of sales. Then next year's expected EBITDA will be $1000 \times (1.06) \times (1 - 0.80) = \$212$. However, since we recognize that our forecasts are not entirely accurate, we want to get an idea of the range of possible values for EBITDA. From past experience, we think that the slowest sales growth we are likely to experience is 3%, and the most rapid is 9%. We do not think that operating expenses will exceed 85% of sales, and they could be as low as 75%. If we consider the nine possible values associated with all the combinations of the three different

Operating	Sales Growth		
Expense	3%	6%	9%
75%	$258	$265	$273
80%	$206	$212	$218
85%	$155	$159	$164

EXHIBIT 5.1   EBITDA with various growth rates and operating expense ratios.

extremes of growth and costs, we see a range of possible levels for EBITDA for the coming year. These are shown in Exhibit 5.1.

EBITDA ranges from $155 with low growth and high expense to $273 with high growth and low expense, and an apparently most likely outcome in the middle of the range at $212. Scenario analysis gives us an idea of the different outcomes, but it does not tell us much about how likely these different combinations are. More disturbingly, as we are contemplating this range of outcomes, Ms. Grimaldi looks over our shoulders and tells us that we need to try Monte Carlo simulation to gain insight into the distribution of next year's EBITDA. To placate her, we will try out Monte Carlo.

In Monte Carlo simulation, we view variables as being drawn randomly from their own probability distributions. The result of each draw is run through the spreadsheet model, producing a different spreadsheet for each random draw. With many random draws from the probability distributions, we essentially have many different spreadsheet results with the relative frequency of the results based on the probabilities. For example, suppose we make one random draw from the distributions for growth and expense. Suppose that, by chance, the growth rate is −2% and the expense ratio is 90%. Then the resulting EBITDA forecast is $98. This is just one out of many possible outcomes that might occur. Certainly, it would be regarded as a disastrous outcome, but then we need to recognize that negative growth and very high expenses are not likely to occur together very often. Of the many spreadsheet results possible, having EBITDA this low will occur on very few spreadsheets. Monte Carlo allows us to consider the frequency of occurrences.

The crucial inputs for our simulations are the probability distributions of the relevant random variables. In our example, the rate of sales growth and the proportional operating costs are the random variables. After examining past data we have concluded that sales growth should be

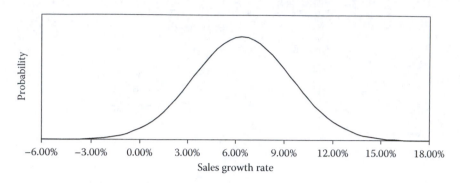

EXHIBIT 5.2 Probability distribution of sales growth.

modeled from a normal distribution with a mean of 6% and a standard deviation of 3% per year. A graph of this frequency distribution is shown in Exhibit 5.2. We also think that operating costs are normally distributed with an expected value (mean) of 80%, with standard deviation of 5%. This distribution is shown in Exhibit 5.3.

Armed with these distributions, we want to see what the distribution of EBITDA would look like. We can make draws from each of the input distributions and trace out the results for EBITDA. To do this, we will use @Risk software, which is an add-in for Excel® that allows us to make these random draws from the specified probability distributions. Using @Risk, we make 1000 random draws, so we have 1000 different values for EBITDA that are distributed as shown as a histogram in Exhibit 5.4. The histogram shows the relative frequency of the EBITDA outcomes for the 1000 random draws; it represents the simulated probability distribution of EBITDA.

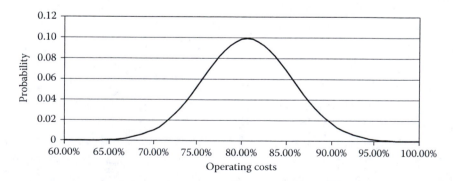

EXHIBIT 5.3 Probability distribution of percent operating costs.

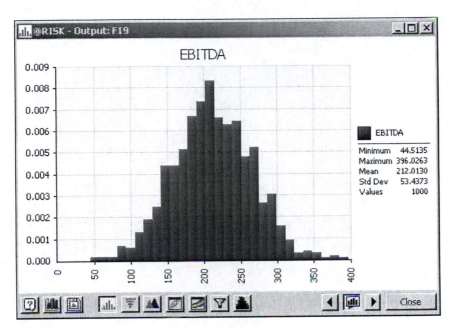

EXHIBIT 5.4   Frequency distribution of EBITDA based on growth and cost uncertainty. 1000 random draws.

The EBITDA values actually drawn ranged from a low value of 44.5 to a high of 396. The mean of the distribution was 212 and the standard deviation was 53.4. The histogram gives us a good idea of the likelihood of different possible EBITDA outcomes because it takes account of the uncertainty of the growth rate and operating costs. This is a richer information set than we got from scenario simulation. Although scenario simulation gave us an idea of the range of outcomes, it did not give us a good idea of the probabilities of the outcomes. Nor did scenario simulation provides a good estimate of the full range of outcomes, because we considered only three possible values for growth and cost, and these values were not at the extreme ends of their ranges. With Monte Carlo sampling we get more information about what might happen.

5.3 THE EBITDA MODEL AND @RISK

Our simple EBITDA problem demonstrated the idea of Monte Carlo simulation and let us see some of the output. Now, we take a closer look at the model and see how @Risk was used. Exhibit 5.5 shows our simple EBITDA model set up to perform as an @Risk Monte Carlo simulation. The input to

	A	B	C	D	E	F	G	H	I	J
1										
2				Simulation of EBITDA						
3				With Random Sales Growth and Operating Costs						
4										
5										
6		Probability Assumptions								
7				Distribution	Parameters		Actual			
8				Type	Mean	Standard	Value	=RiskNormal(E10,F10)		
9						Deviation				
10		Sales Growth %		Normal	6%	3%	6.0%			
11		Operating Costs %		Normal	80%	5%	80.0%			
12								=RiskNormal(E11,F11)		
13										
14				Period	0	1				
15							=E16*(1+G10)			
16			Sales		1,000	1060				
17			Operating Costs		800	848				
18							=F16*G:1			
19			EBITDA		200	212				
20										
21							= F16-F:7			

EXHIBIT 5.5   EBITDA model with uncertain growth and cost.

our model is the initial (period 0) sales level of $1000 (cell E16), the sales growth rate (G10), and operating cost (G11). The model in lines 16–19 is

	Cell formulas
$Sales_1 = Sales_0(1 + Growth)$	$F16 = E16 \times (1 + G10)$
$Operating\ Costs_1 = Sales_1(Operating\ Costs\ \%)$	$F17 = F16 \times G11$
$EBITDA_1 = Sales_1 - Operating\ Costs_1$	$F19 = F16 - F17$

We have assumed that the values of two variables in this model—sales growth and expense in cells G10 and G11—are uncertain. These cells contain @Risk formulas that define these variables as random, with values to be drawn from normal probability distributions. The form for the @Risk normal probability function is =RiskNormal(mean,standard deviation). The formula in G10 is =RiskNormal(E10,F10), where cells E10 and F10 contain our estimates for the mean growth rate and the standard deviation of the growth rate. For operating costs we have G11 =RiskNormal (E11,F11), where E11 and F11 are the mean and standard deviation of the percentage operating costs.

Putting the @Risk distribution formula in a cell makes the contents of that cell a random variable that will be drawn from the specified distribution. On the first iteration of the simulation, a single random draw will be made from each of the specified distributions (growth and cost) in the model. These random values are inserted in the cells (G10 and G11), and the results fed through the rest of the model. This is one random draw or iteration, so it is one spreadsheet result. A second random draw will be made with the new results fed through the model. We get as many different spreadsheet results as the number of draws we specify with the simulation software. If we set the sample size (number of iterations) at 1000, we will make 1000 draws from each distribution in the model, so it is as if we have 1000 different spreadsheets, each with a different value for each of the random variables. Out of the 1000 draws, values that are unlikely will be drawn infrequently, and values with high probabilities will be drawn more frequently. The end result is as shown in the frequency distribution shown in Exhibit 5.4. In this case, we have 1000 different EBITA values, with most occurring in the middle, near the mean, or expected value, of EBITDA, and relatively few occurrences that are at the extremes, far from the middle of the range.

Monte Carlo simulation results are not a magic bullet or panacea. Knowing the shape of the distribution of EBITDA or any other output variable does not magically solve our decision problem. However, it does provide us with more information about what might happen. With this additional information, we may be able to make a better decision because we have a better idea of which outcomes are likely and which are unlikely. In this example, we know a little more about the range of EBITDA and how likely it is to be at any given level. Such information gives us some confidence about what might happen.

One way to measure that confidence is with confidence intervals. If we know the details about the distribution of outcomes, we can determine the likelihood of the outcomes falling within a specified range. For example, having run our EBITDA simulation, we now have estimates of the parameters that describe the probability distribution of EBITDA. These parameters are the mean and the standard deviation of EBITDA, which are 212 and about 53, respectively. Assuming that EBITDA is normally distributed,* we can use our knowledge of the normal distribution to calculate or confidence interval for EBITDA. We know that 95% of the time, a normally distributed random variable will occur within the interval,

---

* EBITDA is normally distributed in this case because both the distributions in the model were normal. However, in other cases, our input distributions will not necessarily be normal. Moreover, when we have more complex models, the output may not have the same distribution type as the inputs. In these more general cases, we will have to examine the output statistics to determine the type of distribution and estimate a confidence interval.

mean ±1.96σ. Applying this, we have a 95% confidence limits for EBITDA of 212 ± 1.96(53) = 212 ± 103.9; that is, the lower limit is 212 − 103.9 = 108.1 and the upper limit is 212 + 103.9 = 315.9. So we estimate that there is a 95% chance that EBITDA will fall between 108.1 and 315.9. Conversely, there is just a 2.5% chance that EBITDA will be less than 108.1 and a 2.5% chance it will exceed 315.9.

An example of another way in which information about the distribution might help us is in evaluating our ability to meet our obligations. Suppose that in the next year we will have to make debt payments of $150, and the funds for the payment will have to come from earnings (EBITDA). What is the likelihood that we will not be able to make the payment? With information about the distribution of EBITDA we can calculate the probability that EBITDA will be less than the required debt payment: Pr{EBITDA < 150}. The debt payment of 150 is 1.17 standard deviation units below the EBIT mean:

$$\frac{150 - \text{mean}}{\sigma} = \frac{150 - 212}{53} = -1.17.$$

Using a standard normal probability table, we determine that the probability of a standard normal variable, x, occurring 1.17σ below the mean is

$$Pr\{x \le \text{mean} - 1.17\sigma\} = 12\%.$$

In this case, it helps us understand our situation when we can estimate that there is about a 12% chance that we will be unable to service our debt from our earnings. Alternatively, as was the case for confidence intervals, simulation data can be used to estimate the simulated likelihood of an adverse event.

Now that we have introduced the basic idea of Monte Carlo simulation, we will provide a short primer on the software we use for Monte Carlo modeling.

5.4 MONTE CARLO SOFTWARE: @RISK

The Monte Carlo software we use in this text is @Risk, for which free student access is provided with this text.* We use @Risk in this chapter,

---

* @Risk is available from Palisade Corporation (www.palisade.com). There are two competing Monte Carlo software packages that are very similar, @Risk and Crystal Ball. Both are add-ins for Excel. Each one has some advantages and disadvantages, although we tend to have a slight preference for @Risk. That is why we have used it in this text. Generally, both packages are useful and easy to use.

beginning with a quick overview of the software. Readers desiring a more detailed explanation should consult the extensive @Risk literature. The books by Wayne Winston, cited in the references at the end of the book, are particularly helpful.

@Risk enables Excel to carryout Monte Carlo simulation by providing functions that are the probability distributions from which we make our random draws. Of course, there is a lot of user-friendly software surrounding the basic functions, but the essence of @Risk is a portfolio of probability distribution functions with which to describe and draw random variables. These functions can be entered into a cell via an @Risk point and click dialog box, or they can be entered much like any other Excel function. Most of our discussion will focus on entering the distribution functions manually. A short list of the common functions includes

Normal distribution	=RiskNormal(mean,standard deviation)
Uniform	=RiskUniform(minimum,maximum)
Triangular	=RiskTriang(minimum,most likely,maximum)
Binomial	=RiskBinomial(draws,probability of success)
Lognormal	=RiskLognorm(mean,standard deviation)

When you enter the function in the appropriate cell in your spreadsheet model, it makes the content of that cell the random variable that will be drawn from the distribution. There are various ways to insert parameters into the functions. Exhibit 5.6 shows several of them. Parameters can be entered directly as numbers (as in Example A), as cell references (as in Example B), or even as functions of cell references and numbers (as in Example C). The software provides the model builder with remarkable flexibility.

Cell E6 in Example A shows EBITDA being calculated with the mean and standard deviations specified directly in the =RiskNormal function. In Example B, the mean and standard deviation are set as constants in cells E10 and E11, and the =RiskNormal function in E13 references those cells. Example C shows a more complicated entry of the parameters. The price is set as a constant in E17. The quantity in E18 is based on a demand function with a random error drawn from a uniform distribution ranging from −5 to +7 units. Sales in E19 is a normal random variable with a mean that is the product of price (E17) and the random quantity (E18). The standard deviation is also the product of price and quantity. This makes both the mean and standard deviation of sales random parameters. % Operating

	A	B	C	D	E	F	G	H	I
1									
2		**Alternative Ways to Enter Parameters in @ Risk Functions**							
3									
4	A	Parameters Entered Directly Into Distribution Function							
5									
6			EBITDA		212	=RiskNormal(212,53)			
7									
8	B	Parameters Specified in Other Cells							
9									
10			Mean		212				
11			Standard Deviation		53				
12									
13			EBITDA		212	=RiskNormal(E10,E11)			
14									
15	C	Parameters Computed Elsewhere & Random Parameters							
16						=200 - 10*E17 + RiskUniform(-5,7)			
17			Price		10				
18			Quantity Sold		101	=RiskNormal(E17*E18,0.05*(E17*E18))			
19			Sales		1010				
20			% Operating Costs		81.7%				
21						=RiskTriang(0.75,0.8,0.9)			
22			EBITDA		185.1667				
23									

EXHIBIT 5.6   Parameter entry in @Risk distribution functions.

Costs (E20) are drawn from a triangular distribution. The result is that EBITDA is random variable based on draws from a uniform, a normal, and a triangular distribution, where some of the parameters of the distributions are themselves random variables. This shows some of the flexibility that is available for portraying the randomness and uncertainty of variables in @Risk.

Another way to enter a distribution function in your model is to use @Risk's graphic "Pop Up Window" that is accessed via the Define Distribution button ⌨. This graphic approach allows you to see a graph of the distribution and set the parameters in a dialog box. For example, Exhibit 5.7 shows the Define Distribution dialog box that pops up when you click the Define Distributions icon. This will allow us to define the distribution for cell E13 in our model because that was the highlighted cell when we clicked the icon. You can click on the picture of the distribution you want to use, and then click on the Select Distribution button at the bottom. This brings up the Define Distribution box shown in Exhibit 5.8.

In Exhibit 5.8, the center panel shows a picture of the distribution and the left side shows the parameters entered to describe the distribution. The mean box (μ) references the model cell where the mean is located (E10),

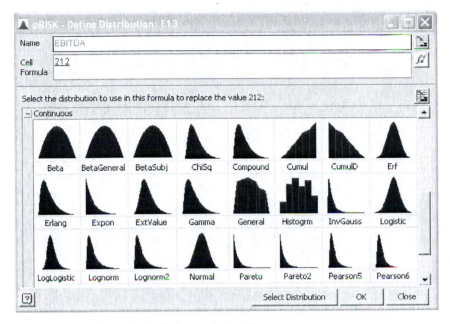

EXHIBIT 5.7 Define distributions dialog box.

and the standard deviation box (σ) shows where its parameter is located (E11). You can change the entries in this section, including the type of distribution, the parameters (mean and standard deviation) and make other adjustments such as truncate the distribution. For example, you can

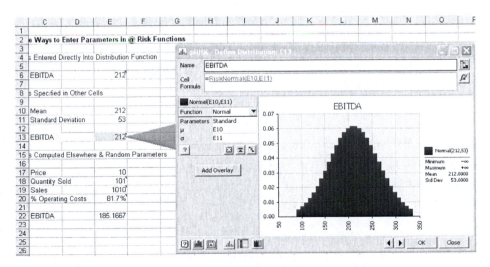

EXHIBIT 5.8 @Risk define distribution dialog box.

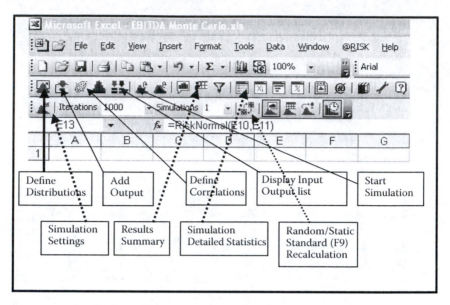

EXHIBIT 5.9   @Risk tool bar.

enter parameters directly in the parameter box [i.e., as the number (212)], change the cell reference, or even enter a formula that would compute the parameter. If you want to change the type of distribution, say from normal to uniform, you click the Function arrow to see the list distributions from which you can choose. After entering the parameters, click the OK button and the distribution formula is entered into the cell.

@Risk is controlled with the set of icon buttons on the @Risk tool bar at the top of the Excel screen as shown in Exhibit 5.9. The crucial buttons to get started are the Define Distributions, Add Output, Simulation Settings, and Start Simulation. Once the input probability distributions are defined, only a few short steps remain before running the simulation.

5.4.1 Steps for Running an Initial Simulation

Returning to the example in Exhibit 5.5, follow the next steps to carry out the simulation.

1. Tell the simulation which variables are the object, or target of the simulation. This is the variable for which you want to see the simulation statistics, frequency distribution, and so on, which in this case is EBITDA.

(a) Select the cell or range of cells that are the target of the simulation: in this case, the EBITDA cell F19.

(b) Click the Add Output button, ![icon], and give the output a name—EBITDA. @Risk will modify the formula in F19, so it becomes =RiskOutput("EBITDA")+F16−F17.

2. Specify the sample size, or number of iterations for the simulation. Clicking the Simulations Settings button brings up the Simulation Settings dialog box shown in Exhibit 5.10.

The default #Iterations is 1000. This is fine for your initial run. For important problems that are sensitive to extreme outcomes, you probably will need a larger sample. The Sampling tab allows you to choose the options for Simulation Type. Latin Hypercube sampling is the default with @Risk because it is a more efficient way to sample. We will stick with that.

At the bottom of the box you can choose "Random Values," or "Static Values." This determines what is shown in your spreadsheet when the

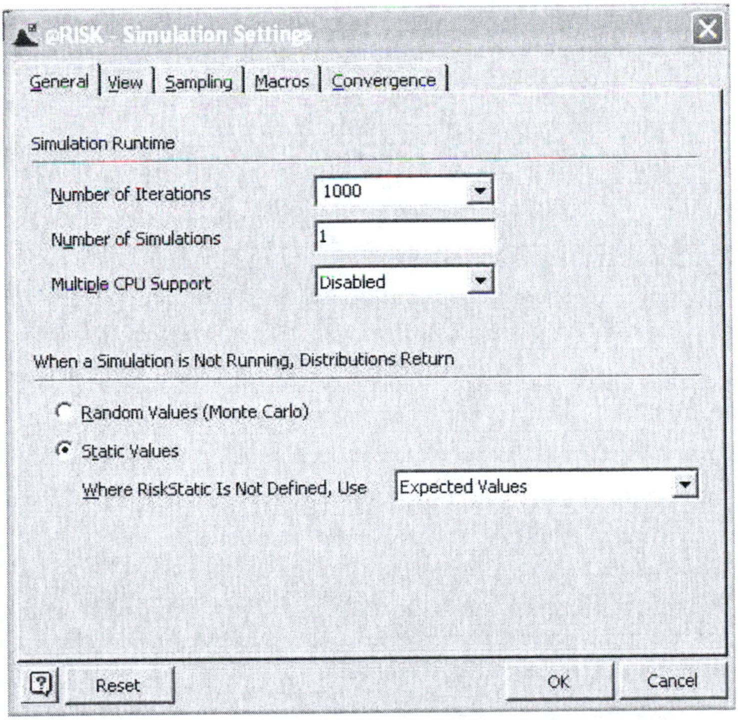

EXHIBIT 5.10   @Risk simulations settings dialog box. Iterations.

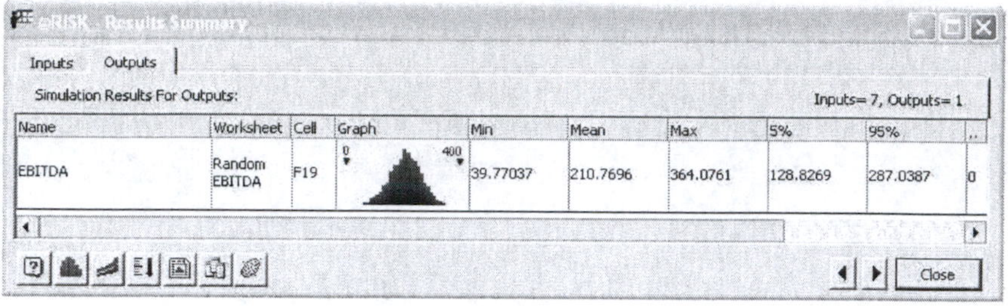

EXHIBIT 5.11   Summary statistics from simulation run.

simulation ends. If you choose Random Values, @Risk returns the last number that was drawn.* If you choose Static with the Expected Value as shown, the number shown in the spreadsheet is the expected (mean) value.

All that remains is to run the simulation and analyze the output. The simulation is started by clicking the Start Simulation button: . When @Risk finishes the specified number of iterations, it automatically opens the Results Window and shows the frequency distribution of the outcomes. You can see more of the results by clicking the "Results Summary" icon, which brings up a display like Exhibit 5.11.

The Results Window reports the summary statistics for the input and output variables for the simulation: the minimum, mean, maximum, and the upper and lower limits of the confidence intervals. For more detail, click the Detailed Statistics icon ![]. This brings up the output shown in Exhibit 5.12.

In addition to the material shown in the Summary Statistics window, the Detailed Statistics window shows standard deviation, skewness, kurtosis, and the variable values at percentile points in 5% intervals.

If your @Risk did not automatically show the frequency distribution, you can click the "Browse Simulation Results" icon, ![], to bring up a frequency distribution for the simulation as seen earlier in Exhibit 5.4.

5.4.2 Putting the Output into the Spreadsheet

Although all this output is interesting in its own right, its sheer quantity can be overwhelming and not always germane to the problem under consideration. Drilling down through the various menus also can be confusing. Sometimes, too, a permanent record of each simulation will be useful. @Risk addresses

---

* For example, suppose a cell contains a normally distributed random variable with a mean of 10. The Expected Value option would return a 10 in this cell; the Monte Carlo option would return a number drawn from the normal distribution defined in that cell.

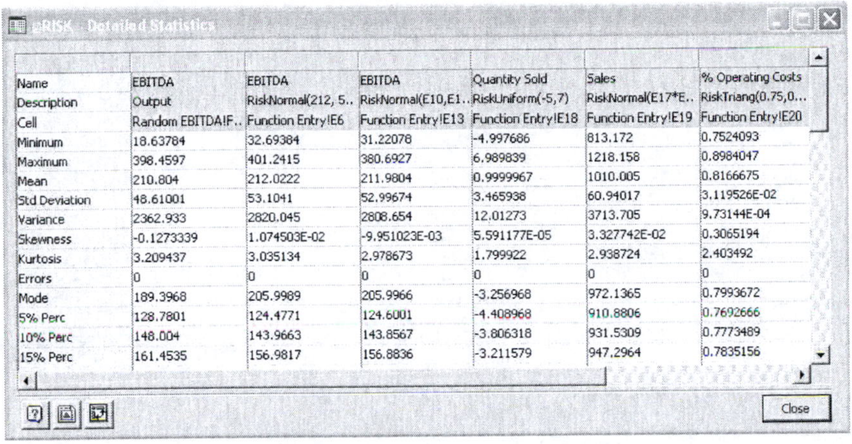

Name	EBITDA	EBITDA	EBITDA	Quantity Sold	Sales	% Operating Costs
Description	Output	RiskNormal(212, 5..	RiskNormal(E10,E1..	RiskUniform(-5,7)	RiskNormal(E17*E..	RiskTriang(0.75,0...
Cell	Random EBITDA!F..	Function Entry!E6	Function Entry!E13	Function Entry!E18	Function Entry!E19	Function Entry!E20
Minimum	18.63784	32.69384	31.22078	-4.997686	813.172	0.7524093
Maximum	398.4597	401.2415	380.6927	6.989839	1218.158	0.8984047
Mean	210.804	212.0222	211.9804	0.9999967	1010.005	0.8166675
Std Deviation	48.61001	53.1041	52.99674	3.465938	60.94017	3.119526E-02
Variance	2362.933	2820.045	2808.654	12.01273	3713.705	9.73144E-04
Skewness	-0.1273339	1.074503E-02	-9.951023E-03	5.591177E-05	3.327742E-02	0.3065194
Kurtosis	3.209437	3.035134	2.978673	1.799922	2.938724	2.403492
Errors	0	0	0	0	0	0
Mode	189.3968	205.9989	205.9966	-3.256968	972.1365	0.7993672
5% Perc	128.7801	124.4771	124.6001	-4.408968	910.8806	0.7692666
10% Perc	148.004	143.9663	143.8567	-3.806318	931.5309	0.7773489
15% Perc	161.4535	156.9817	156.8836	-3.211579	947.2964	0.7835156

EXHIBIT 5.12   Detailed statistics output.

these issues with its Excel Reports option. Excel Reports allows the user to place the @Risk output into a spreadsheet so that it can be printed or manipulated in the spreadsheet. You access this option with the Excel Report icon, ⬛, which brings up the Excel Reports dialog box shown in Exhibit 5.13.

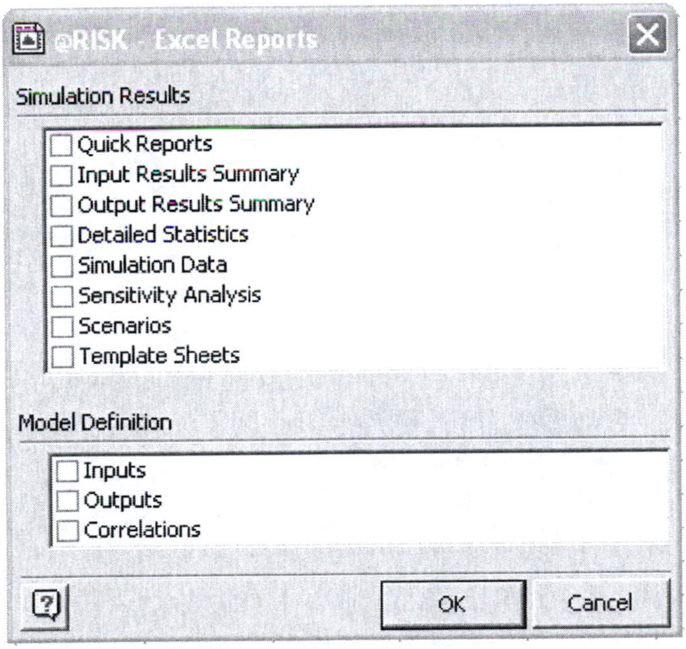

EXHIBIT 5.13   Excel reports dialog box.

Checking off the options such as "Quick Reports" and clicking OK will place the report in a new spreadsheet. The options that you select determine which output report is generated and inserted into Excel. Quick Reports was the option that generated the Excel output shown in Exhibit 5.14.

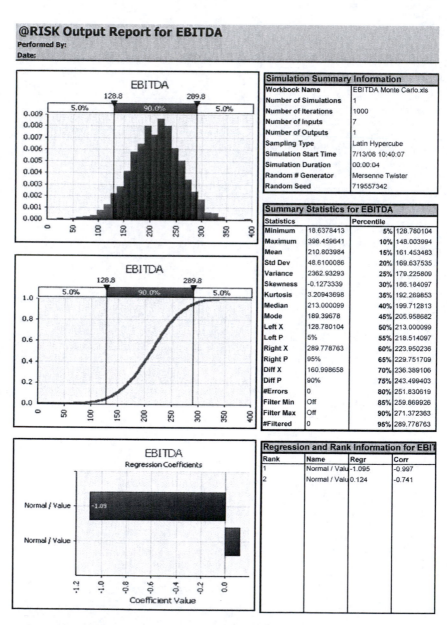

EXHIBIT 5.14   Quick output report from Excel sheet.

5.4.3 Correlating the Variables

The two random variables in our EBITDA model, growth of sales and operating cost, were drawn independently. That is, the operating cost was drawn from its normal distribution without regard for the value of growth that was drawn. It may be more realistic to assume that these variables are correlated. For example, we would expect that sales would grow more rapidly when the economy is doing well. We also would expect that our costs would tend to be higher during periods of prosperity. We can build this kind of linkage into our model by correlating the random variables.

Two random variables are correlated when there is a systematic relationship between the values drawn from each of their distributions. Suppose that whenever the economy declines, sales decline, and raw material prices also tend to be lower. In that case, we say that sales and prices seem to be correlated with the economy, and sales and prices are correlated with each other.

We measure correlation with the correlation coefficient. The correlation coefficient is an index number between −1 and +1; +1 means perfect positive correlation and −1 means perfect negative correlation. Other values between −1 and +1 represent the different degrees of imperfect correlation. Positive correlation (between 0 and +1) means that two variables tend to move with each other, or, more accurately, when one variable is drawn randomly from (say) the upper end of its probability distribution, the other variable also will tend to be drawn from the upper end of its distribution. Negative correlation (between −1 and 0) is when the variables tend to move in opposite directions, so that when one is high, the other tends to be low.

@Risk allows us to specify the correlation coefficients between random variables in our model. This is done either by filling in a matrix of correlation coefficients or with the correlation function =RiskCorrmat. To see how this is done in our EBITDA example, assume that the correlation between sales growth and the operating cost ratio has been estimated to be +0.80. That is, growth and costs are highly positively correlated, but not perfectly so. Most of the time when growth is high, the cost ratio will also be high. However, with imperfect correlation, growth and costs will occasionally move in the opposite directions.

To put the correlation into our model, click the "Define Correlations" icon, 🖼. The correlation window will show, as in Exhibit 5.15. This exhibit is already filled in with the correlation coefficient of 0.80. When it first pops up, it shows a blank correlation matrix for you to fill in. To add the

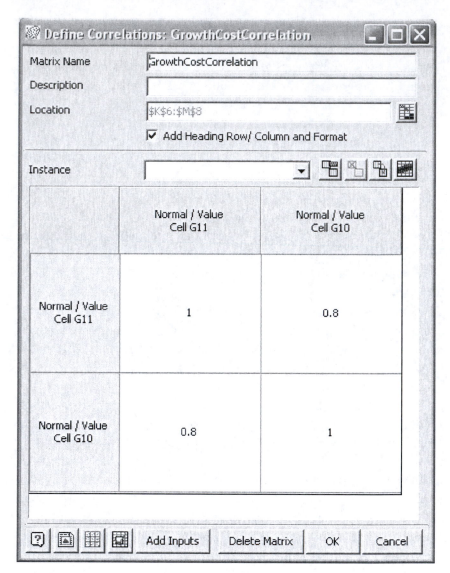

EXHIBIT 5.15   Correlation dialog box in model window.

variables that you want to correlate, click the "Add Inputs" button at the bottom. This brings up the Add Inputs box shown as Exhibit 5.16. Select the cells with the probability distribution functions in them—G10:G11 in this case, click OK. This inserts these cell addresses into the correlation matrix. You can then double click into the matrix to fill in the correlation coefficients—the 0.80 shown in Exhibit 5.15. Click OK, and you are ready to go with the correlated model.

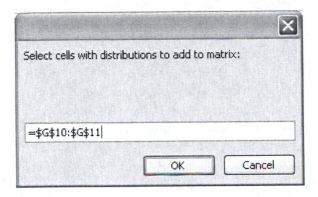

EXHIBIT 5.16   Add inputs box.

At this stage, what @Risk has done is (1) inserted a matrix of correlation coefficients into the spreadsheet; and (2) modified the probability distribution formulas for the correlated variables. The correlation matrix shown in Exhibit 5.17 has been placed in cells K6:M8, which is the cell location indicated in the "Location" line of the Define Correlations box. The probability distribution functions have been modified by adding RiskCorrmat(·) functions.

Before these changes, the distribution formulas for growth (G10) and cost (G11) were the =RiskNormal(mean,standard deviation) functions. These have now been modified to be

Growth(G10): =RiskNormal(E10,F10, RiskCorrmat(GrowthCostCorrel,2))
   Cost(G11): =RiskNormal(E11,F11, RiskCorrmat(GrowthCostCorrel,1))

The =RiskCorrmat function specifies the correlations between the variables by referencing the correlation matrix that is put into your model. In this case the name of the correlation matrix is GrowthCostCorrel that we entered in the Define Correlation box. You can provide your own name for the matrix in the correlation dialog box. As an alternative to using the correlation dialog, you can put the =RiskCorrmat function directly into your model. Consult the @Risk documentation for more detailed information about the function and how to use it.

@RISK Correlations	Normal / Value in $G$11	Normal / Value in $G$10
Normal / Value in $G$11	1	0.8
Normal / Value in $G$10	0.8	1

EXHIBIT 5.17   Correlation matrix "GrowthCostCorrelation" inserted into cells K6:M8.

5.4.3.1 Model Results with Correlated Variables

Now we will take a look at the results with the correlations. Comparing the output shown in Exhibit 5.18 with the uncorrelated case shown in Exhibit 5.4, we find that when cost and growth are positively correlated, the distribution of EBITDA changes. Mean EBITDA is now 210.8 versus 212 before. The distribution is tighter, with a smaller standard deviation (48.6 versus 53.4) and a narrower range between the minimum and maximum. That is, with positively correlated cost and growth, the expected EBITDA has decreased slightly, but there is less uncertainty. The decreased uncertainty is because when sales are low, costs are also low, so earnings do not decline as much as they would otherwise. If the correlation was negative, we would get the opposite result: increased uncertainty because when sales are low, costs are more likely to be high and earnings will be even lower than otherwise. Another difference in this case is the skew of the distribution. The skewness parameter is −0.1273. This indicates that the distribution is slightly skewed to the left, or negative direction so that here is a higher probability of low outcomes than we would get if the distribution was not skewed (skewness parameter of 0).

This brief introduction to @Risk should be sufficient for you to begin to run a Monte Carlo simulation. There are many additional options and a

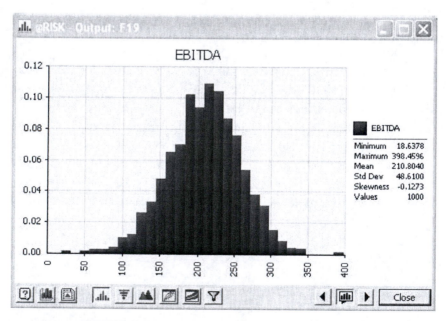

EXHIBIT 5.18   EBITDA output with correlated growth and cost.

lot of capability in @Risk that we have not discussed here. Obviously, you need to consult the @Risk documentation to learn more and use it to its full capability. The next section looks at some additional examples and applications of Monte Carlo simulation.

5.5 USING MONTE CARLO SIMULATION FOR CAPITAL INVESTMENT DECISIONS

An important application of Monte Carlo simulation is in the evaluation of capital investment projects. In normal capital budgeting analysis we discount the future cash flows to determine the NPV of the investment project. However, like anything else that is projected to happen in the future, we do not really know what the cash flows will actually turn out to be. Since there are many different possible cash flows that we could earn, there are potentially many different possible NPVs, depending on which particular cash flows actually occur.

Given the many different NPVs that a project might assume, we need some way to simplify the information so that we can evaluate the project. The risk-adjusted discount rate approach (RAD approach) is one way to do this. With the RAD approach, we estimate the expected or most likely cash flows and discount the flows at the required return for the project. If the cash flows are more uncertain, we use a higher discount rate to penalize the project for its uncertainty. For example, if two projects have the same expected cash flows, but the second project's cash flows are more uncertain, we adjust for that uncertainty by using a higher discount rate. The higher discount rate generates a lower NPV for the riskier project, so we conclude that the more risky project is less valuable than the less risky project. In this way we are adjusting for risk with the discount rate—hence the method's name, RAD.

To apply the RAD approach, we need to be able to estimate the investment's expected, or mean cash flows. Monte Carlo simulation can help us to do that. In addition, Monte Carlo simulation can give us information about the distribution of possible NPVs so that we can make better decisions. For example, the distribution might tell us that even though the expected NPV is attractive, there is a 30% chance that it will be negative. The following example will help us to understand some of these issues.

5.5.1 Example: Capital Investment for Placidia Corporation

The Placidia Corporation is considering a proposal for an investment in new machinery. The project would require an immediate investment in machines of $8000 (figures are in thousands of dollars) and a $2000 commitment of additional NWC. The total capital investment is $10,000 at $t = 0$. If the project is undertaken, Placidia will be able to produce and sell more products, so its sales will increase. It is *expected* that incremental annual sales from the project will be $15,000 for five years. At the end of the fifth year, the equipment will be worn out, and the project will end. Obviously, the production of new output will involve operating costs. It is *estimated* that operating costs will be 75% of sales. At the end of the project, the investment in NWC will be recovered, and the worn out machinery will be sold. The NWC recovery will be $2000, and the sale of the machinery is *expected* to generate about $4000 after taxes. Assume that we use the weighted average cost of capital (WACC) of 9.6% as the discount rate for evaluating this project.

5.5.1.1 Conventional NPV Analysis

Using the information about the project, the income for the project is estimated as

Sales	$15,000
Operating Costs @ 75%	11,250
EBITDA	3750
Depreciation	1600
EBIT	2150

When we use WACC* as the discount rate we calculate the incremental operating cash flow with the formula

$$EBIT(1 - T) + Depreciation.$$

Assuming a tax rate of 40%, the annual incremental cash flow in each of the five years is

$$\$2150(1 - 0.40) + 1600 = \$2890.$$

The initial cost of the project, including both the purchase of the equipment and the investment in NWC, is $10,000. When the project ends in five years, terminal cash flows will be generated from the recovery of

---

* We will discuss capital budgeting in Chapter 10, where the use of WACC in capital budgeting is explained in greater detail.

investment in NWC and the sale of the machinery. This terminal flow is $2000 for the NWC and $4000 for the salvage of the machinery, for a total of $6000.

For this simple project, the evaluation is easy. Letting PV(.) denote the present value of the flow, the NPV is calculated as follows:

$$NPV = -\text{Initial Cost Outlay} + PV(\text{Operating Flows}) + PV(\text{Terminal Flow})$$
$$= -10,000 + PV(\$2890 \text{ per year for 5 years @ } 9.6\%)$$
$$+ PV(\$6000 \text{ @ } t = 5)$$
$$= -10,000 + 2890(3.8298) + 6000(0.6323)$$
$$= -10,000 + 14,862$$
$$= \$4862.$$

(The 3.8298 is the present value of an annuity of $1 for five years at a 9.6% discount rate, and the 0.6323 is the present value of $1 received at the end of five years at 9.6% discount.)

Given its positive NPV, this project is acceptable under traditional financial management practice. However, the positive NPV tells us little about the project's uncertainty. The project's sales, operating costs, and terminal cash flows are unknown. We used their expected values in calculating NPV, but if actual sales and terminal cash flows are lower and operating costs higher than anticipated, the project's actual NPV could be substantially smaller than expected, perhaps even negative. Monte Carlo simulation allows us to dig deeper into the uncertainty underlying our assumptions.

5.5.1.2 Monte Carlo Simulation of the Placidia Project

After discussions with Placidia's sales manager, production manager, and engineers, we have assembled the following information about sales, costs, and terminal flows.

5.5.1.2.1 Sales.   Future sales are expected to be close to $15,000 per year, with actual values becoming less and less certain for the more distant periods. This means that the likelihood that actual sales will come in higher or lower than $15,000 increases each successive period. We model this uncertainty by assuming that sales will be normally distributed with a mean of 15,000 with standard deviations that increase each period. The standard deviations are 1000, 1050, 1100, 1150, and 1200 for periods 1–5, respectively. Sales are specified for each year using the Define Distributions dialog box, as shown in Exhibit 5.19.

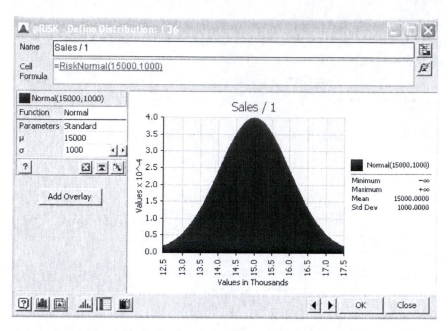

EXHIBIT 5.19   The @Risk Define Distributions dialog box: Normal distribution of sales in the first year.

5.5.1.2.2 Operating Costs.   Our expectations for operating costs are less clearly defined than those for sales. For sales, we had a fairly strong belief that the observations would center around $15,000, hence we used the normal distribution. For operating costs, we have no expectations for a central tendency. Instead, we have strong beliefs concerning the minimum and maximum values these costs might assume: 70% and 80% of sales, respectively—but we have no clear notion for what the actual value within these bounds will be. Therefore, we model operating costs as a uniformly distributed random variable with minimum and maximum values of 0.70 and 0.80, respectively. This distribution is shown in Exhibit 5.20.

5.5.1.2.3 Salvage Value.   The salvage value of the machinery at the end of the project is relatively uncertain. We expect it to be $4000, but it could be as little as $2000 or as much as $4500. Further, we have no clear expectation for its distribution. The triangular distribution is frequently used to model this type of uncertainty. We use it here, specifying the minimum, most likely, and maximum values of 2000, 4000, and 4500, respectively. This distribution is shown in Exhibit 5.21.

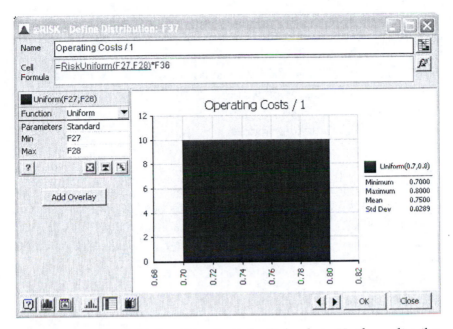

EXHIBIT 5.20 @Risk Define Distributions dialog box: Uniform distribution of percent operating costs.

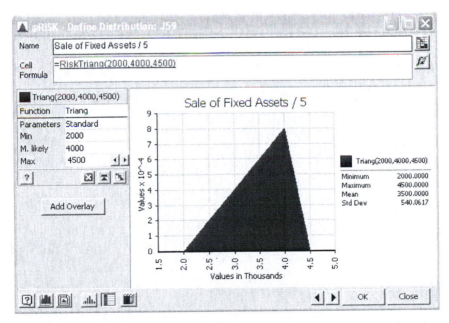

EXHIBIT 5.21 @Risk Define Distributions dialog box: Triangular distribution of salvage value.

Having defined the uncertainty in our expectations for the future, we are ready to develop a model that incorporates and simulates its effect on the project's NPV and internal rate of return (IRR). Exhibit 5.22A shows the section of our NPV model in which we specify the parameters for the model—our assumptions about the state of the firm, the state of the economy, and our forecasts of the future. Lines 22–32 contain the parameters of the probability distributions.

Exhibit 5.22B shows the computational section of the model with the income statement, cash flows, NPV, and IRR. The random variables are sales (F36:J36), operating costs (F37:J37), and Sale of Fixed Assets (salvage value) (J59). Each of these cells has the @Risk distribution formulas as pointed out in the comment boxes. NPV is in E71 and IRR is in E72. Both of these cells are specified as the @Risk output cells that are the objects of our simulation.

We are now ready to run a Monte Carlo simulation. We set the number of iterations to 1000 and click the Start Simulation icon. Within seconds, @ Risk runs the simulation. It shows us the histogram of the NPV, and we choose Excel Reports as the output, which is shown in Exhibit 5.23. Note the much richer information we have to work with. Instead of a single

	A	B	C	D	E	F	G	H	I	J
1										
2					Placidia Corporation					
3					Investment Evaluation					
4										
5	Data Input									
6		Interest Rate			7.0%		Cost of Equity Capital			15%
7		Term To Maturity			5		Tax Rate			40%
8		Proportion of Cost Financed With Debt			50%		Depreciable Life			5
9										
10	Initial Investment									
11				Period:	0					
12										
13		Investment in Fixed Assets			8,000					
14		Net Working Capital Investment			2,000					
15		Total Project Investment			10,000		Costs of Capital			
16							Proportion of Debt			50.00%
17	Financing						Proportion of Equity			50.00%
18			Amount Borrowed		5,000		Cost of Debt			7.00%
19			Equity Capital Investment		5,000		Cost of Equity			15.00%
20			Project Capital Investment		10,000		WACC			9.60%
21										
22	Probability Parameters									
23		Sales								
24		Normal	Mean			15,000	15,000	15,000	15,000	15,000
25			Standard Deviation			1,000	1,050	1,100	1,150	1,200
26		Operating Costs % of Sales								
27		Uniform	Lower Limit			70%	70%	70%	70%	70%
28			Upper Limit			80%	80%	80%	80%	80%
29										
30		Terminal Value of Fixed Assets						Lower	Middle	Upper
31		Triangular							2,000	4,000	4,500
32										

EXHIBIT 5.22A NPV model. Data input section.

	A	B	C	D	E	F	G	H	I	J	K
32											
33	**Income Forecast**										
34				Period	0	1	2	3	4	5	
35											
36		Sales				15,173	17,391	14,664	15,759	14,246	
37		Operating Costs	=RiskNormal(F24,F25)			11,103	13,182	11,073	12,086	10,067	
38		EBITDA				4,070	4,209	3,591	3,673	4,179	
39		Depreciation				1,600	1,600	1,600	1,600	1,600	
40		EBIT				2,470	2,609	1,991	2,073	2,579	
41		Interest				350	350	350	350	350	
42		EBT				2,120	2,259	1,641	1,723	2,229	
43		Tax	=RiskUniform(F27,F28)*F36			848	904	657	689	892	
44		Net Income				1,272	1,355	985	1,034	1,338	
45											
46	**Cash Flow**										
47				Period	0	1	2	3	4	5	
48		To Invested Capital									
49		EBIT (1 - Tax Rate)				1,482	1,565	1,195	1,244	1,548	
50		+ Depreciation				1,600	1,600	1,600	1,600	1,600	
51		= Earnings Flow to Invested Capital				3,082	3,165	2,795	2,844	3,148	
52		Investment in Net Working Capital		2,000							
53		+ Investment in Fixed Assets		8,000							
54	-	= Project Investment		10,000							
55		= Cash Flow to Invested Capital		(10,000)		3,082	3,165	2,795	2,844	3,148	
56											
57		Terminal Value									
58		Net Working Capital Recovery				=RiskTriang(H31,I31,J31)				2,000	
59		Sale of Fixed Assets								3,360	
60		Terminal Flow Net of Taxes								5,360	
61									=J55+J60		
62	**Investment Evaluation**										
63				Period	0	1	2	3	4	5	
64		Invested Capital									
65		Total Net Flow to Invested Capital			(10,000)	3,082	3,165	2,795	2,844	8,508	
66		PV(Initial Cost)			(10,000)						
67		PV(Op Flow)			11,532	=NPV(J20,F51:J51)					
68		PV(Terminal Flow)			3,390						
69		PV(Op Flow + Terminal Flow)			14,921	=J60/(1+J20)^J63					
70											
71		**Net Present Value**			4,921	=RiskOutput("NPV") + E69+E66					
72		**Internal Rate of Return**			24%						
73											
74					4,921	=RiskOutput("IRR") + IRR(E65:J65)					
75											
76											

EXHIBIT 5.22B   NPV model. Computation section.

NPV of $4546, we have a frequency distribution of NPVs (top left graph in the exhibit) ranging between $2507 and $6622, with an expected value of $4546.6. The bar at the bottom of the distribution graph shows the 90% confidence interval limits, indicating that there is a 90% chance that NPV will be between $3520 and $5521. Even after taking the project's uncertainty into account, all the NPV outcomes are positive. We conclude that there does not appear to be any chance that NPV will be negative.

The bottom graph of the Output Report shows a "tornado graph" of the correlations between the random inputs and the output. The bars on each side of the centerline indicate the correlations. The purpose of this graph is to help us understand the sensitivity of the output variable, NPV, in response to the random input variables. At the top of the graph is the Sale of Fixed Assets (salvage value) with the highest correlation at 0.53. This tells us that the NPV is most sensitive to changes in the project's salvage value. Errors in

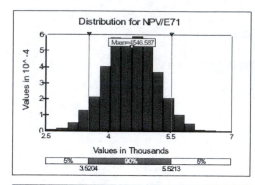

Summary Information	
Workbook Name	Placidia.xls
Number of Simulations	1
Number of Iterations	1000
Number of Inputs	11
Number of Outputs	2
Sampling Type	Latin Hypercube
Simulation Start Time	10/26/2004 14:25
Simulation Stop Time	10/26/2004 14:25
Simulation Duration	00:00:02
Random Seed	675336724

Summary Statistics			
Statistic	Value	%tile	Value
Minimum	2,507	5%	3,520
Maximum	6,622	10%	3,779
Mean	4,547	15%	3,923
Std Dev	618	20%	4,027
Variance	382318.3036	25%	4,117
Skewness	-0.029542393	30%	4,199
Kurtosis	3.089190955	35%	4,294
Median	4,554	40%	4,375
Mode	4,958	45%	4,471
Left X	3,520	50%	4,554
Left P	5%	55%	4,623
Right X	5,521	60%	4,724
Right P	95%	65%	4,806
Diff X	2,001	70%	4,894
Diff P	90%	75%	4,971
#Errors	0	80%	5,061
Filter Min		85%	5,171
Filter Max		90%	5,320
#Filtered	0	95%	5,521

Sensitivity			
Rank	Name	Regr	Corr
#1	Sale of Fixed Ass	0.553	0.517
#2	Operating Costs /	-0.383	-0.309
#3	Operating Costs /	-0.351	-0.294
#4	Operating Costs /	-0.319	-0.302
#5	Operating Costs /	-0.289	-0.306
#6	Operating Costs /	-0.266	-0.217
#7	Sales / $F$36	0.222	0.228
#8	Sales / WACC / $	0.215	0.255
#9	Sales / 2000 / $H	0.205	0.189
#10	Sales / Most Likel	0.193	0.208
#11	Sales / Maximum	0.183	0.167
#12			
#13			
#14			
#15			
#16			

EXHIBIT 5.23 @Risk Quick Output Report. Simulation results for NPV/E71.

forecasting salvage value will lead to greater errors in estimating NPV. This suggests that we may need to put more effort into forecasting the salvage value. The operating costs correlation bars are all on the left side of the centerline, indicating negative correlation. This simply means that higher costs lead to lower NPV. Of course, we already know this, but in more complex models, the link between the input and output may be less obvious, and this graph can help us to understand the relationships. Now, we will look at a different kind of application of Monte Carlo simulation.

5.5.2 Example: Success and Failure of Research and Development

This example shows you how to model a situation where the ultimate success is contingent on the successful completion of one or more intermediate phases.

Genetic Systems Corporation is engaged in the R&D of a new vaccine to treat tumors. This type of biotech research has to go through three phases of development and testing before reaching its market. In the first two phases, the prospective new vaccine undergoes development and testing. If the development and testing is successful at one phase, the product goes on to the next phase where it will face further development, refinement, and testing. If it fails the earlier phase, the product is considered a failure and does not pass on to the next phase and the project is abandoned. In this example, the third phase is the last hurdle, where the vaccine has to receive approval by the Food and Drug Administration (FDA). After that, the company can market its product where the demand for life-saving products seems to be insatiable and price inelastic.

At each phase there is uncertainty regarding the success of the phase and the costs and revenues. However, before describing these uncertainties, let us take a general look at the costs and revenues.

Phase 1 is the initial research. It will last for one year, and it is expected to cost about $10 million to pursue the initial research. If phase 1 is successful, phase 2 will proceed. Phase 2 tests the product in the lab and on animals and humans. It is expected to take about two years, and will cost about $15 million per year. Success at the testing phase will lead to phase 3, the application for FDA approval to market the drug. Phase 3 will cost about $8 million, and will take a year to get a decision. Upon receiving approval, the company will go to market with the vaccine. It is expected that the market will consist of about one million people who will purchase the product each year. The treatment will require 12 monthly injections for the year, with each injection costing the patient $5000. The company will receive about $4000 for each monthly dose, with the balance going to the doctors and hospitals that administer the medicine. Once the product reaches the market, the marginal cost of producing the vaccine is expected to be small relative to the sales price, costing only about $1000 per dose. When we convert the production costs and revenues to annual figures, the company thinks that it is most likely to earn about $36 million in gross profit each year. They expect to market the vaccine for about six years until it is displaced by a competing product.

5.5.2.1 Conventional Analysis of the NPV of Project

Exhibit 5.24 shows a spreadsheet with the forecasts of the most likely costs and revenues. Should the company be so lucky as to wend its way successfully through this process and market the product, it would appear to be a very attractive prospect.

In Exhibit 5.24 the NPV of the vaccine project is shown at the bottom in B23, based on a discount rate of 20%, which is regarded as appropriate for risky drug R&D programs. The costs of the R&D phase are shown in C7:C10, and the gross profits from the marketing program for years 5–10 are shown in D11:D16. The net cash flows for the project are summarized in column E, with the present values of costs and cash inflows in B21 and B22. The NPV would be a very attractive $26.446 million. If we stopped our analysis here, we would conclude that this is a great project, and we should proceed immediately. However, the NPV of $26.4 million assumes the product succeeds in all three phases of development—something that is by no means certain. Therefore, we need to take a closer look at the risks as reflected in the costs and probabilities of successfully completing each phase of the process.

5.5.2.2 Risk in the R&D Program

R&D programs in the pharmaceutical and biotech industries are notoriously risky. At each phase of the R&D process there is a substantial risk of failure. These risks of failure accumulate so that there is high probability that any given drug development project will fail. Some estimates suggest that only about 1 in 10 drug development projects ever get to market. With that in mind, we dig into the probabilities for the Genetic Systems project. As a result of our closer examination, we make the following estimates of the various risks.

Phase 1: Costs in phase 1 are uncertain. We portray them using a triangular probability distribution with a minimum of $9 million, most likely at $10 million, and a maximum of $12 million. The probability of phase 1 research being successful is considered to be only 40%, with a 60% chance of failure.

Phase 2: Costs during the first year of phase 2 are uncertain, and we think the lowest cost would be $12 million, most likely at $15 million, and a maximum at $20 million. We portray this uncertain cost with

	A	B	C	D	E	F	G
1			**Genetic Systems Corporation**				
2			Investment Evaluation With Monte Carlo Simulation				
3			($ thousands)				
4							
5		Year	Cost	Cash In Flow	Net Cash Flow		Success?
6							
7	Phase 1: Initial Developmer	1	10,000.0	0.0	-10,000.0		0
8	Phase 2: Testing	2	15,000.0	0.0	-15,000.0		
9		3	15,000.0	0.0	-15,000.0		0
10	Phase 3: FDA Approval	4	8,000.0	0.0	-8,000.0		0
11	Phase 4: Market	5		36,000.0	36,000.0		
12		6		36,000.0	36,000.0		
13		7		36,000.0	36,000.0		
14		8		36,000.0	36,000.0		
15		9		36,000.0	36,000.0		
16		10		36,000.0	36,000.0		
17							
18	**Present Values**			**Market**			
19	Discount Rate	20.0%		Patients Per Year (thousand			1,000
20	Present Value			Revenue Per Patient			48
21	Cost	-31,288.6		Cost Per Patient			12
22	Cash Flow	57,734.6		Gross Profit Per Patient			36
23	Net Present Value	**26,446.0**		Expected Gross Profit			36,000

EXHIBIT 5.24   Vaccine project costs, revenues, and NPV based on most likely outcomes.

a triangular distribution. Costs are forecasted to increase in the second year, and we think the rate of increase will be normally distributed with mean of 8% and standard deviation of 1%. If phase 1 is successful, the chance of phase 2 succeeding is estimated to be 70%, with a 30% chance of failure. Success or failure of the second phase will be determined when the phase ends in year 3 of the program.

Phase 3: The costs of applying to the FDA are uncertain, with a triangular distribution with minimum of $7 million, most likely of $8 million, and a maximum of $10 million. If we have succeeded at phases 1 and 2, we think we have an 80% chance of success with the FDA application, and a 20% chance that it will be rejected.

Phase 4: Success with the FDA will allow us to go to market. Even then, nothing is certain. The number of patients who will use the vaccine is uncertain. We think that the number of patients per year will be normally distributed with a mean of 1 million and a standard deviation of 100,000. Revenue and cost per patient are also uncertain. We estimate that the Revenue per patient will be normally distributed with mean of $48,000 annually and standard deviation of $5000, and the Cost per patient per year will be normally distributed with mean of $12,000 and standard deviation of $2000.

5.5.2.3 Monte Carlo Simulation of Project

Having identified the risky variables and defined their distributions, we put the probability distributions into our model and run a Monte Carlo simulation. Exhibit 5.25 shows the Excel model with the @Risk probability distribution functions. Note the RiskDiscrete(·) functions in G7:G10, which are of the form

$$=RiskDiscrete(\{0,1\},\{p1,p2\}),$$

where the 0 and 1 represent 0 or 1 outcomes (0 is failure and 1 is success), and the p1 and p2 the probabilities of the 0 and 1 outcomes. In G7 the formula is =RiskDiscrete({0,1},{0.6,0.4}). This is the failure or success probability function for the initial development phase. When we run the Monte Carlo simulation, this function will work like flipping a coin where tails is failure, heads is success, and there is a 60% chance of tails. On any one draw in the simulation this cell will either be 0 or 1 depending on the chance of the draw. But over many draws, 60% of them will be 0 and 40% will be 1.

	A	B	C	D	E	F	G
1			Genetic Systems Corporation				
2			Investment Evaluation With Monte Carlo Simu				
3			(thousands)				
4							
5	Year		Cost	Cash In Flow	Net Cash Flow		Success?
6							
7	Phase 1: Initial Developmen 1		=RiskTriang(9000,10000,12000)	0	=D7-C7		=RiskDiscrete({0,1},{0.6,0.4})
8	Phase 2: Testing 2		=RiskTriang(12000,15000,20000)*G7	0	=D8-C8		=RiskDiscrete({0,1},{0.3,0.7})*G7
9	3		=C8*(1+RiskNormal(0.08,0.01))	0	=D9-C9		=RiskDiscrete({0,1},{0.2,0.8})*G9
10	Phase 3: FDA Approval 4		=RiskTriang(7000,8000,10000)*G9	0	=D10-C10		
11	Phase 4: Market 5			=G23*G10	=D11-C11		
12	6			=D11	=D12-C12		
13	7			=+D12	=D13-C13		
14	8			=+D13	=D14-C14		
15	9			=+D14	=D15-C15		
16	10			=+D15	=D16-C16		
17							
18	Present Values			Market			
19	Discount Rate	0.2		Patients Per Year (thousand			=RiskNormal(1000,100)
20	Present Value			Revenue Per Patient			=RiskNormal(48,5)
21	Cost	=NPV(B19,E7:E10)		Cost Per Patient			=RiskNormal(12,2)
22	Cash Flow	=NPV(B19,D7:D16)		Gross Profit Per Patient			=G20-G21
23	Net Present Value	=RiskOutput("NPV") + B21+B22		Expected Gross Profit			=G22*G19
24							

EXHIBIT 5.25   Monte Carlo model for vaccine program.

Success or failure is carried through the model with the 0s and 1s. By multiplying any subsequent cell formula with the 0 or 1, we can make that cell a 0 or the value determined by the formula in the cell. For example, the cost of testing in phase 2 (cell C8) is to be drawn randomly from the distribution =RiskTriang(12000,15000,20000). However, if we have failed at phase 1, the research program would end, and there would be no costs (or revenues) in any period after the failure. Note in C8 the formula is

$$=RiskTriang(12000,15000,20000) \times G7.$$

If we failed at the end of phase 1, G7 will be 0 and the contents of C8 will have to be 0. If we succeeded at phase 1, G7 will be 1 and C8 will be the result of a draw from the triangular distribution. Thus, at each phase, by multiplying by the success or failure of the previous phase, we can carry success or failure through to the end. If we fail at any phase, all subsequent phases will show 0, representing failure at that phase. If we succeed in phases 1 and 2, and then fail in phase 3, we will incur the costs of phases 1, 2, and 3, but we will not be able to go to the market, and we will not earn any of the income.

Now, let us see how the simulation model works. We specify NPV as the output variable that we want to see, and run the simulation. Exhibit 5.26

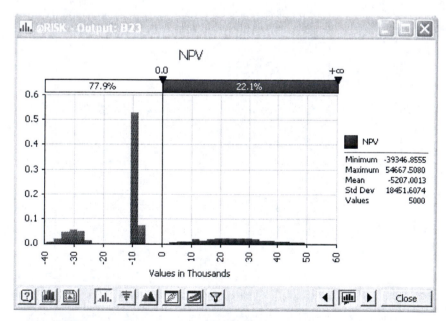

EXHIBIT 5.26   Frequency distribution for NPV.

shows the frequency distribution of the NPVs from a simulation with 1000 iterations. This simulation shows a mean NPV of −5.207 million, with outcomes ranging from −39.3 to +54.6 million. In 77.9% of the iterations, the NPV was negative. This does not look like a great project if we have 77.9% chance that the NPV will be less than zero. This no longer looks like a great project after we have considered the risks of failing at each stage.

Note that the results do not look anything like the distributions we are accustomed to, such as the normal or even the triangular distribution. We get several peaks or modes in this simulation—we call this a multi-modal frequency distribution. Each of the peaks corresponds to the value of the program if we fail at a given phase. For example, if we fail at the end of phase 1, we will have incurred the cost of the research at that phase. The phase 1 cost is a random variable from a triangular distribution with a most likely value of $10 million, but a mean of $10.33 million. The present value of this cost is shown as the tallest spike in the frequency distribution because there is a 60% chance that we will incur these costs and never proceed to the next phase. The outcomes furthest to the left on the graph represent the present values of the greatest losses. These are the outcomes when we incur the costs and succeed at phases 1 and 2, then incur the costs of going to the FDA, and they reject the application. The outcomes that are positive—on the right side of the graph—correspond to the cases where we are successful clear through to the marketing program, and we are then able to sell the product. However, there is only a 22% chance that this success will happen.

The results of this Monte Carlo analysis are a far cry from the attractive NPV that was suggested by our preliminary analysis shown in Exhibit 5.24, where the NPV was an attractive $26.45 million. This helps to point out how Monte Carlo simulation can enhance our planning and decision making. Monte Carlo simulation is an important addition to our collection of tools for financial planning and modeling.

5.6 SUMMARY

This chapter has provided an introduction to the ideas and software for Monte Carlo simulation. Monte Carlo simulation is a tool to help us to understand the impacts of risk and uncertainty on our decisions. Monte Carlo simulation does not provide answers or quick solutions to our questions. It only helps to give us some perspective at to what might happen, given the variety of risks that we face.

The output from the simulation is typically a frequency distribution that tells us how likely the different outcomes are. Armed with that information, we should be able to make a more informed decision.

@Risk software was introduced. This is an add-on to Excel that enables us to specify the probability distribution for a variable in an Excel spreadsheet. This allows us to analyze the impacts of many different types of uncertainty and see the results of combining the uncertainties. In future chapters we will see other ways that Monte Carlo simulation and the @Risk software can help us to analyze difficult problems.

PROBLEMS

1 Sanbull Trombones

Sanbull Corporation is evaluating a proposed project to purchase old student trombones, melt them down to create new brass sheet, and then produce new professional caliber instruments. The data relating to this project are as follows:

New machine	
Initial cost at $t = 0$	$35,000
Depreciation (eight-year straight line)	$4375 per year
Salvage value (after taxes)	$5000
Investment in working capital	
Inventory increases by	$7500 at $t = 0$
Investment in working capital is recovered at the end	$7500 for $t = 8$
Expected income	
Incremental increase in sales	$85,000 for $t = 1, \ldots, 8$
Growth rate of sales	0%
Incremental costs	
CGS	65% of sales
Selling, general and administrative	20% of sales
Other operating expenses	$2500 per year
Income tax rate	34%
Cost of capital:	
Required return on equity investment	$k_E = 0.12$ (12%)

Assignment

    A. Develop a spreadsheet with which you can evaluate the acceptability of this proposal. Calculate the NPV and the IRR on the project. Note that this part of the assignment does not treat uncertainty. You are asked to develop a deterministic spreadsheet model. Briefly summarize your results.

B. Now assume that Sanbull's sales and costs are random variables with distributions characterized as follows:

Sales are distributed normally with mean (average) of $85,000 per year, and standard deviation of $17,000 (Try =RiskNormal (85000,17000)).

You are uncertain about CGS as a proportion of sales. You think the lowest this proportion could go is 60%, the most likely is 65%, and it could go as high as 70% (Try =RiskTriang(0.60,0.65,0.70)).

SGA is uncertain, and past data indicate that the SGA as a percent of sales is normally distributed with a mean of 20% and standard deviation of 4%.

Other operating costs could be as low as $2000 per year, or as high as $3000 per year. You are very uncertain about these costs, and you think all values between these limits are equally likely (Try =RiskUniform(2000,3000)).

Salvage value for the machinery at time 8 is uncertain, and is distributed as a normal distribution with a mean of $5000 and standard deviation of $1000. Treat the recovery of inventory at $t = 8$ as nonrandom: $7500.

Include the effects of these uncertainties in your analysis by the following methods:

(a) First, using scenario analysis to analyze the possible outcomes for the NPV and IRR of the project. With scenario analysis you explore the results of different outcomes without explicitly running a Monte Carlo simulation. That is, simply make your evaluation based on low, most likely, and high outcomes. For most likely values, use the expected values for each distribution. For the low and high values, use the end points of the triangular and uniform distributions, and use the mean $\pm 2 \times$ standard deviations for the normal distributions. Report your results in a table for the worst, expected, and best case scenarios.

(b) Use Monte Carlo simulation to analyze the distribution of NPV and IRR for each of scenarios 1, 2, and 3 listed below. Briefly summarize your results, including answers to the following two questions. What is the likelihood that NPV will turn out to be negative? What is the range of outcomes for IRR? Be sure to include a table that displays your numerical results. Place the area histograms for NPV and IRR under each scenario in an appendix.

1. In specifying the random variables, set the first period's sales or cost as a distribution (e.g., Normal etc.), and make the

subsequent periods' sales or costs equal to the first period. This makes the variables perfectly correlated. Simulate the problem and answer the questions. For example,

Period 1	Period 2
Sales =RiskNormal(85000,17000)	=Sales in period 1

2. Next, make each period's sales or cost an independent random variable by specifying the distribution in each column (year). Simulate and answer the questions. For example,

Period 1	Period 2
Sales =RiskNormal(85000,17000)	= RiskNormal(85000,17000)

3. Give the sales a correlation between periods. Sales in period 2 have a correlation of 0.70 with sales in period 1, and sales in period 4 have a correlation of 0.20 with sales in period 3. Simulate and give answers.

(c) Sanbull intends to locate its smelters in a foreign country and there is some likelihood that their plant will be expropriated in the future. You have estimated that there is a 15% probability that it will be expropriated in year 5, and a 20% chance it will be expropriated in year 6. It is unlikely that the plant would be expropriated in any other year. Obviously, if it is expropriated, all subsequent cash flows will cease, starting with the year of the expropriation. However, since inventories will be warehoused in the United States, they will be recovered in the year of expropriation. Further, if the plant is expropriated in year 5, it cannot be expropriated in year 6. Use the assumptions of b.3 above for other parameters. Briefly summarize your results.

Suggestion

First build the spreadsheet model entirely in Excel, without considering the uncertainty. Make sure it works well. Having completed the basic Excel model, start @Risk and open your Excel model. Change the relevant cells to have the @Risk distribution functions in them. Then, run the @Risk simulation model. In running @Risk, if it fails because of the IRR function, remove the IRR and run it to examine NPV.

2 Cerro Gordo Mining

The Cerro Gordo Mining Corporation is considering an investment in a new machine for use in its silver mine. Your task is to develop a spreadsheet so that you can evaluate this investment proposal. The data relating to this investment prospect are as follows:

New machine	
Initial cost at time $t = 0$	$20,000
Straight line depreciation over the 10-year life	$2000 per year
Salvage value (after taxes)	$5000 at $t = 10$
Investment in working capital	
Inventory increased by	$3000 at $t = 0$
Investment in working capital is recovered at the end	$3000 at $t = 10$
Expected income	
Incremental sales	$50,000 for $t = 1, \ldots, 10$
Growth rate of sales	0% per year
Incremental costs	
CGS	75% of sales (revenue)
SGAs	10% of sales
Other operating expenses	$870 per year
Income tax rate	40%
Cost of capital	
Required Return on Equity	$k_E = 0.18$ (18%)

Assignment

A. Develop a spreadsheet with which you can evaluate the acceptability of this proposal. Calculate the NPV and the IRR on the project.

B. Now assume that Cerro Gordo's sales and costs are random variables with distributions characterized as follows: Sales are distributed normally with a mean (average) of $50,000 per year and standard deviation of $7000 (Try =RISKNormal(50000,7000)).

You are uncertain about CGS as a proportion of sales. You think the lowest this proportion could go is 70%, the most likely is 75%, and it could go as high as 85% (Try =RISKTriangular(0.70,0.75,0.80)).

SGA is uncertain, and past data indicate that the SGA as a percent of sales is normally distributed with a mean of 10% and standard deviation of 2%.

Other operating costs could be as low as $750 per year, or as high as $1200 per year. You are very uncertain about these costs, and you think all values between these limits are equally likely (Try =RiskUniform(750,1200)).

Salvage value for the machinery at time 10 is uncertain, and is distributed as a normal distribution with a mean of $5000 and standard deviation of $1000. Treat the recovery of inventory at $t = 10$ as non-random: $3000.

Include the effects of these uncertainties in your analysis by

(a) Use scenario analysis to analyze the possible outcomes for the NPV and IRR of the project. That is, simply make your evaluation based on low, most likely, and high outcomes. For most likely values, use the expected values for each distribution. For the low and high values, use the end points of the triangular and uniform distributions, and use the mean $\pm 2 \times$ standard deviations for the normal distributions.

(b) Use Monte Carlo simulation to analyze the distribution of NPV and IRR. What is the likelihood that NPV will turn out to be negative? What is the range of outcomes for IRR?

1. In specifying the random variables, set the first period's sales or cost as a distribution (e.g., Normal etc.), and make the subsequent periods' sales or costs equal to the first period. Simulate the problem and answer the questions. For example,

Period 1	Period 2
Sales =RISKNormal(50000,7000)	=Sales in period 1

2. Next, make each period's sales or cost an independent random variable by specifying the distribution in each column (year). Simulate and answer the questions. For example,

Period 1	Period 2
Sales =RiskNormal(50000,7000)	=RiskNormal(50000,7000)

Suggestion

First build the spreadsheet model entirely in Excel, without considering the uncertainty. Make sure it works well. Having completed the basic Excel model, start @Risk and open your Excel model. Change the relevant cells to have the @Risk distribution functions in them. Then, run the @Risk simulation model. In running @Risk, if it fails because of the IRR function, remove the IRR and run it to examine NPV.

3 VitalRitha Pharmaceuticals

VitalRitha Pharmaceutical Company is a biotechnology company engaged in research, development, manufacture, and marketing of pharmaceuticals developed from genetic research. The company was founded in 1976 by Vitali Kopych, a venture capitalist and Saowarin Sritha, a scientist, each of whom invested $50,000 to start the company. Today the company employs 1300 and has annual revenues of about $400 million.

The company has many new applications of its technology that could lead to significant breakthroughs in medicine, and to very profitable medical products. However, there is great risk in pursuing any one of its research projects because of the high costs of development and the likelihood that the product will not be successful, or will not get the necessary approvals. One of the decision problems faced by VitalRitha is to correctly choose which new research project to support, and how long to continue with this support.

Each potential new drug goes through a four-phase R&D process before it can be marketed, as illustrated by the following table that lists each phase, the average number of years for completion of the phase, and the average annual cost of completing that phase, along with the minimum and maximum annual costs of completing the phase.

ValRitha Pharmaceuticals
Development Cost Estimates

Phase	Years	Average Annual Cost ($ million)	Minimum Cost ($ million)	Maximum Cost ($ million)
1. Preclinical development	2	5	4	9
2. Phase II testing	1	15	12	18
3. Phase III testing	3	8	6	10
4. Regulatory filing	2	5	4	8
5. Marketing (Marketing expenditure is a decision variable in Part 2 of the problem)	1	14	10	18

Each different product would have a different cost structure, but these figures are representative. In addition, each product would have a different market potential. For the typical product, if it reaches the marketing stage and is successful, the annual net cash flow generated by the product's sales would be about $60 million, and these cash flows would be earned for about eight years, at which point they would terminate.

To evaluate this type of investment, the cash flows should be discounted at the cost of capital adjusted for the riskiness of the project.

This type of investment is considered to be very risky, with its cash flows about twice as volatile as the average stock market investment. Consequently, in the context of the CAPM, the beta is regarded as about 2.

Assume that the risk-free interest rate is 4% and the risk premium (in excess of the risk-free rate) on the market portfolio (or the average stock market investment) is about 7%.

It is now the end of year 2010. If a new program is started, it will begin in 2011, and its first costs will be incurred at the end of 2011. The NPV should be evaluated as of this date, the end of 2010.

Assignment 1

Develop a spreadsheet model based on the most likely costs (average annual costs) and cash flows with which you can evaluate the prospect of investing in a new research project.

Your model should show the NPV and the IRR for the typical new research project.

Modeling Uncertainty

There is considerable uncertainty about new research projects that should be considered in the evaluation. We can only guess about the possible success of each phase of the research, and there is considerable variation in the possible cash flows. To take account of the uncertainty of a project, assume the following: Each phase has a probability of failure or success:

Phase	1	2	3	4	5
Probability of success (%)	40	60	70	80	80 if high marketing expenditures 50 if low expenditures

If the project is unsuccessful at any of the first four phases, the project will be terminated, and no additional costs will be incurred. If it is successful at the end of a given phase, the next phase will be carried to completion. Note also that failure is recognized only at the end of the phase, so if, for example, a phase lasts three years, costs for that phase are incurred each year for the full three years, and then, with failure, the project ends and no future costs or revenues will occur.

The costs of development at each phase are random variables drawn from a probability distribution. Assume that the development costs are from a triangular distribution with minimum, most likely, and maximum value as shown on the table above. For example, the annual costs for Phase 3 are drawn from a triangular distribution with a minimum of $6 million, a most likely value of $8 million, and a maximum value of $10 million. However, keep in mind that these costs would only be incurred if Phase 2 is successful; in which case the Phase 3 costs will be incurred for three years.

Marketing and Sales

If a drug makes it through the R&D process and receives FDA approval, then it can be marketed to the medical industry and the public. Phase 5 is a marketing program that will begin after phase 4 is successful and which lasts just one year. After the end of that year, the product will be sold to the public, with the first cash flow being received one year after the end of the marketing effort. These flows will continue for eight years. That is, the earnings from the product will be received from years 10 through 17.

The success of the marketing program depends on the amount spent on the program. If the company spends at least $15 million in stage 5, the probability of success will be 80%. If it spends less than that, its probability of success will be 50%. There are two alternative marketing programs available, the less expensive program involves marketing expenditures of $10 million, and the more expensive program requires that $18 million be spent at stage 5.

If the marketing effort is successful, it will have a substantially different market than if it is less successful. Projections and data relating to the markets for successful and less successful marketing programs are as follows:

The Percentage of the Total U.S. Population Who Will Use the Treatment Will Be a Normally Distributed Random Variable with the Following Parameters	Successful	Less Successful
Mean percent of population who use the medication	1%	0.5%
Standard deviation of % usage	0.15%	0.10%
Days per year taken by each patient	365	365
Wholesale price per dose ($)	0.148	0.12
Profit margin	40%	35%

The profit margin represents the rate of profit earned from the wholesale price, and represents the incremental cash flow to the company that will be generated from the sales.

Assume that the population in the United States in 2010 is 300,000,000. It is expected that the population will grow at an annual rate of 0.8% (8/10 of 1%) for the next 30 years.

Assignment 2

Expand your model to take account of the uncertainty of the R&D and marketing processes. Evaluate the attractiveness of the investment in a typical pharmaceutical product in terms of NPV and IRR. If you have difficulty with IRR, you may eliminate it from your random model.

PART III

Introduction to Forecasting Methods

CHAPTER 6

Forecasting I: Time Trend Extrapolation

Forecasting is the substitution of error for chaos.

—ANONYMOUS

6.1 AN INTRODUCTION TO FORECASTING

Forecasts are a major input to financial plans and the models within those plans. They are basic to solving forward-looking models. Certainly, models can be built without forecasts. Still, we need forecasts to make the models work. This and the following three chapters are concerned with forecasting. The intent is to expose you to a variety of forecasting methods, so you will be able to develop and use basic forecasts. We recognize that a few short chapters can barely introduce you to forecasting methods. We direct readers interested in going beyond this brief introduction to any of the many fine texts devoted to forecasting.*

---

* Useful texts that deal with forecasting include DeLurgio (1998), Hanke, Wichern, and Reitsch (2004), Levine, Berenson, and Stephan (2007), and Pindyck and Rubinfeld (1998).

There are many different methods and techniques for generating the forecasts necessary for financial models. The best technique to use depends on the problem being modeled. The choice will be determined by such factors as the time horizon over which the forecast will be made, the level of detail and accuracy required, the costs of being wrong, the type of data available, and the expertise of the forecaster.

The forecasting technique selected should suit the task, and its benefits should outweigh its costs. For example, consider the choice between highly sophisticated and simple methods. Highly sophisticated techniques may be too costly for the task, too complex to engender the confidence of the people in the organization, or too detailed in their output for the task at hand. Less sophisticated forecasting methods may produce equally reliable results at a lower cost and, because they are easier to understand, gain a higher level of acceptance. This chapter and the two that follow are intended to give the reader a sufficient perspective to make this type of evaluation.

6.1.1 Qualitative versus Quantitative Forecasts

Forecasting methods can be either qualitative or quantitative. Before computers, qualitative forecasts were the primary means of forecasting. "Humans possess unique knowledge and inside information not available to quantitative methods. However, empirical studies and laboratory experiments have shown that their forecasts are not more accurate than those of quantitative methods."[*] Quantitative forecasting methods have distinct advantages over qualitative methods—they are consistent, replicable, and testable. These attributes make them amenable to careful evaluation and academic study. Ultimately, perhaps, the best technique may a combination of both qualitative and quantitative methods.

This being a modeling text, we focus on replicable and testable quantitative methods—not because they are inherently superior, but because they are more easily taught and evaluated. This chapter introduces the use of linear regression for extrapolating a time trend. Chapter 7 extends our treatment of regression-based models by introducing the use of econometric models for forecasting. Chapter 8 discusses smoothing methods.

6.2 STEPS FOR DEVELOPING A FORECASTING MODEL

The development of your forecasting model begins with a hypothesis about what factors influence the quantity to be predicted. Even if you are not using a quantitative approach and your sales forecast is based on judgment

---

[*] Makridakis (1986, vol. 2, p. 17).

and an intuitive understanding of your company and industry, your projections will be based on an idea of the factors and events that will affect your company. Similarly, with a quantitative model, you begin with an idea of the influential factors. These ideas form the basis of the hypothesis that supports the forecast.

With the hypothesis in mind, the next steps are to gather and examine data and then use that data to test and refine your hypothesis. In the example to follow, we have assembled past sales data for a company called The Speckled Band, Inc., with the intent to forecast its sales for the next several quarters. We have some ideas about what affects our company's sales. At the top of our list is a time trend. We know that past sales generally have been increasing over time, although with considerable variation from period to period. This suggests that sales depend on time. Now we want to take a more systematic look at the sales data to refine and test our initial hypothesis that sales are following a time trend.

The first step in data analysis is looking at the data—drawing a graph. A visual image of the data can help identify possible patterns in the data and perhaps suggest alternative hypotheses. After examining the data, we specify a model that allows us to evaluate our hypotheses and develop a forecast. Finally, we take a test drive and see if our forecasting model can actually give us a usable forecast.

Forecasting can be a discouraging task. The forecaster almost invariably will be wrong. One of the lessons from this chapter is that forecasts are always imperfect. In the example to follow, we find that our first forecasting model does not provide us with as accurate a forecast as we would like. We improve upon this forecast in the following chapter. Part of the lesson about the imperfection of forecasts is to understand the limitations of both the data and the methods for building forecasting models. The statistical methods we use frequently require a lot of well-behaved data that conforms to the assumptions that underlie statistical analysis. Economic and business data seldom is that well behaved. Compounding our difficulty is the fact that we rarely have as much data as we would like. These, and many other problems, can accumulate to bewilder the forecaster. Nevertheless, our advice is to forge ahead, but always with an understanding of where the potential pitfalls lie and the various things that can go wrong.

6.3 TIME TREND EXTRAPOLATION

We begin by introducing linear regression to extrapolate a trend. Our forecasting discussion use sales as an example, but the methods discussed can be used to forecast most business and economic variables. We use quarterly sales for Speckled Band shown in Exhibit 6.1, in which the Roman numerals denote quarters and the quarters are numbered 1–44. Our goal is to forecast Speckled Band's quarterly sales for year 12 (quarters 12.I–12.IV or quarters 45–48).

The first step is to examine the data and get an intuitive sense of any patterns in the data (e.g., trends and/or cycles). Exhibit 6.2 shows the sales data in a spreadsheet with sales shown in a graph, so we can see the trend and make a visual check for other patterns. Over the 11 years, sales have increased at an average rate of 4.2% per quarter and an average compound rate of 1.42% per quarter.* The graph shows that Speckled Band's sales are trending upward, but with considerable random variation around the trend.

Though the average growth rate is frequently thought of as a descriptive statistic rather than a forecasting method, it can be used to generate a simple short-term forecast. For example, to forecast the next quarter's sales, take the average percentage quarterly change and apply it to the current quarter's sales. Speckled Band's sales were $7220 in the fourth quarter of year 11 (11.IV). The sales forecast for the first quarter of year 12 is

$$\text{Sales}_{12.I} = \text{Sales}_{11.IV} \times 1.0421$$
$$= 7.220 \times 1.0421$$
$$= 7524.$$

---

* The average percentage change is what is called an arithmetic average—the simple average of the percent change for each of the 43 quarters. The compound growth, also called the geometric average growth, is based on the compound growth expression $\text{Sales}_N = \text{Sales}_1 \times (1 - g)^{N-1}$, where g is the compound rate of growth over $N - 1$ periods. Over $N - 1$ periods, the average compound growth in sales is calculated as

$$g = \left(\frac{\text{Sales}_N}{\text{Sales}_1}\right)^{1/(N-1)} - 1.$$

With 44 periods of data and 43 periods of growth, Speckled Band's geometric average growth is

$$g = \left(\frac{\text{Sales}_{44}}{\text{Sales}_1}\right)^{1/43} - 1 = \left(\frac{7,219.5}{3,942.8}\right)^{1/43} - 1 = 0.0142$$

$$= 1.42\% \text{ average compound growth per quarter.}$$

Year	Quarter	Time Period	Sales $000		Year	Quarter	Time Period	Sales $000
1	I	1	3,943		7	I	25	4,698
	II	2	2,823			II	26	4,057
	III	3	2,328			III	27	5,074
	IV	4	2,458			IV	28	6,598
2	I	5	2,792		8	I	29	6,563
	II	6	3,256			II	30	7,226
	III	7	3,177			III	31	6,423
	IV	8	3,726			IV	32	6,409
3	I	9	3,805		9	I	33	6,400
	II	10	2,246			II	34	6,217
	III	11	3,047			III	35	4,931
	IV	12	2,106			IV	36	4,987
4	I	13	1,339		10	I	37	5,338
	II	14	3,190			II	38	5,087
	III	15	3,023			III	39	4,908
	IV	16	3,498			IV	40	5,021
5	I	17	2,948		11	I	41	5,122
	II	18	3,237			II	42	5,392
	III	19	3,917			III	43	7,239
	IV	20	3,230			IV	44	7,220
6	I	21	3,985					
	II	22	3,826					
	III	23	4,159			Average Growth		4.21%
	IV	24	4,747			Compound Growth		1.42%

EXHIBIT 6.1 Speckled Band Corporation: Quarterly sales.

The sales forecast for the second quarter of year 12 takes the forecast sales for the first quarter as its input, thus

$$Sales_{12.II} = E[Sales_{12.I}] \times 1.0421$$
$$= 7.524 \times 1.0421$$
$$= 7851$$

Forecasts using the geometric average growth rate are calculated in the same way, but use the geometric average quarterly growth rate instead of the simple average quarterly growth rate. Though this process can be continued indefinitely, it is not recommended. There is little reason to expect either average growth rate to continue indefinitely into the future.

Excel® makes drawing the graph in Exhibit 6.2 relatively easy. The first step is to enter the data in your spreadsheet, as shown on the left side of Exhibit 6.2. Next, click on the Graph icon and select XY-Scatter as the "Chart type." Click on "Next," then "Series," then "Add" and fill in the boxes as shown in Exhibit 6.2 (filling in the boxes can be automated by clicking on the small squares to the right of the larger boxes). Click on "Next," fill in the titles and legends, then click on Finish and you get the basic graph.

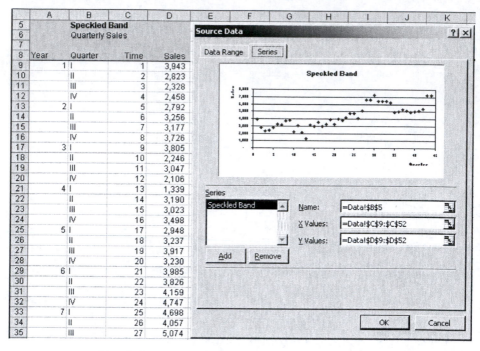

	A	B	C	D	E	F	G	H	I	J	K
5		**Speckled Band**									
6		Quarterly Sales									
7											
8	Year	Quarter	Time	Sales							
9	1	I	1	3,943							
10		II	2	2,823							
11		III	3	2,328							
12		IV	4	2,458							
13	2	I	5	2,792							
14		II	6	3,256							
15		III	7	3,177							
16		IV	8	3,726							
17	3	I	9	3,805							
18		II	10	2,246							
19		III	11	3,047							
20		IV	12	2,106							
21	4	I	13	1,339							
22		II	14	3,190							
23		III	15	3,023							
24		IV	16	3,498							
25	5	I	17	2,948							
26		II	18	3,237							
27		III	19	3,917							
28		IV	20	3,230							
29	6	I	21	3,985							
30		II	22	3,826							
31		III	23	4,159							
32		IV	24	4,747							
33	7	I	25	4,698							
34		II	26	4,057							
35		III	27	5,074							

Source Data dialog box fields:
- Name: =Data!$B$5
- X Values: =Data!$C$9:$C$52
- Y Values: =Data!$D$9:$D$52
- Series: Speckled Band
- Buttons: Add, Remove, OK, Cancel

EXHIBIT 6.2   Excel chart wizard dialog box.

6.3.1 Estimation Period versus Hold-Out Period

Having looked at the sales data, we now develop our forecasting models. Once the forecasting models are developed, we will need to decide which ones seem to generate the best predictions. Obviously, we will not know which model yields the best prediction until we see what actually happens. However, since we do not want to wait to see how accurate our forecasts are, we will apply our model to past data and measure how well it would have worked in the past. We do this by partitioning the data into an *estimation period* and a *hold-out period*. We use data from the estimation period to estimate our model's parameters. Then, we use that model to predict the values for the hold-out period. This way we can check the forecasting accuracy of our model by comparing its forecasts to their respective actual values in the hold-out period. The model that appears to generate the best forecasts for the hold-out period is then re-estimated using data from the entire sample—both the original estimation period and the hold-out period. Finally, the re-estimated model is used to forecast the variables for the still unknown future. For Speckled Band, we use the first eight years as the initial estimation period and the last three years as the hold-out period. The next section uses linear regression to extrapolate the trend in Speckled Band's sales.

6.3.2 Time Trend Extrapolation Using Linear Regression

We consider two time trend regression models using one independent variable, "Time." The time variable simply numbers the periods in the estimation period (1–44 in this example). To apply the model, the series to be forecast is regressed on this time sequence. The two models we consider differ in their assumptions about growth. The constant change model assumes that the series being modeled increases by a constant dollar amount each period. The compound growth model assumes that the series grows at a constant percentage rate. We begin with the constant change model.

6.3.2.1 The Constant Change Model

The general form of the constant change model is shown in Equation 6.1

$$\text{Sales}_t = a + b \times \text{Time}_t + e_t, \tag{6.1}$$

where our task is to estimate the intercept parameter a, the slope parameter b, and e_t denotes the random error term. We use the regression tool from Excel to estimate this model.* You access the regression tool in Excel through "Tools > Data Analysis."† In the Data Analysis list of Analysis Tools, click Regression. The Regression dialog box pops up and you fill it in as shown in Exhibit 6.3. For the Input Y Range select the column (or row) of dependent variable data (Sales in D8:D40). Note that D8 is the label "Sales." The Input X Range is the independent variable data (C8:C40 which includes the label "Time"). With variable name labels "Sales" and "Time" included in the data range, check the Labels box, telling the program that the data range includes the label. F1 is entered in Output Range to indicate that you want the regression output to start at F1. Click OK and the output is inserted in the sheet as shown in Exhibit 6.4.

---

* We are using the Excel regression routine because it is accessible in Excel. There are numerous other statistical software programs that are well suited to developing a forecasting model. These include StatTools from Palisade Corporation (www.Palisade.com), access to which is included with this text, EViews from Quantitative Micro Software (www.eviews.com), SPSS from SPSS Corporation (www.spss.com), and STATA (www.stata.com).

† The DataAnalysis ToolPak can be found on the "Data" ribbon in Excel 2007 or on the "Tools" dropdown menu in Excel 2003. The DataAnalysis ToolPak is an Excel Add-In. To add it into Excel 2007, click on the "Office Button," select "Excel Options," then "Add-Ins" and select DataAnalysis ToolPak. To add it into Excel 2003, click on "Tools," then "Add-Ins" and check the DataAnalysis box.

Speckled Band Quarterly Sales

Year	Quarter	Time	Sales
1	I	1	3,943
	II	2	2,823
	III	3	2,328
	IV	4	2,458
2	I	5	2,792
	II	6	3,256
	III	7	3,177
	IV	8	3,726
3	I	9	3,805
	II	10	2,246
	III	11	3,047
	IV	12	2,106
4	I	13	1,339
	II	14	3,190
	III	15	3,023
	IV	16	3,498
5	I	17	2,948
	II	18	3,237
	III	19	3,917
	IV	20	3,230
6	I	21	3,985
	II	22	3,826
	III	23	4,159
	IV	24	4,747

Regression

Input
- Input Y Range: $D$8:$D$40
- Input X Range: $C$8:$C$40
- ☑ Labels   ☐ Constant is Zero
- ☐ Confidence Level: 95 %

Output options
- ◉ Output Range: $F$1
- ○ New Worksheet Ply:
- ○ New Workbook

Residuals
- ☐ Residuals   ☐ Residual Plots
- ☐ Standardized Residuals   ☑ Line Fit Plots

Normal Probability
- ☐ Normal Probability Plots

[OK] [Cancel] [Help]

EXHIBIT 6.3   Dialog box for regression.

At the bottom of the output are the estimates of the regression coefficients. The intercept, a, estimated to be 1886.2, is shown in cell G17, and the slope coefficient, b, estimated to be 120.26, is shown in cell G18. The rest of the output are the diagnostic statistics that we discuss later. For now, we note that the estimated coefficients are statistically significant with t-statistics of 5.61 and 6.77 in cells I17 and I18 and p-values of 0.00 in cells J17 and J18. The adjusted R^2 of 0.59 in cell G6 tells us that 59% of the variability of sales is accounted for by its relation with Time. Substituting

	F	G	H	I	J	K	L
1	SUMMARY OUTPUT						
2							
3	Regression Statistics						
4	Multiple R	0.78					
5	R Square	0.60					
6	Adjusted R Square	0.59					
7	Standard Error	928.12					
8	Observations	32					
9							
10	ANOVA						
11		df	SS	MS	F	Significance F	
12	Regression	1	39,453,831	39,453,831	45.80	0.00	
13	Residual	30	25,842,303	861,410			
14	Total	31	65,296,134				
15							
16		Coefficients	Standard Error	t Stat	P-value	Lower 95%	Upper 95%
17	Intercept	1,886.20	335.99	5.61	0.00	1,200.02	2,572.37
18	Time	120.26	17.77	6.77	0.00	83.97	156.55

EXHIBIT 6.4   Regression output.

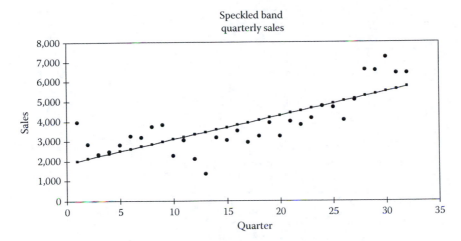

EXHIBIT 6.5   Regression trend line of sales.

the estimated parameters into the general form of the model gives us our forecasting model:

$$Sales_t = 1886.2 + 120.26 \times Time_t. \qquad (6.2)$$

In the regression dialog box we checked the Line Fit Plots. This generates a graph of the regression relationship as shown in Exhibit 6.5, where the dots are the actual sales and the line is the trend line that represents the Sales that would be predicted based on the estimated regression equation. Naturally, our model does not fit the data perfectly. The deviations of the actual sales around the regression line represent the errors in the regression.

To forecast sales in future quarters, substitute the appropriate period's number for $Time_t$. Our estimation period ended with 32. To forecast the first two quarters of the hold-out period, we substitute 33 and 34 for $Time_t$:

$$Sales_{33} = 1886.2 + 120.26 \times 33 = 5854.8$$
$$Sales_{34} = 1886.2 + 120.26 \times 34 = 5975.0. \qquad (6.3)$$

If you want to bypass the detail of the regression and are concerned only with the trend and the most basic statistics, the Excel functions INTERCEPT, SLOPE, and TREND provide the basics as shown in Exhibit 6.6. Both the INTERCEPT and SLOPE functions require only the selection of the column of dependent (D9:D40) and independent (C9:C40) variables, and

	A	B	C	D	E	F	G
29	6	I	21	3,985			
30		II	22	3,826			
31		III	23	4,159			
32		IV	24	4,747		=INTERCEPT(D9:D40,C9:C40)	
33	7	I	25	4,698			
34		II	26	4,057			
35		III	27	5,074			
36		IV	28	6,598		=SLOPE(D9:D40,C9:C40)	
37							
38	=TREND($D$9:$D$40,$C$9:$C$40,C43)						
39						Intercept	1886.20
40		IV	32	6,409		Slope	120.26
41							
42	**Hold-Out Period**					Trend Forecast	
43	9	I	33	6,400		5,854.8	
44		II	34	6,217		5,975.0	
45		III	35	4,931		6,095.3	
46		IV	36	4,987		6,215.6	
47	10	I	37	5,338		6,335.8	
48		II	38	5,087		6,456.1	
49		III	39	4,908		6,576.4	
50		IV	40	5,021		6,696.6	
51	11	I	41	5,122		6,816.9	
52		II	42	5,392		6,937.1	
53		III	43	7,239		7,057.4	
54		IV	44	7,220		7,177.7	

EXHIBIT 6.6   Trend with linear regression using Excel functions.

output is estimated parameters as shown in G39 and G40. The TREND function generates the predicted future values based on the regression trend as shown in F43:F54.

One problem with the constant change model is that the percentage rate of growth declines as time passes and the level of sales increases. This is because the change is the same dollar amount in each period. A compound growth regression model may be better suited when growth is proportional to the level of sales.

6.3.2.2 The Compound Growth Model

A firm's sales may grow by a given percent each period rather than by a constant dollar amount. If so, forecasts made by the constant change model will significantly underestimate future sales. The compound growth model addresses this problem.

If sales grow at a compound rate, g, sales in year t will be

$$\text{Sales}_t = \text{Sales}_0(1 + g)^t, \tag{6.4}$$

where $Sales_0$ is the level in the base period $t = 0$. To use linear regression to estimate the growth rate, transform the expression into the logarithmic form

$$\ln(Sales_t) = a + b \times t,$$

where t is the number of the period and $\ln(\;)$ denotes the natural logarithm. When we regress the log of sales against time as the independent variable, the slope coefficient, b, is the estimate of the compound growth rate, and the intercept, a, corresponds to $\ln(Sales_0)$.*

To run this regression in Excel, the dependent variable, Sales, is transformed with the function =LN(\;); Time remains the independent variable. Using data for the 32-quarter estimation period, our estimated equation for Speckled Band is

$$\ln(Sales_t) = 7.715 + 0.029 \times Time_t$$
$$ (307.9) \quad (7.868)$$
$$R^2 = 51.7\%. \tag{6.5}$$

where the numbers in parentheses are the t-statistics. The estimated compound growth rate is 2.9% per quarter.

To generate forecasts, substitute the appropriate period's number for Time. This yields the forecast for the natural log of sales for that period. Exponentiating this value yields the dollar forecast.† For the 33th quarter (Time = 33), the forecast is

$$\ln(Sales_{33}) = 7.715 + 0.029 \times 33$$
$$= 8.672 \tag{6.6}$$

so

$$Forecast\ Sales_{33} = e^{8.672}$$
$$= 5837.2.$$

The compound growth model addresses the decreasing growth-rate criticism of the constant change assumption. Unfortunately, we cannot know a priori which model generates the better forecasts. To make this

---

* When $Sales_t = Sales_0(1 + g)^t$, the log transformation is

$$\ln(Sales_t) = \ln(Sales_0 \times (1 + g)^t)$$
$$= \ln(Sales_0 + [\ln(1 + g)] \times t$$
$$= a + b \times Time_t$$

† The inverse function of $\ln(x)$ is the exponential function e^y, so $e^{\ln(x)} = x$ transforms the logarithmic forecast back into dollars. The respective Excel functions are =LN(\;) and =EXP(\;).

determination, we must measure and compare the accuracy of each model's forecasts. This is the purpose for the hold-out segment of the data. We take up this subject next.

6.3.3 Assessing Model Validity and Accuracy

Thus far we have considered two regression models to forecast quarterly sales. This section considers how to determine which model is most likely to produce the best forecasts. Our assessment proceeds in two steps. First, we evaluate which forecasting model fits the data best for the initial estimation period. Then, we determine which model generates the best forecasts of sales for the hold-out period.

6.3.3.1 Evaluating Model Validity

This section reviews how to use the diagnostic statistics from Excel's regression output to evaluate how well the regression model fits the data.* Keep in mind, however, that evaluating statistical validity is not the same as forecasting. A model can fit past data very well but still not be a very good forecasting tool. Nevertheless, the first step is to evaluate the model against past data in terms of its statistical validity and fit.

The first thing to understand is the *error* term, also called the *residual* or *deviation*, denoted by e_t. The error is the difference between the actual, observed value of the dependent variable and the value predicted by the regression model. For example, the constant change regression was estimated as

$$\text{Sales}_t = 1886.2 + 120.26 \times \text{Time}.$$

For the first quarter, the estimated or predicted sales would be

$$\text{Sales}_t = 1886.2 + 120.26 \times 1$$
$$= 2006.5$$

and the actual value was 3942.8. The error in the prediction is

$$\text{Error}_t = e_t = \text{Actual Value}_t - \text{Predicted Value}_t. \tag{6.7}$$

---

* An in-depth explanation of statistical methods and regression techniques is beyond the scope of this text. For greater detail and depth, consult texts such as DeLurgio (1998), Hanke, Wichern, and Reitsch (2004), Levine, Berenson, and Stephan (2007), and Pindyck and Rubinfeld (1998).

For quarter 1 the error is

$$e_t = 3943 - 2006$$
$$= 1936.$$

That is, actual sales exceeded the prediction by 1936.3. Exhibit 6.7 shows the calculation of the predicted values and error terms for the constant

	A	B	C	D	E	F	G	H
1						Model 1:	Sales = a + b*Time	
2								
3						Model 1		
4						Constant		
5						Change		
6		**Speckled Band**						Squared
7		Quarterly Sales				Predicted	Error	Error
8	Year	Quarter	Time	Sales				
9	1	I	1	3,942.8		2,006.46	1,936.3	3,749,395.5
10		II	2	2,822.6		2,126.72	695.9	484,301.6
11		III	3	2,328.2		2,246.98	81.3	6,601.9
12		IV	4	2,457.8		2,367.24	90.6	8,203.4
13	2	I	5	2,792.3		2,487.50	304.8	92,882.3
14		II	6	3,256.4		2,607.76	648.7	420,776.4
15		III	7	3,177.5		2,728.02	449.4	201,995.6
16		IV	8	3,726.1		2,848.28	877.8	770,570.5
17	3	I	9	3,805.3		2,968.54	836.7	700,087.1
18		II	10	2,245.9		3,088.80	-842.9	710,477.1
19		III	11	3,047.3		3,209.06	-161.8	26,180.8
20		IV	12	2,105.6		3,329.32	-1,223.7	1,497,435.4
21	4	I	13	1,339.3		3,449.58	-2,110.3	4,453,318.8
22		II	14	3,189.8		3,569.84	-380.0	144,423.9
23		III	15	3,022.8		3,690.10	-667.3	445,327.6
24		IV	16	3,498.1		3,810.36	-312.3	97,506.4
25	5	I	17	2,947.6		3,930.62	-983.1	966,394.2
26		II	18	3,237.3		4,050.88	-813.5	661,853.9
27		III	19	3,917.1		4,171.14	-254.0	64,525.9
28		IV	20	3,230.1		4,291.41	-1,061.3	1,126,378.2
29	6	I	21	3,985.3		4,411.67	-426.4	181,824.3
30		II	22	3,825.6		4,531.93	-706.4	498,955.5
31		III	23	4,159.0		4,652.19	-493.2	243,244.1
32		IV	24	4,746.9		4,772.45	-25.6	653.0
33	7	I	25	4,698.2		4,892.71	-194.5	37,829.1
34		II	26	4,056.8		5,012.97	-956.2	914,312.5
35		III	27	5,074.0		5,133.23	-59.2	3,504.2
36		IV	28	6,597.9		5,253.49	1,344.4	1,807,408.9
37	8	I	29	6,562.9		5,373.75	1,189.2	1,414,175.7
38		II	30	7,226.5		5,494.01	1,732.5	3,001,473.8
39		III	31	6,423.4		5,614.27	809.2	654,759.1
40		IV	32	6,409.5		5,734.53	674.9	455,526.2
41								
42								
43						Total (sum)	0.0	25,842,302.9
44						Mean	0.0	807,572.0
45						Root Mean Square Error		898.65
46						Standard Error		928.12

EXHIBIT 6.7   Errors from the constant change model.

change model (Model 1) for the estimation period. Column D shows actual sales, column F is the value predicted by the regression model, and column G are the errors, or residuals. At the bottom are the totals and averages.

The mean error (ME) in cell G44 is the simple average of the errors over the estimation period.

An ME of zero (ME = 0) tells us that the model is unbiased; that is, the forecasts from the regression are not consistently in error one way or the other, and the over-predictions are balanced by the under-predictions. If the ME is positive, the actual sales tend to be greater than the predicted values, and if the ME is negative, the forecasts tend to be above the actual values. We can have a model with very large errors, but they could average zero. So an ME of zero does not mean the model is accurate—only that it is unbiased. To get a better idea of accuracy we need to look at other statistics like the mean absolute deviation (MAD), mean squared error (MSE), root mean square error (RMSE), and standard error (SE).

Least squares regression focuses on the squared errors that are shown in column H, where, for example, the squared error for quarter 1 is $1936.3^2 = 3,749,395.5$ in cell H9. When the squared errors are totaled, we get the "sum of the squared errors" in cell H43. This is the same as in cell H13 of the regression output in Exhibit 6.4. The mean (simple average) of the squared errors is $25,842,303/32 = 807,572$ in cell H44, which we call the MSE. The square root of the MSE is called the RMSE, shown as 898.65 in H45.

The last item is the SE (a.k.a. residual standard error, RSE), 928.12 in cell H46. This corresponds to cell G7 in the regression output of Exhibit 6.4. The SE is the standard deviation of the error terms around the regression line. This standard deviation is also called the standard error (SE) of the estimate or the regression. The distinction between the SE and the RMSE is that the SE is adjusted for degrees of freedom and the RMSE is not.*

Exhibit 6.8 summarizes the main diagnostic statistics for the two regression models. The coefficient of determination, R^2, measures the

* The adjustment for degrees of freedom takes account of the number of observations and the number of variables in the regression. The divisor of the sum of squared residual errors (cell G13 in the regression output) is $(n - k)$, where n is the number of observations and k the number of variable coefficients estimated. For a regression with one independent variable, $k = 2$ to account for the slope coefficient and intercept term.

	Model 1 Constant Change	Model 2 Compound Growth
Dependent variable	Sales	Ln (Sales)
Independent variable	Time	Time
Intercept	1886.2	7.715
t-Statistic for: intercept	5.61	82.1
Slope coefficient	120.26	0.029
t-Statistic for: slope coefficient	6.77	5.846
Coefficient of determination R^2 (adjusted)	59.1%	51.7%
Standard error of the estimate	928.12	0.26
F-Statistic	45.8	34.2

EXHIBIT 6.8   Diagnostic statistics for time trend regressions.

proportion of variation of the dependent variable that is accounted for by the regression. An R^2 of 1 (100%) means that 100% of the variation of the dependent variable (around its mean) is attributable to its linear relationship with the independent variable, and the regression is a perfect fit. The R^2 for model 1 of 59.1% shows that 59.1% of the variability of sales is accounted for by the trend line and 40.9% is unrelated to the trend. This is a slightly better fit than model 2 with its R^2 of 51.7%. However, this small difference, by itself, is not enough to say that model 1 is really a better model of the data.

Whether or not the R^2 is strong or weak depends on the form of the regression and the type of data. In some cases, such as a regression using cross-sectional data, an R^2 of 59% might be considered a relatively good fit. In other cases, 59% might be considered low. Regressions using time series data frequently yield very high R^2, often in the 90% range. In this case, with time series data, we would assess the R^2 of 59% as being relatively low. However, as the remaining diagnostic statistics will indicate, there is strong evidence that there is a time trend in the data. The problem is that there may be other forces at work in addition to just the time trend. We will deal with these other factors in the following chapter.

For model 1, the coefficient of the time variable is estimated to be 120.26, meaning that each quarter sales are expected to increase by 120.26. We assess the significance of this estimate with the t-statistic. The t-statistic tells us whether the estimated coefficient is significantly different from zero.* The t-statistic of 6.77 indicates that it is very unlikely that we could have gotten an estimated coefficient of 120.26 if, in fact, the true value was zero. Thus, we say the coefficient is significantly different from zero. It seems obvious that 120.26 is significantly different from zero. However, when we look at the time coefficient of 0.029 for model 2, we might be inclined to conclude that this is so small that it might as well be zero. The t-statistic of 5.85 tells us that, despite the coefficient's small magnitude, it is very unlikely that the real value is zero. That is, the time coefficient of 0.029 in the logarithmic regression is almost as significant as the 120.26 in the constant change model. Our conclusion for either regression model is that there is a statistically significant link between sales and time.

The SE of the estimate (SE of the regression) is the estimate of the standard deviation of the error terms around the regression line. This allows us to calculate a confidence interval for the dependent variable around the line. For example, for quarter 16, regression model 1 predicts sales to be

$$\text{Predicted Sales}_{16} = 1886.2 + 120.26 \times 16$$
$$= 3810.4.$$

Due to random error, however, the actual sales will be drawn from a distribution of possible values around the predicted value. With the SE of $\sigma_e = 928.12$, the 95% confidence interval would be

$$\text{Confidence Interval} = \text{Predicted Sales} \pm t \times \sigma_e$$
$$= \text{Predicted Sales} \pm 2.04 \times 928.12$$
$$= 3810.4 \pm 1893.4,$$

---

* The rough rule of thumb (depending on sample size) is that t should be greater than 2 for the coefficient to be significantly different from zero. The t-statistic tests the hypothesis of whether the true value of the regression coefficient is zero. For example, for a large sample, if $t \geq 1.98$, the probability of having obtained the estimated coefficient is <5% when in fact the true value is zero. The t tells us, in standard deviation units, how far the estimated coefficient is from zero. The $t = 6.77$ indicates that our estimated coefficient is 6.77 standard deviations from zero. If the true value was zero, it would be almost impossible to obtain an estimate that was 6.77 standard deviations from zero, so we say that the estimated coefficient is significantly different from zero. The output from most regression packages also reports the likelihood the estimated parameter is the result of chance alone. It is reported as the p-value.

where $t = 2.04$ is the "critical statistic" that represents the number of standard deviations around the mean wherein we expect to observe the actual dependent variable 95% of the time. In this case, with the large SE, the 95% confidence interval is very wide, ranging from about 1917 to 5704.

The last line of Exhibit 6.8 shows the F-statistics of 45.8 and 34.2 for regression models 1 and 2. The F-statistic is an overall significance test of the regression, and is more relevant when we have multiple independent variables. The F-statistic is analogous to the t-statistic in that it tests the hypothesis that there is no relation between the dependent variable and the independent variables. If F is large, it supports the hypothesis that there is in fact a significant relation between the dependent variable and the set of independent variables. The critical value of the F-statistic depends on the sample size, n, and the number of independent variables, k. In this example, with 32 observations and two independent variables (including the intercept), the critical value for F at the 5% significance level is $F_{k-1,n-k} = F_{1,14} = 4.6$. For either model, the F-statistic is far greater than the critical level, supporting the hypothesis that either regression model is valid.*

These diagnostic statistics show slightly greater support for model 1 than model 2, indicating that model 1 fits the data from the estimation period better than model 2. The next step is to see which model seems to give us better predictions.

6.4 EVALUATING FORECAST ACCURACY

6.4.1 Diagnostic Measures

In the previous section we evaluated how well the two regression models fit data from the 32-quarter estimation period. We now measure how well the models forecast sales for the 12-quarter hold-out period.

Exhibit 6.9 shows actual sales, forecast sales, and error terms for the constant change and compound growth regression models for quarters 33–44. The error terms (Actual Sales – Predicted Sales) are used to calculate the statistics for evaluating forecast accuracy. The statistics we consider are the ME, MAD, RMSE, and the RSE. Each statistic is calculated in the same way as for the estimation period, except now we look only at the hold-out period.

---

* As with the t-statistic, most regression packages report the p-value for the F-statistic.

	A	B	C	D	E	F	G	H
1					Speckled Band			
2					Constant-Change		Compound-Growth	
3					Model		Model	
4						Error		Error
5	Year	Quarter	Time	Sales	Forecast	Term	Forecast	Term
6	9	I	33	6400.5	5,855	546	5,850	551
7		II	34	6216.7	5,975	242	6,022	195
8		III	35	4930.6	6,095	-1,165	6,200	-1,269
9		IV	36	4986.7	6,216	-1,229	6,382	-1,396
10	10	I	37	5338.0	6,336	-998	6,570	-1,232
11		II	38	5087.5	6,456	-1,369	6,764	-1,677
12		III	39	4907.6	6,576	-1,669	6,964	-2,056
13		IV	40	5021.5	6,697	-1,675	7,169	-2,147
14	11	I	41	5122.1	6,817	-1,695	7,380	-2,258
15		II	42	5392.3	6,937	-1,545	7,598	-2,205
16		III	43	7238.7	7,057	181	7,822	-583
17		IV	44	7219.5	7,178	42	8,052	-833
18								
19				ME		-861.1		-1242.5
20				MAD		1029.5		1366.8
21				RMSE		1189.4		1529.2
22				RSE		1302.9		1675.2

EXHIBIT 6.9   Forecasts and errors for hold-out period from two regression models.

6.4.1.1 Mean Error

The most obvious measure of forecasting accuracy is an average of the error terms, known as the ME. It is calculated as a simple average of the errors. The MEs over the hold-out period for both of the regression models are shown in line 19 of Exhibit 6.9. Both models have negative MEs, indicating that, on the average over the hold-out period, the forecast is biased with the actual sales less than the forecast. Because the compound growth model's ME is larger (−1242.5 versus −861.1), we are getting a more biased forecast from the compound growth model.

Even if the forecast is unbiased, it may not be accurate. We could have a model with very large errors, but they could average zero. To get a better idea of the accuracy of our forecasts, we look at the MADs.

6.4.1.2 Mean Absolute Deviation

The MAD provides another perspective on the accuracy of our model. The MAD tells us the magnitude of the average error, in either direction, without considering the sign. The absolute values of the error terms are summed and divided by the number of observations, as shown in Equation 6.8

$$\text{MAD} = \sum_{i=1}^{n} \frac{|e_t|}{n}. \qquad (6.8)$$

The constant change model's MAD of 1029.5 versus 1366.8 for the compound growth model, telling us the constant change model makes smaller average forecasting errors.*

6.4.1.3 Root Mean Square Error

The RMSE is the square root of the average squared error term. It is a better indicator of the accuracy of the forecast model than ME or MAD when avoiding the occasional large error is important. Squaring the error term over-weights large errors. Hence, RMSE penalizes models with a few large errors versus models with multiple small- to mid-sized errors. The formula for RMSE is shown in Equation 6.9

$$\text{RMSE} = \sqrt{\sum_{i=1}^{n} \frac{e_t^2}{n}}. \qquad (6.9)$$

The fact that RMSE for the constant change model is smaller than for the compound growth model (1189.4 versus 1529.2) is additional evidence that the constant change model yields better forecasts for the hold-out period.

6.4.1.4 Residual Standard Error

The RSE, also known as the SE of the estimate, is a more robust standard for assessing the forecasting accuracy of a model. Up to now, the goodness-of-forecast statistics have been simple averages. None has taken into account the number of parameters estimated in the model. This is important because as the number of explanatory variables (hence, estimated parameters) in a model increases, the model's forecasting accuracy can appear to increase even if the additional variables have little or no ability

---

* Calculating the MAD is quite simple in Excel if an array function is used. Excel's array functions operate on arrays of numbers rather than single numbers. In Exhibit 6.9, the code for the MAD of the constant change model is =AVERAGE(ABS(F6:F17)). Array functions are entered by pressing Ctrl+Shift+Enter after typing in the formula.

238 ■ Introduction to Financial Models for Management and Planning

to explain the data. The goal of forecasting is to find the most parsimonious model, that is, to find the model that gives the best forecasts with the fewest parameters. The RSE tells us if adding more variables produces better forecasts.

The first two steps in calculating the RSE are identical to those in calculating the RMSE. The error terms are squared and then summed. However, instead of dividing by the number of observations (n), we divide by the number of observations less the number of estimated parameters (k), that is, ($n - k$). As the number of parameters in a model increases, the denominator decreases, thereby increasing the RSE. This penalizes more complex models: Unless the additional variables produce better forecasts, the RSE increases. Equation 6.10 gives the formula; Equation 6.11 shows the calculations for the RSE of the constant change model.

$$\text{RSE} = \left[\sum_{i=1}^{n} \frac{e_t^2}{(n-k)} \right]^{1/2}, \tag{6.10}$$

where k is the number of estimated parameters in the model

$$\text{RSE} = \sqrt{\frac{\left(546^2 + 242^2 + -1165^2 + -1229^2 + -998^2 + -1369^2 + -1669^2 + -1675^2 - 1695^2 - 1545^2 + 181^2 + 42^2 \right)}{12 - 2}}$$

$$= 1302.9. \tag{6.11}$$

Once again, the constant change model yields better forecasts than the compound growth model. In this case, all our statistics for evaluating the forecasts agree: the constant change regression model works better for Speckled Band in the hold-out period. Note however that the goodness-of-forecast statistics will not necessarily always agree. We could find that one model is less biased, but does not generate as good a forecast (a smaller ME but larger MAD); or, a model might yield smaller RMSE but the results are due to spurious correlations associated with using many irrelevant independent variables.

6.4.2 Combining the Estimation Period and the Hold-Out Period

To complete the forecasting process, re-estimate the parameters for best model using the entire data set. For example, assume that we have tested

	F	G	H	I	J	K	L
1	SUMMARY OUTPUT						
2							
3	*Regression Statistics*						
4	Multiple R	0.80					
5	R Square	0.64					
6	Adjusted R Square	0.64					
7	Standard Error	928.7					
8	Observations	44					
9							
10	ANOVA						
11		*df*	*SS*	*MS*	*F*	*Significance F*	
12	Regression	1.00	65,405,429	65,405,429	75.84	0.00	
13	Residual	42.00	36,220,565	862,394			
14	Total	43.00	101,625,995				
15							
16		*Coefficients*	*Standard Error*	*t Stat*	*P-value*	*Lower 95%*	*Upper 95%*
17	Intercept	2,196.92	284.84	7.71	0.00	1,622.1	2,771.7
18	Time	96.01	11.02	8.71	0.00	73.8	118.3

EXHIBIT 6.10   Regression output for the whole sample period.

several models including both the constant change and the compound growth regression models, and we have concluded that the constant change model is our best forecaster. We now re-estimate the constant change model using all 44 quarters. The output from this regression is shown in Exhibit 6.10. Our re-estimated regression for sales as a function of time is now estimated as

$$\text{Sales}_t = 2196.9 + 96.0 \times \text{Time}_t$$
$$\quad\quad\quad\; (7.71) \quad\quad (8.71)$$

$$R^2_{\text{adj}} = 0.64\ (64\%). \tag{6.12}$$

The diagnostic statistics are slightly improved over the shorter estimation period, so the regression seems robust. The regression parameter estimates are different from those of the original estimation period shown in Exhibit 6.4. It is expected that they will vary a little with the addition of the data from the hold-out period. If the estimated parameters are substantially different it may indicate that the link between sales and time had changed between the two periods. Or, perhaps there is something else besides the time trend that is important for our forecast.

6.4.3 The Last Step: Making the Forecast

Having re-estimated the relationship between sales and time using data from the entire sample period, we now use the estimated equation to forecast future sales. Unfortunately, we will not know how accurate the forecasts are until the future unfolds. We want to forecast the next four

Period	45	46	47	48
Sales forecast	6516.9	6612.9	6708.9	6804.9
95% confidence interval				
Upper bound	8477	8580	8682	8784
Lower bound	4557	4647	4737	4827

EXHIBIT 6.11 Forecast of speckled band sales based on time trend regression.

quarters: periods 45–48. Applying the re-estimated regression to the 45th quarter, our forecast is

$$\text{Sales}_{45} = 2196.9 + 96.0 \times 45$$
$$= 6516.9.$$

Each of the following quarters is forecasted similarly as shown in Exhibit 6.11. The second line of the table shows the confidence intervals for our forecasts, explained next.

6.4.4 Assessing Forecast Accuracy: Confidence Intervals

One advantage of quantitative over qualitative methods is that they provide a way to estimate the accuracy of the forecasts—a margin of error. This is particularly useful in financial modeling, where the goal is to determine the joint consequences of forecast events and decisions. Because forecasts almost never will be correct, it is important to know how far off the mark they might be. This is what the confidence intervals tell us.

Even when the model is correctly specified, there are three inter-related sources of error. To see why, consider the regression model with one independent variable X_t, and dependent variable Y_t:

$$Y_t = a + b \times X_t + e_t. \tag{6.13}$$

First, even when the estimates of a and b are correct, there is the error term e_t. We assume that it is normally distributed with zero mean and standard deviation σ_e. This error term means that the actual Y will be distributed around the forecast Y due to pure chance. However, this assumes that our estimates of a and b are correct, bringing us to the second and third sources of error. We only estimated the regression coefficients a and b. The true coefficients fall within some confidence interval around the estimates. Given all of this opportunity for error, we know that our forecasts probably will be wrong. The question is, how wrong? Confidence intervals tell us the interval around our forecast in which we would expect the actual value to occur.

Confidence intervals are based on the SE of the estimate, a statistic reported regression software. The SE or RSE is converted to the SE of the forecast ($SE_{forecast}$). In turn, the $SE_{forecast}$ is used to calculate the confidence intervals for the forecast.

When the estimated regression has considerable explanatory power (a high adjusted R^2 and small SE), the confidence interval is a narrow band. We can be reasonably confident in our forecast. However, when the regression lacks explanatory power (a low adjusted R^2 and large SE), the confidence interval is a wide band. We cannot have much confidence in our forecast.

The general expression for a confidence interval when the error terms are normally distributed is given by Equation 6.14, where Z is the standard unit-normal variable.*

$$\text{Confidence Interval}_t = \text{Forecast}_t \pm Z \times SE_{forecast_t}. \qquad (6.14)$$

The confidence interval forms a band around the forecasts within which the true value of a predicted variable should fall at a specified percentage of the time.

Three elements are needed to calculate confidence intervals: the forecast, the $SE_{forecast}$, and the critical statistic. The forecast is derived in the usual way. The other two arguments require some discussion. To explain the confidence interval, we will use the one-factor (one independent variable) model for Speckled Band. When we have more than one independent variable it becomes more complicated, and the reader should consult a more specialized source such as Neter (1996).

6.4.5 The Standard Error of the Forecast

The $SE_{forecast}$ measures the scatter of individual actual observations about the sample regression line. It is the estimated error when making a forecast for Y given X. The $SE_{forecast}$ is the mechanism that adjusts the width of

---

* For smaller samples, we use the t-distribution instead of the standardized unit normal distribution, Z, to calculate the confidence interval.

the confidence interval to reflect the distance from the center of the data. The formula is given in Equation 6.15

$$SE_{forecast} = \sigma_e \left[1 + \frac{1}{n} + \frac{(X_f - \bar{X})^2}{\sum_{i=1}^{n}(X_i - \bar{X})^2} \right]^{\frac{1}{2}}, \quad (6.15)$$

where σ_e is the SE of the regression or the residual SE from fitting the predicted values and X_f the forecast for the independent variable.

The confidence interval is narrowest at the center of the data, where the independent variable, X, and the dependent variable, Y, are both near their means. As the forecasts move away from the data's center (i.e., as X moves further from its mean, $\bar{X}$), the numerator of the third term inside the brackets (in Equation 6.15) increases at an increasing rate. This increases the confidence interval and induces the curvature of the confidence bands shown in Exhibit 6.12. This curvature means that as we move away from our range of experience as represented by the center of the data, our forecasts become less and less accurate. That is, if we extrapolate Y for values of X that are far from the values of X and Y that were used to estimate our model, we are less and less confident that our forecast will be accurate. If we use recent data to generate our forecast, we are more confident about our forecast for next year than we are forecasting 20 years into the future.

For our Speckled Band forecast, we want to calculate the $SE_{forecast}$ for period 45 (quarter 9.I), the quarter following the last one in our sample

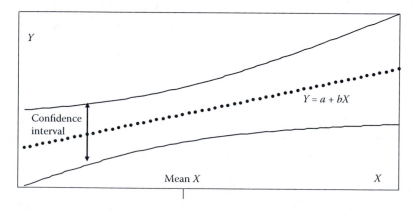

EXHIBIT 6.12   Forecast confidence interval.

period. The first step is to gather the following numbers from the regression output and a spreadsheet program

$$n = 44$$
$$\sigma_e = 928.65$$
$$\overline{X} = 22.5 \ (X \text{ is Time})$$
$$\Sigma(X_i = \overline{X})^2 = 7095.$$

Next, use these numbers to calculate the $SE_{forecast}$ from Equation 6.15. The $SE_{forecast}$ for period 45 (where Time = X = 45) is calculated as

$$SE_{forecast} = 928.65 \times \left[1 + \frac{1}{44} + \frac{(45 - 22.5)^2}{7095}\right]^{\frac{1}{2}}$$

$$= 971.4. \tag{6.16}$$

6.4.5.1 The Critical Statistic

The last argument in the expression for the confidence interval is the critical statistic that specifies the width of the confidence interval. The confidence interval is the range, or band, around the value predicted by the regression model, within which we say the actual sales will be likely to occur. If we are willing to tolerate a higher likelihood of having the actual sales occur outside this band, we would set a narrow confidence interval. For example, our model predicts that sales in period 45 will be $6516.9. However, we know that sales will not be exactly this amount. Suppose we predict that sales will be within $100 of this amount—that is, between $6416.9 and $6616.9. Given the large SE of the regression, there is a very low likelihood that sales will actually fall within this narrow interval. If we want greater confidence that sales will actually turn out to be within the band we specify, then we need to set a wider band. For example, suppose we predict period 45 sales to be 6516.9 ± $2000—that is, between $4516.9 and $8516.9. Clearly, we will be more confident that actual sales will be within this wide band than within the narrower $100 band.

The critical statistic sets the width of the confidence band based on our specification of our desired probability of being right and the acceptable probability of being wrong. The link between the target probability and the width of the confidence band is the standard deviation (SE). The critical statistic specifies the width of the confidence interval in standard deviation units. For example, suppose we can tolerate a 5% chance that we are wrong. This means we want a 95% chance that sales will actually be within the specified interval around the predicted value. If sales are normally distributed around the prediction, we can consult a normal probability table to determine the 95% bounds. For a normally distributed variable, the 95% bounds are (mean − 1.96σ) and (mean + 1.96σ). The 1.96 is the critical statistic, specifying the number of standard deviations above and below the expected value such that there is a 95% chance that the actual value will be within this range. If the sample is small or the distribution is unknown, we use the t-statistic and the t-distribution table instead of the standard normal table to determine the critical statistic.

The current goal is to find the interval that is likely to contain the actual sales for period 45 95% of the time. In this example, we use the t-statistic because we have a relatively small sample of 44 observations. The critical value of the t-statistic is determined by the degrees of freedom in our model and the level of confidence we desire. We used 44 observations to estimate our model's two parameters. This leaves 42 (44 − 2) degrees of freedom. Our desired level of confidence is 95% (2.5% of the observations should fall in either tail of the distribution). Therefore, we find the t-statistic for a two-tailed test and a 5% level of significance. From a t-distribution table or Excel's =TINV( ) function, the t-statistic is 2.018.* All the pieces necessary to calculate the confidence interval are now in place. All that remains is to calculate the confidence interval:

$$E[\text{Sales}_{45}] = 2197 + 96.0 \times 45$$

$$= 6517$$

$$\text{Confidence Interval} = E[\text{Sales}_{45}] \pm t_{\substack{df=42 \\ \alpha=0.05}} \times \text{SE}_{\text{forecast}}$$

$$= 6517 \pm 2.018 \times 971.4$$

$$= 6517 \pm 1960$$

$$= \text{lower limit of 4557 to upper limit of 8477.} \quad (6.17)$$

This confidence interval is very wide, providing little confidence in our ability to forecast with the simple time trend (single-factor) model. So what is wrong with our forecasting model? In the next section we discuss some of the problems with our model.

---

* For $n > 30$ or so, t- and Z-statistics are virtually identical.

6.4.6 Problems with the Forecasting Model

We have used a simple time trend regression model to show the method and steps for developing a forecasting model. However, we have seen that this model generates a forecast within a very wide range, so it is not as accurate as we might like. There are myriad reasons why we may not be able to generate a forecast that is as accurate as we would like. We discuss two of them. First, the data may be inconsistent with the assumptions of the regression model. Second, we may have the wrong model in that (a) sales may not really follow a trend, or (b) there may be other influences in addition to the trend.

6.4.6.1 Assumptions of the Regression Model

The reliability of linear regression depends on whether the data are consistent with the assumptions underlying the mathematics of regression. The major assumptions are

1. The form of the relationship between the dependent variable, Y_t, and the independent variable, X_t, is a linear function of the form

$$Y_t = a + bX_t + e_t,\qquad(6.18)$$

where:

2. The error terms, e_t, are normally distributed with mean zero and a standard deviation, σ_e, which is constant over time; and

3. The error terms are uncorrelated between periods.

Assumption 1 is that we have the right model. If the real relationship is actually $Y_t = a + bX_t^2$, then Equation 6.18 is simply the wrong model. Similarly, if there are other variables, W and Z, that have an important influence on Y, and the actual relation is of the form

$$Y_t = a + bX_t + cW_t \times dZ_t + e_t,$$

then Equation 6.18 may not yield reliable estimates of coefficients a and b, and may not provide a good forecast.

Assumption 2 is that the error terms are homoskedastic, that is, the dispersion of error terms around the regression line is the same at all values of the independent variable. If the spread of error terms around the regression line changes systematically with the independent variable, then the errors are said to be heteroskedastic. In that case the regression can still yield valid estimates of the regression coefficients, but the estimated SEs will be biased.

Assumption 3 is that there is no serial correlation in the error terms. Serial correlation of error terms means that the error or deviation away from the regression line in one period is related to the size and direction of the error in a previous period. For example, our time series regression exhibits serial correlation. In Exhibit 6.5, in the early period the data points tend to cluster above the regression line, in the middle periods they are all below the regression line, and in the last four periods they are above the line. This is visual evidence of serial correlation of the error terms. The statistical evidence of serial correlation is the Durbin–Watson (DW) statistic. A DW statistic near 2 indicates that serial correlation is not a problem. Positive serial correlation will result DW being below 2. DW will be between 2 and 4 if there is negative serial correlation. For Speckled Band sales for the 32-quarter estimation period, the DW statistic is 0.676.* This low value of DW is a strong indicator of positive serial correlation in the residuals.

Like heteroskedasticity, the presence of serial correlation does not necessarily invalidate estimates of the regression coefficients, but it does bias the estimates of their SEs. This means, for example, that while our estimate of slope coefficient b is consistent and unbiased, the estimated SE of b may be smaller than the actual SE. In this case, we would be led to believe that the confidence interval for b is narrower than it really should be, and that the forecast is more accurate than it really is. We have already seen that the confidence interval for our forecast is very wide. With the presence of serial correlation, we conclude that this already wide confidence interval may be even wider than we think it is.

There are several reasons why we may have serial correlation in the regression. Frequently, we get serially correlated errors when we have left out of the model other factors that influence the dependent variable. That is, we have omitted variables. If we add these omitted variables to the regression model, we will tend to get a better fitting model, and hopefully a better forecast. This is the subject of the next chapter.

---

* Regression in Excel does not compute the DW statistic. Other regression software such as EViews routinely shows this statistic.

6.5 SUMMARY

This chapter is an introduction to using linear regression to develop a simple time trend forecasting model. We used two simple models to demonstrate the basic methods of developing and testing a forecasting model. Though more complex models might be able to provide better forecasts, the methods for developing and testing the models would still follow the same basic steps.

The steps recommended for developing and testing the forecasting model include:

- Develop hypotheses about what factors and variables affect the forecast variable.

- Gather data to test your hypotheses.

- Develop the models for testing.

- Separate the time series of data into an estimation period and a hold-out period.

- Test your alternative models on data from the estimation period.

- Evaluate the validity and fit of the models for the estimation period.

- Apply the superior models to forecasting the hold-out period.

- Assess which models generate the best forecast results for the hold-out period.

- Re-estimate the relationships for the whole sample period for the models that provide the best forecasts.

- Use the re-estimated model to forecast period following the sample period.

Evaluating a forecasting model can differ from the standard methods for testing statistical hypotheses and evaluating regressions. The regression model that provides the best fit of the data may not necessarily provide the best forecast. For comparing different forecasting models, several summary measures were discussed:

- Mean error (ME)

- Mean absolute deviation (MAD)

- Root mean square error (RMSE)

- Residual standard error (RSE)

In the next chapter we extend these methods to consider relationships more complex than a simple time trend.

PROBLEMS

1 Antsy Corporation

The following table reports the annual sales for Antsy Corporation over the past nine years. Use this information to perform tasks a–d. Estimate your models using all nine years of data.

Antsy Corp.	
Year	Revenue
1	26.1
2	14.7
3	20.1
4	29.5
5	23.1
6	37.9
7	57.4
8	91.1
9	120.7

(a) Forecast Antsy's sales for years 10–13 using the simple average percentage increase method.
(b) Forecast Antsy's sales for years 10–13 using the geometric average percentage increase method.
(c) Forecast Antsy's sales, along with their 95% confidence intervals, for years 10–13 using the constant change linear model.
(d) Forecast Antsy's sales for years 10–13 using the compound growth log-linear model.

2 Zephyr Corporation

The Zephyr Corporation markets wingbrats in the United States, which it imports from Zongthia. Its wingbrats are of the highest quality, and the market is limited. However they are gaining in acceptance and popularity, and Zephyr's sales have been growing over the last eight years. The chief financial officer (CFO) of Zephyr has asked you to analyze their sales data and develop a simple forecasting model. Zephyr's quarterly sales for the last 7.5 years are shown in the table below.

			Zephyr Corp.			
Year	Quarter	Sales		Year	Quarter	Sales
1	I	303		5	I	467.7
	II	278.9			II	560.6
	III	354.8			III	628.8
	IV	411.1			IV	556.5
2	I	422.9		6	I	587.4
	II	459.3			II	562.4
	III	317.7			III	571.1
	IV	410.6			IV	620.9
3	I	478.8		7	I	644.1
	II	406.5			II	605.3
	III	453			III	682.3
	IV	428.9			IV	703.8
4	I	504.8		8	I	740.4
	II	561.1			II	735.4
	III	572.9					
	IV	609.3					

To develop your forecasting model, you will use the simple trend methods discussed in Chapter 6. Begin by graphing the series. Next, determine which of the following methods yields the best fit with the historical data. The methods to consider are

i. Average percentage increase per quarter
ii. Geometric mean percentage increase
iii. Constant change linear model
iv. Compound growth log-linear model

Evaluate the goodness-of-fit of each method using ME, MAD, RMSE, and RSE. Determine the method that yields the best fit, and then use that method to predict sales for the next four quarters (8.III–9.II). Add your predictions to the initial graph of Zehpyr's sales.

What You Turn In

Submit a report in memo format that describes the results of your work, including which goodness-of-fit statistic determined your choice of model and why you chose that statistic. Include a time-series graph of sales, actual and predicted, in your memo. Your graph should contain a title and axis labels. Also include a table that summarizes all of goodness-of-fit statistics similar to Exhibit 6.9.

3 G&A Acquisitions

Imagine that you work for an econometric consulting firm. Imagine, too, that you have a client named G&A Acquisitions who is interested in acquiring Lindberg, Inc., a small consumer staples firm that has been growing steadily for 10 years. G&A has asked your firm to build a model to forecast Lindberg's quarterly sales over the next year and a half. They have supplied you with the data shown below.

Year	Quarter	Time	Sales	Year	Quarter	Time	Sales
1 I		1	759.2	6 I		21	1,491.2
	II	2	761.4		II	22	1,674.4
	III	3	863.1		III	23	1,790.2
	IV	4	629.5		IV	24	1,704.6
2 I		5	838.8	7 I		25	1,825.2
	II	6	917.6		II	26	1,975.0
	III	7	962.0		III	27	1,795.8
	IV	8	978.8		IV	28	1,667.4
3 I		9	1,272.0	8 I		29	1,732.8
	II	10	1,051.4		II	30	1,956.0
	III	11	1,325.0		III	31	2,030.7
	IV	12	1,249.8		IV	32	1,855.0
4 I		13	1,378.2	9 I		33	2,088.7
	II	14	1,269.0		II	34	2,264.3
	III	15	1,258.8		III	35	2,057.7
	IV	16	1,301.5		IV	36	2,180.6
5 I		17	1,514.0	10 I		37	2,101.0
	II	18	1,602.0		II	38	2,417.4
	III	19	1,397.5		III	39	2,364.2
	IV	20	1,556.3		IV	40	2,116.3

Lindberg, Inc. Quarterly Sales

Your task is to identify the model that best forecasts Lindberg's sales—the constant change model or the compound growth model. Begin by graphing Lindberg's sales. Estimate your preliminary models using the first seven years of data and test your models using the last three years of data. Measure their forecasting ability using the goodness-of-forecast statistics discussed in Chapter 6 (RMSE, ME, MAD, and RSE).

After identifying the model that produces the best forecasts, re-estimate that model using all 10 years of data. Label your graph appropriately. Finally, use this newly estimated model to forecast sales for the next six quarters. Add the six forecasts to your original graph as a separate series so that your forecasts are clearly identified.

What You Turn In

Submit a report in memo format that describes the purpose and results of your work, including the criteria you used to select your forecasting model. Place the graphs of sales (including predicted sales) in the body of your memo. Also include a table in your memo that reports the model used (e.g., $\text{Sales}_t = 45 + 4 \times \text{Time}_t$) and the sales predicted. In an appendix, place a well-organized printout of the spreadsheet showing the goodness-of-forecast calculations you used to identify the best model. Use Exhibit 6.9 as your template.

CHAPTER 7

Forecasting II: Econometric Forecasting

C HAPTER 6 INTRODUCED FORECASTING using linear regression to extrapolate a time trend. For most companies, though, that model is too simple. Other forces besides a simple trend can influence a company's sales. Most businesses are subject to the cycles of the economy, the vagaries of interest rates, and other economic and demographic factors. Our knowledge of these links may allow us to improve our forecasts. In this chapter we explain how to account for these other economic influences and include them in a forecasting model. We call the resulting model a "structural econometric model" to reflect the fact that our company is affected by the structure and performance of its industry and the economy.

7.1 DEVELOPING A STRUCTURAL ECONOMETRIC MODEL

The first step in developing a structural econometric model is to state our hypotheses about the linkages between our company and economic variables that influence it. A good starting point is a description of our company and how we think it is related to other parts of the economy.

Armed with the description, we can refine it until we have captured the most important influences.

In this regard, it is important to have as short a list as possible because we want our model to be *parsimonious*. We want only the most important variables in our model, each of which captures a different aspect of the company's relationships with the economy. Indiscriminately including many different variables can distort our forecast with spurious correlation, meaning have added variables that add little real explanatory power to our model. In addition, we do not want the variables to overlap, that is, to measure the same influence. Highly correlated independent variables in a regression model reduce the reliability of the estimated relationships in the model. Finally, we need to keep in mind that we are building a forecasting model. Any variable that we use as an input to the forecast must, itself, be known or more easily forecasted than our company variable. Including too many variables also could overwhelm our forecasting ability.

Once the hypotheses are formulated about the variables that influence our company, we gather the data, estimate the models, and then test the hypotheses. Based on our statistical analysis, some models will be eliminated immediately. The few remaining are then tested for their forecasting ability.

7.1.1 Example: Speckled Band, Inc.

Our task once again is to develop a forecasting model for Speckled Band, Inc., first introduced in Chapter 6. Its quarterly sales data are reprised in Exhibit 7.1. In Chapter 6 we used simple linear regression to provide a time trend forecast of Speckled Band's sales. However, the time trend forecast is unlikely to be reliable because the confidence intervals are so wide. Moreover, we found that the residuals were serially correlated. This led us to suspect that we had omitted other important variables from our regression. We should be able to improve on our time trend forecasting model if we take account of the variables that we did not consider in the time trend model.

The next step is to decide what variables we might have overlooked. This is where our hypothesis building comes into play. We need to think about the industry and economic environment of the company and specify the factors that we think affect Speckled Band's sales.

7.1.1.1 The Economic and Industry Context

Speckled Band manufactures industrial efficiency equipment used by other manufacturers. Since the products it sells constitute capital investment for

Year	Quarter	Time Period	Sales $000	Year	Quarter	Time Period	Sales $000
1	I	1	3,943	7	I	25	4,698
	II	2	2,823		II	26	4,057
	III	3	2,328		III	27	5,074
	IV	4	2,458		IV	28	6,598
2	I	5	2,792	8	I	29	6,563
	II	6	3,256		II	30	7,226
	III	7	3,177		III	31	6,423
	IV	8	3,726		IV	32	6,409
3	I	9	3,805	9	I	33	6,400
	II	10	2,246		II	34	6,217
	III	11	3,047		III	35	4,931
	IV	12	2,106		IV	36	4,987
4	I	13	1,339	10	I	37	5,338
	II	14	3,190		II	38	5,087
	III	15	3,023		III	39	4,908
	IV	16	3,498		IV	40	5,021
5	I	17	2,948	11	I	41	5,122
	II	18	3,237		II	42	5,392
	III	19	3,917		III	43	7,239
	IV	20	3,230		IV	44	7,220
6	I	21	3,985				
	II	22	3,826				
	III	23	4,159		Average Growth		4.21%
	IV	24	4,747		Compound Growth		1.42%

EXHIBIT 7.1   Speckled Band Inc.: Quarterly sales.

the buyers, we would expect that Speckled Band's sales are cyclical, tracking the overall level of industrial equipment purchases in the economy. When the economy turns up, manufacturers increase investment in additional equipment; when the economy turns down, they decrease those investments. Consequently, we would expect that Speckled Band's sales to be highly correlated with the general level of economic activity. In addition, if sales of their equipment follow the pattern of other capital equipment, we would expect that their sales would tend to decline if interest rates increase, because higher rates decrease the present value of their benefits for the purchasers.

However, discussions with the sales manager of Speckled Band suggest that this relationship might go the other way. That is, he has observed that the company's sales cycle seems to follow the interest rate cycle. His observation is that rates increase when demand for industrial investment is high, so that even though high rates tend to decrease the NPV of customer purchases, this happens when demand is coincidentally high. Thus high rates are a measure of high demand for capital goods, and this effect tends to overshadow the NPV-depressing effect of interest rates. Given these

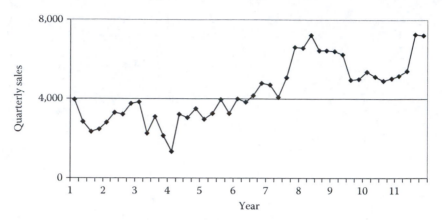

EXHIBIT 7.2   Speckled Band Inc.: Time pattern of quarterly sales.

observations, we should be able to use the links between economic activity, interest rates, and company sales to develop a forecasting model. Our hypothesis is that Speckled Band's sales will increase with the level of economic activity as measured by gross domestic product (GDP) and the level of interest rates as measured by the rate on long-term U.S. Treasury bonds.

The starting point of the analysis is to plot Speckled Band's sales, shown in Exhibit 7.2, then look for possible influences in the data. In the last chapter we confirmed a time trend in sales, with sales growing from about $4000 to more than $7200 over 11 years. But the plot in Exhibit 7.2 does not show purely random variation about a trend. Rather, sales tend to remain flat for the first six years, grow rapidly between years 6 and 9, only to fall back and then grow again in years 9–11. This pattern could indicate the presence of a cyclical influence.

A cyclical element, if present, can be made more visible by graphing the residual error terms that result from fitting a simple trend line. Most regression packages offer the option of reporting the residual errors from a regression. (In Excel®, simply check the "Residuals" box in the regression wizard.) The residual plot for Speckled Band is shown in Exhibit 7.3. The error terms do not appear to be random. They first decrease, then remain negative beginning in year 3, then increase beginning in year 7, and then decrease again in year 9. Each error appears to be related to the errors that immediately precede or follow it, producing a smooth, non-periodic wave-like pattern, a pattern similar to that identified in Exhibit 7.2. Such a pattern may indicate the presence of a long-term cycle.

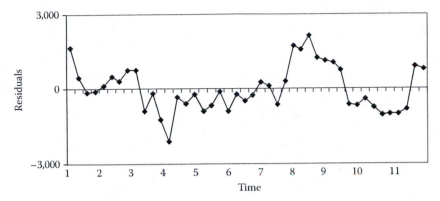

EXHIBIT 7.3 Speckled Band Inc.: Residual plot (trend as explanatory variable).

Given our previous conjecture about the relationship between Speckled Band's sales and the business cycle, Exhibit 7.3 suggests that adding variables that capture the business cycle might improve our model's forecasts. We hypothesize the following model, using GDP and the 10-year Treasury bond rate to capture the influence of the business cycle.

$$\text{Sales}_t = a + b_1 \times \text{GDP}_t + b_2 \times \text{T-bond rate}_t + b_3 \times \text{Time}_t \qquad (7.1)$$

The data needed to estimate and test this model are given in Exhibit 7.4. As we did in Chapter 6, we partition the data into a 32-observation estimation

Speckled Band
Quarterly Sales & Economic Data

Year	Quarter	Time	Sales	GDP	10-year TB	Year	Quarter	Time	Sales	GDP	10-year TB
1	I	1	3,943	6542.7	6.60	7	I	25	4,698	9066.6	4.72
	II	2	2,823	6612.1	5.97		II	26	4,057	9174.1	5.18
	III	3	2,328	6674.6	5.81		III	27	5,074	9313.5	5.79
	IV	4	2,458	6800.2	5.33		IV	28	6,598	9519.5	6.11
2	I	5	2,792	6911.0	5.75	8	I	29	6,563	9629.4	6.66
	II	6	3,256	7030.6	6.97		II	30	7,226	9822.8	5.99
	III	7	3,177	7115.1	7.30		III	31	6,423	9862.1	6.05
	IV	8	3,726	7232.2	7.74		IV	32	6,409	9953.6	5.74
3	I	9	3,805	7298.3	7.78	Hold Out Period:					
	II	10	2,246	7337.7	7.06	9	I	33	6,400	10024.8	5.16
	III	11	3,047	7432.1	6.28		II	34	6,217	10088.2	5.14
	IV	12	2,106	7522.5	6.04		III	35	4,931	10096.2	5.24
4	I	13	1,339	7624.1	5.65		IV	36	4,987	10193.9	4.57
	II	14	3,190	7776.6	6.51	10	I	37	5,338	10329.3	5.04
	III	15	3,023	7866.2	6.87		II	38	5,087	10428.3	5.21
	IV	16	3,498	8000.4	6.53		III	39	4,908	10542.0	4.65
5	I	17	2,948	8113.8	6.58		IV	40	5,021	10623.7	3.94
	II	18	3,237	8250.4	6.89	11	I	41	5,122	10735.8	4.05
	III	19	3,917	8381.9	6.22		II	42	5,392	10846.7	3.96
	IV	20	3,230	8471.2	6.03		III	43	7,239	11107	3.98
6	I	21	3,985	8586.7	5.54		IV	44	7,220	11262	4.29
	II	22	3,826	8657.9	5.64						
	III	23	4,159	8789.5	5.46						
	IV	24	4,747	8953.8	4.53						

EXHIBIT 7.4 Speckled Band Inc.: Quarterly sales and economic data.

SUMMARY OUTPUT						
Regression Statistics						
Multiple R	0.969					
R Square	0.939					
Adjusted R Square	0.933					
Standard Error	376.724					
Observations	32					
ANOVA						
	df	SS	MS	F	Significance F	
Regression	3	61322341.75	20440780.58	144.03	0.00	
Residual	28	3973792.279	141921.1528			
Total	31	65296134.02				
	Coefficients	Standard Error	t Stat	P-value	Lower 95%	Upper 95%
Intercept	-61514.41	5178.47	-11.88	0.00	-72122.0	-50906.8
Time	-970.30	92.85	-10.45	0.00	-1160.5	-780.1
GDP	9.81	0.82	11.90	0.00	8.12	11.50
10-year TB	255.36	95.80	2.67	0.01	59.11	451.61

EXHIBIT 7.5   Regression results from model (7.1).

segment and a 12-quarter hold-out period. Regression results for model (7.1) are shown in Exhibit 7.5.

All the statistics in the regression output look good. The adjusted R^2 is 93%, as compared to 59% for the time trend regression, and the t-statistics confirm the link between sales and GDP and interest rates. However, there are other statistics that we should check. First, as with the time trend model, we should check for serial correlation of the error terms. Second, we want to make sure that the independent variables are not too closely related to each other.

We check for serial correlation of the residuals with the DW statistic. If the DW statistic is near 2.0, it indicates that serial correlation among the error terms is not a problem. For this regression, the DW statistic is 1.96.* The inclusion of the two economic variables has eliminated the serial correlation that was present in the time trend model.

The second issue is whether there is multicollinearity in the model— that is, whether the independent variables are highly correlated. If they are, estimates of their regression coefficients are unreliable. Multicollinearity is a common problem in regressions with time series data where some or all of the variables are following a similar time trend. This model has two variables that could be highly related: Time and GDP. If GDP tends to follow a trend and increase with time, GDP and Time will be highly correlated. In Excel we quickly check for correlation with the

* Regression in Excel does not provide the DW statistic, whereas, most statistical software provides it. The EViews program was used to check these other statistics such as DW.

	Sales	Time	GDP	10-year TB
Sales	1.000			
Time	0.802	1.000		
GDP	0.833	0.998	1.000	
10-year TB	-0.461	-0.734	-0.726	1.000

EXHIBIT 7.6   Correlation matrix.

correlation tool in Tools > Data Analysis > Correlation, yielding the correlation matrix as shown in Exhibit 7.6. We see that all the independent variables are correlated. But the relation that draws our attention is the 99.8% correlation between GDP and Time, which is considerably higher than the correlation between Sales and Time. This very high correlation means that the estimated coefficients may not be reliable.

To obtain more reliable estimates, we can eliminate one of the highly correlated independent variables, or perhaps change the form of the model. For example, we could eliminate Time and use GDP to capture both the trend and cyclical aspects of Sales. However, we would like to separate these two components in our model so that we can see the effects of the business cycle separate from the trend. One way to do this is to keep the trend variable, Time, and measure the cyclical component of GDP that is independent of the time trend. Another way to do this is to estimate the trend of GDP with the regression

$$GDP_t = a + b \times Time_t + u_t \qquad (7.2)$$

and to use the residuals from that regression, u_t, to measure the GDP cycle around its trend. To do this in Excel, click "Residuals" in the Regression dialog box when setting up this regression. The regression output will then report the residuals, u_t, for each of the 32 periods in our sample. These residuals then are used as a new independent variable that measures the cyclical component of GDP, denoted by $GDPu_t$. Our revised sales regression becomes

$$Sales_t = a + b_1 \times GDPu_t + b_2 \times T\text{-bond rate}_t + b_3 \times Time_t + e_t. \qquad (7.3)$$

The output for the three-factor least squares regression is shown in Exhibit 7.7 for the 32-quarter estimation period.

The fitted model is

$$Sales_t = 430.5 + 9.81 \times GDPu_t + 255.4 \times T\text{-bond rate}_t + 108.2 \times Time_t. \qquad (7.4)$$

SUMMARY OUTPUT

Regression Statistics	
Multiple R	0.97
R Square	0.94
Adjusted R Square	0.93
Standard Error	376.7
Observations	32

ANOVA

	df	SS	MS	F	Significance F
Regression	3	61,322,342	20,440,781	144.03	0.00
Residual	28	3,973,792	141,921		
Total	31	65,296,134			

	Coefficients	Standard Error	t Stat	P-value
Intercept	430.5	657.37	0.65	0.52
GDPu	9.81	0.82	11.90	0.00
Treasury	255.4	95.80	2.67	0.01
Time	108.2	8.06	13.42	0.00

EXHIBIT 7.7   Regression output for regression model (7.3).

Its adjusted R^2 of 0.93 indicates a very good fit: the three-factor model explains 93% of the variation in Speckled Band's quarterly sales. In addition, the t-statistics for each of the explanatory variables are significant at the 1% level of confidence, telling us that each variable contributes important information to our forecasts. Despite the apparent good fit, the SE of the regression, 376.7, is about 10% of the mean sales level. This is a relatively large SE for a forecasting model, and it will lead to wide confidence intervals around the forecasts. Statistics not shown are the DW, which at 1.96 shows no evidence of serial correlation, and the correlations between the independent variables that show no evidence of multicollinearity. We conclude that adding GDPu and the T-bond rate to the time trend model appears to capture much of the cyclical component in Speckled Band's sales.

These statistics, however, do not tell us which model (the time trend model or the three-factor model) will produce the most accurate forecasts. To answer this question, we use the models to forecast sales in the hold-out segment of our data (observations 33–44), then calculate goodness-of-forecast measures for both sets of forecasts. To calculate forecasts, plug historical values for Time, the T-bond rate, and GDPu, into each model, as appropriate. For example, during the first quarter of the year 9, Time was 33, the T-bond rate was 5.16%, and the GDPu was 84.3. Plugging these values into the three-factor model (Equation 7.5) produces the predicted sales for the first quarter of year 9, $6145. Plugging the value for Time into the one-factor model (Equation 7.6) produces the predicted sales for the first quarter of year 9 of $5855.

$$\begin{aligned} \text{Sales}_{33} &= 430.5 + 9.8 \times \text{GNP}u_{33} + 255.4 \times \text{T-bond rate}_{33} + 108.2 \times \text{Time}_{33} \\ &= 430.5 + 9.8 \times 84.3 + 255.4 \times 5.16 + 108.2 \times 33 \\ &= 6145, \end{aligned} \qquad (7.5)$$

and for the one-factor model,

$$\begin{aligned} \text{Sales}_t &= 1886.2 + 120.3 \times \text{Time}_t \\ &= 1886.2 + 120.3 \times 33 \\ &= 5856. \end{aligned} \qquad (7.6)$$

Repeating the calculations for each quarter through quarter 44, then calculating the error terms and goodness-of-forecast statistics, produces the table shown in Exhibit 7.8. Comparing the goodness-of-forecast statistics shows that the three-factor model produces the more accurate forecasts. Its RMSE and MAD are smaller, suggesting that the three-factor model produces fewer large forecast errors; its ME is smaller (on an absolute basis), suggesting that it is less biased; and its RSE is smaller, suggesting that the two additional variables (GDPu and T-bond rate) significantly increase the model's explanatory ability.*

The three-factor model is clearly the superior forecasting model. However, before using it to forecast the sales for year 12, we re-estimate it using all 11 years of data. Doing so yields our final forecasting model (Equation 7.7)

$$E[\text{Sales}_t] = 624.7 + 9.95 \times E[\text{GDP}u_t] + 223.8 \times E[\text{T-bond}_t] + 108.8 \times \text{Time}_t. \quad (7.7)$$

Note the terms $E[\text{GDP}u_t]$ and $E[\text{T-bond}_t]$. To forecast future sales, we must first forecast future values of GDPu and T-bond rates, adding still more uncertainty to our forecasts.† In this case, suppose we consult various

---

* The RSEs were calculated using $(n-2)$ for the one-factor and $(n-4)$ for the three-factor models.
† The forecast of the GDP residual, GDPu, warrants some explanation. The GDP in quarter 44 was $11,262 (billion). Economists' consensus forecast for each of the next four quarters is shown in the table below as the "Actual GDP Forecast." The time trend regression of GDP (Equation 7.2) was estimated to be GDP$_t = 6313.3 + 109.9 \times \text{Time}_t$. This time trend model yields the Time Trend GDP Forecast shown in the table below. The difference between the amounts actually forecast and the time trend regression forecast provides the residuals, GDPu, for the next four quarters that are the input to our forecasting model:

	Quarter 45	Quarter 46	Quarter 47	Quarter 48
Actual GDP forecast	11,374.6	11,488.4	11,603.3	11,719.3
Time trend GDP forecast	11,259.5	11,369.4	11,479.3	11,589.3
Difference (residual)	115.1	119.0	124.0	130.0

Speckled Band, Inc.
Forecast Sales

Year	Quarter	Sales	Time	GDP Residual	10-year Treasury	1-Factor Model Forecast	1-Factor Model Error	3-Factor Model Forecast	3-Factor Model Error
9	I	6,400.5	33	84.3	5.16	5,855	546	6,145	255
	II	6,216.7	34	37.8	5.14	5,975	242	5,792	425
	III	4,930.6	35	-64.1	5.24	6,095	-1,165	4,925	5
	IV	4,986.7	36	-76.4	4.57	6,216	-1,229	4,743	244
10	I	5,338.0	37	-50.9	5.04	6,336	-998	5,221	117
	II	5,087.5	38	-61.8	5.21	6,456	-1,369	5,265	-178
	III	4,907.6	39	-58.0	4.65	6,576	-1,669	5,268	-360
	IV	5,021.5	40	-86.2	3.94	6,697	-1,675	4,918	104
11	I	5,122.1	41	-84.0	4.05	6,817	-1,695	5,075	47
	II	5,392.3	42	-83.0	3.96	6,937	-1,545	5,170	222
	III	7,238.7	43	67.3	3.98	7,057	181	6,759	480
	IV	7,219.5	44	112.4	4.29	7,178	42	7,389	-169

	1-Factor Model	3-Factor Model
ME	-861	99
MAD	1,030	217
RMSE	1,189	258
RSE	1,303	317

Model Parameters

	1-Factor	3-Factor
Intercept	1,886.20	430.47
GDP	---	9.81
TB		255.36
Time	120.26	108.18

EXHIBIT 7.8   Speckled Band Inc.: Forecast sales.

			GDP	10-year	
Speckled Band, Inc. Forecast sales and cyclical variables					
Year	**Quarter**	**Sales**	**Residual**	**Treasury**	**Time**
12 I		7,593	115	4.15	45
	II	7,726	119	4.08	46
	III	7,855	124	3.95	47
	IV	7,979	130	3.75	48

EXHIBIT 7.9   Speckled Band Inc.: Forecast sales and cyclical variables.

services that provide forecasts of national economic data. We use these forecasts as input to our sales forecasting model. The resulting forecasts are shown in Exhibit 7.9 for the next four quarters.

7.1.2 Confidence Interval of the Forecast

The next step in our forecasting process is to calculate the confidence intervals for our forecast, so we have a better idea how far from the forecast the actual sales might turn out to be. In this multivariate case, we use the SE of the regression to calculate the confidence interval.* The SE of the regression associated with estimating Equation 7.4 for the whole estimation period of 44 quarters was 342.06, with 40 degrees of freedom $[(n - k) = (44 - 4)]$. The 95% confidence interval for each forecast period, t, will be

$$E[\text{Sales}_t] \pm t_{\alpha,n-k} \times \text{SE} = E[\text{Sales}_t] \pm 2.021 \times 342.06$$
$$= E[\text{Sales}_t] \pm 691.3,$$

where $t_{\alpha,n-k} = t_{0.05,40} = 2.021$, obtained with the Excel function =TINV (0.05, 40). The forecasts and the upper and lower 95% confidence bounds are shown in Exhibit 7.10.

---

* In Chapter 6 we calculated the confidence interval for the forecast when we had 1 independent variable. Based on the standard error of the forecast calculated according to expression (6.15), we saw that the confidence interval widened as we forecasted further into the future. However, when we have several independent variables the calculation and application of the equivalent standard error of the forecast is beyond the scope of this book. Consequently, in this example we use the standard error of the regression as given by the Excel regression routine. The resulting confidence intervals are slightly narrower than their true values. However, the differences are quite small.

Quarter	Lower Bound	Sales Forecast	Upper Bound
45	$6902	$7593	$8284
46	7035	7726	8417
47	7164	7855	8546
48	7288	7979	8670

EXHIBIT 7.10   Forecast and confidence interval for Speckled Band sales.

These confidence intervals are considerably narrower than those for the time trend model. Nevertheless, at about 9% on each side of the forecast value they still are quite wide. And, of course, these confidence intervals are slightly understated because they do not take account of possible errors in the regression coefficients.

If we are not satisfied with the accuracy of this forecasting model, we could try to improve the model, perhaps by looking for independent variables with greater explanatory ability. However, the costs in time and effort of finding such variables are likely to exceed the benefit of the additional accuracy. As with nearly everything else in business, the costs of finding a better solution (in this case, a model) must be balanced against the benefits of that better solution (in this case, improved accuracy).

7.2 SUMMARY

This chapter presented the basic steps for developing a forecasting model based on statistical relationships between the forecast variable and other economic data. We used linear regression to estimate these relationships, captured by the estimated parameters for the models, and then used the models to forecast future values of the forecast variable. The forecast normally requires that we also forecast the independent variables that are input into our forecast model. Usually, we can get these forecasts from forecasting services that follow the economy.

This and the previous chapter considered how to build forecasting models for time series that exhibit increasing or decreasing trends through time. Sometimes, however, a time series exhibits no trend whatsoever, but instead varies randomly about a central tendency. Sometimes, too, more recent data contains more information about the future than data further in the past. In these situations, smoothing techniques are useful short-term forecasting tools. We introduce several of these in the next chapter.

PROBLEMS

1 Speckled Band

In Chapters 6 and 7 we presented quarterly sales data for Speckled Band Inc. and developed two regression models. The one-factor model used time as the single independent variable, and the three-factor model used time, interest rates, and GDP. However, we did not assess the effectiveness of basing our forecasts on other combinations of these three variables. For this exercise, you are asked to develop, test, and compare regression forecast models that use: (a) GDP only, (b) interest rates (only), (c) GDP and time, and (d) interest rates and time.

2 Corrigan, Inc.

The forecasting firm for which you work has been hired by Corrigan, Inc., to forecast Corrigan's sales for the next four quarters. Corrigan supplies industrial equipment to small and mid-sized businesses. Corrigan's management suspects its sales are influenced by the macro economy and general level of interest rates. To this end, they have supplied you with quarterly time series for the percentage change in hours worked (to capture the overall economy) and the annualized yield on the 20-year Treasury bond (to capture interest rates).

One aspect of the data complicates the model selection process. The economy experienced robust growth over the early years the data were collected. This could produce spurious results in your regression models. It means that you will need to rely more on goodness-of-forecast measurements and less on regression statistics to identify the best model for forecasting.

Your task is the following. Estimate the hypothesized models for Corrigan's sales shown below using the first seven years of data and test their forecasting accuracy using the last four years of data.

Hypothesized forecasting models for Corrigan, Inc.

$$\text{Sales}_t = a + b_1 \times \text{Trend}_t$$
$$\text{Sales}_t = a + b_1 \times \text{Trend}_t + b_2 \times \text{Hours\_worked}_t$$
$$\text{Sales}_t = a + b_1 \times \text{Trend}_t + b_3 \times \text{T\_bond\_yield}_t$$
$$\text{Sales}_t = a + b_1 \times \text{Trend}_t + b_2 \times \text{Hours\_worked}_t$$
$$+ b_3 \times \text{T\_bond\_yield}_t$$

After identifying the model that produces the best forecasts, re-estimate that model using all 11 years of data. Have Excel report

Corrigan, Inc.

Year	Quarter	Trend	Quarterly Sales	Economic Data Change in Worked Hours	T-bond	Year	Quarter	Trend	Quarterly Sales	Economic Data Change in Worked Hours	T-bond
1	I	1	680.2	-6.02	6.60	7	I	25	778.2	2.97	4.72
	II	2	683.9	-6.18	5.97		II	26	837.4	2.52	5.18
	III	3	761.8	-6.33	5.81		III	27	788.4	2.46	5.79
	IV	4	664.9	-5.09	5.33		IV	28	806.8	1.91	6.11
2	I	5	632.8	-4.34	5.75	8	I	29	776.3	1.05	6.66
	II	6	648.1	-3.89	6.97		II	30	922.2	0.30	5.99
	III	7	670.7	-3.35	7.30		III	31	1,004.5	-0.16	6.05
	IV	8	689.9	-3.00	7.74		IV	32	1,013.4	-0.71	5.74
3	I	9	746.1	-1.86	7.78	9	I	33	1,032.4	-0.37	5.16
	II	10	536.9	-0.01	7.06		II	34	1,082.8	-0.22	5.14
	III	11	600.5	1.53	6.28		III	35	1,060.1	-0.77	5.24
	IV	12	523.6	3.08	6.04		IV	36	1,131.6	-0.43	4.57
4	I	13	551.7	3.62	5.65	10	I	37	1,035.1	-0.78	5.04
	II	14	434.4	4.57	6.51		II	38	1,202.3	-1.44	5.21
	III	15	435.9	4.92	6.87		III	39	1,238.2	-1.89	4.65
	IV	16	424.1	5.46	6.53		IV	40	1,279.9	-3.45	3.94
5	I	17	579.4	5.31	6.58	11	I	41	1,224.0	-3.60	4.05
	II	18	619.1	4.95	6.89		II	42	1,327.7	-4.26	3.96
	III	19	477.1	4.60	6.22		III	43	1,434.7	-4.41	3.98
	IV	20	623.6	4.04	6.03		IV	44	1,491.8	-5.16	4.29
6	I	21	542.4	4.29	5.54	12	I	45		-5.44	4.16
	II	22	655.6	3.63	5.64		II	46		-5.92	4.01
	III	23	732.0	3.38	5.46		III	47		-6.15	3.89
	IV	24	744.1	3.13	4.53		IV	48		-5.91	4.05

the residuals so that you can graph them and look for evidence of a missing factor. Label your graph appropriately. Finally, forecast sales for the next four quarters. In Excel, draw a graph that shows Corrigan's actual and predicted quarterly sales. Draw actual and predicted sales as separate series so that your forecasts are clearly identified. Be sure the company name appears in the graph title.

What You Turn In

Submit a report in memo format that describes the purpose and results of your work, including the criteria you used to select your model. Place the graph of actual and predicted sales in the body of your memo. Include a table in your memo that reports the model you used to predict sales (e.g., $Sales_t = 45 + 4 \times Time_t$) and the sales predicted. In your appendix, place clearly labeled graphs of the error terms, before and after model identification, where the "before" error terms are those from a simple time trend regression. In addition, include a well-organized printout of the spreadsheet showing the goodness-of-forecast calculations you used to identify the best model.

A Suggestion

To help organize your work, we suggest you use three worksheets: one to run the initial time trend regression and plot the resulting error terms; another to determine the best-forecasting model; and a third to re-estimate the final model, forecast sales, and draw the graphs for sales and the residuals from your final model.

3 Extravagant Goodies, Inc.

You have been assigned to forecast the next four quarters of sales for Extravagant Goodies, Inc., (hereafter EGI). EGI is a retailer of high-end products—fancy toys for individuals with more disposable income than common sense. Its sales have been growing fairly steadily over the past seven years. EGI's CFO believes their sales are closely related to overall economic activity and/or inflation. Sales tend to decrease during recessions, increase during expansions, and increase during periods of inflation. As a result, their sales are highly cyclical. Management believes that sales respond to expectations for economic growth (measured by the Treasury spread).* In addition, casual observation suggests that EGI's sales increase with the Consumer Price Index (CPI).

Below, find 7.5 years of quarterly sales data and concurrent time series for the Treasury spread and the CPI. The data also include forecasts for both the Treasury spread and CPI over the next four quarters.

					Extravagant Goodies, Inc. (EGI)						
				Change	Quarterly					Change	Quarterly
Year	Quarter	Time	T_sprd	CPI	Sales	Year	Quarter	Time	T_sprd	CPI	Sales
1	I	1	1.54	0.71	274.7	5	I	17	4.13	0.54	321.4
	II	2	0.52	0.95	320.7		II	18	4.22	0.86	341.8
	III	3	0.24	1.05	303.6		III	19	3.91	0.96	346.6
	IV	4	-0.07	0.69	291.4		IV	20	3.09	0.79	373.3
2	I	5	0.50	0.98	328.6	6	I	21	2.44	0.26	351.1
	II	6	1.91	0.46	319.7		II	22	1.97	1.36	423.3
	III	7	2.24	0.57	270.5		III	23	1.26	0.62	398.7
	IV	8	3.18	0.11	256.5		IV	24	1.03	2.10	472.3
3	I	9	4.04	0.06	262.3	7	I	25	0.41	-0.10	413.9
	II	10	4.13	0.90	247.7		II	26	0.62	0.90	429.7
	III	11	3.83	0.39	255.5		III	27	0.30	1.20	
	IV	12	3.42	0.67	267.5		IV	28	0.02	-0.69	
4	I	13	3.85	0.61	296.2	8	I	29	-0.03	0.67	
	II	14	3.78	0.55	298.7		II	30	0.08	1.40	
	III	15	4.02	0.27	308.4						
	IV	16	4.29	0.60	326.9						

Your job is to determine which of the four hypothesized linear models shown below produces the best forecasts and then use that

* The Treasury spread for this problem is the difference between the yield on 20-year Treasury bonds less the yield on 6-month Treasury bills.

model to forecast EGI's sales. Hypothesized models for EGI quarterly sales:

Model 1:
   Quarterly Sales$_t$ = $a + b \times$ Time$_t$
Model 2:
   Quarterly Sales$_t$ = $a + b \times$ Time$_t$ + $d \times$ CPI$_t$
Model 3:
   Quarterly Sales$_t$ = $a + b \times$ Time$_t$ + $c \times$ T_Spread$_t$
Model 4:
   Quarterly Sales$_t$ = $a + b \times$ Time$_t$ + $c \times$ T_Spread$_t$ + $d \times$ CPI$_t$.

After identifying the model that produces the best forecasts, re-estimate it using all 7.5 years of data and use it to forecast EGI's sales for the next four quarters. Have Excel report the residuals of your final regression so that you can graph them and look for evidence of a missing factor. In Excel, draw a graph that shows EGI's actual and predicted quarterly sales. Draw actual and predicted sales as separate series so that your forecasts are clearly identified. Label your sales and residuals graphs appropriately.

What You Turn In

Submit a report in memo format that describes the purpose and results of your work, including the criteria you used to select your model. Identify the model that is the least biased, the model that produces the smallest average error regardless of direction, and the model that is the most efficient with respect to the number of independent variables used. Place your graph of EGI's sales (including predicted sales) in the body of your memo. Include a table in your memo that reports the model used to predict EGI's sales (e.g., Sales$_t$ = $45 + 4 \times$ Time$_t$) as well as the predicted sales. Place the following in an appendix: A clearly labeled graph of the error terms from the time series regression used to estimate your forecasting model and a well-organized printout of the spreadsheet showing the goodness-of-forecast calculations you used to identify the best model.

4 Aetius Equipment Corporation

It is late in the year, 1992. You are an analyst with the Aetius Equipment Corporation. The CFO of the company has asked you to develop a forecast of the company's sales for each quarter of 1993 and the first quarter of 1994.

The Aetius Equipment Corporation has been in continuous operation since 1924. It manufactures and markets a line of farm implements including plows, harrows, rakes, and balers. These are all items that are either pulled by farm tractors or are fitted to the tractor. In addition to farm equipment, it manufactures equipment that can be fitted to tractors for use in the construction industry, such as add-on forklifts, scoops and shovels, and grading blades. The company has been relatively successful in tapping the export market because its equipment is sufficiently adaptable to be used with farm and construction equipment manufactured in foreign countries.

Its sales have grown slowly but steadily as the economy has expanded. Superimposed on the long-term growth in sales is a very cyclical pattern of sales. Aetius has been vulnerable to the business cycle, with sales turning down sharply during recessions. The official dates of recent recessions were: 11/73–3/75, 1/80–7/80, 7/81–11/82, and 7/90–3/91, and it is clear that there were declines in sales during these periods. This cyclical pattern may be related to the fact that sales of farm equipment are sensitive to the prosperity of farmers. Farmers tend to invest in new equipment when they have a good year and can afford the purchase. In addition, because some of Aetius' sales are to the construction industry, the company is sensitive to the construction cycle, which is, in turn, very sensitive to the business cycle and the level of interest rates.

To help you develop your forecast, you have gathered some data that you hope will help. The company's sales are shown in the graph below. Sales and economic data are contained in an Excel spreadsheet called "Aetius Sales Data S" that is included with text disk. Quarterly data for 1980–1992 are shown below. The spreadsheet below shows the company's sales for each quarter starting with the first quarter 1980 (1980.1) through the last quarter of 1992 (1992.4), and the data on the disk start in 1973.1. The sales data are shown in millions of dollars, and are seasonally adjusted; quarterly data shown at an annual rate.

You plan on developing a forecasting model using linear regression that shows the relationship between Aetius' sales and broader measures of economic activity. The spreadsheet shows various series of economic data that you have gathered that you think might possibly relate to the company's sales. These quarterly data (on the disk) start in 1973.1 and end with the current quarter, 1992.4. Forecasts of these other economic series were obtained from an econometric forecasting service and are shown for the next five quarters (1993.1–1994.1). The aggregate economic activity data are in billions of dollars, and are seasonally adjusted, at an annual rate. Interest rates are expressed as a percent, and prices are indices.

Year & Quarter	AETIUS EQUIPMENT Sales Quarterly @Annual Rate	TBILLS 3 MONTH RATES	CORP BOND AAA RATE	DOW JONES INDUST AVG	S&P 500 AVG	CONSUMER PRICE INDEX	GROSS DOMESTIC PRODUCT BILL $	PERSONAL CONSUMPT. EDNEFX	PERSONAL CONSUMPT EXPEND DURABLES	PERSONAL CONSUMPT EXPEND NONDUR	PERSONAL CONSUMPT EXPEND SERVICES	GROSS PRIVATE DOMEST. INVEST	FIXED INVEST	RESIDENT INVEST	GROSS DOMEST INVEST: CHANGE BUSINESS INVENTOR	EXPORTS	GOVRMNT PURCHASES	GROSS FARM PRODUCT	EMPLOYEE WAGES & SALARIES	FARM INCOME	CORP PROFITS	INDEX OF FARM PRODUCT PRICES
1980.1	118.146	15.20	12.96	803.56	104.69	80.1	2650.1	1701.5	218.7	667.1	815.7	495.3	488.2	134.6	7.1	267.5	490.5	57.5	1337.7	14.2	100.4	128.0
1980.2	96.407	7.07	10.58	869.86	114.55	82.5	2643.9	1704.9	198.2	673.8	832.9	451.5	453.8	111.2	-2.2	276.2	504.1	45.1	1353.9	0.9	89.1	127.0
1980.3	115.826	10.27	12.02	946.67	126.51	83.9	2705.3	1762.3	211.3	686.2	864.9	432.1	468.0	115.9	-35.9	282.7	507.4	56.2	1379.9	11.3	87.3	141.0
1980.4	133.065	15.49	13.21	945.96	133.48	86.4	2832.9	1823.6	221.8	704.6	897.2	491.5	498.4	131.3	-6.8	290.4	526.4	65.6	1434.9	19.7	94.8	145.0
1981.1	142.601	13.36	13.33	987.18	133.19	88.6	2953.6	1876.0	230.8	731.3	913.9	548.5	515.6	131.9	32.9	303	545.4	69.9	1473.4	22.8	98.6	143.0
1981.2	144.682	14.73	13.75	996.27	132.28	90.5	2993.0	1908.9	225.5	741.6	941.7	543.3	529.5	128.7	13.9	305.8	556.8	71.7	1500.3	23.3	96.8	142.0
1981.3	144.003	14.70	15.49	853.38	118.27	93.1	3079.6	1952.1	236.3	748.5	967.2	575.4	538.5	120.1	36.9	299.9	562.2	71.7	1532.8	22.3	108.4	134.0
1981.4	134.795	10.85	14.23	878.28	123.79	94.1	3096.3	1968.0	221.4	755.5	991.1	564.7	546.6	109.5	18.1	303.4	579.9	66.3	1556.0	16.4	99.7	128.0

Data for 1975 - 1989 is Omitted for Display. Missing Data is included in Student Data Disk

Year & Quarter	AETIUS EQUIPMENT Sales Quarterly @Annual Rate	TBILLS 3 MONTH RATES	CORP BOND AAA RATE	DOW JONES INDUST AVG	S&P 500 AVG	CONSUMER PRICE INDEX	GROSS DOMESTIC PRODUCT BILL $	PERSONAL CONSUMPT. EDNEFX	PERSONAL CONSUMPT EXPEND DURABLES	PERSONAL CONSUMPT EXPEND NONDUR	PERSONAL CONSUMPT EXPEND SERVICES	GROSS PRIVATE DOMEST. INVEST	FIXED INVEST	RESIDENT INVEST	GROSS DOMEST INVEST: CHANGE BUSINESS INVENTOR	EXPORTS	GOVRMNT PURCHASES	GROSS FARM PRODUCT	EMPLOYEE WAGES & SALARIES	FARM INCOME	CORP PROFITS	INDEX OF FARM PRODUCT PRICES
1990.1	223.759	7.90	9.37	2700.13	338.47	128.7	5461.9	3679.3	479.8	1201.7	1997.8	828.9	819.3	233.2	9.6	542	1027.7	86.9	2689.2	49.9	250.6	150.0
1990.2	219.662	7.73	9.26	2894.82	360.39	130.0	5540.8	3727.0	466.0	1213.6	2047.5	837.8	804.5	222.4	33.3	553.5	1037.3	87.1	2739.1	42.5	269.5	152.0
1990.3	209.793	7.36	9.56	2550.69	315.41	132.6	5583.8	3801.7	467.3	1241.0	2093.4	812.5	804.1	209.9	8.4	555.3	1048.3	84.8	2770.6	31.6	221.8	147.0
1990.4	212.851	6.74	9.05	2610.92	328.75	134.2	5597.9	3836.6	459.5	1260.7	2116.4	756.4	780.3	195.8	-23.9	577.6	1076.5	81.5	2781.3	43.8	225.8	142.0
1991.1	205.348	5.91	8.93	2920.11	372.28	135.1	5631.7	3843.6	448.9	1252.3	2142.4	729.1	749.0	182.2	-19.9	576.5	1093	79.2	2782.2	37.2	243.9	148.0
1991.2	214.760	5.57	9.01	2968.14	378.29	136.1	5697.7	3887.8	452.0	1259.2	2176.6	721.5	744.5	183.6	-23	600.7	1099.9	83.0	2800.6	42.6	242.8	153.0
1991.3	203.668	5.22	8.61	3010.35	387.20	137.1	5758.6	3929.8	465.1	1260.0	2204.8	744.5	745.0	192.4	-0.5	603	1104	78.7	2823.4	29.8	245.4	148.0
1991.4	205.800	4.07	8.31	2958.64	388.51	138.2	5803.7	3964.1	465.2	1260.0	2239.0	752.4	743.5	200.3	8.9	625.7	1100.2	74.5	2853.6	37.6	246.5	137.0
1992.1	225.052	4.04	8.35	3247.42	407.36	139.3	5908.7	4046.5	484.0	1278.2	2284.4	750.8	755.9	208.9	-5.1	633.7	1118.5	84.8	2892.2	45.6	262.9	144.0
1992.2	225.102	3.66	8.22	3337.79	408.27	140.2	5991.4	4099.9	487.8	1288.2	2323.8	799.7	786.8	220.6	12.9	632.4	1125.8	83.4	2933.6	44.9	258.7	140.0
1992.3	224.654	2.91	7.92	3293.92	418.48	141.3	6059.5	4157.1	500.9	1305.7	2350.5	802.2	792.5	223.3	9.7	641.1	1139.1	85.8	2970.7	36.8	237.4	138.0
1992.4	231.987	3.22	7.98	3303.15	435.64	141.9	6194.4	4256.2	516.6	1331.7	2407.9	833.3	821.3	241.8	12	654.7	1143.8	83.6	3015.8	47.6	284.5	137.0
Future Forecast																						
1993.1		2.95	7.58	3440.74	450.16	143.6	6261.6	4296.2	515.3	1335.3	2445.5	874.1	839.5	244.9	34.6	651.3	1139.7	83.8	3054.3	55.7	271.2	141.0
1993.2		3.07	7.33	3513.81	448.06	144.4	6327.6	4359.9	531.6	1344.8	2483.4	874.1	861.0	241.9	13.1	660	1158.6	83.3	3082.7	47.0	284.8	140.0
1993.3		2.95	6.66	3592.28	459.24	145.1	6395.9	4419.1	541.9	1352.4	2524.8	884.0	876.3	251.3	7.7	653.2	1164.8	73.2	3115.4	24.8	299.1	145.0
1993.4		3.06	6.93	3736.23	465.95	145.8	6526.5	4492.0	562.8	1367.5	2561.8	934.5	927.6	271.6	6.9	682.4	1169.1	89.2	3149.6	56.4	315.4	145.0
1994.1		3.50	7.48	3816.98	463.81	147.2	6609.4	4549.4	577.4	1376.1	2595.9	978	943.8	279.1	34.2	668.8	1164.4	98.1	3200.7	60.0	-100.0	148.0

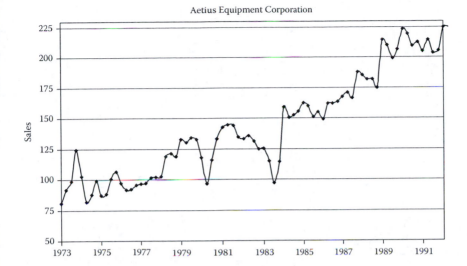

Your mission, should you choose to accept it, is to develop an econometric forecasting model for Aetius' quarterly sales for the next five quarters.

You should develop several alternative forecasting equations that relate Aetius' sales to broader measures of economic activity. Estimate the parameters of these models with an estimation sample using data from 1973.1 through 1985.4. When you have narrowed your models down to your best three or four models, apply your model to the data from the hold-out period, 1986.1–1992.4. Evaluate the ability of each model to forecast by comparing your model's forecast against the actuals for the hold-out period. Determine which is your best model, and use it to forecast Aetius' sales for 1993.1–1994.1.

Your report should briefly explain your task, your method, and your results. Explain the various models you have tried and their results. Show how you have compared the alternative forecasting models, and explain how you decided on your final model.

CHAPTER 8

Forecasting III: Smoothing Data for Forecasts

8.1 INTRODUCTION

Economic and business data are not always well-behaved. It frequently exhibits extreme variations, regular patterns, or both that can mask underlying trends or cycles. There are a variety of techniques to smooth extreme variation or remove regular patterns. These techniques result in better behaved data series that in turn produce more precise models and improved forecasts. Several of these techniques are the focus in this chapter—moving averages, exponential smoothing, and seasonal decomposition. We finish with a brief overview of time series models. All these techniques share a common element—past observations of a variable are used to forecast future values of that same variable.

8.2 MOVING AVERAGE

The simplest smoothing technique is a moving average forecast. In the simplest approach, the forecaster decides upon the length of the horizon over which the average is to be taken (the number of periods, n, in the average). The average value for the observations within this horizon becomes the forecast for the next period. As the series moves forward from period to period, the earliest observation is dropped, the most recent observation added, and a new average computed. This new average becomes the forecast for the next period. The average moves through time, hence its name. In general, the n-period moving average forecast for period t is given by Equation 8.1.

$$\text{Forecast Sales}_t = \frac{\sum_{i=t-n}^{t-1} \text{Sales}_i}{n} \tag{8.1}$$

The forecast can also be thought of as a weighted average of past values, where the weight, $1/n$, is the same for each period. The moving average forecast can be an effective method for very short-term (one-period) forecasts of simple series that do not exhibit cyclical or seasonal patterns or a trend.

To see how this works, consider the quarterly sales data for Tudor Corporation, shown in Exhibit 8.1. The chart of Tudor's sales is shown in Exhibit 8.2. From the chart we see that Tudor's sales do not exhibit any obvious cyclical pattern or growth trend.

Exhibit 8.3 illustrates the calculations for a 4-period moving average in Excel®. The sales forecast for quarter 5 in cell F12, 872, is the average of sales in quarters 1–4, as shown in Equation 8.2.

$$E\left[\text{Sales}_5\right] = \left(\frac{822 + 900 + 860 + 905}{4}\right)$$
$$= 872 \tag{8.2}$$

The sales forecast for quarter 6, 887, is the average of sales in quarters 2–5, as shown in Equation 8.3.

$$E\left[\text{Sales}_6\right] = \left(\frac{900 + 860 + 905 + 882}{4}\right)$$
$$= 887 \tag{8.3}$$

The average moves forward with each new period.

Tudor Corp. Quarterly Sales			
Year	Quarter	Time	Sales ($000)
1	I	1	822
	II	2	900
	III	3	860
	IV	4	905
2	I	5	882
	II	6	881
	III	7	883
	IV	8	885
3	I	9	872
	II	10	822
	III	11	817
	IV	12	812
4	I	13	856
	II	14	862
	III	15	840
	IV	16	904
5	I	17	861
	II	18	904
	III	19	872
	IV	20	824
6	I	21	895
	II	22	871
	III	23	866
	IV	24	822

EXHIBIT 8.1   Tudor Corporation quarterly sales.

Moving averages smooth a series. The idea is to remove the period-to-period variation to better discern the long-term tendency in a time series. The greater the number of periods in the moving average, the greater the smoothing effect. Consider Exhibits 8.4A through 8.4D. The number

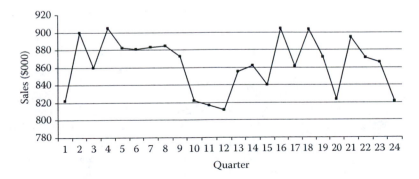

EXHIBIT 8.2   Tudor Corporation sales.

	B	C	D	E	F	G	H
3		Tudor Corp. Quarterly Sales					
4					4-period		
5					Moving Average		
6					Error		
7	Year	Quarter	Time	Sales ($000	Forecast	Term	
8	1 I		1	822			
9		II	2	900			=AVERAGE(E8:E11)
10		III	3	860			
11		IV	4	905			
12	2 I		5	882	872	11	=AVERAGE(E9:E12)
13		II	6	881	887	-6	
14		III	7	883	882	2	
15		IV	8	885	888	-3	
16	3 I		9	872	883	-11	
17		II	10	822	880	-58	
18		III	11	817	866	-49	
19		IV	12	812	849	-37	
20	4 I		13	856	831	25	
21		II	14	862	827	36	
22		III	15	840	837	3	
23		IV	16	904	842	62	
24	5 I		17	861	866	-4	
25		II	18	904	867	37	
26		III	19	872	877	-5	
27		IV	20	824	885	-61	
28	6 I		21	895	865	29	
29		II	22	871	874	-3	
30		III	23	866	865	0	
31		IV	24	822	864	-42	

EXHIBIT 8.3 Tudor Corporation: Four-period moving average sales forecasting model.

of quarters used to smooth sales increases from 0–4 to 4–6 to 6–8 periods. Note that the longer the moving average period, the smoother the series.

Moving average forecasting suffers from several problems. It can be difficult to determine the appropriate number of periods. Often, though, a graph of the series helps the forecaster to determine the appropriate number.* In addition, moving averages give equal weight to all observations in the average. Sometimes, however, more recent observations contain more information, whereas old observations contain less, if any, information. In this case, the solution is to give more weight to recent observations. This leads us into a discussion of exponential smoothing.

* You can also calculate goodness-of-fit statistics for these models. Goodness-of-fit statistics are identical to goodness-of-forecast statistics discussed in Chapters 6 and 7. The difference is that goodness-of-fit statistics are used when the model is fit to the entire data set instead of just the hold-out segment.

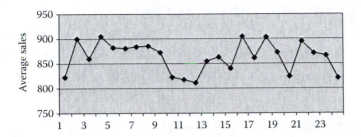

EXHIBIT 8.4A   Tudor Corporation quarterly sales.

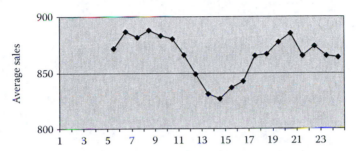

EXHIBIT 8.4B   Tudor Corporation quarterly sales. Four-period moving average.

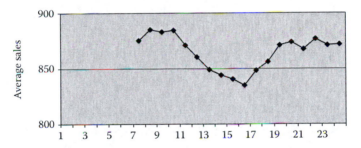

EXHIBIT 8.4C   Tudor Corporation quarterly sales. Six-period moving average.

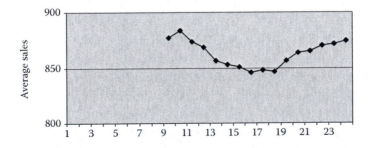

EXHIBIT 8.4D   Tudor Corporation quarterly sales. Eight-period moving average.

8.3 EXPONENTIAL SMOOTHING

Exponential smoothing is a moving average in which some observations are more important than others. It is based on the notion that the most recent observations of the variable being forecast convey more information about the future values of that variable than observations in the more distant past. To capture this information, exponential smoothing weights each observation in the moving average, typically assigning the greatest weight to the most recent observation and the smallest weight to the most distant observation. It is calculated according to Equation 8.4

$$\text{Forecast Sales}_{t+1} = a \times \text{Sales}_t + a(1-a) \times \text{Sales}_{t-1}$$
$$+ a(1-a)^2 \times \text{Sales}_{t-2} + \cdots + a(1-a)^j \times \text{Sales}_j, \quad (8.4)$$

where $0 < a < 1$. Because $0 < a < 1$, the terms $a(1-a)^j$ decrease monotonically as j increases. Although this expression looks daunting, it reduces to the simple expressions shown below:

$$\text{Forecast Sales}_{t+1} = a \times \text{Sales}_t + (1-a) \times \text{Forecast Sales}_t \quad (8.5)$$

or, alternatively,

$$\text{Forecast Sales}_{t+1} = \text{Forecast Sales}_t + a \times (\text{Sales}_t - \text{Forecast Sales}_t). \quad (8.6)$$

In Equation 8.5, the forecast for period $t + 1$ is a weighted average of the sales observed in period t, Sales_t, and the forecast for sales in period t that was made in period $t - 1$, Forecast Sales_t. Equivalently, Equation 8.6 shows the forecast for period $t + 1$ as a function of the forecast for period t sales, Forecast Sales_t, and the error in that forecast, $\text{Sales}_t - \text{Forecast Sales}_t$.

There are two issues to consider when building an exponential smoothing model—the initial forecast and the value of the smoothing constant. The forecast for period t depends on the forecast for period $t - 1$, raising the question of where the initial forecast comes from. Forecasters frequently use either of two methods for the initial forecast. One method assumes a random walk and uses the observation for period 1, Sales_1, as the forecast for period 2, Forecast Sales_2. The other method uses a moving average to get the first forecast. For example, the average of the first four observations can be used to forecast the fifth observation. In the example that follows, we use the first approach. As for the smoothing parameter, a, an arbitrarily chosen value is used to build the model. The final value used is the one that minimizes some measure of the model's goodness-of-fit, such as the RMSE.

Return to Tudor Corporation's quarterly sales. To start the exponential smoothing algorithm, we need a value for the smoothing constant and the actual and forecast sales for a given period. Let the smoothing constant, a, be 0.10. Observed sales in quarter 2 were 899.9. Finally, the first, or "seed" forecast will be for quarter 2. To start the algorithm we set the forecast for period 2 equal to the actual sales in period 1, 822.1. With these numbers in place, we now proceed to forecast quarter 3 using Equation 8.5.

$$\text{Forecast Sales}_3 = a \times \text{Sales}_2 + (1 - a) \times \text{Forecast Sales}_2$$
$$= 0.10 \times 899.9 + 0.90 \times 822.1$$
$$= 829.8 \tag{8.7}$$

Alternatively, Equation 8.6 can also be used to calculate the forecast for quarter 3.

$$\text{Forecast Sales}_3 = \text{Forecast Sales}_2 + a \times (\text{Sales}_2 - \text{Forecast Sales}_2)$$
$$= 822.1 + 0.10 \times (899.9 - 822.1)$$
$$= 829.8 \tag{8.8}$$

Because successive forecasts combine a with the sales and forecast sales from the previous period, all that is needed to complete this model is to make the cell address of the smoothing constant an absolute reference ("dollar sign" it) and copy your code into the remaining rows of your model. The final model, assuming $a = 0.10$, is shown in Exhibit 8.5.

The value of the smoothing constant, a, determines how much influence the most current observation has on the forecast value. To see this, we repeat Equation 8.5.

$$\text{Forecast Sales}_{t+1} = a \times \text{Sales}_t + (1 - a) \times \text{Forecast Sales}_t$$

When a is close to one, $(1 - a)$ is close to zero, so the most recent observation has a significant influence on the next period's forecast. Very little smoothing of the series takes place. When a is close to zero, the most recent observation has little influence on the new forecast. There is a substantial

	A	B	C	D	E	F	G	H	I
1		Tudor Corp. Quarterly Sales							
2						Exponential			
3		Smoothing Constant		0.100		Smoothing			
4						Error			
5	Year	Quarter	Time	Sales ($000)	Forecast	Term			
6	1	I	1	822.1					
7		II	2	899.9	822.1	77.9			
8		III	3	859.8	829.8	30.0	=D6		
9		IV	4	904.7	832.8	71.8			
10	2	I	5	882.3	840.0	42.2			
11		II	6	880.8	844.3	36.6			
12		III	7	883.5	847.9	35.6	=$D$3*D7+(1-$D$3)*E7		
13		IV	8	884.7	851.5	33.2			
14	3	I	9	872.3	854.8	17.5			
15		II	10	822.4	856.5	-34.1			
16		III	11	817.0	853.1	-36.1			
17		IV	12	811.7	849.5	-37.8			
18	4	I	13	855.6	845.7	9.8			
19		II	14	862.2	846.7	15.5			
20		III	15	840.1	848.3	-8.2			
21		IV	16	904.2	847.5	56.8			
22	5	I	17	861.4	853.1	8.3			
23		II	18	903.7	854.0	49.8			
24		III	19	872.2	858.9	13.2			
25		IV	20	824.0	860.3	-36.3			
26	6	I	21	894.6	856.6	38.0			
27		II	22	871.1	860.4	10.7			
28		III	23	865.9	861.5	4.4			
29		IV	24	821.6	861.9	-40.4			
30									
31					ME	15.6			
32					MAD	32.3			
33					RMSE	37.7			
34									
35			=SQRT(SUMSQ(F7:F29)/COUNT(F7:F29))						
36									
37									

EXHIBIT 8.5   Sales smoothed with smoothing constant $a=0.10$.

degree of smoothing. The new forecast will be very similar to the old forecast.

From another perspective, the smoothing constant determines the speed at which past observations lose their importance to the forecast. When a is small, random variations are smoothed and the influence of distant observations on the forecast is reasonably large. Forecasts will be stable from period to period. When a is large, the influence of distant observations will be small. The forecasts will respond rapidly to recent observations and will exhibit more variation from period to period.

We now consider how to determine the best smoothing constant, a. The standard approach is to find the smoothing constant, a, that minimizes the RMSE using a non-linear optimization program such as Excel's Solver.

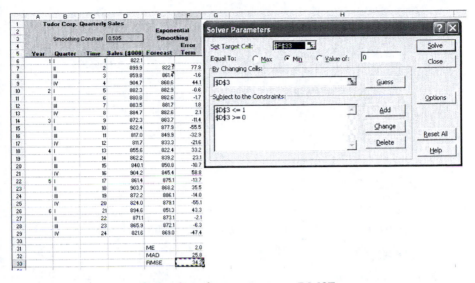

	A	B	C	D	E	F	G	H
1		Tudor Corp. Quarterly Sales						
2					Exponential		Solver Parameters	
3		Smoothing Constant		0.505	Smoothing			
4						Error	Set Target Cell:	$F$33
5	Year	Quarter	Time	Sales ($000)	Forecast	Term	Equal To: ○ Max ● Min ○ Value of: 0	
6	1	I	1	822.1			By Changing Cells:	
7		II	2	899.9	822.1	77.9		
8		III	3	859.8	861.4	-1.6	$D$3	
9		IV	4	904.7	860.6	44.1		
10	2	I	5	882.3	882.9	-0.6	Subject to the Constraints:	
11		II	6	880.8	882.6	-1.7	$D$3 <= 1	
12		III	7	883.5	881.7	1.8	$D$3 >= 0	
13		IV	8	884.7	882.6	2.1		
14	3	I	9	872.3	883.7	-11.4		
15		II	10	822.4	877.9	-55.5		
16		III	11	817.0	849.9	-32.9		
17		IV	12	811.7	833.3	-21.6		
18	4	I	13	855.6	822.4	33.2		
19		II	14	862.2	839.2	23.1		
20		III	15	840.1	850.8	-10.7		
21		IV	16	904.2	845.4	58.8		
22	5	I	17	861.4	875.1	-13.7		
23		II	18	903.7	868.2	35.5		
24		III	19	872.2	886.1	-14.0		
25		IV	20	824.0	879.1	-55.1		
26	6	I	21	894.6	851.3	43.3		
27		II	22	871.1	873.1	-2.1		
28		III	23	865.9	872.1	-6.3		
29		IV	24	821.6	869.0	-47.4		
30								
31					ME	2.0		
32					MAD	25.8		
33					RMSE	34.2		

(Solver dialog box buttons: Solve, Close, Guess, Options, Add, Change, Delete, Reset All, Help)

EXHIBIT 8.6 Solver dialog box for minimizing RMSE.

In Excel 2003, Solver is accessed through the Tools menu: Tools > Solver; in Excel 2007, Solver is found on the "Data" ribbon.* Clicking on Solver opens the dialog box shown in Exhibit 8.6.

The steps to fill out the Solver dialog box are:

1. *Set Target Cell.* This is the cell that we want to optimize. In this case we want to make the RMSE as small as possible. RMSE is in cell F33, so select and enter F33 in the Set Target Cell box.

2. *Equal To.* Click Min because we want to minimize the RMSE that is the Target Cell.

3. *By Changing Cells.* The decision variable that we want to adjust is the smoothing constant, *a*, that is in cell D3. Click in the By Changing Cells box, and select and enter D3.

4. *Subject to the Constraints.* The constraints limit the range of values of the decision variable. In this case, we consider values of *a* between 0 and 1. To enter these constraints, click the Add button to bring up the constraint dialog box shown in Exhibit 8.7. The dialog box is shown filled in with the constraint D3 <= 1, which says that the smoothing

* Solver is an Excel Add-In. To add it into Excel 2007, click on the "Office Button," select "Excel Options," then "Add-Ins" and select Solver. To add it into Excel 2003, click on "Tools," then "Add-Ins" and check the Solver box.

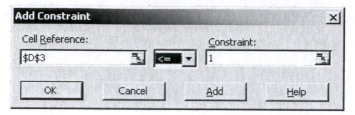

EXHIBIT 8.7   Constraint dialog box.

constant cannot exceed 1. Fill in a similar constraint saying that the smoothing constant cannot be less than zero (D3 >=0). Click the down arrow: [<= ▾] to choose the direction of the inequality.

5. *Solve.* Click the Solve button and Solver follows your directions to find the value of the smoothing constant (the Changing Cell) that minimizes RMSE (Target Cell). In this example, the optimal solution is $a = 0.505$, yielding the minimum RMSE of 34.2, and the forecasts as shown in column E of Exhibit 8.6.

In this example, we entered the exponential smoothing formula (8.6) in the spreadsheet to develop our forecast. Excel can do this for us with the Exponential Smoothing tool. Access exponential smoothing via Tools > Data Analysis > Exponential Smoothing. This brings up the dialog box that you fill in as shown in Exhibit 8.8.

To use the Exponential Smoothing tool, you specify the input data by selecting the column of actual sales, D6:D29, and indicate the Output Range where the program should insert the forecasted values, E6 (which implicitly matches the length of the input data to E29). The exponential smoothing routine inserts the formulas for the forecast in the Output Range column. The formula inserted in the Output cells is

$$\text{Forecast Sales}_t = (1 - \text{Damping Factor})\text{Sales}_{t-1} + (\text{Damping Factor})\text{Forecast Sales}_{t-1}, \qquad (8.9)$$

where the Damping Factor that is entered in the dialog box is $(1 - a)$. Note that Excel's Damping Factor is *not* the same as our smoothing constant. It is the complement of the smoothing constant. Our optimal smoothing constant is $a = 0.505$, so the Excel Damping Factor is $(1 - a) = (1 - 0.505) = 0.495$, as shown in the dialog box.

	A	B	C	D	E	F	G	H	I	J	K	L
1		Tudor Corp. Quarterly Sales										
2						Exponential						
3		Smoothing Constant		0.505		Smoothing						
4						Error						
5	Year	Quarter	Time	Sales ($000)	Forecast	Term						
6	1 I		1	822.1	#N/A	#N/A						
7	II		2	899.9	822.1	77.9						
8	III		3	859.8	861.4	-1.6						
9	IV		4	904.7	860.6							
10	2 I		5	882.3	882.9							
11	II		6	880.8	882.6							
12	III		7	883.5	881.7							
13	IV		8	884.7	882.6							
14	3 I		9	872.3	883.7							
15	II		10	822.4	877.9							
16	III		11	817.0	849.9							
17	IV		12	811.7	833.3							
18	4 I		13	855.6	822.4							
19	II		14	862.2	839.2							
20	III		15	840.1	850.8							
21	IV		16	904.2	845.4							
22	5 I		17	861.4	875.1							
23	II		18	903.7	868.2							
24	III		19	872.2	886.1							
25	IV		20	824.0	879.1							
26	6 I		21	894.6	851.3	43.3						
27	II		22	871.1	873.1	-2.1						
28	III		23	865.9	872.1	-6.3						
29	IV		24	821.6	869.0	-47.4						
30												
31					ME	-3.7						
32					MAD	23.5						
33					RMSE	34.2						

Exponential Smoothing dialog box:
- Input
  - Input Range: $D$6:$D$29
  - Damping factor: 0.495
  - ☐ Labels
- Output options
  - Output Range: $E$6
  - New Worksheet Ply:
  - New Workbook
  - ☐ Chart Output
  - ☐ Standard Errors
- OK
- Cancel
- Help

EXHIBIT 8.8   Exponential smoothing tool in Excel.

The other thing to note about the Exponential Smoothing tool is that it enters #N/A for the first period's forecast. It sets the second period forecast equal to the first period's actual sales. The result is the same as our earlier manual approach.

8.4 EVALUATING THE MODEL

We now have two models for the quarterly sales of Tudor Corporation, raising the question of which one is likely to produce more accurate forecasts? To answer this question in the previous chapters, we measured forecasting accuracy. A portion of the data set was used to estimate the models and then the models were used to generate forecasts for the values of the remaining hold-out period. Error terms were calculated and then used to calculate four goodness-of-forecast statistics—ME, MAD, RMSE, and RSE. Unfortunately, we cannot use this same procedure to rank the two smoothing models for Tudor Corporation because smoothing models can only forecast a limited number of periods, typically just one. This means that there is insufficient data to measure forecasting accuracy. Instead, we measure how well each model "fits" the data.

Return to the 4-period moving average model of Tudor Corporation's quarterly sales. It forecasts sales for all but the first four periods in the data

	L	M	N	O	P	Q	R	S
1	Tudor Corp. Quarterly Sales						Exponential	
2						4-period	Smoothing	
3						Moving Average	constant 0.505	
4						Error		Error
5	Year	Quarter	Time	Sales ($000)	Forecast	Term	Forecast	Term
6	1 I		1	822.1			N/A	
7		II	2	899.9			822.1	
8		III	3	859.8			861.4	
9		IV	4	904.7			860.6	
10	2 I		5	882.3	871.6	10.6	882.9	-0.6
11		II	6	880.8	886.7	-5.8	882.6	-1.7
12		III	7	883.5	881.9	1.6	881.7	1.8
13		IV	8	884.7	887.8	-3.1	882.6	2.1
14	3 I		9	872.3	882.8	-10.5	883.7	-11.4
15		II	10	822.4	880.3	-57.9	877.9	-55.5
16		III	11	817.0	865.7	-48.7	849.9	-32.9
17		IV	12	811.7	849.1	-37.4	833.3	-21.6
18	4 I		13	855.6	830.9	24.7	822.4	33.2
19		II	14	862.2	826.7	35.5	839.2	23.1
20		III	15	840.1	836.6	3.5	850.8	-10.7
21		IV	16	904.2	842.4	61.8	845.4	58.8
22	5 I		17	861.4	865.5	-4.1	875.1	-13.7
23		II	18	903.7	867.0	36.7	868.2	35.5
24		III	19	872.2	877.4	-5.2	886.1	-14.0
25		IV	20	824.0	885.4	-61.4	879.1	-55.1
26	6 I		21	894.6	865.3	29.3	851.3	43.3
27		II	22	871.1	873.6	-2.5	873.1	-2.1
28		III	23	865.9	865.5	0.4	872.1	-6.3
29		IV	24	821.6	863.9	-42.3	869.0	-47.4
30								
31					ME	-3.7		-3.7
32					MAD	24.2		23.5
33					RMSE	32.3		30.7

EXHIBIT 8.9   Comparison of forecasts.

sample. The exponential smoothing model forecasts sales for all but the first period that starts the algorithm. To compare forecasts, we exclude the first four periods, thereby considering only the periods for which both methods make forecasts. Subtracting these forecasts from their respective actual values gives us error terms that can be used to calculate each model's ME, MAD, and RMSE. These statistics are calculated in precisely the same way as the goodness-of-forecasting statistics in the previous section. They also convey the same information. Models with lower RMSEs are said to fit the data better; models with lower MEs are less biased; and models with lower MADs have smaller absolute errors, on average.[*]

Exhibit 8.9 reports the forecasts for Tudor made by the 4-period moving average model and an exponential smoothing model with a smoothing

---

[*] Goodness-of-fit statistics also can be used in lieu of goodness-of-forecast statistics if the data set is not large enough to permit partitioning into estimating and testing segments.

constant of 0.505. The error terms and the diagnostic statistics also are shown. The exponential smoothing model produces a lower RMSE (it fits the data better) and a lower MAD (its absolute errors are smaller, on average) than the four-period moving average model. Both models show the same bias with an ME of −3.7. These results indicate that exponential smoothing generates forecasts that fit Tudor's sales slightly better than the four-period moving average.

8.4.1 Making the Forecast

We have evaluated each of these methods relative to past data and found that the exponential smoothing method provides the best fit of past data. The last step is to use the model to forecast the next period's sales. Our best smoothed model is

$$\text{Sales}_{t+1} = 0.505 \times \text{Sales}_t + (1 - 0.505) \text{ Forecast Sales}_t. \qquad (8.10)$$

For quarter 25 the forecast is

$$\begin{aligned}\text{Sales}_{25} &= 0.505 \times 821.6_{24} + (1 - 0.505) \times 869.0_{24} \\ &= 845.1. \qquad\qquad\qquad\qquad\qquad\qquad\qquad (8.11)\end{aligned}$$

We will not know if this is an accurate forecast until the events of quarter 25 unfold.

As previously mentioned, smoothing methods are intended only to provide short-term (one-period) forecasts; they are not well suited for longer-term forecasts. Furthermore, both moving average and exponential smoothing are most effective with stable time series with minimal trend, seasonal, or cyclical components.* If the forecaster wishes to use these models for series that exhibit trends or seasonal components, these components must first be removed. We consider the process of removing seasonal variation next.

8.5 SEASONALITY AND SEASONAL DECOMPOSITION

In the previous sections we considered two ways to smooth a data series. Our purpose was to reduce the influence of extreme observations on

---

* Both moving average and exponential smoothing have been extended to handle trends and seasonal effects. These extensions are beyond the scope of this book. The interested reader can consult a more specialized forecasting text to learn more about methods such as the double moving average, Holt's method of exponential smoothing and the Holt-Winter method, that handle trends and seasonality.

short-term forecasts. Frequently, however, data will also exhibit a periodic variation about its mean or trend. The most common type of periodic variation in economic data is seasonal, in which patterns repeat weekly, monthly, or quarterly. These seasonal swings often mask the other influences, and though predictable, decrease the precision of our forecasts. Therefore, most economic forecasting utilizes data that have had the seasonal pattern removed. That is what you see when you look at a series that is labeled "seasonally adjusted." In this section we introduce another smoothing method, seasonal adjustment, also known as seasonal decomposition.

We are all familiar with the fact that retail stores tend to receive an overwhelming proportion of their sales in the Christmas season, followed by very low sales after the beginning of the year. Suppose we are at the end of the year, and we are using monthly data as input to our forecast for the coming year. We observe that monthly sales have increased at a rate of 20% per month over the last three months. Should we extrapolate that trend and forecast that sales will continue increasing at 20% per month for the coming year? Being familiar with the company, you say "Of course not. It is normal for monthly sales to go up 20% per month during the holiday season, and that based on this data, there is no reason to think that next year's total sales will be any greater than this year's." What you have just done intuitively is deseasonalize the data to help you make a forecast.

8.5.1 Sources of Variation in Data

Many economic and business data series reflect several sources of variation: seasonal, trend, business cycle, firm- or industry-specific variation, and random fluctuations. Exhibit 8.10 displays the sales for a company that exhibit four different influences: a simple trend with cyclical, seasonal, and random variation. The data we actually observe are the dots on the graph. The underlying trend is the straight line; the long wave indicates the presence of a long-term economic cycle; the short wave indicates the presence of seasonal variation; and the displacements of the observations from the other three influences represent a random component.

When such influences are present, they must be removed or captured by the modeling process if reliable forecasts are to be made. As we saw in the previous chapter, econometric models can be used to capture cyclical components. Seasonal decomposition uses the data to identify and remove the systematic components of season. Any remaining trend or economic

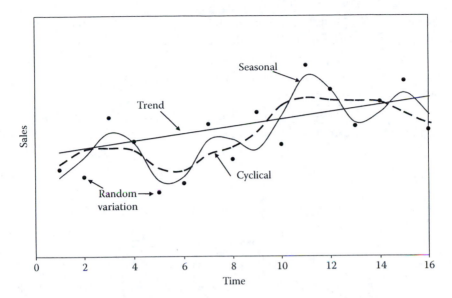

EXHIBIT 8.10   Components of sales.

cycle effects are then used to develop a forecast with a method such as smoothing or linear regression. The process is sequential. First, remove seasonality. Next, look for a trend. What is left should be either random or the result of a long-term cycle. After generating the forecast that excludes the seasonal component, we add back the seasonal factor to obtain a complete forecast. Through using this method, our forecast takes into account all of the influences: trend, cycle, and seasonal.

8.5.2 Seasonal Adjustment Factors

In its most general form, seasonal decomposition characterizes a given period's sales as a function of the four influences discussed in the previous section. This is shown formally in Equation 8.12

$$Y = f(T, C, S, e), \tag{8.12}$$

where Y is the actual value of the time series, T the trend, C the cyclical influences, S the seasonal influences, and e the random error term. The multiplicative form of this model is

$$Y_t = TC_t \times S_t \times e_t. \tag{8.13}$$

In Equation 8.13, the trend term "T" represents the level we would expect without taking account of the cyclical, seasonal, or random effects. C and S are proportional indexes centered on 1. Ignoring the error term, when the product CS is >1, sales will be above the trend line; when it is <1, sales will be below the trend line.

Consider the following example. Suppose the trend component of sales is shown by the trend regression equation

$$T_t = \text{Sales Trend}_t = 4761 + 178.4 \times \text{Time}_t. \qquad (8.14)$$

Before accounting for the stage of the economic cycle or the fact that it is the holiday season, the expected sales for quarter 16 are

$$T_{16} = \text{Sales Trend}_{16} = 4761 + 178.4 \times 16$$
$$= 7645.4 \qquad (8.15)$$

Now suppose the economy is in a recession, so sales are expected to be 3% below normal; then C would be 0.97. Suppose, too, that even though we are in a recession, this is the holiday season and sales are expected to be 20% higher than the "average" for the year; then S would be 1.20. The combined effect of cyclical and seasonal effects is that sales are $CS = (0.97)(1.20) = 1.164$ times higher than would be indicated by the trend by itself. Taking account of trend, cycle, and season, sales are expected to be

$$\text{Forecast Sales}_{16} = T_{16} C_{16} S_{16}$$
$$= 7615.4 \times C_{16} S_{16}$$
$$= 7615.4 \times 0.97 \times 1.20$$
$$= 8864.3. \qquad (8.16)$$

In this example, sales are 20% higher in the holiday season. But how did we arrive at the 20% figure? We first determined the normal level of sales without the seasonal effects, and then calculated how much sales exceed the normal level during the holiday season. To estimate the normal level, we use a special type of moving average.

Consider the quarterly sales of Warwick Corporation in Exhibit 8.11, shown graphically in Exhibit 8.12. The seven years of quarterly data suggest a seasonal component: sales tend to rise in the first quarter, and then fall back over the next three quarters.

Warwick Corp. Quarterly Sales		
Period	Quarter	Reported Sales
1	I	6,284
	II	5,796
	III	5,084
	IV	3,734
2	I	6,763
	II	6,048
	III	5,359
	IV	5,496
3	I	6,321
	II	6,775
	III	6,222
	IV	6,383
4	I	7,563
	II	8,081
	III	7,199
	IV	6,571
5	I	8,066
	II	8,586
	III	6,758
	IV	6,085
6	I	8,854
	II	9,084
	III	7,693
	IV	7,312
7	I	9,793
	II	9,054
	III	8,870
	IV	8,984

EXHIBIT 8.11   Warwick quarterly sales.

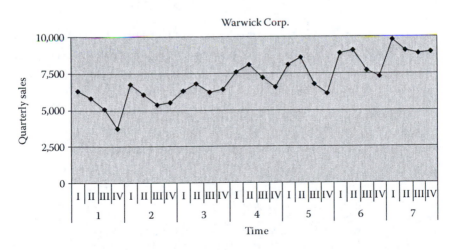

EXHIBIT 8.12   Graph of Warwick sales.

Our goal is to develop a model to forecast quarterly sales for Warwick Corporation. Following the procedure used in the previous chapter, we partition the data into two segments—one to estimate the model, the other to test it. First we use Warwick's quarterly sales from the years 1 through 4 to estimate seasonal adjustment factors (S_t) for each quarter. We then use these factors to remove the seasonality from the quarterly sales figures. Next, we graph the deseasonalized data and look for any trend or cycle that might be present. If a trend or cycle is present, we fit a model to capture those effects and then use it to estimate the seasonally adjusted (or deseasonalized) quarterly sales for years 5–7. Finally, we use the seasonal adjustment factors to restore seasonality to our forecasts. This last step produces forecasts for the actual sales in each quarter of years 5–7.

8.5.2.1 Estimating the Seasonal Adjustment Factors

The first step is to determine the seasonal adjustment factors for each period.* For Warwick, each period is one quarter. While there are various approaches, we use seasonal variation relative to an annual moving average to estimate each quarter's seasonal adjustment factor, S_t. The model is shown in Equation 8.17

$$S_t = \frac{Y_t}{TC_t},$$

(8.17)

where S_t is the seasonal adjustment factor for period t, Y_t the actual sales for period t, and TC_t the centered moving annual sales centered around period t.

The first step in calculating the quarterly seasonal adjustment factors is to calculate the centered four-quarter average sales for each quarter t, TC_t. Why a centered average? The center of a year that runs from quarters 1 through 4 lies midway between quarters 2 and 3. Similarly, the center of a year that runs from quarters 2 through 1 of the next year lies between quarters 3 and 4. We want our seasonal adjustment factor to be centered precisely on a given quarter, say quarter 3. Doing so aligns the numerator, Y_t, with the denominator, TC_t, thereby reducing any estimation error

---

* Our explanation of seasonal adjustment is relatively simplified. More complex algorithms and programs are available for more complex data sets. A very widely used program is the X-12 ARIMA program from the U.S. Census Bureau, available via the internet at www.census.gov.

	P	Q	R	S	T
80		Warwick Corp.			
81		Quarterly Sales		Centered	Seasonal
82			Reported	Moving	Adjustment
83	Period	Quarter	Sales	Average	Factor
84	1	I	6,284		
85		II	5,796		=R86/S86
86		III	5,084	5,284	0.96
87		IV	3,734		
88	2	I	6,763		
89		II	6,048		
90		III	=AVERAGE(R84:R87,R85:R88)		
91		IV	5,496		

EXHIBIT 8.13   Calculating a third quarter seasonal adjustment factor.

caused by asynchronous estimates.* By averaging quarterly sales for the years that run from quarters 1 through 4 and 2 through 1, we achieve an average quarterly sales figure that is precisely centered on quarter 3. This centered average becomes our estimate for TC_t for quarter 3. Dividing each quarter's actual sales (Y_t) by its respective centered average (TC_t) yields estimates of the seasonal adjustment factors, four in this example because the data are quarterly. This process is shown in Excel in Exhibit 8.13. Note how the centered moving average in cell S86 falls precisely in between the values in cells R84:R88.[†]

To calculate the centered four-quarter average sales for quarter 1.III, calculate the average quarterly sales for the year that runs from 1.I through 1.IV and the average quarterly sales for the year that runs from 1.II through 2.I. These are 5225 and 5344, respectively. Taking the average of these two averages gives us the average quarterly sales centered on quarter 1.III, thus

$$TC_3 = \frac{5225 + 5344}{2}$$
$$= 5284.$$

(8.18)

* Certainly, we could accomplish this using a five-quarter average (e.g., quarter 1 through the next year's quarter 1). But doing so would overweight one quarter, quarter 1 in this case.

† You can calculate a centered moving average directly in Excel by using arrays as arguments in Excel's AVERAGE function. For example, in the above example the centered moving average for quarter 1.III is =AVERAGE(R84:R87,R85:R88).

The seasonal adjustment factor for quarter 3 (SAF$_3$) now can be calculated as shown in Equation 8.19.

$$\text{Seasonal Adjustment Factor}_3 = \frac{\text{Sales for 1.III}}{\text{Centered Four Quarter Average}_{1.\text{III}}}$$

$$SAF_3 = \frac{Y_3}{TC_3}$$

$$= \frac{5084}{5284}$$

$$= 0.96 \tag{8.19}$$

This process is repeated for each subsequent period (quarters in the current example). For example, for the fourth quarter of year 1, calculate the average quarterly sales for the quarters 1.II–2.I and the quarters 1.III–2.II, average the two averages, then divide quarter 1.IV's sales by this centered average. Continuing the process for the subsequent 10 quarters yields three estimates for each seasonal adjustment factor. Finally, the estimated adjustment factors for each quarter are averaged to give us the four seasonal adjustment factors that we will use to deseasonalize the data. The final results are shown in Exhibit 8.14. The final four adjustment factors should sum to four (there are four quarters), providing a check on your calculations.

The seasonal adjustment factors tell us the portions above or below average seasonal sales that the actual seasonal observations should fall. For example, Warwick's first quarter seasonal adjustment factor is 1.11, telling us that Warwick's first quarter sales tend to be 111% of its average quarterly sales. Similarly, Warwick's fourth quarter sales tend to be 85% of its average quarterly sales.

8.5.3 Removing Seasonality

Once estimated, these seasonal adjustment factors are used to deseasonalize the series' observed sales. Divide each quarter's actual sales by its respective average seasonal adjustment factor. This grosses up the sales for quarters with traditionally low sales and grosses down the sales for quarters with traditionally high sales. For example, actual quarter 1.I sales were 6284 and the first quarter seasonal adjustment factor is 1.11, so seasonally adjusted 1.I sales are 6284/1.11 = 5657; similarly, actual 1.IV sales were 3734 and the fourth quarter seasonal adjustment factor is 0.85, so seasonally

	G	H	I	J	K	L
1						
2		**Warwick Corp.**				**Seasonal**
3		**Quarterly Sales**		**4-period**	**Centered**	**Adjustment**
4			**Reported**	**Moving**	**Moving**	**Factor**
5	**Period**	**Quarter**	**Sales**	**Average**	**Average**	**(SAF)**
6	1	I	6,284			
7		II	5,796			
8		III	5,084	5,225	5,284	0.96
9		IV	3,734	5,344	5,376	0.69
10	2	I	6,763	5,407	5,442	1.24
11		II	6,048	5,476	5,696	1.06
12		III	5,359	5,916	5,861	0.91
13		IV	5,496	5,806	5,897	0.93
14	3	I	6,321	5,988	6,096	1.04
15		II	6,775	6,204	6,314	1.07
16		III	6,222	6,425	6,580	0.95
17		IV	6,383	6,736	6,899	0.93
18	4	I	7,563	7,062	7,184	1.05
19		II	8,081	7,306	7,330	1.10
20		III	7,199	7,353		
21		IV	6,571			
22						
23					**Average SAF**	
24					I	1.11
25					II	1.08
26					III	0.94
27					IV	0.85

EXHIBIT 8.14   Seasonal adjustment of Warwick sales.

adjusted 1.IV sales are 3734/0.85 = 4390. Exhibit 8.15 shows the seasonally adjusted (smoothed) monthly sales for Warwick Corporation.

Once the data have been deseasonalized, look for other influences that might be present, for example, trends or business cycles. The actual and deseasonalized quarterly sales for Warwick Corporation are graphed in Exhibit 8.16. With seasonality removed, we observe an upward trend in Warwick Corporation's quarterly sales.

The next step is to fit a trend model to the data. We use the constant change model* of the form

$$\text{Forecast Sales}_t = a + b \times \text{Time}_t. \qquad (8.20)$$

* In actual application, both the constant change and constant growth models should be estimated, goodness-of-forecast statistics calculated and compared, and the model that yields the best forecasts selected.

Period	Warwick Corp. Quarterly Sales Quarter	Reported Sales	Average Seasonal Adjustment Factor	Deseasonalized Sales
1	I	6,284		5,657
	II	5,796	=AG84/$AH$88	5,372
	III	5,084		5,405
	IV	3,734		4,390
2	I	6,763	1.11	6,088
	II	6,048	1.08	5,605
	III	5,359	0.94	5,697
	IV	5,496	0.85	6,461
3	I	6,321		5,691
	II	6,775		6,278
	III	6,222	=AG85/$AH$89	6,615
	IV	6,383		7,504
4	I	7,563		6,808
	II	8,081		7,489
	III	7,199		7,653
	IV	6,571		7,725

EXHIBIT 8.15   Warwick Corporation: Seasonally adjusted sales.

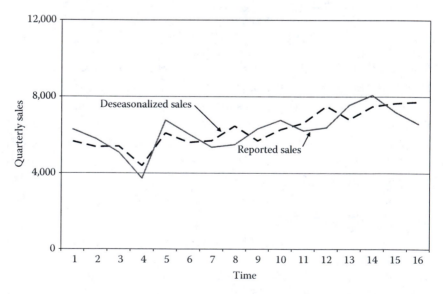

EXHIBIT 8.16   Warwick Corporation: Reported and deseasonalized sales.

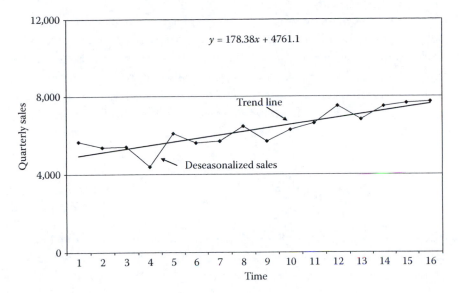

EXHIBIT 8.17   Warwick Corporation: Sales with fitted trend line.

The resulting forecasting model is

$$\text{Forecast Sales}_t = 4761 + 178.4 \times \text{Time}_t. \qquad (8.21)$$

Warwick's deseasonalized quarterly sales data with a fitted trend line are shown in Exhibit 8.17.

8.5.4 Forecasting Sales

Everything is now in place to forecast Warwick Corporation's quarterly sales for years 5–7, the goal started at the beginning of this extended process. We do this by combining the seasonal adjustment factors in Exhibit 8.14 and our model of the growth trend (Equation 8.21). We assume that both the seasonality and trend will continue through years 5–7. The forecast for seasonally adjusted sales in quarter 5.I is shown in Equation 8.22.

$$E(\text{Seasonally Adjusted Sales}_{5.I}) = 4761 + 178.4 \times \text{Time}_{5.I}$$
$$= 4761 + 178.4 \times 17$$
$$= 7794 \qquad (8.22)$$

Repeating the calculations for the next 11 quarters yields the seasonally adjusted quarterly sales forecasts for years 5–7. Multiplying these forecasts by their respective average seasonal adjustment factors yields the actual quarterly sales forecasts. For example, the seasonally adjusted forecast for 5.I is $7794 and the first quarter seasonal adjustment factor is 1.11. Multiplying the two yields the actual 5.I sales forecast of $8657, as shown in Equation 8.23.

$$E[\text{Actual Sales}_{5.I}] = E[\text{Seasonally Adjusted Sales}_{5.I}] \times SAF_I$$
$$= 7794 \times 1.11$$
$$= 8657 \tag{8.23}$$

The seasonally adjusted and actual quarterly sales forecasts for years 5–7 are reported in Exhibit 8.18. Exhibit 8.19 shows the forecast added to the actual sales for quarters 1–16 from Exhibit 8.11.

The penultimate step is to calculate the goodness-of-forecast statistics and compare them with those for other models being considered, such as the compound growth model. The statistics for the constant change model are shown in Exhibit 8.20. Recall that the RSE adjusts for the number of

	BG	BH	BI	BJ	BK
80			Warwick Corp.		
81			Quarterly Sales		
82			Seasonal	Forecast	Forecast
83			Adjustment	Deseasonalized	Actual
84	Period	Quarter	Factor	Sales	Sales
85	5 I			7,794	8,657
86		II		7,972	8,602
87		III		8,150	7,667
88		IV		8,329	7,084
89	6 I		1.11	8,507	9,450
90		II	1.08	8,685	9,372
91		III	0.94	8,864	8,338
92		IV	0.85	9,042	7,691
93	7 I			9,221	10,243
94		II		9,399	10,142
95		III		9,577	9,009
96		IV		9,756	8,298
97					

EXHIBIT 8.18 Forecast seasonally-adjusted and seasonal sales for Warwick Corporation.

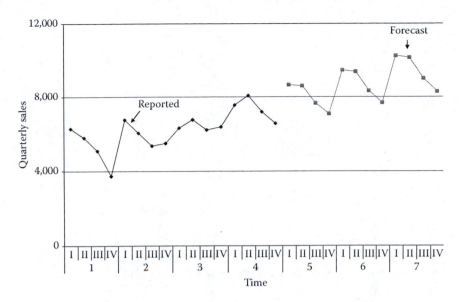

EXHIBIT 8.19   Warwick Corporation: Actual and forecast sales.

		Warwick Corp. Quarterly Sales		
Period	Quarter	Actual Sales	Forecast Actual Sales	Error Term
5 I		8,066	8,657	-592
	II	8,586	8,602	-16
	III	6,758	7,667	-909
	IV	6,085	7,084	-999
6 I		8,854	9,450	-596
	II	9,084	9,372	-288
	III	7,693	8,338	-645
	IV	7,312	7,691	-379
7 I		9,793	10,243	-450
	II	9,054	10,142	-1,088
	III	8,870	9,009	-140
	IV	8,984	8,298	686

Goodness of Forecast Statistics	
RMSE	649
ME	-451
MAD	566
RSE (denominator: 12 - 6)	917

EXHIBIT 8.20   Warwick Corporation: Goodness-of-fit statistics for constant change model.

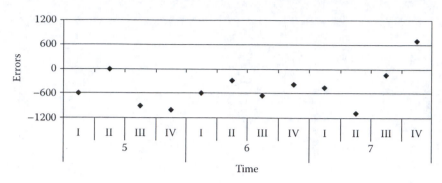

EXHIBIT 8.21   Warwick Corporation: Forecasting errors.

parameters estimated. In the current situation, the number of estimated parameters is six—the two regression coefficients and the four seasonal adjustment factors.

Finally, look at the forecasting errors. If they do not display any obvious cyclicality, your model has probably done an acceptable job of capturing the structure of the data. The forecasting errors are shown in Exhibit 8.21. In this case, a substantial number are consecutively below the 0 line, indicating that there may be serial correlation in the series. This suggests we look for another variable that, when added to the model, will improve its predictive power. But assume for our present purposes, we have accomplished our seasonal adjustment. It remains as further work for the forecaster to look at this remaining issue.

8.5.5 Review of Seasonal Adjustment

The steps to seasonal adjustment and forecasts are summarized below:

1. Calculate a centered moving average equal to the length of the seasons (12 months if the pattern is monthly, four quarters if quarterly, etc.).

2. Calculate the actual observation as a proportion of the centered moving average to obtain the seasonal index for each period. If you have enough data, consider calculating factors over several cycles, then averaging them.*

---

* DeLurgio (1998) recommends adjusting outlying observations and seasonal adjustment factors, both of which are beyond the scope of this text.

3. Deseasonalize the time series by dividing each observation by its respective average seasonal adjustment factor.

4. Estimate the trend and/or cyclical components using the deseasonalized data.

5. Check the model's goodness-of-forecast.

   (a) Use the model to forecast seasonally adjusted values.

   (b) Multiply the seasonally adjusted values by their appropriate average seasonal factors to compute the seasonal values.

   (c) Calculate errors and goodness-of-forecast statistics using known actuals.

6. Check to see if cyclical factors are important by plotting the error terms of your model. They should exhibit a random pattern. If they do not, add explanatory variables and repeat steps 5 and 6.

   Having removed the seasonality from Warwick Corporation's quarterly sales, the series in Exhibit 8.17 exhibits no obvious cyclicality. However, we do observe a possible cyclical effect in the forecast errors in Exhibit 8.21. If cyclicality is present, consider using an econometric model as discussed in Chapter 7.

8.6 TIME SERIES MODELS

The last forecasting method we consider, albeit briefly, goes by many names, including ARMA, ARIMA, Box–Jenkins methods, or simply, time series forecasting. ARIMA has come to be used to refer generally to the methods for forecasting all these processes. It takes its name from the acronyms for processes that are assumed to generate the time series. AR stands for autoregressive, MA for moving average, ARMA means autoregressive, moving average, and ARIMA for autoregressive, integrated moving average. George Box and Gwilym Jenkins (1976) were authors who developed some of the statistical methods for forecasting these series. As with the other smoothing methods considered in this chapter, this forecasting method uses the past observations and forecasting errors of a time series to predict the future values of that same time series; hence, the name "time series forecasting." It also shares another trait with the simpler smoothing methods previously considered—it only is capable of producing very short-term forecasts.

Time series models are a logical extension of the processes we considered earlier in this chapter. Moving averages assume a stationary series (a series whose mean and standard deviation remain constant across time) in which recent observations contain more information than do distant observations. Exponential smoothing also assumes a stationary series, but one in which both recent observations and forecast errors contain information about future values. However, neither of these methods assumes the presence of a formal systematic relationship between the past observations and forecast errors of a time series and its future observations.* Therefore, neither is capable of modeling a systematic undulating time series or capturing turning points in that series.

ARIMA models can model undulating times series. They do so by assuming some underlying structure, or process, in which current and future observations are dependent upon past observations. Statistical methods are used to identify or "fit" a model to the patterns in the past historical time series. The fitted model then is used to forecast future values for the series. There is no underlying theory that states why such models should adequately describe the movement of a given economic time series. ARIMA models do not attempt to explain why a variable behaves in a particular way. Rather, they try to detect patterns in a time series, use those patterns to identify an underlying process, and then use the identified process to predict future values.

These models are considerably more difficult to understand than what has come before, and the modeling processes required to identify them are beyond the scope of this book. We introduce them because they can be effective short-term forecasting tools for certain types of time series, and because easy to use, low cost software has greatly simplified the estimation of these models and their forecasts. The method also is frequently included in many commercial forecasting software programs.† The wide availability and ease of use of this software creates the danger of using a method about which you know very little. Because you will likely have access to such

---

* By "formal relationship" we mean a relationship that can be expressed mathematically. For example, a simple trend line such as Sales $= a + b \times$ Time is a formal relationship between sales and time.

† ARIMA methods are available in numerous software packages including most widely used statistical programs such as EViews and STATA. StatTools by Palisade that is included with this text has some capability to handle time series analysis. Other software packages devoted to forecasting include ForecastX by John Galt Solutions, and a freeware package, FreeFore by Autobox.

programs, it is important for you to have some familiarity with the methods. Our goal is to help the reader understand in a non-technical way what ARIMA forecasting programs are doing. When finished with this section, the reader should have a general idea of how these models work and what they do. After all, our intent in Chapters 6, 7, and 8 is only to expose you to some of the forecasting methods that are available and to give you sources to consult if you want to dig deeper.

8.6.1 A Description of Time Series Models

A time series is a time-ordered, equally spaced series of observations of a random variable. Examples include the variation in crime rates by day of the month, the pulsations of light from distant stars, the daily fluctuations in the stock market, and, of course, the daily, weekly, monthly, or quarterly variations in firm's sales or costs.

We have already looked at a variety of ways to describe or characterize time series with the idea of forecasting the next few observations. For example, in Chapters 6 and 7 we treated a series of observations as independent random draws from some larger population and then looked for independent factors (e.g., a time trend) to explain the variation in the series and predict future values. ARIMA models assume a time series itself contains information that can be used to forecast the next several values. We have seen one simple example of this in exponential smoothing, where we used previous forecasting errors to predict the next observation. ARIMA models extend this concept by adding formal structure to a time series' relationship with itself. The models we consider view the entire series as a single, self-contained sample created by some underlying process, then try to identify that process and use it to explain the variation in the series and predict future values. These linkages are depicted in Exhibit 8.22.

To see how the underlying logic and assumptions of ARIMA models differ from macroeconomic models, consider a model to predict housing

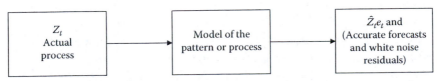

| Z_t Actual process | Model of the pattern or process | $\hat{Z}_t, e_t$ and (Accurate forecasts and white noise residuals) |

EXHIBIT 8.22   The link between an actual ARIMA process and an observed time series.

starts. A macroeconomic model might use interest rates and GDP (economic activity) as independent variables, the underlying logic being that rising interest rates slow the rate of housing construction, while increasing economic activity increases it. However, an ARIMA model only uses the past data on housing starts to predict future housing starts.

In general mathematical terms, the current observation in a time series (Z_t) is a function of past observations (Z_{t-k}), and past random error terms (e_{t-k}), and a current random error (e_t), expressed as follows:

$$Z_t = f(Z_{t-k}, e_{t-k}) + e_t, \qquad (8.24)$$

where Z_t is the current observation in a time series that is t periods long, Z_{t-k} the observation in the time series k periods earlier, e_t the current random error in the process, and e_{t-k} the random error of the time series k periods earlier.

An example of a specific realization of this general process is shown in Equation 8.25

$$Z_t = a_0 + b_1 Z_{t-1} - a_1 e_{t-1} + e_t, \qquad (8.25)$$

where a_0, b_1, and a_1 are least squares regression coefficients; and e_t and e_{t-1} the error term in periods t and $t-1$. Equation 8.25 tells us that the current observation in a time series is influenced by the observation that immediately precedes it and the forecast error to the previous period. The next section considers the most frequently encountered processes.

8.6.2 Two Processes: AR and MA

Two possible processes can be at work in ARIMA models (assuming a stationary series): autoregressive processes and moving average processes. These two processes can occur separately or together. This section looks more closely at these two processes.

8.6.2.1 Autoregressive Processes

An autoregressive process is one in which the value of the variable today is correlated with the value of the same variable in the past. This means past values give us information about future values. For example, suppose the variable Z_t follows an autoregressive process with a k-period lag. We can express the relationship as a simple regression equation of the form

$$Z_t = a + b Z_{t-k} + e_t. \qquad (8.26)$$

Equation 8.26 tells us that an observation at time t is closely related to an observation at time $t - k$ (a time k periods in the past). Suppose that when the value at time $t - k$ is large, the value at time t is also large. If so, we would expect the correlation between these k-lagged variables to be close to +1. Alternatively, suppose that an observation at time t is inversely related to an observation k time periods in the past, so that when the value at time $t - k$ is large, the value at time t is small. In this scenario, we would expect the correlation between these k-lagged variables to be close to −1. In either case, knowing the value of the observation in period $t - k$ helps us anticipate or predict the value in period t.

8.6.2.1.1 First-Order Autoregressive Processes

8.6.2.1.1.1 Example of a First-Order Autoregressive Process. To see how an autoregressive process works, consider the monthly consumer sentiment data released by the University of Michigan. Suppose that your firm has developed a forecasting model for sales that is a function of consumer sentiment. To implement your model, you first will need a forecast of consumer sentiment. Applying a forecasting software program to the monthly consumer sentiment time series gives you the following model.

$$\text{Consumer\_sentiment}_t = 36.53 + 0.577 \times \text{Consumer\_sentiment}_{t-1} + e_t$$

$$(8.27)$$

Equation 8.27 tells us that consumer sentiment this month is related to consumer sentiment last month. This is an autoregressive model. To estimate it, the software regressed current consumer sentiment on the previous month's consumer sentiment. It regressed the consumer sentiment series on itself, hence the name, "autoregressive."

We now use this model to forecast future consumer sentiment. Suppose consumer sentiment for August was 89.3. Substituting 89.3 for $\text{Consumer\_sentiment}_{t-1}$ in Equation 8.27 yields the forecast for September shown in Equation 8.28.

$$\text{Consumer\_sentiment}_{\text{SEP}} = 36.53 + 0.577 \times \text{Consumer\_sentiment}_{\text{AUG}}$$

$$= 36.53 + 0.577 \times 89.3$$

$$= 88.06 \qquad\qquad (8.28)$$

To forecast consumer sentiment in October, we substitute our September forecast into our model, thus

$$\text{Consumer\_sentiment}_{\text{OCT}} = 36.53 + 0.577 * \text{Consumer\_sentiment}_{\text{SEP}}$$
$$= 36.53 + 0.577 * 88.06$$
$$= 87.34. \tag{8.29}$$

8.6.2.1.1.2 General Form of a First-Order Autoregressive Process. In the above example, the observation one period back (at a lag of one period) explains some of the variance in the current observation. This is known as a first-order autoregressive process, or AR(1), where "(1)" denotes the lag. Mathematically, it is given as

$$Z_t = a_0 + bZ_{t-1} + e_t, \tag{8.30}$$

where a_0 and b are least squares regression coefficient, and e_t the error term in period t. This is a simple linear relationship between Z_t and Z_{t-1}.

8.6.2.1.2 Second-Order Autoregressive Processes

8.6.2.1.2.1 Example of a Second-Order Autoregressive Process. In second-order autoregressive processes (AR(2)), the two most recent observations are used to predict the next several. Consider the following model of monthly total business inventories.

$$\text{Inventories}_t = 20{,}600 + 1.52 \times \text{Inventories}_{t-1} - 0.573$$
$$\times \text{Inventories}_{t-2} + e_t \tag{8.31}$$

Equation 8.31 tells us that inventories in this month are related to inventories in the previous two months. Suppose the two most recent observations for inventories are 1,179,980 (in millions of dollars) and 1,180,343 for July and June, respectively. Then the forecasts for August and September inventories are calculated as follows:

$$\text{Inventories}_{\text{AUG}} = 20{,}600 + 1.52 \times \text{Inventories}_{\text{JUL}} - 0.573 \times \text{Inventories}_{\text{JUN}} + e_t$$
$$= 20{,}600 + 1.52 \times 1{,}179{,}980 - 0.573 \times 1{,}180{,}343$$
$$= 1{,}180{,}325$$
$$\text{Inventories}_{\text{SEP}} = 20{,}600 + 1.52 \times \text{Inventories}_{\text{AUG}} - 0.573 \times \text{Inventories}_{\text{JUL}} + e_t$$
$$= 20{,}600 + 1.52 \times 1{,}180{,}325 - 0.573 \times 1{,}179{,}980$$
$$= 1{,}181{,}045. \tag{8.32}$$

Note that the September forecast is based on the actual July number and the August forecast.

8.6.2.1.2.2 General Form of an AR(2) Process. An AR(2) process assumes that the value of a variable (Inventories in this example) at time t depends on the observed or expected values of that variable at times $t - 1$ and $t - 2$. This model is a linear regression that regresses each observation in the time series beginning at $t = 3$ on the two previous values. Mathematically, AR(2) processes are characterized as follows:

$$Z_t = a_0 + b_1 Z_{t-1} + b_2 Z_{t-2} + e_t \qquad (8.33)$$

where a_0, b_1, and b_2 are least squares regression coefficients, and e_t the error term in period t. This is a linear relationship between Z_t and the two most recent observations, Z_{t-1} and Z_{t-2}.

8.6.2.1.3 Indentifying the Processes. How does the forecaster (or computer program) know what model to try first? For example, in the first example consumer sentiment was modeled as an AR(1) process. In the second example, inventories were modeled as an AR(2) process. How did we know to fit an AR(1) process to the consumer sentiment data and an AR(2) process to the inventories data? We knew because each process leaves behind a distinct footprint related to the autocorrelations between current and past observations. We use the term autocorrelation for the same reason we use the term autoregressive. We are looking at the correlation between current values of a time series with past value of the same series—the correlation of a time series with itself.

8.6.2.1.3.1 Autocorrelation and Partial Autocorrelation. Two types of autocorrelations, known as autocorrelation and partial autocorrelation, are used to generate the footprints needed to identify the different processes. To see the difference between these two terms, consider the following time series, (u, v, w, x, y, z). We use letters instead of numbers because the sequence order is readily observable. Suppose we wish to find the correlation of current observations with those from two periods ago, for example, the influence of observation x on observation z. Then we would calculate the autocorrelation between series (w, x, y, z) and series (u, v, w, x). This tells us the influence of x on z.

Note, however, that x could be influencing z in two different ways: directly (denoted as $x \rightarrow z$), and indirectly, though x's influence on y and y's influence on z. Autocorrelation includes both paths of influence. However, additional information useful to process identification can be gleaned by measuring observation x's direct influence on z, the influence of x on z after removing influences from y. Partial autocorrelation measures this relationship, the part of the autocorrelation related only to the direct effect.*

We use bar graphs to show the autocorrelations and partial autocorrelations between current and successively lagged values of the time series. These graphs are called the autocorrelation function (ACF) and partial autocorrelation function (PACF). These graphs provide the footprints used by forecasting software to identify the models to be used to fit the time series. We look at the footprints for the AR(1) and AR(2) processes next.

8.6.2.1.3.2 The Footprints. When a time series is generated by an AR(1) process, the plot of its autocorrelations for successive lags shows the correlations dying away. Exhibit 8.23 shows an ACF when the b in Equation 8.30 equals 0.9.

Similar to the ACF, the PACF is a plot of the partial autocorrelations at successive lags. The PACF for a simple AR(1) process shows a single spike

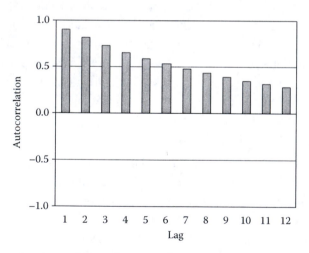

EXHIBIT 8.23 Autocorrelation function for an AR(1) process, where $b = 0.9$.

---

* The reader interested in the formulas for the ACFs and PACFs in this chapter is referred to O'Donovan (1983); Box and Jenkins (1976).

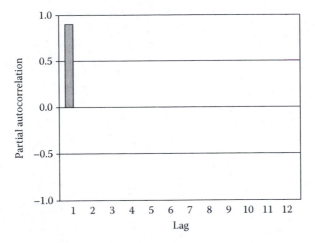

EXHIBIT 8.24   Partial autocorrelation function for an AR(1) process, where $b = 0.9$.

at the first lag; partial correlations at all other lags are zero. Exhibit 8.24 shows the PACF for $b = 0.9$.

AR(2) processes exhibit different ACF and PACF footprints. The ACF shows the correlations of successive lags dying away. The PACFs for an AR(2) process show two distinct spikes at the first two lags, then zero for all subsequent lags. The respective ACF and PACF for an AR(2) process when the first two partial autocorrelations are positive are shown in Exhibits 8.25 and 8.26.

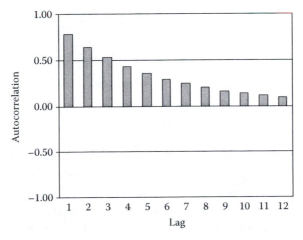

EXHIBIT 8.25   Autocorrelation function for an AR(2) process, where $b_1 = 0.7$ and $b_2 = 0.1$.

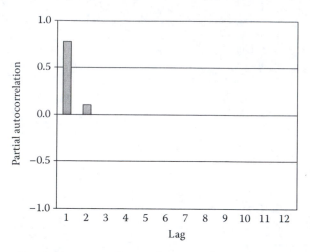

EXHIBIT 8.26   Partial autocorrelation function for an AR(2) process, where $b_1 = 0.7$ and $b_2 = 0.1$.

Exhibits 8.25 and 8.26 present a different footprint, than do Exhibits 8.23 and 8.24. Similarly, other processes exhibit different footprints. These distinct footprints tell the analyst or the software program where to begin the model-fitting process.

8.6.2.2 Moving Average Processes

Moving average processes are less common, and typically, have less explanatory power than autoregressive processes (Makridakis and Hibon, 1997). They also suffer from the problem that they can only forecast one period into the future. For these reasons, we only describe a general first-order moving average (MA(1)) process in this section. In a moving average process, the value of a random variable depends on a moving average of past random error terms. Consider the following simple MA(1) process:

$$Z_t = a_0 - a_1 e_{t-1} + e_t. \qquad (8.34)$$

Equation 8.34 tells us that the current observation Z_t is a function of a moving weighted average of the current and previous random error terms. At first, the concept of a current value being related to a previous forecast error may seem odd. However, we have seen this before in the exponential smoothing technique. Exponential smoothing used the previous period's forecast error to forecast the next period's value. The error terms in Equation 8.24 come about in much the same way.

All autoregressive and moving average processes have distinctive ACF and PACF footprints. ARIMA modeling software begins by calculating the ACFs and PACFs for the series under consideration, and then uses those footprints to identify the processes at work in that particular series.

8.6.3 ARMA and ARIMA Processes

There is nothing to prevent more complex combinations of processes. For example, an autoregressive process and a moving process can exist simultaneously. In the case of two first-order processes, the general form is as follows:

$$Z_t = a_0 + b_1 Z_{t-1} - a_1 e_{t-1} + e_t, \qquad (8.35)$$

where a_0, b_1, and a_1 are least squares regression coefficients, and e_t and e_{t-1} the error terms in periods t and $t-1$. As was the case for less complex processes, this process (notated as ARMA(1,1)) has its own distinctive footprint that is used to identify it.

Autoregressive and moving averages processes can also be used to model the changes in a time series instead of the time series itself. For example, suppose that you create a time series by calculating the changes from one observation to the next, a process known as first differencing, and then looked for underlying processes in this newly created time series of changes. One such model might be

$$(Z_t - Z_{t-1}) = a_0 + b_1 (Z_{t-1} - Z_{t-2}) - a_1 e_{t-1} + e_t, \qquad (8.36)$$

where a_0, b_1, and a_1 are least squares regression coefficients, and e_t and e_{t-1} the error tetms in periods t and $t-1$. Such a process is known as an ARIMA process, where the I (for integrated) identifies the number of times the data is differenced. Equation 8.36 contains an AR(1) process, an I(1) process (first differences), and an MA(1) process. It would be notated as an ARIMA(1,1,1) model.

8.6.4 An Overview of the Box–Jenkins Method

We have now considered several of the more basic processes that could be at work in a time series. But correctly identifying the underlying processes is not easy, and afterwards you must still estimate the parameters in those relationships. This problem was solved by Box and Jenkins (1976). They devised an iterative method to identify the appropriate model, estimate its parameters, and then generate forecasts. Their method is to extract all possible information (patterns) from a time series so that any remaining variation in the series, the error terms (e_t's), are distributed as white noise.* As a result of their work, the method for estimating ARIMA models is called the Box–Jenkins method and ARIMA models themselves tend to be called Box–Jenkins models.

The Box–Jenkins method is a highly refined curve-fitting algorithm that uses current and past values of the dependent variable to produce accurate short-term forecasts. The Box–Jenkins method:

- Analyzes the patterns of correlations between current and past observations on a variable (autocorrelations and partial autocorrelations). These relationships create distinctive footprints for the various possible models.

- Based on these footprints, the analyst or forecasting software determines the appropriate structure for a model.

- A computer program then fits the parameters to the structure via an iterative process.

- Once the parameters of the model are estimated, the distribution of the residual errors is examined. If the model is correctly identified, the error terms will be white noise. If they are not, the process begins again with a slightly different model. The process iterates until it finds a solution.

- The fitted model then is used to generate forecasts.

As with most other econometric methods, the Box–Jenkins method has positives and negatives. On the positive side, the Box–Jenkins method can produce very accurate short-term forecasts. For certain types of economic data these forecasts can be more accurate than those produced by many other methods. The Box–Jenkins method allows for a wide range of models and offers a strategy for selecting a model from that class of models that best represents the data. Finally, the process can be used to model a wide variety of variables. On the negative side, the process of identifying the

---

* White noise refers to a process where the sequence of error terms is distributed randomly with zero mean and zero correlation.

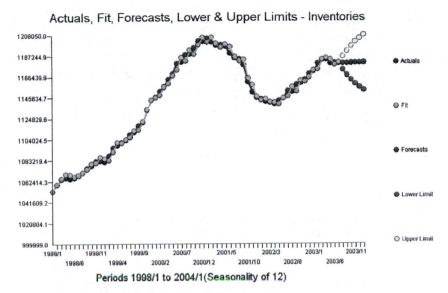

EXHIBIT 8.27 Output from using the Box–Jenkins method to fit a model for inventories.

appropriate model is difficult to understand and sometimes produces different models that fit the data equally well.

Exhibit 8.27 shows the output from an ARIMA model for inventories. It highlights both the strengths and weaknesses of this forecasting method. The Box–Jenkins algorithm found a model that closely tracks the inventories time series and forecasts inventories for the next six quarters. However, note the rapidly widening confidence intervals around the forecasts. The accuracy of the first forecast may be reasonably good, but it deteriorates for longer forecasts.

What does this complexity buy us? Box–Jenkins models are better than simpler models at some things, such as predicting turning points and short-term forecasts. However, econometric models are better for long-term forecasts and identifying relationships between variables. So as always, the model builder must assess her forecasting needs and horizons, then select the forecasting method that best addresses those needs.

8.7 SUMMARY

These three chapters have presented only the most basic methods of forecasting. The goal was to introduce future model builders to a few of the wide variety of forecasting techniques available. If you are going to do a lot of forecasting, and the forecasts are crucial input to your modeling, you will need to develop considerably more expertise as a forecaster than we have space to discuss here. Readers interested in pursuing the subject further should consult a text devoted exclusively to forecasting, such as DeLurgio (1998) or Pindyck and Rubinfeld (1998) listed in our references.

PROBLEMS

1 Zeplin Corporation

Use the following data for Zeplin Corporation to answer the following two questions.

Zeplin Data

Year	Quarter	Sales	Time	Year	Quarter	Sales	Time
1	I	853.0	1	5	I	1017.7	17
	II	828.9	2		II	1110.6	18
	III	904.8	3		III	1178.8	19
	IV	961.1	4		IV	1106.5	20
2	I	972.9	5	6	I	1137.4	21
	II	1009.3	6		II	1112.4	22
	III	867.7	7		III	1121.1	23
	IV	960.6	8		IV	1170.9	24
3	I	1028.8	9	7	I	1194.1	25
	II	956.5	10		II	1155.3	26
	III	1003.0	11		III	1232.3	27
	IV	978.9	12		IV	1253.8	28
4	I	1054.8	13	8	I	1290.4	29
	II	1111.1	14		II	1285.4	30
	III	1122.9	15				
	IV	1159.3	16				

(a) Fit four- and six-period moving average models to Zeplin's sales. Calculate the ME, MAD, and MSE for each model. Based on these statistics, which model best fits the data? Use the best-fitting model to forecast Zeplin's sales for the next two quarters.

(b) Fit an exponential smoothing model to Zeplin's sales. Calculate the ME, MAD, and MSE for each model. Use the average of the first four periods for your initial forecast and the smoothing constant that minimizes the MSE. Forecast Zeplin's sales for the next quarter using your model.

2 Boulder Brass Works, Inc.

Boulder Brass Works, Inc. (BBW) is a struggling young company that manufactures toy replicas of Civil War brass instruments. The owners realize that their sales move with the seasons, but lack the necessary skills to analyze these seasonal fluctuations. In an effort to better plan their production levels, they have asked you to build a model that tells them, on average, how much each quarter's sales depart from BBW's average quarterly sales. They have given you BBW's sales data for the last 7.5 years. It is reported below.

Boulder Brass Data

Boulder Brass Works, Inc.					
Year	Quarter	Sales	Year	Quarter	Sales
1	I	923.0	5	I	1,456.8
	II	1,015.3		II	1,783.8
	III	289.2		III	404.9
	IV	434.6		IV	607.6
2	I	1,071.1	6	I	1,300.3
	II	1,420.3		II	2,009.9
	III	318.2		III	458.4
	IV	471.8		IV	721.4
3	I	1,020.7	7	I	1,621.4
	II	1,294.0		II	1,866.5
	III	375.0		III	533.2
	IV	545.0		IV	753.9
4	I	1,118.1	8	I	1,772.9
	II	1,449.7		II	2,913.2
	III	336.6			
	IV	592.4			

Using the simple seasonal decomposition method discussed in this chapter earlier, estimate the seasonal adjustment factors for BBW's sales. Next, use those factors to remove the seasonality from BBW's sales. Finally, on the same chart, plot BBW's sales with and without seasonality.

3 Comfort Food Corporation

Comfort Food Corporation (hereafter CFC) is a wholesaler of high-end food products—food for individuals who miss their mothers' cooking. As part of their production planning, their management has asked you to forecast quarterly sales for the next six quarters.

CFC sales are highly seasonal. Each year, they grow steadily from the first quarter through the fourth quarter, then fall back at the beginning of the first quarter of the next year and repeat the process. In addition to a year-over-year increasing trend, casual observation suggests that their sales increase with consumer sentiment and, perhaps, disposable income. Some in management also believe that sales may increase with positive expectations for economic growth (measured by the Treasury spread) and decrease with the overall level of risk aversion (measured by the risk spread).* CFC has supplied 11 years of quarterly data consisting of the change in consumer sentiment, the percentage change in disposable income, the Treasury and risk spreads, and CFC's quarterly sales. The data set also contains CFC's forecasts for the next six quarters of macroeconomic data. It is shown on next page.

Your job is to determine which of four hypothesized linear models produces the best forecasts and then use that model to predict CFC's sales for the next six quarters. The four models are shown below.

Hypothesized models for CFC's sales

Model 1:
$$\text{Quarterly Sales}_t = a + b \times \text{Consumer Sentiment}_t$$
$$+ c \times \text{Disposable Income}_t + d \times \text{Treasury Spread}_t$$
$$+ e \times \text{Risk Spread}_t + g \times \text{Time}_t$$

Model 2:
$$\text{Quarterly Sales}_t = a + b \times \text{Consumer Sentiment}_t$$
$$+ c \times \text{Treasury Spread}_t + d \times \text{Time}_t$$

Model 3:
$$\text{Quarterly Sales}_t = a + b \times \text{Consumer Sentiment}_t + c \times \text{Risk Spread}_t$$
$$+ d \times \text{Time}_t$$

Model 4:
$$\text{Quarterly Sales}_t = a + b \times \text{Consumer Sentiment}_t + c \times \text{Time}_t$$

---

* The Treasury spread in this case is the difference between the yield on 20-year Treasury bonds and the yield on 6-month Treasury bills. The risk spread is the difference between the yields on Baa and the Aaa 20-year corporate bonds.

CFC Data

Year	Quarter	Time	Comfort Foods Change Sentiment	% change Disp. Inc.	Treasury Spread	Risk Spread	Quarterly Sales
1	I	1	-1.7	1.83	3.31	0.64	2,248
	II	2	-3.6	0.66	2.92	0.69	2,668
	III	3	3.7	1.39	2.69	0.63	3,111
	IV	4	4.9	0.61	1.76	0.62	3,298
2	I	5	-5.1	-0.08	1.68	0.57	2,182
	II	6	1.9	0.52	1.37	0.63	2,594
	III	7	-4.2	0.53	1.13	0.63	2,893
	IV	8	-0.9	1.12	1.19	0.66	3,411
3	I	9	3.4	0.83	1.92	0.69	2,394
	II	10	2	0.85	1.84	0.7	2,527
	III	11	1.8	0.51	1.79	0.68	3,247
	IV	12	0.9	0.96	1.81	0.67	3,838
4	I	13	4	0.82	1.83	0.61	3,013
	II	14	5.7	1.14	1.44	0.61	2,687
	III	15	-1.5	1.32	1.29	0.57	3,290
	IV	16	1	2.14	0.85	0.58	3,461
5	I	17	2.1	1.45	0.94	0.64	2,567
	II	18	-3.5	1.09	0.75	0.6	2,681
	III	19	-7.8	0.78	1.25	0.81	3,396
	IV	20	6.5	0.74	1.12	1.05	4,050
6	I	21	0.7	0.14	1.45	0.84	2,636
	II	22	1.4	0.50	1.7	0.76	3,076
	III	23	-2.8	1.45	1.68	0.83	3,325
	IV	24	8.8	2.22	1.36	0.55	3,825
7	I	25	-2.8	0.67	0.37	0.76	2,792
	II	26	-0.9	1.28	0.2	0.7	2,878
	III	27	-2.5	0.14	0	0.79	3,322
	IV	28	-11.1	0.32	0.7	0.78	3,714
8	I	29	-6.3	-0.42	1.93	0.87	2,879
	II	30	4	2.76	2.3	0.84	3,088
	III	31	-9.7	-1.43	3.22	0.88	3,196
	IV	32	10.3	2.78	3.96	1.32	4,641
9	I	33	0	0.47	3.92	1.27	3,411
	II	34	-4.9	-0.39	3.81	1.37	3,496
	III	35	-7.5	0.05	3.44	1.41	3,364
	IV	36	1.8	0.43	3.82	1.18	4,247
10	I	37	3.6	1.24	3.77	1.11	3,112
	II	38	4.9	1.55	3.97	1.13	3,848
	III	39	-1.3	0.42	4.21	1.03	4,261
	IV	40	14.2	0.96	4.04	0.9	4,616
11	I	41	-9.6	0.59	4.07	0.73	3,099
	II	42	2.5	0.68	3.58	0.8	3,564
	III	43	-5	1.82	2.85	0.74	3,878
	IV	44	3.8	-1.01	2.16	0.66	4,177
12	I	45	-7.8	0.11	1.7	0.72	
	II	46	8.8	-0.14	1.06	0.89	
	III	47	-22.3	1.35	0.75	0.95	
	IV	48	17	1.14	0.34	0.95	
13	I	49	-3.8	-0.38	0.5	0.84	
	II	50	-2.7	1.02	0.19	0.91	

Begin with a simple time-trend regression (seasonal sales on Time). Have Excel's regression package report the residual errors. Plot the errors to confirm the presence of seasonal and cyclical variation.

Next, identify which of the four hypothesized models best forecasts future quarterly sales. Estimate your preliminary models using the first seven years of data and test their forecasting accuracy using the last four years of data. Base your goodness-of-forecast statistics (RMSE, ME, MAD, and RSE) on seasonal sales forecasts. (Because you will estimate your models using deseasonalized data, you will need to restore seasonality to your forecasts before calculating the error terms.)

After identifying the model that produces the best forecasts, re-estimate that model using all 11 years of data. Have Excel report the residuals from this regression so that you can graph them and look for evidence of a missing factor. Label your graph appropriately. Finally, forecast sales for the next six quarters. Add the six forecasts to your original graph of CFC's quarterly sales. Add them to your graph as a separate series so that your forecasts are clearly identified. Be sure the company name appears in the graph title.

What You Turn In

Submit a report in memo format that describes the purpose and results of your work. Include the criteria you used to select your model. Identify the model that is the least biased, the model that produces the smallest average error regardless of direction, and the model that is the most efficient with respect to the number of independent variables. Place the graph of sales for CFC (including predicted sales) in the body of your memo. Also include a table in your memo that reports the model used (e.g., $Sales_t = 45 + 4 \times Time_t$) and the sales forecasts. Place the following in your appendix: Clearly labeled graphs of the error terms, before and after model identification; and a well-organized printout of the spreadsheet showing the goodness-of-forecast calculations you used to identify the best model.

4 Brass Wind and Wood Wind

This problem asks you to apply the techniques of seasonal decomposition, trend analysis, and econometric time series models to forecast quarterly sales for two companies, Brass Wind and Wood Wind. Both companies' sales exhibit seasonality. In addition, one exhibits a cyclical component related to the treasury spread (long-term treasury rate less T-bill rate) and possibly to changes in the GDP. The data you need to perform for this analysis is reported on the next page.

Brass & Wood Data

| | | Reported Sales | | | | | | Reported Sales | | | |
| | | Brass | Wood | Change in | Treasury | | | Brass | Wood | Change in | Treasury |
Year	Quarter	Wind	Wind	GDP (%)	Spread	Year	Quarter	Wind	Wind	GDP (%)	Spread
1	I	209.4	105.8	1.50%	3.04	6	I	294.4	268.4	1.20%	0.67
	II	260.8	140.5	1.80%	3.02		II	716.1	354.4	0.70%	1.01
	III	247.4	238.7	1.10%	2.63		III	409.6	685.9	1.50%	1.23
	IV	121.5	275	1.70%	2.32		IV	294.9	580.2	2.40%	0.87
2	I	161.5	159.8	1.10%	1.34	7	I	366.3	329.7	1.50%	-0.22
	II	503.2	277.6	0.60%	1.31		II	688.6	510.7	1.90%	1.06
	III	295.1	320	1.20%	1.1		III	426.9	819.1	0.80%	-0.42
	IV	171	435.7	1.30%	0.81		IV	314.9	670.9	0.90%	-0.52
3	I	197.2	232.4	1.30%	0.92	8	I	429.9	355.9	1.10%	0.39
	II	473.5	301	2.00%	1.52		II	863.4	314	0.60%	1.66
	III	203	528.6	1.00%	1.51		III	450.7	594.5	0.40%	2.05
	IV	195.8	363.3	1.50%	1.49		IV	279.9	268.2	0.00%	3.15
4	I	332.4	198.5	1.80%	1.52	9	I	370.6	101.4	-1.00%	3.68
	II	606.1	270.1	1.90%	1.49		II	878.8	179.6	0.00%	3.85
	III	285.1	487.7	1.30%	1.18	Forecast	III			0.20%	4
	IV	218.8	406.3	1.00%	1.03		IV			0.50%	3.75
5	I	279.5	270.3	1.70%	0.58	10	I			0.50%	3.35
	II	696.5	383.9	0.80%	0.66		II			1.00%	2.75
	III	364.9	782.1	1.40%	0.45						
	IV	243.8	530.3	1.90%	0.81						

Quarterly Sales for Brass Wind and Wood Wind

Begin by identifying which series displays seasonality only and which displays seasonality and cyclicality. To do this, deseasonalize both series, fit a simple trend line, and plot the deseasonalized errors. One error term series will display a somewhat random pattern (below left), the other a clear long-term cycle (below right). The former is the series that displays only seasonality.

Residual plot graph

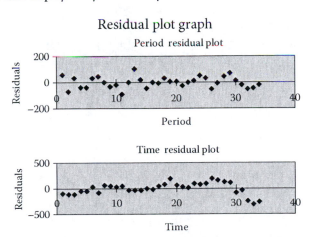

Period residual plot

Time residual plot

A. For the company that displays seasonality but not cyclicality:
   1. Determine the model—the constant change or compound growth model—that generates the most accurate forecasts.

Use the first 20 observations to estimate each model, the last 14 to test each model's forecasting ability. Measure forecasting accuracy using the RSE.

2. Re-estimate the model with the lowest RSE using all 34 observations. Use this new model to predict the sales for the next four quarters.

3. Create a graph that displays actual and predicted sales (as separate series).

B. For the company whose sales display both seasonality and cyclicality:

1. Determine the linear model (trend + Treasury spread or trend + Treasury spread + GDP) that generates the best forecasts as measured by the RSE. As before, estimate each model using the first 20 observations and test them using the last 14.

2. Re-estimate the model with the lowest RSE and use it to predict sales over the next four quarters.

3. As in part A, create a graph that displays actual and predicted sales (as separate series).

What You Turn In

Submit a report in memo format that describes the results of your work. Include the graphs for each firm's actual and forecast sales, with the company name appearing in the title of the graph. Also include a table in your memo that lists, by company, the model used (e.g., $Sales_t = 45 + 4 \times Time_t$) and the forecast sales. Include the following in your appendix: Graphs of the error terms, clearly labeled; and well-organized printouts of the spreadsheets showing your RSE calculations.

PART IV

A Closer Look at the Details of a
Financial Model

CHAPTER 9

Modeling Value

T HROUGHOUT THIS BOOK WE have emphasized that you need to link the decisions in your model to an objective, which is assumed to be the maximization of the value of the equity of the firm. This chapter digs deeper into the details of modeling equity value. Although many of these details were touched on in previous chapters, we now weave them together into a complete valuation module.

9.1 AGGREGATE EQUITY VALUE

A basic equity valuation module was presented in Chapter 4 as part of the O&R model. It expressed the aggregate value of equity as

$$\text{Equity Value}_0 \equiv E_0 = \sum_{t=1}^{h} \frac{D_t}{(1+k_E)^t} + \frac{E_h}{(1+k_E)^h} \qquad (9.1)$$

where D_t is the aggregate dividend paid at time t, k_E the discount rate, h the number of periods in the forecast horizon, and E_h the terminal value of the equity at the end of the forecast horizon.

Equation 9.1 is an overly simple model of valuation. Each of its components (dividends, discount rate, and terminal value) has its own nuance or complication. Moreover, this model does not handle the problem of dilution created when new equity is issued. This chapter considers each of these components and explains how to handle the complications. We begin with a discussion of dividends, followed by discussions of terminal value and the cost of equity. When these sections are complete, we will have a proper model of aggregate equity value and the cost of equity. With these in place, we will be ready to take up the problem of dilution.

9.2 DIVIDENDS VERSUS EQUITY CASH FLOW

If equity value is the present value of future dividends, it seems that computing value should be relatively easy. Equation 9.1 shows equity value as the present value of dividends and a terminal price. All we should have to do is forecast future dividends. Unfortunately, calculating value is more complicated. The complications arise from the manner in which dividends are defined. Besides dividends, we have also discussed other definitions of cash flow, such as equity cash flow and cash flow available for dividends. This section considers the differences between these definitions and how to handle them in valuation.

Like most questions in finance, the answer to the question, "What is the difference between dividends and equity cash flow?" is, "It depends." The most direct interpretation of D_t in Equation 9.1 is that it represents the dividends that are expected to be paid in period t. In the simplest case, the dividends paid would be identical to the equity cash flow available for dividends. However, the simplest case ignores some of the complications, so we consider the more general case.

We begin by reprising the various definitions of cash flow, beginning with Equity Cash Flow. This cash flow can potentially be paid to the owners of the firm after meeting the needs to invest in assets and fulfilling all other financial obligations. Exhibit 9.1 shows a format for calculating equity cash flow.

The intent of the calculations in Exhibit 9.1 is to account for all cash flows into and out of the firm. In the plain vanilla case, the equity cash flow is identical to the dividends that are paid to the stockholders. It is the amount of money left over at the end of the year that the owner of the firm can take home. To understand this "plain vanilla case" and how Equity Cash Flow can become another flavor, we consider some of the components of this cash flow.

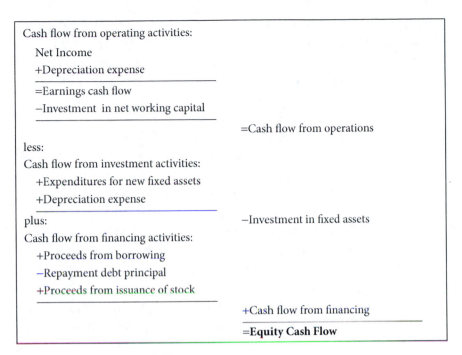

EXHIBIT 9.1   Equity cash flow.

To keep their firm in business and growing, owners need to plow money back into the firm to finance the working capital and fixed assets necessary to support production and sales. This is what the investment in NWC and fixed assets in Exhibit 9.1 represents—the plow back of funds to keep the firm on its trajectory. However, when investments (components of cash flow) are more or less than are necessary, complications arise and a firm's equity cash flow departs from the plain vanilla case. For example, the investment in NWC includes the increase in cash and marketable securities. As a firm grows, these accounts should increase to support the larger operation. However, suppose more funds are invested in marketable securities than is necessary for normal business operations. The amount of funds by which the investment exceeds what is necessary is the amount that could have been paid to the stockholders as dividends but was not.

When the excess investment in securities is included in the investment in NWC, the resulting equity cash flow still represents the funds that are available to be paid as dividends. So equating dividends and equity cash flow is correct. However, the dividends will be less than those that could have been paid had the excess investment not occurred. Actual dividends and dividend-paying ability will differ.

This difference creates a problem for modeling equity value: the model's calculations of investment and cash available for dividends (based on its "necessary investment" assumptions) differ from the firm's actual level of investment and dividends paid. Since investment assumptions differ, the cash available to pay dividends and actual dividends paid differ. As a consequence, the valuations based on these cash flows differ.

To see this, suppose we model current and fixed assets based on our estimate of what is "necessary" for normal operations. Following the format in Exhibit 9.1, our calculation of equity cash flow will represent the dividends that theoretically could be paid. However, suppose the firm is investing more than necessary in marketable securities, the dividends it actually pays will be less than those that the model forecasts (the equity cash flow). Since the model's forecast of equity cash flow and the firm's dividends differ, valuations based on the assumption that equity cash flow equals dividends will be incorrect.

There are several approaches to this type of situation. One is to separate the value of the firm's non-operating assets from the value of its normal business operations, and then sum them. For example, if a firm is making excess investments in cash and securities, estimate the value of the firm based on the dividends expected to be paid, given their low payout, and then add the value of the excess cash and securities to the value of the operating firm. The value obtained for the combined entity (operating firm value + value of non-operating assets) is the value of the firm, given that it continues making excess security investments and paying dividends that are less than its dividend-paying capacity.

Our discussion of excess investment in securities also applies to other types of investments. It applies whether we are dealing with short- or long-term investments. If the firm pays out smaller dividends than it has capacity to pay, the funds are being used elsewhere to purchase unnecessary assets, to repay debt, or maybe even to pay higher expenses than necessary (executive salaries, for example).

In the valuation literature,* the distinction between the two approaches (actual dividends versus dividend-paying capacity) is regarded as the difference between the value to an investor who purchases a controlling interest in the firm and who purchases a minority position with no control, respectively. Controlling value should be based on the dividend-paying

---

* For a discussion of control versus minority value, see Chapter 15, "Control and Acquisition Premiums," Pratt, Reilly, and Schweihs (2000).

	A	B	C	D	E	F	G	H	I
115									
116	CASH FLOW								
117					2005	2006	2007	2008	2009
118									
119	Earnings Cash Flow				210.0	222.5	235.9	250.1	265.1
120	Other Current Asset Investment				32.4	34.3	36.4	38.6	40.9
121	Spontaneous Financing				36.0	38.2	40.4	42.9	45.4
122	NWC Investment				-3.6	-3.8	=((Cash/Sales)* Sales $_t$) - Cash $_{t-1}$		
123	Cash Flow From Operations				213.6	226.3			
124	Capital Expenditures		=F126-F127 = Net Cash Inflow - Increase in Cash		119.0	122.7	130.1	137.9	146.2
125	Capital Financing					3.9	4.2	4.4	4.6
126	Net Cash Inflow					107.5	113.9	120.8	128.1
127	Increase in Cash				10.2	10.8	11.5	12.1	12.9
128	Cash Available for Dividends				91.3	96.7	102.5	108.7	115.2
129									

EXHIBIT 9.2   Cash flow sector of O&R model.

ability of the firm because the control investor can determine the dividend policy and does not have to settle for the present dividend policy. The minority value should be based on the forecast of actual dividends, because the minority investor cannot dictate the dividend or investment policies of the firm and must settle for whatever dividends the directors choose to pay. It may differ from the value based on dividend-paying capacity.

The relevance of this discussion to modeling is that we can change the way we model decisions, such as dividends and investment based on the policy that we think will prevail. For example, Exhibit 9.2 shows the cash flow sector of our O&R model from Chapter 4. Line 126 is the Net Cash Inflow after taking account of all the needs of the business except for increasing the cash balance and paying dividends. It is the flow available to be used either to add to cash and securities or to be paid out as dividends. In line 127, the investment (increase) in the cash balance is set as a percent of sales based on the liquidity needs of the firm. Cash Available for Dividends is the firm's residual cash flow—the cash flow left over after meeting all the business needs of the firm, including increasing the cash balance. This model sets the dividend equal to the residual cash flow, thereby basing it on the firm's ability to pay dividends. The cash balance shown in line 127 is the "necessary" investment used in the model previously discussed. If the actual investment in cash and securities were greater than this amount, the values in line 127 would be larger and the cash available for dividends smaller.

Dividends need not be specified as the residual cash flow. We can easily change the way dividends are determined in a model by setting the dividend according to a specific policy and making cash the residual. For example, we could set the dividend based on a payout ratio, or perhaps even a dollar amount, and add the residual cash flow to the cash balance:

$$\text{Dividends} = \text{Maximum [Minimum(Payout Ratio} \times \text{Net Income,}$$
$$\text{Net Cash Inflow), 0]} \qquad (9.2)$$

$$\text{Increase in Cash} = \text{Net Cash Inflow} - \text{Dividends}$$

The statement

$$\text{Minimum(Payout Ratio} \times \text{Net Income, Net Cash Inflow)}$$

prevents the dividend from exceeding the Net Cash Inflow, and the statement

$$\text{Maximum[Minimum(...), 0]}$$

prevents the dividend from being negative. When using this approach, the payout ratio can be set very low, thereby making more funds available to invest in cash and securities. In this case, the level of cash and securities may not be determined by the needs of the business. Instead, cash absorbs the residual cash flow, and the cash remaining after the target dividend is paid. Even though the dividend is not based on the dividend-paying ability of the firm, it nonetheless equals equity cash flow as long as it is defined as what is left over after the excessive build up of the cash balance.

9.2.1 Value for a Non-Dividend-Paying Firm

Our discussion of dividends raises the question of how to handle the situation where the firm pays no dividends. If we use a dividend valuation model, should not that lead to the result that firms that pay no dividends would have zero value? The answer is usually no.

There are at least four situations when a firm pays no dividends. First, there is the firm that has no earnings and is not expected to have earning or cash flow to payout. The value of this firm's dividends is zero, but the firm may have value in its assets. The value of the firm would be based on the market value of the assets that perhaps is based on the liquidation value of the assets.

Second, there are the small businesses where the owner is an employee who takes the firm's surplus in the form of salary, so the firm earns little or no profits, and pays no dividends. The value would be based on the firm's ability to pay dividends assuming that the owner/manager was paid the going market wage for managerial services and the excess wages are treated as dividends.

Third, and closely related to the second type, are firms, particularly small, closely held firms, that pay no dividends either because the stockholders do not need the funds, or because the owners want to avoid paying taxes on the dividends. This firm retains the earnings and invests them in other assets, including perhaps marketable securities or other assets that are not necessary for the normal operations of the firm. The assets that are not necessary for the ongoing business are called non-operating assets. The value of the firm would be in two parts: the value of the operating business and the value of the non-operating assets. The value of the operating business would be based on its ability to pay dividends assuming normal investment in assets necessary for its normal operations, and the non-operating assets would be valued separately.

Fourth, there is the growth firm that pays no dividends because it needs to plow its earnings back into the firm to finance current operations and growth. This investment enhances the firm's earnings capacity and dividend- paying ability in future years. In terms of our dividend valuation model with a terminal value at the horizon, the value of the zero dividends in the near future would be zero, and all the value would be based on the terminal value. The valuation horizon should be extended out to the point where the firm is expected to have the ability to pay dividends.

9.3 TERMINAL VALUE

The next part of the equity valuation model that we need to consider is the terminal value. Equation 9.1 shows equity value as the present value of the dividends over an h-period horizon plus the value of the equity at the end of the h periods. The value of equity at the end of h periods is called the terminal value. This presents a dilemma. We want to know the equity value today, but this value depends on the value h periods in the future. If it is difficult to estimate the value today, it is much more difficult to determine the value in the future. Yet, today's value depends on our forecast of the future value. There are several ways to handle this.

9.3.1 The Constant Growth Model

The most common way to estimate terminal value is to use the constant growth model of the form

$$E_h = \frac{D_{h+1}}{(k_E - g)}, \tag{9.3}$$

where E_h is the aggregate value of equity at time h, D_{h+1} the aggregate dividend expect at time $h + 1$, g the rate of growth in dividends expected to prevail from time $h + 1$ in perpetuity, and k_E the firm's cost of equity.

The constant growth model is a remarkably useful way to estimate value. However, like almost all financial models, it derives from several strong assumptions. Since these assumptions create potential pitfalls when applying the model, they should be understood before the model is used. As implied by its name, the model assumes that dividends will grow at the same rate in every period, forever. As in the song made famous by Johnny Mathis, "that's a long, long time." Because of this heroic assumption, the model only should be applied to stable growth situations. So long as the growth rate is moderately stable and the rate used (g) is an unbiased estimate* of future growth, the model will yield a reasonable estimate of value.

The model is inappropriate when near term growth rates depart systematically from the average perpetual growth rate. Since the near future has a bigger impact on value than the more distant future, if near-term growth rates are applied in perpetuity they may yield excessively high valuations. The solution to this problem is to forecast dividends (or equity cash flow) period-by-period until the time in the future when the firm's growth is expected to reach stability. At that point in time, we can use the constant growth model to estimate terminal value. A secondary benefit of this approach is that the further into the future we push the terminal value, the smaller will be its impact on the current valuation. When the terminal value calculation is pushed far into the future, errors introduced by the constant growth model will not significantly affect the estimate of current value. Another limitation of the constant growth model that stems from its underlying assumptions is that the growth rate must be less than the discount that is, that is, $g < k_E$, otherwise the constant growth model is invalid.

9.3.1.1 Components of the Terminal Growth Rate

Two important factors influence the long-run growth of the firm: (1) the rate of growth in the economy and the firm's industry; and (2) the limit imposed by the firm's sustainable growth.

9.3.1.1.1 *Growth of the Industry and Economy.* Most industries follow a life cycle with rapid growth early in their life, followed by a slowing of

---

* Unbiased means that actual growth rates are symmetrically distributed around their mean, g.

growth as the industry matures. A firm will tend to grow with its industry, with variations among companies within the industry as they strive for competitive advantage. Consequently, it is normal for a firm's earnings and cash flows to grow rapidly in the early stages of its industrial life cycle, but as the firm and its industry mature and competition increases, the rate of growth will taper off. Eventually, as the industry stabilizes and matures, its growth will be limited by the rate of growth of the economy as a whole. In the long run, neither the industry nor its companies can grow faster than the economy. The future date when we expect the industry and firm to mature and reach stable growth consistent with the economy is the point where we should consider using the constant growth model.

9.3.1.1.2 Sustainable Growth. The company's growth is limited not just by the economy and the industry. It is also limited by internal constraints imposed by its own structure and operating methods. In Chapter 3 we introduced the concept of sustainable growth,[†] which is an excellent indicator of the growth that the firm can handle within the limits of its capital structure and other operating characteristics. When the firm matures, most of its operating ratios will be fairly stable. Its profit margin, asset utilization, and dividend payout would not be expected to fluctuate widely. It also will tend to maintain its debt ratios within a narrow range. This suggests that the sustainable growth rate, g^{**}, serves as a good estimate of the long-run rate of growth that the firm can handle. This does not mean that the firm cannot grow more rapidly in some periods, or that its growth cannot vary from g^{**} from time to time. Rather, it means that the firm's average growth rate over

[†] The sustainable growth formula in Chapter 3 was

$$g^{**} = \frac{\left(\frac{\text{Net Income}}{\text{Sales}}\right)(1 - \text{Payout Ratio})\left(1 + \frac{\text{Debt}}{\text{Equity}}\right)}{\left(\frac{\text{Total Assets}}{\text{Sales}}\right) - \left(\frac{\text{Net Income}}{\text{Sales}}\right)(1 - \text{Payout Ratio})\left(1 + \frac{\text{Debt}}{\text{Equity}}\right)}$$

Note that sustainable growth formula developed by Higgins (1981) is not exact if the various operating ratios are not exact and stable. In the normal situation when the ratios vary from period to period and there are non-linearities in the relationships, the Higgins sustainable growth formula serves only as an approximation of the growth that can be sustained. An alternative to Higgins' sustainable growth is the PRAT, discussed in Chapter 3, which expresses growth in equity flows as

$$(1 - \text{Payout Ratio}_t)\left[\left(\frac{\text{Net Income}_t}{\text{Sales}_t}\right)\left(\frac{\text{Sales}_t}{\text{Total Assets}_t}\right)\left(\frac{\text{Total Assets}_t}{\text{Equity}_{t-1}}\right)\right].$$

long periods cannot exceed g^{**} without it having to change its operating methods and structure. This makes g^{**} something of a speed limit. Just like a car's speed is constrained by its mechanical makeup, the firm is limited by its structural and financial makeup. If the structure is expected to stabilize when the firm reaches maturity, then its growth would also be expected to stabilize at a sustainable level. This makes g^{**} a sensible candidate for use in the constant growth model.

Both the growth of the economy and sustainable growth based on the firm's structure limit the long-run growth of the firm. The one that prevails will be the limit that the firm reaches first, that is, the long-run growth could be considered to be the lower of either the economic growth limit, or the sustainable growth limit imposed by the firm's structure. For example, the structural sustainable growth might be as high as 15%, but the economy is expected to grow at, say, 6% in the very long run. Even though the firm might have the structural capacity to grow at 15%, it would not make sense to forecast that it will grow *forever* at 15% when the economy will grow at just 6%.

9.3.2 Other Models of Terminal Value

There are many cases for which the constant growth model is inappropriate, for estimating the terminal value. For these cases, multipliers provide a simple alternative for calculating terminal value. Typical multipliers include the ratios of price-to-earnings, price-to-revenue, market-to-book, dividends-to-price (dividend multiple), and price-to-cash flow (cash flow multiple). This section considers the price-to-earnings or PE ratio. Other multipliers can be estimated and used in the same or a similar manner.

The PE ratio is probably the most widely used metric of value. It provides an easy way to estimate terminal value that is consistent with market practices. Aggregate terminal value of equity is estimated as

$$E_h = PE_h \times (\text{Net Income}_h),$$

and terminal per share value is estimated as

$$P_h = PE_h \times (\text{Earnings per Share}_h).$$

Although the model provides the forecast of income or earnings, we need to provide the PE ratio. Obviously, we cannot know the PE ratio that will prevail several years into the future. Therefore, a forecast for the PE ratio either must be provided as an input to the model or calculated by a module within the model based on other form of variable in the model.

Two logical forecasting methods are available. One is to use the historical average PE ratio for the industry or company as an indicator of what we expect at the end of the horizon. This method is justifiable because markets exhibit a tendency to "regress to the mean." This means that if PE ratios are currently higher than the average, there will be a tendency for them to work their way down towards the historical mean in future periods; the converse is true if the PE ratio is currently below its historical mean.

The second method for forecasting the future PE ratio is to model it based on its historical relationship to other fundamental variables. For example, we would expect that the PE ratio will be higher for firms that exhibit more stable earnings, less debt, higher growth rates, longer potential duration of growth, or a combination of the above.

Various authors have examined the empirical relation between PE ratios and fundamental company variables. An early example of such a relationship is the simple model by Whitbeck and Kisor (1963). They estimated the empirical relationship between PE ratios and company fundamentals with a linear regression equation, thus

$$PE = a + b_1(\text{Earnings Growth}) + b_2(\text{Dividend Payout})$$
$$+ b_3(\text{Standard Deviation of Earnings Growth})$$

$$= 8.2 + 1.5(\text{Earnings Growth}) + 6.7(\text{Dividend Payout})$$
$$- 0.2\,(\text{Standard Deviation of Earnings Growth}).$$

Cragg and Malkiel (1968) updated the Whitbeck and Kisor study using the stock's beta coefficient as the measure of risk in place of the standard deviation of earnings growth. Their fitted equation using 1965 data was

$$PE = 0.96 + 2.74(\text{Earnings Growth}) + 5.01(\text{Dividend Payout}) - 0.35\beta.$$

Damodaran (2002) updates this model using data from 2000. His estimated equation is

$$PE = -17.2 + 1.55(\text{Earnings Growth}) + 10.9(\text{Dividend Payout}) + 16.44\beta.$$

Using these models to estimate terminal value presents several types of problems. The models fit the data better for some periods than for others and the relationships vary over time. These problems can be addressed in several ways. First, we can re-estimate this type of regression model using recent data from companies in the same industry as the subject firm. Second, recognizing that the relationships change with market conditions and that we do not know what conditions will prevail at the end of the planning horizon, we can estimate the relationship over several periods so as to get a smoothed, normalized relationship, and then use the estimated coefficients as input to our model. By doing this, the terminal PE ratio and resulting equity value are adjusted for the characteristics of our firm at the terminal date.

Other multipliers such as price-to-book, price-to-revenue, or price-to-dividend, can also be used to generate estimates for terminal value. As with the PE ratio, we do not know the ratio that will prevail at the terminal date, so we will have to forecast it. We can use methods similar to those suggested for estimating the PE ratio to model our forecast of these future multipliers.

9.4 THE DISCOUNT RATE: LEVERAGE, CIRCULARITY, AND CONVERGENCE

When we presented a simple valuation model in Chapter 4, the cost of equity, k_E, was based on the CAPM, but the beta coefficient was assumed to be fixed, and did not vary as the capital structure changed. However it is well-known that the systematic risk of the equity should vary with the financial leverage of the firm. In this section, we explain how to deal with a changing capital structure in the valuation model.

As the firm's financial leverage increases, the beta for the stock should also increase. This adjustment is made with Hamada's (1969) formula that expresses the leveraged beta as a function of the unleveraged beta and the debt-to-equity ratio:

$$\beta_L = \beta_U \left[1 + \frac{D}{E}(1 - T) \right] \tag{9.4}$$

where β_U is the beta for the firm with no debt, β_L the leveraged beta, $\frac{D}{E}$ denotes the debt-to-equity ratio, and T the firm's tax rate.

The proper use of this formula requires that the debt-to-equity ratio be based on the market values of the firm's debt and equity. This creates a circularity problem:

- We wish to estimate market value
- To estimate market value we need to know β_L

		Empruntay Corporation		
		Balance Sheet		
		Year 0		
Assets		Liabilities		
		Current Liabilities		595.4
		Long Term Debt		420.0
		Stockholders' Equity		1,118.6
Total Assets	2,134.0	Total Liabilities & Equity		2,134.0

EXHIBIT 9.3   Empruntay Corporation: Balance sheet.

- To calculate β_L we need to know the debt-to-equity ratio based on market value

- But market value is what we are trying to estimate.

This problem can be solved using an iterative method that converges on the correct value of equity.* The iterative process follows the following steps:

Iteration 1: Use an initial assumed value of equity (usually book value) to calculate the components in the Hamada model: D/E, β_L, k_E, and the market value of equity, V_E.

Iteration 2: Use V_E from iteration 1 to recalculate, D/E, β_L, k_E, and the market value of equity, V_E.

Iteration 3: Use V_E from iteration 2 to recalculate, D/E, β_L, k_E, and the market value of equity, V_E.

These iterations continue until the value of equity in the Debt/Equity ratio is the same as the calculated value of equity V_E. This usually takes from three to five iterations to converge.

To see how this works, consider the following example. The Empruntay Corporation's current balance sheet is shown in Exhibit 9.3.

Iteration 1: Exhibit 9.4 shows the equity valuation module from Empruntay's planning model.† The assumed equity value is the book value of equity, $1118.60 (cell B127). The book value debt-to-equity ratio is

---

* The iterative method that we explain here can be circumvented by using the iterative calculation feature of Excel. We will explain that later.

† Empruntay's planning model is identical to the O&R model introduced in Chapter 4. Empruntay's financial structure is very similar to O&R except that it has more initial debt for this example. The unleveraged beta is 1.0 and the five-year growth in sales is 10%.

	A	B	C	D	E	F	G	H
119				Empruntay Corporation				
120				EQUITY VALUATION MODULE				
121								
122	**Cost of Equity**			Initial value:				
123	Risk Free Interest Rate	5.0%		Book value of equity				
124	Market Risk Premium	7.0%						
125	Beta (unleveraged)	1						
126	Assumed Equity Value	1,118.60						
127	Debt/Equity	37.55%		Debt/Equity in B127 is based on the book value of				
128	Beta (leveraged)	1.225		equity at t = 0, shown in B126.				
129	Discount Rate	13.58%						
130						Period		
131	**Dividend Valuation Model**		0	1	2	3	4	5
132	Dividends			$ 84.38	$ 91.28	$ 98.82	$ 107.08	$ 116.13
133	Present Value of Dividends		$ 338.29	$ 299.84	$ 249.27	$ 184.30	$ 102.24	$ -
134	Terminal Value							
135	Long Term Growth Rate							7.38%
136	Dividend t = 6							$ 124.70
137	Terminal Value of Stock at t = 5							$ 2,012.21
138	PV of Terminal Value		$ 1,064.69	$ 1,209.24	$ 1,373.42	$ 1,559.88	$ 1,771.67	$ 2,012.21
139								
140	**Aggregate Value of Equity**		$ 1,402.98	$ 1,509.08	$ 1,622.69	$ 1,744.18	$ 1,873.91	$ 2,012.21

EXHIBIT 9.4 Equity valuation module.

$$\frac{\text{Debt}}{\text{Equity}} = \frac{420}{1118.6} = 37.55\% \quad \text{(cell B128)}.$$

The leveraged beta is calculated in cell B129 based on the debt-to-equity in cell B128. If Empruntay had no debt, its unleveraged beta would be 1.0. However, with the debt shown in the balanc e sheet, the leveraged beta is estimated using the Hamada equation as

$$\beta_L = \beta_U \left[1 + \frac{D}{E}(1-T) \right]$$
$$= 1.0 \times [1 + 0.3755 \times (1 - 0.40)]$$
$$= 1.225, \qquad \text{(cell B129)}$$

where D is the book value of debt. Given this beta, the cost of equity is

$$k_E = 0.05 + (0.07) \times 1.225 = 13.58\%. \qquad \text{(cell B130)}$$

Using these values, the estimated value of equity is $1402.98 (C140). This computed value of equity differs from the assumed (book) value that was used to calculate the debt-to-equity ratio, beta, and cost of equity. Because of this inconsistency, we need to go to the next iteration.

Iteration 2: For the second iteration, we use the computed value of equity from iteration 1 to calculate the debt-to-equity, leveraged beta, and cost of equity:

$$\frac{D}{E} = \frac{420}{1402.6} = 29.94\%$$

Row		Iteration						
		1	2	3	4	5	6	7
1	Assumed Equity Value	$1,118.60	$1,402.98	$1,497.60	$1,496.24	$1,499.67	$1,500.37	$1,500.55
2	Debt/Equity	37.55%	29.94%	28.39%	28.07%	28.01%	27.99%	27.99%
3	β_L	1.225	1.18	1.17	1.168	1.168	1.168	1.168
4	k_E	13.68%	13.29%	13.19%	13.18%	13.18%	13.18%	13.18%
5	Derived Equity Value	$1,402.98	$1,479.60	$1,496.24	$1,499.67	$1,500.37	$1,500.55	$1,500.55

EXHIBIT 9.5   Iterations to converge on equity value.

$$\beta_L = 1.0 \times [1 + 0.2994 \times (1 - 0.40)]$$
$$= 1.18$$
$$k_E = 13.358\%.$$

Using these values, the estimated value of equity is $1479.60. Now the assumed equity value used in the debt-to-equity ratio and the derived value of equity differ by just $76.62. However, they are still inconsistent. The iterations continue until the equity values converge on a solution in which the assumed value of equity used to calculate the cost of equity agrees with the derived value. Exhibit 9.5 shows the results for seven iterations. Equity values converged by the seventh iteration to an equity value of $1500.55.

Exhibit 9.6 shows the valuation model with the converged values. The sequence of iterations was performed by manually and repeatedly inserting the derived equity value (C142) from the previous iteration into cell B127.

	A	B	C	D	E	F	G	H
121			Empruntay Corporation					
122			EQUITY VALUATION MODULE					
123	Cost of Equity							
124	Risk Free Interest Rate	5.0%						
125	Market Risk Premium	7.0%						
126	Beta (unleveraged)	1						
127	Assumed Equity Value	1,500.55	Debt/Equity in B128 is based on the assumed value of					
128	Debt/Equity	27.99%	equity of $1,500.6 (B127) that was computed in the					
129	Beta (leveraged)	1.168	previous iteration.					
130	Discount Rate	13.18%						
131								
132			Period					
133	Dividend Valuation Model		0	1	2	3	4	5
134	Dividends			$84.38	$91.28	$98.82	$107.08	$116.13
135	Present Value of Dividends		$341.79	$302.44	$251.02	$185.27	$102.61	$0.00
136	Terminal Value		Converged Solution					
137	Long Term Growth Rate		Computed value of equity (C142) agrees with the assumed					7.38%
138	Dividend t = 6		value of equity (B127).					$124.70
139	Terminal Value of Stock at t = 5							$2,151.57
140	PV of Terminal Value			$1,158.76	$1,311.43	$1,484.22	$1,679.77	$1,901.09	$2,151.57
141								
142	Aggregate Value of Equity		$1,500.55	$1,613.87	$1,735.24	$1,865.05	$2,003.70	$2,151.57
143								

EXHIBIT 9.6   Equity valuation module with converged equity value.

As a practical matter, we do not need to converge down to the last cent. Nonetheless, we do want the assumed value and the derived value to be quite close. Rarely are more than four or five iterations required to bridge the gap between the initial and converged values. However, when there is proportionally more debt or a bigger difference between book value and the initial derived value, more iterations may be needed. Regardless of the number of iterations, this kind of iterative process is necessary to produce an estimate of the value of equity that is consistent with the equity value used to calculate the cost of equity. However, we do not necessarily have to do the iterations manually as shown in this example. If our model is not too complex, involving numerous circularities, we can use the iterative calculation capability of Excel®.

Instead of manual iterations for the Empruntay model, we can set the Assumed Equity Value (cell B127) equal to the computed equity value (cell C142). This will bring about a warning of circular reference. We solve this problem with Excel's iterative recalculation by selecting Tools > Options > Calculation, and then checking the "Automatic" and "Iteration" boxes. With "Iterations" checked, Excel will follow automatically the process that we have just carried out manually. Clearly, this is a lot easier. However, this procedure can fail if our model is complex with many different circular variables. Generally, it is better to try to minimize the number of circular references so that our model does not fail or freeze-up.

9.4.1 Cost of Equity with a Changing Capital Structure

The calculations just discussed assume a stable capital structure over the estimation period. If the debt–equity mix changes over time, the cost of equity also should change. For example, if the proportional amount of debt increases over the investment horizon, so too will financial risk, and the return investors will be required to bear the increased risk. The converse is also true if debt decreases. An equity valuation model should account for the changes in the capital structure and the stockholders' financial risk. We use the Empruntay Corporation as an example to show how to handle this problem.

Empruntay's sales are expected to grow at 10% per year for the next five years, after which growth is expected to stabilize at 7.38% for the indefinite future. New asset investment will be required to support this growth. Empruntay will not generate enough cash to fund these investments, so external financing will be required. Management has decided to use the debt to meet their financing needs. As a result, the firm's

Period	0	1	2	3	4	5
Required external financing	—	$88.5	$99.9	$110.9	$123.1	$136.5
Total interest bearing debt	$420	$508.5	$608.4	$719.13	$842.4	$978.9
Debt/equity (book)		37.5%	43.3%	49.2%	55.3%	61.3%	67.5%

EXHIBIT 9.7   Empruntay Corporation: Schedule of external financing, debt, and debt-to-equity.

debt-to-equity ratio will increase over the next five years as shown in Exhibit 9.7.

The valuation model of Empruntay in the previous section calculated the cost of equity for the whole planning period based on the debt-to-equity at the beginning of the horizon. It did not take into account the changing capital structure. However, the cost of equity depends on the debt-to-equity ratio, so ignoring the changing capital structure introduces error into the valuation process. The discount rate should account for the fact that the stockholders' risk will change over the planning period due to changing financial leverage.

In the previous section we used the CAPM and the Hamada adjustment for leverage to calculate the cost of equity. The CAPM, though, is a single period model and the Hamada formula assumes that the capital structure will not change during the single period of valuation. To apply the CAPM correctly, the leveraged beta and cost of equity should be calculated for each period based on the capital structure prevailing during that period. Exhibit 9.8 shows the beta and the cost of equity for Empruntay for each period based on the respective book value debt-to-equity ratios. Note that as the debt-to-equity ratio increases, so too do Empruntay's beta and cost of equity.

Ignoring for the moment the problem of the inconsistency of book value versus market value discussed in the previous section, it is easy to take

Period	0	1	2	3	4	5
Debt/equity (book)	37.5%	43.3%	49.2%	55.3%	61.3%	67.5%
Beta	1.225	1.26	1.295	1.332	1.368	1.405
Cost of equity	13.58%	13.82%	14.07%	14.32%	14.58%	14.84%

EXHIBIT 9.8   Empruntay Corporation: Schedule of beta and cost of equity.

into account the changing capital structure and cost of equity in our valuation model. Simply base each period's cost of equity on the capital structure prevailing at the start of that period.

When the discount rate, k_j, is different in each period j, the discount factor for period t becomes

$$\text{Discount Factor}_t = \frac{1}{[(1 + k_0)(1 + k_1)(1 + k_2) \ldots (1 + k_{t-1})]}.$$

For example, using the equity costs in Exhibit 9.8, the discount factor for period 2 is

$$\text{Discount Factor}_2 = \frac{1}{[(1 + k_0)(1 + k_1)]}$$

$$= \frac{1}{[(1.1358)(1.1382)]}$$

$$= 0.7735.$$

We account for the changing capital structure and use this method of discounting in the calculations of equity value for Empruntay shown in Exhibit 9.9. The book value of equity at each date is shown in line 127, the book value debt-to-equity ratio in line 128, and the leveraged beta in line 129. The cost of equity in line 130 is different each period, reflecting the changing debt-to-equity ratio.

	A	B	C	D	E	F	G	H
121			Empruntay Corporation					
122			EQUITY VALUATION MODULE					
123	Cost of Equity		Assumed equity value is equal to book value, with the					
124	Risk Free Interest Rate	5.0%	amount of book equity and the Debt / Equity changing each					
125	Market Risk Premium	7.0%	period.					
126	Beta (unleveraged)	1						
127	Assumed Equity Value		$ 1,118.65	$ 1,174.90	$ 1,235.75	$ 1,301.63	$ 1,373.02	$ 1,450.43
128	Debt/Equity		37.55%	43.28%	49.23%	55.26%	61.35%	67.48%
129	Beta (leveraged)		1.225	1.260	1.295	1.332	1.368	1.405
130	Discount Rate		13.58%	13.82%	14.07%	14.32%	14.58%	14.83%
131								
132			Period					
133	Dividend Valuation Model		0	1	2	3	4	5
134	Dividends			84.4	91.3	98.8	107.1	116.1
135	Present Value of Dividends		$335.56	$296.74	$246.46	$182.32	$101.35	$0.00
136	Terminal Value							
137	Long Term Growth Rate		=(D134+D135)/(1+C130)					7.38%
138	Dividend t = 6				=H140/(1+G130)			124.70
139	Terminal Value of Stock at t = 5							1672.8
140	PV of Terminal Value		$866.08	$983.67	$1,119.59	$1,277.09	$1,459.98	$1,672.79
141								
142	Aggregate Value of Equity		$1,201.64	$1,280.40	$1,366.05	$1,459.41	$1,561.33	$1,672.79

EXHIBIT 9.9 Empruntay equity valuation. Changing capital structure: Beta based on book equity.

Since the debt-to-equity ratio increases over time, the average cost of equity is higher than the cost of equity shown in Exhibit 9.4 that reflected only the initial book value debt-to-equity ratio. Given the higher cost of equity, the estimated value of equity is $1201.64 versus the converged value of $1500.55 in Exhibit 9.6. Of course, these two values are not directly comparable. The $1500.55 was calculated using market values, whereas $1201.64 was calculated using different book values of equity for each period of the planning horizon. Neither estimate is correct. One more step is needed to complete our valuation model. We need to combine the two methods by accounting for the changing capital structure and using market values in the cost of equity calculations. We do this next.

The valuation model in Exhibit 9.9 needs to be modified so that the debt-to-equity ratio, beta, and the cost of equity in each period are based on market values. As we did previously, we use an iterative method. But because the debt–equity mix changes over time, we have to perform a separate iteration for each period in the planning horizon. We start at the end of the planning horizon and calculate a converged value of equity where the equity value in the debt-to-equity ratio agrees with the derived value of equity. Then we step back one period in time and compute a converged value of equity. This process continues until we reach the beginning of the planning period.

In the Empruntay example, the starting point for our period-by-period iteration is the fifth period valuation. Consider the results of Exhibit 9.9 as Iteration 1. At time 5, the book value of equity is $1450.43 (H127), and the derived value of equity is $1672.79 (cell H142). The second iteration is shown in Exhibit 9.10. We plug the estimated equity value of $1672.80 (H142 in Exhibit 9.9) into the Assumed Equity Value for period 5 (cell H127 in Exhibit 9.10). This generates a time 5 equity-derived value of $E_5 = \$1761.84$(H142 in Exhibit 9.10). Since the assumed value and derived values still do not agree, we need to perform additional iterations. After several iterations, we get a converged value of equity at time 5 of $E_5 = \$1809.21$.

We then step back to $t = 4$ for the next stage of the iterative process shown in Exhibit 9.10. The initial derived value of equity at time $t = 4$ is calculated using the converged $t = 5$ value in cell H142 with the formula

$$E_4 = D_5 \left[\frac{1}{(1+k_4)} \right] + E_5 \left[\frac{1}{(1+k_4)} \right]$$

$$= 116.13 \left[\frac{1}{(1+k_4)} \right] + 1809.21 \left[\frac{1}{(1+k_4)} \right].$$

	A	B	C	D	E	F	G	H
121			Empruntay Corporation					
122								
123	Cost of Equity		Iteration #2:			Iteration #2:		
124	Risk Free Interest Rate		5.0%	Assumed equity values for times 1		Equity value at t = 5 is assumed to be equal to		
125	Market Risk Premium		7.0%	to 4 are set equal to book values		value computed in Interation #1 (cell H142)		
126	Beta (unleveraged)		1			that was based on book value.		
127	Assumed Equity Value		$ 1,118.65	$ 1,174.90	$ 1,235.75	$ 1,301.63	$ 1,373.02	$ 1,672.80
128	Debt/Equity		37.55%	43.28%	49.23%	55.26%	61.35%	58.51%
129	Beta (leveraged)		1.23	1.26	1.30	1.33	1.37	1.35
130	Discount Rate		13.58%	13.82%	14.07%	14.32%	14.58%	14.46%
131								
132			Period					
133	Dividend Valuation Model		0	1	2	3	4	5
134	Dividends			$ 84.38	$ 91.28	$ 98.82	$ 107.08	$ 116.13
135	Present Value of Dividends		$ 335.56	$ 296.74	$ 246.46	$ 182.32	$ 101.35	$ -
136	Terminal Value							
137	Long Term Growth Rate							7.38%
138	Dividend t = 6							$ 124.70
139	Terminal Value of Stock at t = 5							$ 1,761.84
140	PV of Terminal Value		$ 912.19	$ 1,036.03	$ 1,179.19	$ 1,345.07	$ 1,537.70	$ 1,761.84
141								
142	Aggregate Value of Equity		$ 1,247.75	$ 1,332.77	$ 1,425.65	$ 1,527.39	$ 1,639.05	$ 1,761.84

EXHIBIT 9.10 Equity value estimation. Changing capital structure iteration starting at time 5.

Using book value as the assumed value at $t = 4$ for the first iteration yields

$$E_4 = 116.13\left(\frac{1}{1.14547}\right) + 1809.21\left(\frac{1}{1.14547}\right)$$

$$= \$1679.62.$$

This is not a correct value because we used book equity as the assumed equity value. To get a correct value, we must iterate until the assumed and derived equity values at time $t = 4$ converge. We repeatedly substitute the derived estimate of E_4 from cell G142 into the Assumed Equity Value in G127. When the assumed and the derived values agree, we have converged. Our converged time 4 value is $E_4 = \$1687.46$.

Having obtained the converged value at time 4, we step back to time 3 and repeat the iterative process, then step back successively to times 2, 1, and 0. Exhibit 9.11 shows the valuation model with converged value for each period. The objective of the calculation is the value at time $t = 0$. Our converged estimate is $E_0 = \$1288.74$. This value takes into account the different capital structures in each of the six periods ($t = 0, 1, \ldots, 5$), with each period's debt-to-equity ratio based on the estimated market value of equity at each date.

We now have generated four different estimates of equity value: $1402.98 (Exhibit 9.4), $1500.55 (Exhibit 9.6), $1201.64 (Exhibit 9.9), and $1288.74 (Exhibit 9.11). Of these estimates, only the last one, $1288.74 is correct,

	A	B	C	D	E	F	G	H
121			Empruntay Corporation					
122			EQUITY VALUATION MODULE					
123	**Cost of Equity**							
124	Risk Free Interest Rate	5.0%						
125	Market Risk Premium	7.0%						
126	Beta (unleveraged)	1						
127	Assumed Equity Value		$ 1,288.74	$ 1,376.65	$ 1,471.93	$ 1,575.29	$ 1,687.46	$ 1,809.21
128	Debt/Equity		32.59%	36.94%	41.33%	45.66%	49.92%	54.10%
129	Beta (leveraged)		1.196	1.222	1.248	1.274	1.300	1.325
130	Discount Rate		13.37%	13.55%	13.74%	13.92%	14.10%	14.27%
131								
132			**Period**					
133	**Dividend Valuation Model**		0	1	2	3	4	5
134	Dividends			$ 84.38	$ 91.28	$ 98.82	$ 107.08	$ 116.13
135	Present Value of Dividends		$338.05	$298.86	$248.08	$183.34	$10.78	$0.00
136	**Terminal Value**							
137	Long Term Growth Rate							7.38%
138	Dividend t = 6							$ 124.70
139	Terminal Value of Stock at t = 5							$ 1,809.21
140	PV of Terminal Value		$950.70	$1,077.79	$1,223.85	$1,391.96	$1,585.68	$1,809.21
141								
142	**Aggregate Value of Equity**		$1,288.74	$1,376.65	$1,471.93	$1,575.29	$1,687.46	$1,809.21
143								
144			The cells in row 127 are iteratively matched to equity values in row 142.					
145			Alternatively, set Excel to Iterative calculations and point the values in row 127					
146			to those in row 142 (e.g., the code in H127 is = H142).					
147								
148								
149								

EXHIBIT 9.11 Converged equity values. Changing capital structure: Based on market value debt-to-equity ratios in each period.

because it is the only estimate that properly takes into account the capital structure at each date, and it does so based on market values instead of book values. However, it took 24 iterations (averaging four iterations at each date) of the model in Exhibit 9.11 to reach a converged value. The difference between the highest and the most accurate estimate is $1500.55 − $1288.74 = $211.81, an error of about 16% of the correct value, which is a large error. In addition, we do not know ahead of time how big an error we may make when we do not carry out all the steps required for a properly converged value estimate.

Is the more accurate estimate worth the effort of working through these many iterations? It depends on the purpose of the model. If an accurate estimate for the value of equity is the objective of our modeling effort, it is probably worth the extra effort. However, if the objective of the modeling effort is to analyze decisions for which we are only concerned with relative values, then we probably do not need this level of accuracy. For example, suppose we are considering alternative financial strategies and we need to know which strategy yields the highest value. There is a good chance the question can be answered by the simpler, non-iterative approach. However, if we do not do the iterations, we may always wonder whether we might have gotten a different answer had we done the extra work. Fortunately,

Excel can perform the iterations quickly, rendering this question nearly moot. Armed with a proper model of aggregate equity value and the cost of capital, we are now ready to consider how to model per share value when new equity is issued.

9.5 VALUE PER SHARE AND ISSUING NEW EQUITY

The objective of financial decisions should be to make the current owners of the firm as wealthy as possible—that is, to maximize the value of their investment in the firm. We operationalize this objective by maximizing the value per share of the stock owned by the current owners rather than maximizing the aggregate value of the equity. Consequently, we need to discuss how to handle calculating value per share in a model.

Modeling value per share seems to be easy; simply divide the aggregate value of equity by the number of shares. However, this is not always correct. When new equity is issued, the number of shares outstanding during the planning period may change. This creates a circularity problem. When shares are sold to raise new equity capital, the number of shares to be sold will depend on the price per share, and the price per share, in turn, depends on the number of shares issued. Unless the model is carefully structured, this problem results in a seemingly endless cycle of calculations. This section shows how to address this problem.

The problem being addressed is how to account for dilution when raising new equity capital by issuing new shares. The apparent simultaneity of the share price and the number of shares to be issued is demonstrated by the following equations:

$$\text{Shares Issued} = \frac{\text{Equity Issued}}{\text{Price per Share}}$$

$$\text{Price per Share} = \frac{\text{Equity Value}}{(\text{Initial Shares} + \text{Shares Issued})}, \tag{9.5}$$

where Equity Value is the aggregate value of equity, Equity Issued is the aggregate amount of new capital to be raised, Shares Issued the number of shares sold, Initial Shares the number of shares outstanding prior to selling new shares, and Initial Shares + Shares Issued the number of shares outstanding after the new issue.

In the first Equation of 9.5, the number of shares issued depends on the price per share, whereas in the second equation the price per share depends on the number of shares issued. The problem is clearly circular.

One solution to this circularity, or simultaneity, problem derives from Miller and Modigliani (1961). It expresses the price per share as

$$\text{Price per Share}_0 = \frac{\sum_{t=1}^{h} \left[\frac{(\text{Net Income}_t) - (\text{Internal Equity Financing}_t) - (\text{Equity Issued}_t)}{(1+k)^t} \right] + \frac{E_h}{(1+k)^h}}{\text{Shares}_0}.$$

$$(9.6)$$

This equation splits Net Income between dividends and reinvestment in the firm. The share price on a given date depends on the shares outstanding at that date, prior to the issuance of new shares. Alternatively, Price per Share can be expressed as

$$\text{Price per Share}_0 = \frac{(\text{Equity Value}_0) - \sum_{t=1}^{h} [\text{Equity Issued}_t / (1+k)^t]}{\text{Shares}_0}, \quad (9.7)$$

where

$$\text{Equity Value}_0 = \sum_{t=1}^{h} \frac{[(\text{Net Income}_t) - (\text{Internal Equity Financing}_t)]}{(1+k)^t} + \frac{E_h}{(1+k)^h}.$$

$$(9.8)$$

The term "$\sum_{t=1}^{h} [(\text{Equity Issued}_t)/(1+k)^t]$" in Equation 9.7 represents the present value of the new equity to be issued in future periods.

These expressions avoid the circularity problem that frequently occurs in calculating the price per share when we issue new equity. Since the price per share is expressed in both Equations 9.6 and 9.7, in terms of the number of shares outstanding at time $t = 0$, Shares$_0$, we do not need to know the number of shares to be issued at each future date. However, we do need to know the aggregate value of new equity issued at each future date, Equity Issued$_t$. Note that Equity Issued$_t$ appears in Equations 9.6 and 9.7 with a negative sign. This captures the dilution that occurs for the original stockholders (the stockholders at time $t = 0$) when new shares are sold. It does this by reducing the total value of equity in a given period by the amount of new equity financing used in that period. Subtracting the aggregate value of the new shares issued implicitly accounts for the dilution that occurs when more shares are going to be outstanding in future periods.

9.5.1 Example: Mythic Corporation

To see how we use the Miller and Modigliani relationship to model the issuance of new shares, consider the following simple example. It has two scenarios: (A) in which no new equity is issued; and (B) in which new equity is issued.

Scenario A: The Mythic Corporation is financed entirely with equity, has 100 shares of common stock outstanding, and a cost of equity $k_E = 12\%$. At the end of five years it will cease operations and liquidate. The first scenario assumes that Mythic will make no capital investment over the next five years, and therefore will not have to issue any new shares. The equity cash flow available to be paid as dividends is $1000 for each of the next five years, as shown in Exhibit 9.12.

Given that all equity cash flow is paid as dividends, the aggregate value of the stock at the end of each period is simply the present value of future dividends, as shown in line 23. The aggregate value of the stock at $t = 0$ is $3604.78. With 100 shares outstanding, the value per share is $36.05 at $t = 0$ (line 26). The value shown for $t = 5$ is zero because these values are ex-dividend and there are no future cash flows after the last dividend has been paid.

Scenario B: This scenario considers what happens when new shares are sold to raise additional equity capital. Assume that $200 must be spent for

	B	C	D	E	F	G	H
2			Mythic Corporation				
3			Equity Cash Flow				
4			Scenario A: No New Equity Issued				
5				Period			
6		0	1	2	3	4	5
7	Net Income		700	700	700	700	700
8	+ Depreciation		300	300	300	300	300
9	Earnings Cash Flow		1,000	1,000	1,000	1,000	1,000
10	- Net Working Capital Investment		0	0	0	0	0
11	- Capital Investment		0	0	0	0	0
12	+ Capital Financing: Equity Issued		0	0	0	0	0
13	Equity Cash Flow		1,000	1,000	1,000	1,000	1,000
14							
15	**Equity Valuation Module**						
16				Period			
17		0	1	2	3	4	5
18	Cost of Equity						
19	Discount Rate	12.0%					
20							
21	Dividend Valuation Model						
22	Dividends		1,000	1,000	1,000	1,000	1,000
23	Aggregrate Value of Stock	$ 3,604.78	$ 3,037.35	$ 2,401.83	$ 1,690.05	$ 892.86	$ -
24							
25	Number of Shares	100	100	100	100	100	100
26	Value per Share	$ 36.05	$ 30.37	$ 24.02	$ 16.90	$ 8.93	$0.00

EXHIBIT 9.12   Equity cash flow and valuation. Scenario A: No new equity issued.

capital equipment in period 2 to maintain Mythic's productive capacity. Assume that this $200 will be financed by issuing new equity. If the firm did not sell new shares, the investment would have to be financed with retained earnings (internal equity), and dividends would be reduced by $200 to a level of $800.

However, since new shares are sold to finance the purchase of the equipment, the equity flow available for dividends remains at $1000, as shown in Exhibit 9.13. Furthermore, since the aggregate equity cash flow remains the same, the aggregate value of the equity is still $3604.78. However, since new shares were issued, the value *per share* will change.

Consider the position of the original stockholders. Under scenario A the investors who owned shares at $t=0$ expect to receive 100% of the equity cash flow. Under scenario B, new shares will be sold at $t=2$ to finance the new investment. The sale of the new shares will increase the number of shares outstanding, thereby diluting the ownership of the original shareholders. Though the aggregate equity cash flow is unchanged, it is now distributed across more shares. Hence, the dividend per share will decrease. Since the dividend is smaller, the NPV of the dividend per share is smaller, making the value per share smaller that it would have been otherwise.

The effect of dilution is shown in the lower panel labelled Scenario B. Exhibit 9.14A which shows the calculations of the aggregate and per share value of equity based on the issuance of $200 of new equity in period 2. (Exhibit 9.14B shows the Excel formulas for two columns of the model.) At $t=0$, the aggregate value of equity is $3604.78 and the value per share is $34.45. This value per share reflects stockholders' knowledge that new shares will be issued at $t=2$.

At time $t=2$, $200 of external equity is raised by selling 9.083 shares (line 37) at a price of $22.02 (line 40). How do we know how many shares

	B	C	D	E	F	G	H
2			Mythic Corporation				
3			Equity Cash Flow				
4			Scenario B: New Equity Issued				
5				Period			
6		0	1	2	3	4	5
7	Net Income		700	700	700	700	700
8	+ Depreciation		300	300	300	300	300
9	Earnings Cash Flow		1,000	1,000	1,000	1,000	1,000
10	- Net Working Capital Investment		0	0	0	0	0
11	- Capital Investment		0	200	200	200	200
12	+ Capital Financing: Equity Issued		0	200	200	200	200
13	Equity Cash Flow		1,000	1,000	1,000	1,000	1,000

EXHIBIT 9.13 Equity cash flow. Scenario B: New equity issued.

	A	B	C	D	E	F	G
14		Mythic Corporation					
15		Equity Cash Flow					
16		Scenario B: New Equity Issued					
17				Period			
18		0	1	2	3	4	5
19	Cost of Equity						
20	Discount Rate	12.0%					
21	**Dividend Valuation Model**						
22	*Scenario A*						
23	Dividends		1,000	1,000	1,000	1,000	1,000
24	Aggregrate value of Stock	$ 3,604.78	$ 3,037.35	$ 2,401.83	$ 1,690.05	$ 892.86	$ -
25							
26	Number of Shares	100	100	100	100	100	100
27	Value per Share	$ 36.05	$ 30.37	$ 24.02	$ 16.90	$ 8.93	$ -
28	*Scenario B*						
29	Equity Financing						
30	Amount of Equity Issued	0	0	200	0	0	0
31	Aggregate Stock Value	$ 3,604.78	$ 3,037.35	$ 2,401.83	$ 1,690.05	$ 892.86	$ -
32	PV Equity Issued	$159.44	$178.57	$200.00	$0.00	$0.00	$0.00
33	Equity Value - PV (Equity Issued)	$ 3,445.34	$ 2,858.78	$ 2,201.83	$ 1,690.05	$ 892.86	$ -
34							
35	Shares Outstanding						
36	Number of Shares - beginning	100.000	100.000	100.000	109.083	109.083	109.083
37	Shares Issued	0.000	0.000	9.083	0.000	0.000	0.000
38	Shares Ending	100.000	100.000	109.083	109.083	109.083	109.083
39							
40	Issue Price per Share	$ 34.45	$ 28.59	$ 22.02	$ 15.49	$ 8.19	
41	Dividends per Share		$ 10.00	$ 10.00	$ 9.17	$ 9.17	$ 9.17
42	PV of Dividends per Share	$ 34.45	$ 28.59	$ 22.02	$ 15.49	$ 8.19	
43	Value per Share - ex dividend	$ 34.45	$ 28.59	$ 22.02	$ 15.49	$ 8.19	

EXHIBIT 9.14A   Equity value: Aggregate and per share. New shares issued.

	A	B	C
21	**Dividend Valuation Model**		
22	*Scenario A*		
23	Dividends		=C12
24	Aggregrate value of Stock	=NPV($B$20,C23:G23)	=NPV($B$20,D23:H23)
25			
26	Number of Shares	=B36	=B26
27	Value per Share	=B24/B26	=C24/C26
28	*Scenario B*		
29	Equity Financing		
30	Amount of Equity Issued	0	0
31	Aggregate Stock Value	=B24	=C24
32	PV Equity Issued	=NPV($B$20,C30:G30)+B30	=NPV($B$20,D30:H30)+C30
33	Equity Value - PV (Equity Issued)	=B31-B32	=C31-C32
34			
35	Shares Outstanding		
36	Number of Shares - beginning	=100	=B38
37	Shares Issued	0	=C30/C40
38	Shares Ending	=B36+B37	=C36+C37
39			
40	Issue Price per Share	=B33/B36	=C33/C36
41	Dividends per Share		=C23/C36
42	PV of Dividends per Share	=NPV($B$20,C41:G41)	=NPV($B$20,D41:H41)
43	Value per Share - ex dividend	=B42	=C42

EXHIBIT 9.14B   Selected formulas for equity financing section of Exhibit 9.14A.

need to be issued or the price at which they will be issued? First, we use Equation 9.9 to determine the price per share.

$$\text{Price per Share}_t = \frac{\text{Equity Value} - PV_t\,(\text{Equity Issued})}{\text{Shares}_{t-1}} \tag{9.9}$$

where Equity Value$_t$ is the aggregate value of equity at t (line 31 repeating line 24), PV_t (Equity Issued) the present value at time t of all future issues of new equity (line 32), and Equity Value$_t$ − PV_t (Equity Issued) the aggregate value of the original 100 shares at time t (line 33).

Line 40 calculates value per share by dividing the net equity value of current shareholders (line 32) by the number of shares at the beginning of the period (line 36), Shares$_{t-1}$. Since price per share at time t depends on Shares at time $t-1$, simultaneity is avoided. Dividing the amount of new equity financing ($200) by the price per share at time $t = 2$ yields the number of shares issued (line 37).

This analysis demonstrates how issuing new shares affects the value of the previously existing shares. The reduction in value due to this dilution will be equal to the present value of the reduction in dividends per share. This has several important implications. First, as Miller and Modigliani emphasize, if the firm increases its dividend, it will have to issue shares to finance the dividend (holding all other variables constant). The reduction in share value from this dilution will exactly equal the value of the additional dividend paid. Hence, there is no net benefit from paying the extra dividend.*

Even though the aggregate value of the equity is the same under both scenarios, $3604.78, the values per share differ. The value per share when no new equity issued is $36.05, and when equity is issued it is $34.45, with a difference of $1.59. The value per share decreases because future dividends per share decrease after new equity is issued at time $t = 2$. Exhibit 9.15 shows the effects of this dilution. Issuing new shares in period 2 causes the dividends per share to drop by $0.833 in periods 3, 4, and 5. The present value, at time $t = 0$, of this reduction in dividends per share is $1.59. For the 100 original shares, this reduction in value amounts to $159.44, which equals the present value at $t = 0$ of the equity issued at $t = 2$.

---

* This can be demonstrated in the scenario B model by supposing that the $200 of equity is issued to increase the dividend in year 2 rather than pay for capital expenditures. Simply change capital expenditures to zero. The resulting price per share is identical to the price per share absent issuing new equity.

	A	B	C	D	E	F	G
48				Period			
49		0	1	2	3	4	5
50	Dividend Per Share w/o Equity Issue		10.000	10.000	10.000	10.000	10.000
51	Dividend Per Share w/ Equity Issue		10.000	10.000	9.167	9.167	9.167
52	Difference in Dividend Per Share		0.000	0.000	0.833	0.833	0.833
53							
54	PV of Difference in Dividends per Share	$1.59	$1.79	$2.00	$1.41	$0.74	$0.00
55							
56	Price Per Share w/o Equity Issue	$ 36.05	$ 30.37	$ 24.02	$ 16.90	$ 8.93	$ -
57	Price Per Share w/ Equity Issue	$ 34.45	$ 28.59	$ 22.02	$ 15.49	$ 8.19	$ -
58	Difference in Price Per Share	$ 1.594	$ 1.79	$ 2.00	$ 1.41	$ 0.74	$ -
59							
60	Equity Issued	$ -	$ -	$ 200.00			
61	PV Equity Issued	$159.44	$178.57	$200.00			

EXHIBIT 9.15 Difference in dividends and equity values. No equity issue versus equity issue.

The second implication of this analysis, and the one most useful to financial planning, is how to handle the problem of issuing new equity while avoiding the problem of simultaneity. The simple Miller and Modigliani solution is to calculate the aggregate value of equity, independent of the number of shares, subtract from that aggregate the value of new equity financing, and divide the difference by the number of shares outstanding prior to the new issue.

9.5.2 Share Issue and Dilution in the Empruntay Model

The Mythic Corporation example was simple, so we could introduce dilution and see how to model share value. Now we will extend the concept in a more complex model such as Empruntay that we used earlier in the chapter.

The modifications needed to calculate value per share when facing the dilution created by issuing new equity are fairly easy. At each date, the present value of future equity issued is subtracted from the aggregate value of equity to yield the value to the stockholders on that date. We divide that total value by the number of shares outstanding on that date:

Price per Share$_t$

$$= \frac{[\text{Aggregate Equity Value}_t] - PV_t(\text{Future Equity Issues})}{\text{Shares Outstanding}_t}$$

which, at $t = 0$ is

$$= \frac{\left[\sum_{t=1}^{h}(D_t/(1+k_E)^t) + (E_h/(1+k_E)^h)\right] - \sum_{t=1}^{h}(\text{Equity Issued}_t/(1+k_E)^t)}{\text{Shares Outstanding}_0}.$$

(9.10)

	A	B	C	D	E	F	G	H
120				Empruntay Corporation				
121								
122				Growth = 10%; Debt portion = 50%				
123	EQUITY VALUATION MODULE			1	2	3	4	5
124	Cost of Equity							
125	Risk Free Interest Rate	5.0%						
126	Market Risk Premium	7.0%						
127	Beta (unleveraged)	1						
128	Assumed Equity Value			1,551.08	1,670.47	1,797.72	1,933.14	2,077.00	2,229.55
129	Debt/Equity			27.08%	27.79%	28.58%	29.39%	30.24%	31.13%
130	Beta (leveraged)			1.162	1.167	1.171	1.176	1.181	1.187
131	Discount Rate			13.14%	13.17%	13.20%	13.23%	13.27%	13.31%
132								
133	Iterative Aggregate Equity Value							
134	Equity Valuation Model			1	2	3	4	5
135	Equity Cash Flow			84.38	92.71	101.88	111.97	123.07
136	Present Value of Equity Flow		$351.72	$313.55	$262.12	$194.84	$108.65	$0.00
137	Terminal Value							
138	Long Term Growth Rate							7.38%
139	Equity Flow t = 6							132.15
140	Terminal Value of Stock at t = 5							2,229.55
141	PV of Terminal Value		$1,199.36	$1,356.93	$1,535.60	$1,738.30	$1,968.35	$2,229.55
142								
143	Aggregate Value of Equity		$1,551.08	$1,670.47	$1,797.72	$1,933.14	$2,077.00	$2,229.55
144								
145	Value Per Share							
146	Accounting For Dilution			1	2	3	4	5
147	Amount of Equity Issued			$44.25	$49.45	$54.43	$59.91	$65.93
148								
149	Aggregate Equity Value		$1,551.08	$1,670.47	$1,797.72	$1,933.14	$2,077.00	$2,229.55
150	PV Equity Issued		187.25	211.85	189.68	158.74	118.11	65.93
151	Aggregate Equity Value - PV Equity Issued		1,363.83	1,458.62	1,608.04	1,774.40	1,958.89	2,163.62
152	Number of Shares at Beginning of Period		100.00	100.00	103.03	106.20	109.46	112.81
153	Value Per Share		13.64	14.59	15.61	16.71	17.90	19.18
154								
155	Value Based on Per Share Data							
156	Number of Shares - beginning		100.0	100.0	103.0	106.2	109.5	112.8
157	Shares Issued			3.0	3.2	3.3	3.3	3.4
158	Shares Ending		100.0	103.0	106.2	109.5	112.8	116.2
159								
160	Issue Price Per Share		13.64	14.59	15.61	16.71	17.90	19.18
161	Dividends Per Share			0.84	0.90	0.96	1.02	1.09
162	PV Dividend / Share		3.32	2.91	2.40	1.75	0.96	0.00
163	Terminal Value / Share							19.18
164	PV Terminal Value		10.32	11.67	13.21	14.95	16.93	19.18
165	Price Per Share -ex div		13.64	14.59	15.61	16.71	17.90	19.18

EXHIBIT 9.16A Empruntay equity valuation. Value per share with dilution from issuing new equity.

Exhibit 9.16A shows the equity valuation section for Empruntay based on the assumptions that sales are growing at 10% and 50% of Empruntay's REF is new equity; the formulas for selected columns are in Exhibit 9.16B. Lines 143 and 149 show the aggregate value of equity. Line 147 is the amount of external financing from issuing new equity, and line 150 is the present value at each date of the future equity that will be issued from that date onward. For example, $187.25 in C150 is the present value at $t = 0$ of all the issues from periods 1 through 5. Line 151 takes account of the dilution due to future stock issues by subtracting the present value of future equity issues from aggregate equity value; and when the values in line 151 are divided by the number of shares outstanding at the beginning of each period, we get the value per share (line 153), as prescribed by Equation 9.10.

	A	C	D
120			**Empruntay Corporation**
121			
122			Growth = 10%, Debt portion = 50%
123	**EQUITY VALUATION MODULE**		=D47
124	**Cost of Equity**		
125	Risk Free Interest Rate		
126	Market Risk Premium		
127	Beta (unleveraged)		
128	Assumed Equity Value	=C143	=D143
129	Debt/Equity	=C73/C128	=D73/D128
130	Beta (leveraged)	=$B$127+C129*$B$127*(1-Tax)	=$B$127+D129*$B$127*(1-Tax)
131	Discount Rate	=RF+C130*Risk_premium	=RF+D130*Risk_premium
132			
133	**Iterative Aggregate Equity Value**		
134	*Equity Valuation Model*		=D47
135	Equity Cash Flow		=D58
136	Present value of Equity Flow	=(D135+D136)/(1+C131)	=(E135+E136)/(1+D131)
137	Terminal Value		
138	Long Term Growth Rate		
139	Equity Flow t = 6		
140	Terminal Value of Stock at t = 5		
141	PV of Terminal Value	=D141/(1+C131)	=E141/(1+D131)
142			
143	**Aggregate Value of Equity**	=C136+C141	=D136+D141
144			
145	**Value Per Share**		
146	**Accounting For Dilution**		=D47
147	Amount of Equity Issued		=D100
148			
149	Aggregate Equity Value	=C143	=D143
150	PV Equity Issued	=C147+((1/(1+C131))*D150)	=D147+((1/(1+D131))*E150)
151	Aggregate Equity Value - PV Equity Issued	=C149-C150	=D149-D150
152	Number of Shares at Beginning of Period	=C156	=D156
153	Value Per Share	=C151/C156	=D151/D156
154			
155	**Value Based on Per Share Data**		
156	Number of Shares - beginning	=H41	=C158
157	Shares Issued		=D147/D160
158	Shares Ending	=C156	=D156+D157
159			
160	Issue Price Per Share	=C151/C156	=D151/D156
161	Dividends Per Share		=D135/D156
162	PV Dividend / Share	=(1/(1+C131))*(D161+D162)	=(1/(1+D131))*(E161+E162)
163	Terminal Value / Share		
164	PV Terminal Value	=D164*(1/(1+C131))	=E164*(1/(1+D131))
165	Price Per Share -ex div	=C162+C164	=D162+D164

EXHIBIT 9.16B   Formulas for share valuation.

The calculations of the number of shares and the issue price per share are in lines 155–165. Circularity is avoided by dividing aggregate values at each date by the number of shares outstanding at the beginning of each period—prior to the issuance of new shares at that date. The exception is terminal value per share in H163, where the aggregate terminal value at the end of the horizon is divided by the number of shares outstanding at the end of the horizon. Note that the end result is a value per share at $t = 0$ of $13.64. Had we not taken the dilution into account and simply divided the aggregate value of $1551 by the original number of shares, we would get a value per share of $15.51. Hence, for exploring the impact of new equity financing, it is worthwhile to consider dilution and the number of shares in the model.

PROBLEMS

1 TellAll, Inc., Valuation

TellAll, Inc., is a mid-cap company engaged in designing and market-ing equipment to telecommunications service providers in the United States and internationally. The company also provides deployment and professional services to support its products. Its income statements and balance sheets for 2006–2010 are shown in Table 9.1. A variety of average financial ratios are also shown in Table 9.2. The ratios are two-year averages, taken over the years 2009–2010. Table 9.3 shows forecast parameters and policy variables that will apply for the next five years.

Your assignment is to

(a) Develop a financial statement simulation model to generate *pro forma* financial statements (income statement, balance sheet, and a statement of cash flows) that cover the years 2011–2015. It should include the following modules/sectors: accounting (income statement and balance sheet), investment, financing, cash flow, and valuation. For the initial model, assume that the forecast parameters and policy variables are as listed in Table 9.3. In addition, assume that any REF will be funded by issuing new debt and that the Notes Payable in 2010 will be retired in 2011. Assume also that the short-term investments in 2010 will be liquidated in 2011 and that no long-term debt will be repaid, nor equity repurchased.

Your model should be constructed so that it balances: assets should equal liabilities (plus shareholders' equity), and sources should equal uses of cash. Put checks into your model to verify that it balances. In addition, your model should be constructed in such a way that any equations that refer to exogenous parameters do not contain numbers. The parameters should be in a separate input section, with the two decision/sensitivity variables—*payout* and *debt portion*—clearly identified.

When the above steps are completed, you will have developed a basic financial statement simulation model. Hopefully, it worked the first time. However, it is important to make sure that it balances and remains logical under a variety of circumstances. Check the balancing by varying the rate of growth of sales. Use high growth rates such as 30% or 50% per year and see how it performs. Then use low growth rates such as −20% and −30% per year. As before, assets should equal liabilities and all cash flows should go to some use. You probably will need to modify your model slightly under these extreme scenarios for all of the accounts to remain logical. In particular, you will need to add conditional statements to prevent NFA from becoming negative and to prevent your model from taking a depreciation deduction once NFA reach zero. If your model balances and remains logical under the various extreme growth assumptions, return the five-year growth rate to the base case (10%). Once your model balances you will be ready to move on to part b.

(b) The focus of this portion of the assignment is upon the valuation module. The parameters needed for the valuation module are

TABLE 9.1   TellAll Financial Statements 2006–2010

	2006	2007	2008	2009	2010
Annual Income Statement, TellAll, Inc. ($ millions)					
Total revenue	1317	980	1232	1883	2041
Cost of revenue, total	831	626	575	1028	1108
SGA	298	243	237	271	293
R&D	335	286	250	344	357
EBITDA	−147	−175	170	240	283
Depreciation/amortization	9	13	18	36	29
Restructuring charge	174	77	231	15	8
EBIT	−330	−265	−79	189	246
Interest expense	0	0	0	10	11
Interest income	33	31	34	53	69
Net income before taxes	−297	−234	−45	232	304
Income tax	−15	−3	20	37	91
Net income after taxes	−282	−231	−65	195	213
Dividends	0	0	0	0	0
Retained earnings	−282	−231	−65	195	213
Annual Balance Sheet, TellAll, Inc. ($ millions)					
Cash and equivalents	454	223	269	876	168
Short-term investments	566	847	914	490	1435
Receivables	217	197	309	319	411
Inventory	175	42	116	109	167
Other current assets	206	137	89	74	66
Total current assets	1618	1446	1697	1868	2247
Property/plant/equipment—Gross	770	644	619	614	638
Accumulated depreciation	349	362	380	416	445
NFA	421	282	239	198	193
Goodwill and intangibles	526	660	1248	1229	1197
Other long-term assets	143	133	133	113	185
Total assets	2708	2521	3317	3408	3822
Accounts payable	188	208	216	250	301
Notes payable/short-term debt	0	0	204	181	289
Other current liabilities, total	0	0	16	38	153
Total current liabilities	188	208	436	469	743
Long-term debt	0	0	0	0	0
Other liabilities, total	158	180	212	175	222
Total liabilities	346	388	648	644	965
Common stock	548	550	1151	1244	1400
Retained earnings (accumulated deficit)	1944	1713	1648	1843	2056
Treasury stock—Common	−130	−130	−130	−323	−599
Total equity	2362	2133	2669	2764	2857
Total liabilities and shareholders' equity	2708	2521	3317	3408	3822
Total common shares outstanding	412	415	464	449	439

TABLE 9.2　TellAll, Inc., Two-Year Average Ratios

	Two-Year Average
Annual Income Statement, TellAll, Inc. (Percent of Sales)	
Total revenue	0.306
Cost of revenue, total	0.544
SGA	0.144
R&D	0.179
EBITDA	0.133
Depreciation/amortization	0.053
Restructuring charge	0.006
EBIT	0.110
Interest expense	0.055
Interest income	0.048
Net income before taxes	0.136
Income tax	0.229
Net income after taxes	0.104
Dividends	—
Annual Balance Sheet, TellAll, Inc. (Percent of Sales)	
Cash	0.274
Short-term investments	0.482
Receivables	0.185
Inventory	0.070
Other current assets	0.036
Total current assets	1.046
Property/plant/equipment—Gross	0.319
Accumulated depreciation	0.219
NFA	0.100
Goodwill and intangibles	0.620
Other long-term assets	0.075
Total assets	1.841
Accounts payable	0.140
Notes payable/short-term debt	0.119
Other current liabilities, total	0.048
Total current liabilities	0.307
Long-term debt	—
Other liabilities, total	0.101
Total liabilities	0.407
Common stock	0.673
Retained earnings (accumulated deficit)	0.993
Treasury stock—Common	(0.233)
Total equity	1.434
Total liabilities and shareholders' equity	1.841

TABLE 9.3   Input Module for TellAll, Base Case

TellAll, Inc., Input Sector

Parameters		Summary Statistics
Cost of revenue, total	0.52	*Master check*
SGA	0.144	Aggregate equity
R&D	0.18	Average D/E
Depreciation	0.053	g^* (2015)
Cash and equivalents	0.1	PRAT (2015)
Receivables	0.19	Terminal growth
Inventory	0.08	
Other current assets	0.03	*Decisions/sensitivities*
Property, plant and equipment	0.32	Growth
Goodwill and intangibles	0.62	Payout
Other long-term assets	0.1	Debt portion
Accounts payable	0.14	Short-term debt portion
Other current liabilities	0.08	
Other long-term liabilities	0.08	*Cash control variables*
		Cash returned
Economy		Cash invested
Interest expense	0.06	Debt retired
Interest earned	0.05	
Risk-free rate	0.05	
Market risk premium	0.045	
Unleveraged beta	1.4	
Tax rate	0.32	
Long-term GDP growth	0.04	

shown in the Economy section of the Input Sector. Use Hamada's (1969) formula to calculate TellAll's leveraged beta in which the debt-to-equity ratio is the average of the forecast $D/E^\dagger$ ratios for years 2007–2011. Calculate TellAll's cost of equity using the CAPM. Calculate the terminal growth rate equal to g^* based on ratios in 2015. Base your forecast for the 2016 dividend on the Cash Available for Dividends in 2015 compounded to 2016. This simple valuation

† The ratio of interest-bearing debt-to-equity.

module returns the estimated value for all equity at the end of 2010 (since we are only using debt financing, you need not worry about dilution).

When you are sure that your valuation module works well, try a variety of growth rates to see the impact of growth on value. Try growth rates of 0%, 5%, 10%, 20%, and 30%. In each case, explain what happens to the capital structure, dividend payments, and the value of equity.

2 CyberCorp, Inc.: Spreadsheet Model for Growth

Antsy Corporation has grown its business via an active and successful acquisition program. It is currently considering several prospective future acquisitions. One of these is CyberCorp. Antsy's management has asked you to build a simple valuation model of CyberCorp that will enable them to estimate the effects of various mixes of debt and equity financing on CyberCorp's fully diluted value per share. In addition, Antsy's acquisition team has estimated the relationships between many of the items contained in CyberCorp's financial statements and its sales, as well as anticipated economic conditions (interest and tax rates). You will find these with the financial statements. You will find CyberCorp's most recent Income Statement and Balance Sheet along with the input parameters you need to build the *pro forma* statements at the end of these instructions and in the Excel workbook Ch9_P2.xls available to the instructor.

Part I

The first question Antsy would like you to answer is what is CyberCorp's stand-alone intrinsic value per share, given CyberCorp's current financial policies and assumptions? These assumptions and policies include:

- A 20% dividend payout rate
- Revenue growth of 30% for the next five years
- Long-term growth beginning in year 6 of 7%
- Sales, general, and administrative SGA costs that decrease at the rate of 5% per year beginning in 2010 (e.g., if 2009s SGA/Sales is 0.62, 2010s will be 0.62 ¥ 0.95)
- 100% equity financing

Calculate CyberCorp's cost of equity using the CAPM with Hamada's adjustment and the average projected D/E ratio. In addition, since CyberCorp is a micro-cap stock, add a small cap risk premium (1.5%) to the leverage-adjusted CAPM cost of equity. Your formula for the cost of equity is as follows:

$$k_E = k_F + [k_M - k_F]\beta_L + [\text{Small Cap Risk Premium}].$$

What is CyberCorp's fully diluted per share intrinsic values when a 50/50 debt equity mix or 100% debt is used to finance Cyber-Corp's growth? Assume that all excess cash is paid out as a special dividend.

Part II

The next question Antsy would like answered is what CyberCorp would be worth under Antsy's management for each of the above financing scenarios? Antsy believes that if they acquire CyberCorp they will be able to

- Accelerate SGA costs savings to a 7.5% reduction each year
- Increase revenue growth to 35% per year for the next five years
- Reduce CyberCorp's cash balance from 40% to 30% of sales

General Comments

This model requires an additional input parameter—a decision variable that allows you to adjust the rate at which SGA expense decreases. As always, be sure that your model balances and makes sense in extreme scenarios. It should not take depreciation when NFA are zero, and NFA and dividends should never become negative.

Finally, the CFO is certain to raise several questions. The first is why dilution does not seem to have a significant effect on per share valuations? The second is why CyberCorp is worth more than its current intrinsic value to Antsy? You should address these questions in your memo.

What You Hand In

Turn in a memo targeted for senior management. Your memo should briefly introduce the problem and questions posed, answer the questions clearly and concisely using tables where appropriate, and, lastly, summarize your findings. In an appendix, attach the *pro forma* statements for the base case scenario (using the initial parameter values).

Format your output such that the income statement and balance sheet fit on one page; the investment, financing, and cash flow modules fit on another; and the formal statement of cash flows and valuation module fit on yet another. There should be only one model in your workbook. You answer the above questions simply by changing the values of your decision/sensitivity variables and observing the effects.

3 Problem 3 CyberCorp, Inc. (continued)

This problem is a continuation of Problem 2. Use your solution to that problem as the base model for this one.

Problem 2 asked you to payout any excess cash generated by CyberCorp as a special dividend. However, Antsy's CFO has pointed out that there are other uses for this excess cash that might create greater value for CyberCorp's shareholders. In particular, she proposes that any excess cash first be applied to retiring outstanding long-term debt. Once all debt has been retired, any remaining excess cash would be used to permanently retire outstanding shares of Antsy's stock. What are CyberCorp's intrinsic values per share under this cash policy for the three financing mixes, assuming Antsy acquires CyberCorp?

To answer these questions, you will need to adapt the financial statement simulation model of CyberCorp that you built in Problem 2. Add one decision variable that allows you to decide how much excess cash to return to shareholders and how much to use to retire debt and equity. (We suggest adding a second line to your financing module directly underneath the REF line.) Have it report excess cash when REF is negative, else 0. Similarly, have your REF line show a balance when it is positive, but be zero when excess cash is generated.* Add another decision variable that determines the portion excess case used to retire debt. Link it to your newly created "Excess Cash" line. When you choose to retire current debt and equity, retire debt first. Once all debt has been retired, use any remaining excess cash to repurchase and retire equity. Because you are retiring the equity repurchased, debit (decrease) the Common Stock account.

Antsy's CFO had a second criticism of your Problem 2 model—it made overly simplistic assumptions about the cost of equity. In particular, the cost of equity was based on the average predicted debt-to-equity ratio. But if CyberCorp retires its outstanding debt,

---

* HINT: You will have to calculate REF in both cells (same formula). Your logic in each cell determines which one is zero and which is positive.

TABLE 9.4 CyberCorp Data

CyberCorp, Inc ($ millions)					
Model Parameters			**Decision / Sensitivity**		
Cost of goods sold	0.150		Payout	0.20
Selling, general & administrative	0.620		Debt portion	0.00
Research & Development	0.160		%SGA decrease	0.05
Depreciation	0.120		Excess cash	0.00
Cash and Short Term Investments	0.400		Growth	0.30
Receivables	0.145		Lt growth	0.07
Inventory	0.070			
Gross PPE	0.230			
Accounts Payable	0.040		**Economy**	
Other Current Liabilities	0.100		Riskfree rate	0.045
			Risk Premium	0.060
			Small Cap risk	0.015
			Unleveraged beta	0.340
			Interest expense	0.070
			Taxes (EBT)	0.400

INCOME STATEMENT ($millions)	
	2009
Revenue	113.5
COGS	16.3
Gross Profit	97.2
SG&A	70.6
Research & Development	17.1
EBITDA	9.5
Depreciation	2.8
EBIT	6.7
Interest expense	0.6
Net Income Before Taxes	6.1
Taxes	0.2
Net Income After Taxes	5.9
Dividends	0.0
Retained Earnings	5.9
Shares outstanding	22.9

BALANCE SHEET ($millions)	
	2009
Cash and Short Term Investments	58.4
Receivables	18.5
Total Inventory	9.0
Total Current Assets	85.9
Gross PPE	23.1
Accumulated Depreciation	14.7
PPE, Net	8.4
Total Assets	94.3
Accounts Payable	4.4
Other Current Liabilities	10.7
Total Current Liabilities	15.1
Debt	10.2
Total Liabilities	25.3
Common Stock, Total	0.2
Additional Paid-In Capital	184.4
Retained Earnings (Accumulated Deficit)	-115.6
Total Equity	69.0
Total Liabilities & Shareholders' Equity	94.3

future cash flows will be less risky and should be discounted at a lower rate. Further, the discount rate in Problem 2 was based on the book values of debt and equity. Antsy's CFO would like to see the discount rate based on market values of equity. To address these concerns, you will need to build a market value-based valuation module similar to the one in Exhibit 9.6. Base your terminal value calculations on the minimum dividend calculated in the financing module. You will need to set Excel's calculation option to "Iterative" for the model work correctly.

We recommend you to incorporate the cash management variable first—it will take a fair amount of trial and error to solve. Once the new decision variables work, modify your valuation module to include market-based time-varying discount rates.

Use your new model to answer the following questions:

- Suppose that Antsy now owns CyberCorp. What is CyberCorp's intrinsic value per share if 100% debt financing is used: 50/50 debt and equity financing, or 100% equity financing?
- How should CyberCorp handle its excess cash? Return it to shareholders, use it to retire debt first, then equity, or use it to retire equity.

What You Hand In

Turn in a memo targeted for senior management. Your memo should briefly introduce the problem and questions posed, answer the questions clearly and concisely using tables where appropriate, and, lastly, summarize your findings. There should be only one model in your workbook. You answer the above questions simply by changing the values of your decision/sensitivity variables and observing the effects.

CHAPTER 10

Modeling Long-Term Assets

I N CHAPTER 4, we developed a model that simulated the financial statements in very general form. To introduce the subject and understand the structure of the model, the number and detail of the different accounts was kept to a minimum. It was noted that once we developed a functioning model, we could always add detail to expand it to meet our needs. In this chapter, we discuss how to develop more detailed treatments of long-term assets. We look at ways to model the investment in fixed assets. First, we consider ways to include fixed assets in the long-run planning model where the assets are driven by the sales forecast; second, in the capital budgeting section, we will develop a model where the investment in assets is the driver for the purpose of evaluating the investment in fixed assets; and third, we look at the decision to acquire and/or merge with another company as a capital budgeting decision.

10.1 FIXED ASSETS IN A LONG-RUN PLANNING MODEL

With a long planning horizon and long periods comprising the horizon, a percent-of-sales approach to fixed assets is probably adequate. In the long run the firm's ability to increase product output is tied fairly closely to its productive capacity, so that any increases in output will require increases in the fixed assets that constitute its productive capacity. Thus, the percent-of-sales relation is a good first approximation for fixed assets, where

$$\text{Fixed Assets}_t = a \times \text{Sales}_t, \tag{10.1}$$

and for investment we have

$$\text{Fixed Asset Investment}_t = a \times (\text{Sales}_t - \text{Sales}_{t-1}). \tag{10.2}$$

This simplified approach leads to immediate, concurrent responses of investment to changes in sales. This approach may be adequate for very aggregative models where the time periods are relatively long. However, it may be misleading in a more detailed model that focuses on shorter time periods. One difficulty is that if sales decline, this model would always show disinvestment in response to the sales decline, whereas, in the short run, a firm can cut production in response to a sales decline without selling off assets. Conversely, most firms have some capacity to increase sales without an immediate increase in assets.

The O&R model discussed in Chapter 4 linked fixed assets directly to sales and would disinvest with declines in sales. The fixed assets were modeled as

$$\text{GFA}_t = \left(\frac{\text{GFA}}{\text{Sales}}\right) \text{Sales}_t, \tag{10.3}$$

depreciation expense as

$$\text{Depreciation Expense}_t = (\text{Depreciation Rate})\,\text{GFA}_{t-1}, \tag{10.4}$$

and NFA as

$$\text{NFA}_t = \text{GFA}_t - \text{Allowance for Depreciation}_t,$$

where the Allowance for Depreciation is the accumulated depreciation. In addition to disinvesting when sales decline, this model may even disinvest when sales are growing. With this model, if the depreciation rate exceeds the growth rate of sales, depreciation expense will exceed the net addition to assets each period and NFA will decline. That is, even though sales are growing, depreciation is so large that the model shows net assets declining despite the sales growth. This may or may not be appropriate for the situation, so this is just a caution for the modeler to be aware of the implications of the interplay between growth and the depreciation rate. Even if it is appropriate for NFA to decline, you need to put checks in your model so that it does not produce nonsense results such as negative NFA. Referring again to the O&R model, we included an IF statement so that depreciation would be zero if it would otherwise cause NFA to be negative:

$$\text{Depreciation Expense}_t = (\text{Depreciation Rate})\text{GFA}_{t-1},$$
$$\text{If } [(\text{Depreciation Rate})\text{GFA}_{t-1} \le \text{NFA}_{t-1}],$$
$$\text{Depreciation Expense}_t = \text{NFA}_{t-1}, \text{ otherwise.} \qquad (10.5)$$

In "Excel-ese," this statement is expressed as

$$\text{Depreciation Expense}_t = \text{IF}[(\text{Depreciation Rate})\text{GFA}_{t-1} <= \text{Net Fixed}$$
$$\text{Assets}_{t-1}, (\text{Depreciation Rate})\text{GFA}_{t-1}, \text{NFA}_{t-1}].$$

There are various ways to modify the link between sales and investment so as to avoid this disinvestment problem. Francis and Rowell (1978) presented a model where investment occurs only if the sales forecast exceeds the level of sales that can be produced with existing assets. For example, assume the minimum stock of assets that must be available to produce a given level of sales is

$$\text{Required Fixed Assets}_t = a \times \text{Sales}_t,$$

then asset investment can be specified as

$$\text{Fixed Asset Investment}_t = \max[0, \text{Required Fixed Assets}_t - \text{Fixed Assets}_{t-1}]$$
$$= \max[0, a \times \text{Sales}_t - \text{Fixed Assets}_{t-1}]. \qquad (10.6)$$

If Fixed Assets$_{t-1}$ exceeds $a \times$ Sales$_t$, no new investment occurs in the model. With this code, fixed assets do not decrease in periods where a decline in sales is fed into the model.

The technology of the production process may help us to model the link between sales and investment in fixed assets. The firm's production function, expressing its maximum output as a function of the factor inputs, can be utilized to derive the level of capital assets required to support a given level of sales. For example, in the planning model of AT&T (Davis, Caccappolo & Chaudry, 1973), the firm's production function is estimated. Based on the production function, the required investment in capital assets can be expressed as a function of sales and the relative prices of the factors of production, including the cost of physical capital services.

One difficulty frequently encountered in structuring the simulation model, and particularly in relating investment to sales, is whether to define output in physical units or in dollars. It can be useful to focus on the physical units of output because it is the capacity to produce physical units that most directly affects the need for new investment, and physical capacity is less likely to be confounded by the effects of inflation. In addition, the separation of units from product prices makes the model better suited to exploring pricing policies and the effects of competition and inflation.

For example, in the Francis and Rowell model the dollar amount of investment is linked to the forecast of unit industry sales with the following four simplified equations:

$$\text{Unit Company Sales}_t = \text{Market Share} \times \text{Industry Unit Sales}_t \quad (10.7)$$

$$\text{Capital Units Required}_t = A \times \text{Unit Company Sales}_t \quad (10.8)$$

$$\text{Required Fixed Assets}_t = P \times \text{Capital Units Required}_t \quad (10.9)$$

$$\text{Fixed Asset Investment}_t = \max[0, \text{Required Fixed Assets}_t - \text{Fixed Assets}_{t-1}], \quad (10.10)$$

where A is the ratio of physical capital to sales, and P the price per unit of physical capital. Another example of focusing on physical units was Gershefski's (1968, 1969) oil company model that defined market share in terms of gallons of gasoline sold, and it determined the investment in new service stations required to sell a target number of gallons for the forecast period.

The problem with defining variables in physical units is that it can lead to a model that is excessively detailed, requiring too much data input and generating more output than is necessary to answer the questions that the model is intended to answer. This is especially true for a firm with many different product lines. The excessive detail demanded by unit definitions can be avoided either by sticking with a model defined in dollar amounts, or by using an index of unit output that summarizes the unit output of all the different product lines.

The modeling of fixed assets demonstrates a major difference between a planning model and an accounting model. An accounting model would usually be oriented toward keeping track of the costs of the assets and would track the accounts for specific classes of assets over their life. This would usually be too much detail to be useful in a planning context. For planning purposes, it is usually better to focus attention on the broad aggregate levels of the asset accounts, rather than develop excessive detail about the different account categories. An exception occurs when the purpose of the simulation is to analyze the acceptability and feasibility of specific investment proposals, as will be discussed in the next section.

10.2 DIRECT INVESTMENT EVALUATION

When the purpose of the simulation is to evaluate specific investment projects, the modeling approach is different from the case where investment reacts passively to the sales forecast. In this case, the focus is on the specifics of the proposed investment, and the modeling problem is how to relate the details of the investment to the rest of the simulation model of the firm.

In the context of financial modeling and planning there are two perspectives to the consideration of investment proposals. The first is the evaluation of the investment itself, and the second is relating the investment to the firm as a whole so that the user can evaluate how the investment affects the firm's overall performance. The evaluation will be considered first, and then we briefly discuss how to integrate the investment into the larger firm model.

10.2.1 Investment Evaluation

There are various techniques that are available for evaluating capital investments by the firm. The two most generally accepted techniques are

NPV and IRR.* The NPV technique will be emphasized here. NPV is calculated according to the equation

$$\text{NPV} = \sum_{t=1,h} \text{Inflow}_t \left[\frac{1}{(1+k)^t} \right] - \sum_{t=1,h} \text{Outflow}_t \left[\frac{1}{(1+k)^t} \right], \quad (10.11)$$

where Inflow$_t$ is the increment in the cash inflow that is generated by the investment project in period t; Outflow$_t$ is the outflow, typically the cost of the project; h the number of periods in the life of the project; and k the rate of return that is required on the investment.

The decision criterion used to evaluate the acceptability of an investment project is that the NPV should *not be less than zero*, that is, NPV $\geq$ 0. If the NPV is exactly equal to zero it means that the cash flows are just sufficient to return to the investor the original cost of the investment (return *of investment*), plus yield a rate of return *on investment* equal to the discount rate k. If NPV > 0, the investment yields a rate of return greater than k in addition to returning the original principal of the investment. Applying the NPV criterion is relatively straightforward so long as one is careful to use the correct definitions of cash flow, cost, and discount rate. In this regard, two alternative applications of NPV will be discussed: first, the WACC method and second, the equity method. The WACC method evaluates the project as a whole, whereas the equity method evaluates just the equity share of the project. The WACC approach seeks to answer the question: Are the total (or overall) cash flows generated by the project sufficient to return the original total cost (return of investment principal) and provide the required returns on investment to *all classes of security holders* who provided capital to finance the cost of the project? Because the WACC method is concerned with all the contributors of capital, it is also referred to as the invested capital method. The equity method asks whether the cash flows to the equity *owners* of the project will be sufficient to return the initial investment by the owners and give the owners their required return on investment. We will explain the WACC approach first and then the equity method.

---

* More detailed discussions of the techniques for investment evaluation can be found in any standard financial management text such as Brigham and Ehrhardt (2008) or Ross, Westerfield, and Jaffe (2008).

10.2.2 The WACC Method

10.2.2.1 Cost of Investment

For the purpose of evaluating the acceptability of a project, the cost of the project should be the present value of all the incremental outlays necessary to complete the project and make it operational. The outlays typically include expenditures for the purchase of plant and equipment and any investment necessary to increase the balances of working capital, net of any spontaneous increase in current liabilities. But how we calculate these costs depends on whether we are using the WACC method or the equity method of evaluation. The total cost of an investment project is financed either entirely or partly with equity capital. The part of the total investment cost that is not financed with equity is financed with debt (or some other hybrid source). The total cost includes not just the investment by equity security holders, but also whatever funds were borrowed to finance the project. For example, suppose we invest $2 million of our own money and borrow $8 million to purchase assets costing $10 million. The total project investment is $10 million, but the equity investment is $2 million. If we use the WACC method, we consider the cost of the project as $10 million, but if we use the equity method, the investment cost is just the $2 million that we will invest as the equity owners of the project.

10.2.2.2 Discount Rate

The discount rate used to calculate NPV with the WACC approach is a weighted average of the return required by the equity investors (cost of equity, denoted by k_E) and the the lenders (cost of debt, denoted by k_D). WACC is calculated as

$$k_A = \left(\frac{\text{Debt}}{\text{Total Capital}} \right) k_D (1-T) + \left(\frac{\text{Equity}}{\text{Total Capital}} \right) k_E,$$

(10.12)

where the subscript "A" denotes average; (Debt/Total Capital) the proportion of interest-bearing debt in the firm's capital structure if the investment is undertaken; (Equity/Total Capital) the proportion of equity used to finance the firm if the project is undertaken; and we multiply the cost of debt by $(1 - T)$ to put our costs on an after corporate income tax basis. If our project can earn a return of at least k_A, the project is expected to satisfy all investors by being able to provide each group of investors with their required return.

For this example with 80% debt and 20% equity, assume that the cost of debt is $k_D = 7\%$, the cost of equity is $k_E = 18\%$, and the tax rate is $T = 40\%$. The WACC will be

$$k_A = \left(\frac{\text{Debt}}{\text{Total Capital}}\right)k_D(1-T) + \left(\frac{\text{Equity}}{\text{Total Capital}}\right)k_E$$
$$= (0.80)7\%(1-0.40) + (0.20)18\% = 6.96\%.$$

The correct application of WACC requires that the proportions of debt and equity be the proportions of these capital sources for the firm as a whole as it would be financed if the project is adopted. It would be incorrect to use only the proportions that apply to the project financing. In this example it is assumed that the project is the whole firm so the 80/20% mix is appropriate. On the other hand, if the project was part of a larger firm, it would be the firm's capital proportions that are relevant, not just the project's. Another important point about applying WACC is that the proportions of capital should be based on market values, not book values. That is, the amounts of debt and equity used in calculating the proportions should be market values at the time the project is undertaken.

10.2.2.3 Cash Flow to Invested Capital

The cash flow that should be matched to the WACC method for calculating the discount rate is the cash flow available to be distributed to those who contributed capital to finance the project. We will refer to this cash flow as the cash flow to invested capital. Since both lenders and equity investors contributed capital, this should be the after-tax cash flow available to be distributed to both lenders and equity investors, where, for purposes of cash flow calculation, interest expense and the tax effects of interest are ignored because they will be taken into account in the calculation of k_A. For example, suppose our project is expected to earn incremental income as follows:

Sales	15,000
Operating costs	11,250
EBITDA	3750
Depreciation	1560
EBIT	2190
Interest	433
EBI	1757
Tax	703
Net income	1054

EBIT $(1 - T) = 2190 (1 - 0.40)$	= 1314
+ Depreciation	+ 1560
= Earnings Flow to Invested Capital	= 2874
− Investment in Net Working Capital	− 500
− Gross Investment in Fixed Assets	− 1000
= Cash Flow to Invested Capital	= $1374

EXHIBIT 10.1 Calculation of cash flow to invested capital.

In the same period that we earn the income shown above, suppose that NWC must be increased by $500, and an additional $1000 must be invested in fixed assets.

For purposes of using the WACC approach, we calculate the cash flow to invested capital on an after-tax basis as shown in Exhibit 10.1.

The method shown above for calculating the cash flow shows neither interest nor principal payments on the debt as deductions from the basic cash flow. Nor does it consider the tax deductibility of interest payments. The reason for this apparent omission is that by using the interest rate on the debt on an after-tax basis, $k_D \times (1 - T)$, in the calculation of WACC, the question being asked is whether the cash flows will be large enough to cover these costs, so we exclude interest from the cash flow to avoid double counting.

It is important to note that we must take care to be consistent in the method for calculating project cost, project cash flows, and the discount rate. In this example, the focus was placed on WACC so the cost, cash flows, and discount rate were all defined so that they would be consistent for this approach. To clarify these issues, we compare WACC approach to the equity method.

10.2.3 The Equity Method

The equity approach to investment evaluation asks whether the cash flows generated by the investment project are sufficient to return the original principal amount invested by the equity investors and provide the return on investment required by the equity investors.

In the context of NPV, the cost of the investment is the amount of capital contributed by the equity investors; the cash flow is that which is available to be distributed to the equity investors; and the discount rate is the minimum expected return required by the investors as compensation for the risk of the investment. Thus, equity NPV is the present value of the cash

Net Income	= 1054
+ Depreciation	+ 1560
= Earnings Flow to Equity	= 2614
− Investment in Net Working Capital	−500
− Gross Investment in Fixed Assets	− 1000
− Repayment of Debt Principal	− 1075
+ Proceeds from New Borrowing	+ 1200
= Cash Flow to Equity	= $1239

EXHIBIT 10.2 Calculation of cash flow to equity.

flows available to stockholders (equity), discounted at the stockholders' required return.

Continuing our numerical example, but now focusing on equity, the initial cost of the project from the stockholders' viewpoint is $2000 (2 million). The discount rate is the required return to equity, $k_E = 18\%$. The operating flow to be discounted at 18% is the flow available to the stockholders. Assume in the period being considered, that an additional $1200 will be borrowed, and because of earlier borrowing attributable to the project, debt principal payment of $1075 will also have to be made. Starting with the net income from the statement shown above, the equity cash flow would be calculated as in Exhibit 10.2.

This equity cash flow represents the amount that could be taken out of the project by the equity investors after paying all expenses, making the necessary investments in new assets, and meeting the debt obligations. Note that the equity flow takes account of the debt payments. Interest has already been deducted and the tax effects of interest are taken into account in the calculation of net income. Then, we subtract the principal payments on debt and add proceeds from borrowing to determine the equity flow. With the WACC approach, we did not consider either the interest expense or the principal flows from borrowing. The major differences between these two approaches are summarized in Exhibit 10.3.

To evaluate a project, we can use either the WACC approach or the equity approach. Each approach is valid, provided the cost, cash flows, and discount rate are defined consistently. Both are discussed here to help the reader understand the differences in approach. The WACC approach to NPV is the method that is most frequently used. Generally, the non-expert would tend to make fewer obvious mistakes in using the WACC

	WACC Method	Equity Method
Discount Rate:	Weighted Average Cost of Capital (k_A)	Cost of Equity (k_E)
Cash Flow:	Total Project Cash Flow	Equity Cash Flow

Investment Cost:	Total Capital Invested (D + E) =
	  Total Fixed Asset Investment
	  + Net Working Capital Investment
		Equity Invested (E) =
		  Total Fixed Asset Investment
		  + Net Working Capital Investment
		  − Amount Borrowed

Earnings	EBIT $(1 − T)$	$(EBIT − I)\,(1 − T)$
Cash Flow	+ Depreciation Expense	+ Depreciation Expense
Calculation	= Earnings Flow to Invested Capital (CF_{IC})	= Earnings Flow to Equity (CF_E)
Output: Value	NPV of the Project	NPV of the Equity of the Project

Definitions and notation:

k_E = required return on equity

k_A = weighted average cost of capital (WACC) = $(D/V)k_D\,(1 − T) + (E/V)k_E$

k_D = required return on the debt

D = amount of interest bearing debt

E = amount of equity

V = D + E, the total capital invested

T = tax rate

EBIT = Earnings Before Interest and Taxes

I = Interest expense

EXHIBIT 10.3   Input and output for the WACC and equity methods of investment evaluation.

approach for investment evaluation. Just be very careful how cash flows are calculated.

10.2.4 Working Capital

Regardless of which approach to investment evaluation is used, it is especially important to use the correct method for calculating the various factors that enter the NPV calculation. One aspect that is frequently handled improperly is the determination of the cost of the project for purposes of calculating NPV. A major source of confusion is the treatment of the investment in working capital.

The cost of undertaking an investment project should include not just the direct costs of purchasing and installing the long-term assets, but also any incremental investment in working capital necessary to make the investment operational. If the purchase and operation of a new plant requires that the plant be provided with an initial inventory of raw materials or an initial balance of cash or other current assets, the funds committed to these current assets must be considered part of the cost of the investment. Any funds provided by increases in spontaneous current liabilities, such as AP, offset the increase in the current asset accounts, so we consider only the increase in the NWC account as part of the cost of the investment that must be financed by the contributors of permanent capital.

We usually assume that we will recover the investment in NWC at the end of the project. This is analogous to salvage value of long-term assets at the end of the project.

The foregoing discussion of methods for evaluating investments was intended to provide the general framework. Now, an example will be presented that will demonstrate some of the detail of modeling a capital budgeting problem.

10.3 EXAMPLE: EVALUATING AN INVESTMENT FOR STILIKO PLASTICS

10.3.1 The Details of the Investment

10.3.1.1 The Opportunity

Flavio Stiliko has developed a new process for manufacturing a wide variety of plastic parts and components that can be used in products ranging from toys to components for automobiles and aircraft. Flavio is considering a project that would supply the plastic body for a new sports car to be marketed by Rocky Mountain Motor (RMM) Company. Supplying the bodies in the quantity and quality required by Rocky Mountain would require that Flavio's nascent company, Stiliko Plastics, construct a new manufacturing plant from scratch. Land would have to be purchased, a new building constructed, and new machinery, moulds, and equipment purchased and installed. In addition, the project would require an investment in working capital. The project would be quite expensive for a small startup company, requiring an investment of about $10 million over two years. While there is little doubt that this opportunity to contract with Rocky Mountain is the break that Flavio has been waiting for, the profitability and value of the project need to be rigorously evaluated.

10.3.1.2 The Objective

Flavio and his other investors agree that they would adopt the project only if it will increase the value of their investment. Laric Amal, one of Flavio Stiliko's co-investors, says that even though the link between their decision to adopt the project and their main objective is intangible, the firm can use as its decision criterion the NPV of the project: It should accept this project only if its NPV is positive. Positive NPV serves as the operational goal that, if attained, would help Stiliko Plastic's investors reach their primary objective of making the value of their equity as large as possible.

10.3.1.3 Financing

Flavio Stiliko believes they can finance the $10 million investment by borrowing $8 million at an interest rate of 7%. The remaining $2 million will come from Flavio and the other equity investors. They have estimated that with this capital mix, they would require an equity return of about 18%.

10.3.1.4 The Planning Horizon

The contract with RMM would last for six years. Hopefully, the life of the process and product will extend well beyond that horizon, but Flavio wants to plan conservatively so that if the project ends with the Rocky Mountain Project, it still would have been profitable. In addition, the equipment for the new process is expected to deteriorate sufficiently rapidly so that by the end of the sixth year any continuation would require substantial new capital investment. Consequently, it is decided that the pro-ject will be evaluated over a six-year planning horizon. Since annual data will provide sufficient detail for the evaluation, the planning horizon is broken up into six sub-periods of one year each. The capital investment in the project is assumed to begin immediately, with the first capital investment occurring at time $t = 0$. Subsequent capital expenditures and cash inflows are assumed to occur at the end of each year.

10.3.1.5 The Constraints

10.3.1.5.1 Sales, Production, and Costs.   The main constraints relevant to this project are the economic conditions and the technological relationships that limit the firm's ability to attain its objectives. The relevant economic conditions are those in Stiliko Plastic's product market, the requirements in their contract with RMM Company, and the costs of the factors of production. Since the Rocky Mountain contract and the success of their car will determine the price of the product and the number of units that can be sold, these factors can be considered as constraints that limit the firm. On the basis of the features of the contract and the forecast of the future of the market for this model of car, Stiliko estimates sales of the plastic car bodies will be $15 million per year, beginning in year 2 and ending in year 6. The firm's production technology and conditions in the markets for labor and materials serve as constraints on production and costs. Estimates of the level of production, the costs of labor and materials, and the technical specification of the new plant lead to a forecast for operating and administrative costs (excluding depreciation) equal to 75% of sales.

10.3.1.6 Project Investment

This project will require Stiliko to have a manufacturing facility built on land that must be purchased. The building will cost $3 million and the land $1 million. The necessary machinery and equipment will cost an additional

	Period		
	1	2	3
Depreciable Assets			
Equipment	3000	1500	
Building	1500	1500	
Non-Depreciable Assets			
Land	1000		
Total Fixed Asset Investment	5500	3000	
Net Working Capital Investment	—	—	1500
Incremental Project Investment	5500	3000	1500
Total Cumulated Capital Investment	5500	8500	10,000

EXHIBIT 10.4   Stiliko plastics: Schedule of investment expenditures (in thousands of dollars).

$4.5 million. In addition, the project will require an investment in working capital of about $1.5 million at the end of period 2. These costs will be incurred according to the schedule shown in Exhibit 10.4 (numbers in thousands of dollars). A total of about $10 million will have to be committed to the project, with $5.5 million required immediately, another $3.0 million at the end of the year 1, and $1.5 million at the end of the year 2.

10.3.1.7 Depreciation

The equipment will be depreciated using the three-year MACRS life. The MACRS depreciation percentages for three-year assets are 33%, 45%, 15%, and 7%, for years 1, 2, 3, and 4, respectively. The building will be depreciated over 40 years using straight-line depreciation. The depreciation will begin in year 2, when the building and equipment are placed in service, and end in year 6. The land will not be depreciated.

10.3.1.8 Terminal (Salvage) Value

Stiliko expects that the building and machinery and equipment will decrease in market value over the life of the project: the building by 4% per year and the machinery and equipment by 15% per year. Hence, the prices at which these assets can be disposed of at the end of the project are expected to be less than their purchase prices. The land is expected to increase in market value by 3% per year.

10.3.2 Constructing the Spreadsheet Model

The purpose for our model is to determine the NPV and IRR for this project from the perspectives of all investors (the invested capital method) and of equity alone (the equity method). To this end, we will build a model consisting of three worksheets, named Data Input, Data Processing, and Evaluation. The first sheet (Data Input) lists the basic input data; the second sheet (Data Processing) processes the data in the Data Input sheet for input into the third sheet; and the third sheet (Evaluation) calculates and evaluates the cash flows to all investors and to equity. The Data Input sheet contains the assumptions, data, and information necessary to construct the evaluation model. The other two sheets only contain the formulas necessary to calculate cash flows, discount rates, NPV, IRR, and other ancillary details.

The analysis begins with the current state of Stiliko Plastics, represented by the data and assumptions in the Data Input sheet. However, there is a large gap between this raw data and the input needed by the Evaluation sheet. The components of future cash flows and the costs of capital used to

	A	B	C	D	E	F	G	H	I	J	K
1						Stiliko Corporation					
2						Investment Evaluation					
3						Data Input					
4											
5	Global Assumptions and Data										
6											
7		Number of periods in Horizon				6		Dollars in $000			
8		Number of periods of Initial Investment				3					
9											
10		Economic and Market Data						Working Capital Assumptions			
11		Riskfree Interest Rate				4.0%		Cash/Sales			1.0%
12		Market Risk Premium				5.0%		Receivables/Sales			5.0%
13		Borrowing Rate				7.0%		Inventory/Sales			6.0%
14		Beta - Unleveraged				1.0		Payables/Sales			2.0%
15											
16	Fixed Asset Assumptions										
17				Period	0	1	2	3			
18		Investment in Assets									
19		Depreciable Assets									
20			Equipment Investment		3000	1500					
21			Total Deprec. Cost - Equipment			4500					
22			Building Investment		1500	1500					
23			Total Deprec. Cost - Building			3000					
24		Non-Depreciable Assets									
25			Land Investment		1000						
26											
27		Depreciation		Period	0	1	2	3	4	5	6
28			Depreciation Rate 3-year MACRS				33.0%	45.0%	15.0%	7.0%	0.0%
29			Depreciation Rate Building				2.5%	2.5%	2.5%	2.5%	2.5%
30		Appreciation in Value per Year									
31			Equipment			-15.0%					
32			Building			-4.0%					
33			Land			3.0%					
34											
35	Income Assumptions & Forecast										
36				Period	0	1	2	3	4	5	6
37			Initial Incremental Sales		0	0	15000				
38			Sales Growth					0	0	0	0
39			Operating Costs/Sales				75.0%				
40			Tax Rate			40.0%					
41											
42	Financing Assumptions										
43				Period	0	1	2	3	4	5	6
44			Debt Financing (%)		80.0%	80.0%	80.0%				
45			Debt Term to Maturity		6	5	4				
46			Interest Rate		7.0%	7.0%	7.0%				
47			Equity Financing (%)		20.0%	20.0%	20.0%				
48											

EXHIBIT 10.5   Investment evaluation sheet: Data input.

evaluate them must be estimated. This is the job of the Data Processing worksheet. Fortunately, both evaluation methods share common inputs, so the raw data need only be processed once. Using the processed data as input, the Evaluation sheet estimates future cash flows to all investors and to equity, and then calculates their NPVs and internal rates of return. All that is left is for the investors to interpret the results.

10.3.2.1 Data Input

The Data Input sheet, shown in Exhibit 10.5, contains four sections. The Global Assumptions and Data section lists the number of periods of initial investment and the horizon (cells F8 and F7), market data for calculating discount rates (cells F11:F14), and current assets and liabilities expressed as a percent of sales (cells K11:K14). The latter ratios allow the model to calculate the investment in NWC associated with any given sales forecast. The Fixed Asset Assumptions section contains the amounts to be invested in equipment, building, and land in each period (lines 18–25), depreciation rates (lines 27–29), and asset appreciation rates (lines 30–33). The latter rates are used to calculate the terminal asset values. The Income Assumptions and Forecast section contains sales and cost forecasts (lines 35–40). Finally, the Financing Assumptions section contains assumptions about financing (lines 42–47). The data from the Data Input sheet feed into the Data Processing and Evaluation sheets for processing, leading to the ultimate goal of determining the project's NPV and IRR from the perspectives of invested capital and equity.

	A	B	C	D	E	F	G	H	I	J	K
1						Stiliko Corporation					
2						Investment Evaluation					
3						Data Processing					
4											
5	Initial Investment										
6					Period	0	1	2	3	4	5	6
7	Depreciable Assets										
8		Equipment			3,000	1,500					
9		Building			1,500	1,500					
10	Non-Depreciable Assets										
11		Land			1,000						
12	Total Fixed Asset Investment				5,500	3,000					
13											
14	Net Working Capital Investment				-	-	1,500				
15	Incremental Project Investment				5,500	3,000	1,500				
16		Total Cumulated Capital Investment			5,500	8,500	10,000				
17											
18	Working Capital Accounts			Period	0	1	2	3	4	5	6
19		Cash					150	150	150	150	150
20		Accounts Receivables					750	750	750	750	750
21		Inventory					900	900	900	900	900
22		Current Assets					1,800	1,800	1,800	1,800	1,800
23		Accounts Payable					300	300	300	300	300
24		Net Working Capital					1,500	1,500	1,500	1,500	1,500
25		Net Working Capital Investment					1,500	-	-	-	-
26											
27	Asset Accounting										
28	Depreciation Schedule			Period	0	1	2	3	4	5	6
29		Equipment					1,485	2,025	675	315	-
30		Building					75	75	75	75	75
31		Total					1,560	2,100	750	390	75
32											
33	Asset Book Value			Period	0	1	2	3	4	5	6
34		Equipment			3,000	4,500	3,015	990	315	-	-
35		Building			1,500	3,000	2,925	2,850	2,775	2,700	2,625
36		Land			1,000	1,000	1,000	1,000	1,000	1,000	1,000
37		Total Book Value of Fixed Assets			5,500	8,500	6,940	4,840	4,090	3,700	3,625
38											
39	Asset Market Value			Period	0	1	2	3	4	5	6
40		Equipment			3,000	4,500	3,825	3,251	2,764	2,349	1,997
41		Building			1,500	3,000	2,880	2,765	2,654	2,548	2,446
42		Land			1,000	1,030	1,061	1,093	1,126	1,159	1,194
43		Total Market Value of Fixed Assets			5,500	8,530	7,766	7,109	6,543	6,056	5,637
44											
45	Terminal Value			Period	0	1	2	3	4	5	6
46		Net Working Capital Recovery									1,500
47		Sale of Fixed Assets									5,637
48		Fixed Asset Book Value									3,625
49		Gain of Sale of Fixed Assets									2,012
50		Tax on Gain									805
51		Terminal Flow net of Taxes									6,332

EXHIBIT 10.6A   WACC investment evaluation. Data processing sheet 1.

	A	B	C	D	E	F	G	H	I	J	K
54	Debt Accounting										
55				Period	0	1	2	3	4	5	6
56		Amount Borrowed			4,400	2,400	1,200				
57		Rate			0.07	0.07	0.07				
58		Term			6	5	4				
59		Debt 0 Payment				923	923	923	923	923	923
60		Interest				308	265	219	170	117	60
61		Principal				615	658	704	754	806	863
62		Balance				3,785	3,127	2,423	1,669	863	0
63		Debt 1 Payment					585	585	585	585	585
64		Interest					168	139	108	74	38
65		Principal					417	447	478	511	547
66		Balance					1,983	1,536	1,058	547	0
67		Debt 2 Payment						354	354	354	354
68		Interest						84	65	45	23
69		Principal						270	289	309	331
70		Balance						930	641	331	-
71		Totals									
72		Debt Service				923	1,508	1,863	1,863	1,863	1,863
73		Interest				308	433	442	342	236	122
74		Principal				615	1,075	1,421	1,521	1,627	1,741
75		Balance				3,785	5,109	4,888	3,368	1,741	0
76											
77	Income Forecast										
78				Period	0	1	2	3	4	5	6
79		Sales				-	15,000	15,000	15,000	15,000	15,000
80		Operating Costs				-	11,250	11,250	11,250	11,250	11,250
81		EBITDA				-	3,750	3,750	3,750	3,750	3,750
82		Depreciation				-	1,560	2,100	750	390	75
83		EBIT				-	2,190	1,650	3,000	3,360	3,675
84		Interest				308	433	442	342	236	122
85		EBT				(308)	1,757	1,208	2,658	3,124	3,553
86		Tax				(123)	703	483	1,063	1,250	1,421
87		Net Income				(185)	1,054	725	1,595	1,875	2,132

EXHIBIT 10.6B   WACC investment evaluation. Data processing sheet 2.

	A	B	C	D	E	F	G	H	I	J	K
89	Financing										
90				Period	0	1	2	3	4	5	6
91		Amount Borrowed			4,400	2,400	1,200				
92		Equity Capital Investment			1,100	600	300				
93		Additional Equity Investment				800	-	-	-	-	-
94		Project Capital Investment			5,500	3,800	1,500				
95											
96	Costs of Capital										
97		Total Invested Capital			10,800						
98		Total Contributed by Equity			2,800						
99		Total Contributed by Lenders			8,000						
100		Proportion of Debt			74.07%						
101		Proportion of Equity			25.93%						
102		Cost of Debt			7.0%						
103		Cost of Equity									
104			Kf		4.0%						
105			Km - Kf		5.0%						
106			Bu		1.0						
107			BL		2.71						
108			Ke		17.57%						
109		WACC			7.67%						

EXHIBIT 10.6C   WACC investment evaluation. Data processing sheet 3.

10.3.2.2 Data Processing Sheet

The Data Processing sheet contains the formulas and links needed to process the raw data from the Data Input sheet for input into the Evaluation sheet. It is organized in six sections and structured, so the calculations flow from the top to the bottom. The sections are shown in Exhibits 10.6A through 10.6C: Part A shows the Initial Investment and Asset Accounting sections (lines 1–51); Part B shows the Debt Accounting and Income Forecast sections (lines 54–87); and Part C shows the Financing and Costs of Capital sections (lines 89–109). Once the Data Processing sheet is complete, the processed data are fed into the appropriate cash flow section in the Evaluation sheet.

10.3.2.3 Initial Investment

This section (lines 5–25) determines the total outlays for fixed assets and the increase in NWC for each period. The outlays for fixed assets (cells E8:F12) are taken from the Data Input sheet. The NWC investment (line 14) links to the Working Capital Accounts sector (lines 18–25). NWC requirements (in thousands of dollars) consist of the following items:

$$Cash_2 = \left(\frac{Cash}{Sales}\right) \times Sales_2 = (1\%) \times 15 \text{ million} = 150 \text{ (line 19)}$$

$$Accounts\ Receivables_2 = \left(\frac{Accounts\ Receivables}{Sales}\right) Sales_2$$
$$= (5\%)15 \text{ million} = 750 \text{ (line 20)}$$

$$Inventory_2 = \left(\frac{Inventory}{Sales}\right) Sales_2 = (6\%)15 \text{ million} = 900 \text{ (line 21)}$$

Less

$$AP_2 = \left(\frac{AP}{Sales}\right) Sales_2 = (2\%)15 \text{ million} = 300 \text{ (line 23)}$$

Yielding

$$NWC_2 = 1500 \text{ (line 25)}$$

Stiliko's project will require an investment in NWC of $1.5 million at time $t = 2$. The investment in net work capital is the change in NWC, calculated as

$$NWC\ Investment_t = (CA_t - AP_t) - (CA_{t-1} - AP_{t-1}), \qquad (10.13)$$

where NWC denotes NWC, CA the current assets, and AP the accounts payable. Since we have assumed that Stiliko has no working capital prior to the start of the project, the total NWC represents the incremental amount that must be invested in year 2. If we were dealing with an ongoing firm with previous working capital balances, the investment in NWC would be the increment required should the project be undertaken.

10.3.2.4 Accounting Sections

The next sections of the Data Processing sheet are Asset Accounting (lines 27–51, Exhibit 10.6A) and Debt Accounting (lines 54–75, Exhibit 10.6B). These sections track the asset and debt values needed by the Terminal Value sector (lines 45–51), the Income Forecast section (lines 77–87), and the operating cash flow sections of the Evaluation sheet.

10.3.2.4.1 Asset Accounting. The Asset Accounting section tracks the depreciation, book values, and market values of the assets used in the project. Depreciation expense (lines 28–31) is based on the MACRS and 40-year straight-line percentages specified in the Data Input sheet. The book value of each type of asset, such as equipment, is tracked as

$$\text{Equipment}_t = \text{Equipment}_{t-1} + \text{Equipment Investment}_t$$
$$- \text{Equipment Depreciation}_t.$$

For period 2 this is

$$G34 = F34 - G29$$

$$3015 = 4500 - 1485.$$

The book value of the building is tracked in the same way. As land is not depreciated, its book value remains equal to the original purchase price.

The market values of the assets are based on their forecast rates of appreciation (a negative appreciation rate means the asset is depreciating). These rates, provided in the Data Input sheet, are −15% for equipment (F31), −4% for the building (F32), and +3% for the land (F33). For example, the market value of the equipment at the end of period t is

$$\text{Equipment Value}_t = \text{Equipment Value}_{t-1} \times (1 + \text{Equipment}$$
$$\text{Appreciation Rate}).$$

For time $t = 2$, this is

$$G40 = F40 \times (1 + \text{"Data Input"!F31}),$$

$$3825 = 4500 \times [1 + (-0.15)].$$

Equations for the building and land are similar.

The last sector in the Asset Accounting section calculates the project's terminal value, the cash flows at the end of the project. The figures in this sector come from the previous sectors of the Asset Accounting section. NWC recovery is the NWC balance at the end of the project. The proceeds from the sale of fixed assets at the terminal date (cell K47) are the sum of the market values of equipment, building, and land (cells K40:K42). It is assumed that each asset category is sold at its appreciated market value at the end of the project. The difference between the sale prices of the assets and their book values (cells K34:K36) is the net gain or loss. If the assets are sold at a gain, the gain is taxed; if they are sold at a loss, the loss reduces taxes. The terminal cash flows, net of taxes, are summarized in cell K51.

10.3.2.4.2 Debt Accounting.   Stiliko will borrow a total of $8 million over three years: $4.4 million at time 0, $2.4 million at time 1, and $1.2 million at time 2. Each issue will bear an interest rate of 7%, mature at the end of year 6, and be serviced by equal annual payments. Borrowing terms are specified on lines 42–48 of the Data Input sheet. The Debt Accounting section (lines 54–75, Exhibit 10.6B) calculates the total payments, interest expense, principal payments, and balance for each of the debt issues, based on the borrowing terms given.

The annual payment for each issue is calculated with the function

$$\text{Payment} = \text{PMT}(\text{Interest Rate, Term to Maturity, Amount of Loan}). \tag{10.14}$$

The interest portion of the payment can be calculated either as

$$\text{Interest}_t = \text{Balance}_{t-1}(\text{Interest Rate}), \tag{10.15}$$

or using Excel®'s interest function as

$$\text{Interest}_t = \text{IPMT}(\text{Interest Rate, Period in the Life of Loan,} \\ \text{Term to Maturity, Amount of Loan}). \tag{10.16}$$

If you use the IPMT function, "Period in the Life of the Loan" is the number of the period when they are numbered in ascending order from beginning to maturity. Since the period changes in each column of the model, the way this is done is to use the COLUMNS function. For example, for the loan taken out at $t = 0$, the interest in period 2 is

$$\text{G60} = \text{IPMT}(\text{E57,E58} - \text{COLUMNS}(\text{H49:K49}),\text{E58,E56}),$$

where COLUMNS(H49:K49) counts the number of columns remaining from H to K (the number of periods remaining after $t = 2$ until the loan matures) and where the other cell references are

$$E57 = \text{Interest Rate,}$$

$$E58 = \text{Term to Maturity,}$$

and

$$E56 = \text{Amount of Loan.}$$

The principal portion of the annual payment is calculated in each case with the relation

$$\text{Principal Payment}_t = \text{Payment}_t - \text{Interest}_t. \qquad (10.17)$$

The principal balance for each debt issue is

$$\text{Balance}_t = \text{Balance}_{t-1} - \text{Principal Payment}_t. \qquad (10.18)$$

The last part of the debt accounting (lines 72–75) totals the total payments, interest expenses, principal payments, and balances for the three debt issues. Total interest (line 73) feeds into the Income Forecast section as interest expense (line 84); Total principal (line 74) feeds into the equity cash flow calculations in the Evaluation worksheet (line 33). The Total balance (line 75) feeds into the terminal flows to equity in the case when, at the end of the project, the remaining debt balance must be repaid before stockholders can receive their terminal cash flow. In this particular project, however, the debt will have been completely repaid by the end of the project.

10.3.2.4.3 Income Forecast.   The Income Forecast section (lines 77–87) calculates expected sales based on the specified initial sales (line 37, Data Input) and sales growth rates (line 38, Data Input). Operating costs are a percent of sales (line 39, Data Input); interest expense is drawn from the Debt Accounting section above (line 73); and depreciation expense is taken from the depreciation schedule in the Asset Accounting section (line 31, Exhibit 10.6A). Taxes are calculated based on the tax rate specified in cell F40 in the Data Input sheet. The Income Forecast section provides almost all the basic information required to calculate the operating cash flows to invested capital and to equity in the Evaluation worksheet.

10.3.2.4.4 Financing Section.   The project's capital investment to be provided by the lenders and the equity investors in each period is shown in the Initial Investment section (line 15 in Exhibit 10.6A). Absent any complications, the capital outlays for Stiliko would be $5.5 million, $3 million, and $1.5 million at times 0, 1, and 2, respectively, for a total of $10 million. With the 80% debt target, the amount of debt would be $8 million and equity would be $2 million. However, there is a complication: a cash shortfall of $0.8 million in period 1. The shortfall results from a negative net income of $185,000 and a debt principal payment of $615,000. The funds to cover this shortfall must be provided by someone. We have assumed that operating losses will be covered by the equity investors, making equity's proportion of the financing the sum of their direct project investment and the cash shortfall. The total funds that will have to be provided by the equity investors at $t = 1$ is calculated (in thousands) as

$$
\begin{aligned}
\text{Total Equity Capital Investment} &= \text{Equity Capital Investment} \\
&\quad + \text{Excess Equity Investment} \\
&= \text{F92} + \text{F93} \\
&= 600 + 800 \\
&= \$1400.
\end{aligned}
$$

Including the additional $0.8 million needed by the project, the project's total capital investment is $10.8 million (cell E97). Because of the additional equity investment, the proportional amounts of debt and equity will differ slightly from the target (80/20)% mix. The proportion of debt is calculated in E100 as

Proportion of Debt

$$
= \frac{\displaystyle\sum_{t=0}^{2}\text{Amount Borrowed}_t}{\displaystyle\sum_{t=0}^{2}\text{Amount Borrowed}_t + \sum_{t=0}^{2}\text{Equity Capital Investment}_t}
$$

$$
\text{E100} = \frac{\text{Sum(E91:G91)}}{\text{Sum(E91:G91)} + \text{Sum(E92:G93)}}
$$

$$
74.07\% = \frac{8000}{8000 + 2800}.
$$

The proportion of equity is calculated similarly in E101 based on the total equity investment at times 0, 1, and 2, and comes to 25.9%.

10.3.2.5 Costs of Capital Section

10.3.2.5.1 Costs of Capital. There are two components to Stiliko's costs of capital: the cost of debt and the cost of equity. Two cash flows will be evaluated in the Evaluation sheet—the cash flow to invested capital and the cash flow to equity. The former uses the WACC; the latter the cost of equity (k_E).

The WACC uses both costs of capital as inputs. The cost of debt is relatively easy. Stiliko expects to borrow its funds at $k_D = 7\%$. The cost of equity, however, is more complex. It is based on the CAPM formula

$$k_E = k_f + (k_m - k_f)\beta. \tag{10.19}$$

The risk-free interest rate is 4% and the market risk premium ($k_m - k_f$) is estimated to be 5%. However, β is more difficult. It should reflect the amount of debt used to finance the project. We assumed the systematic risk of the project equals that of the market, and given the debt–equity mix we have calculated, the beta coefficient for this project is estimated to be 2.714. Therefore, the cost of equity capital is

$$k_E = 4\% + (5\%)2.714 = 17.57\%,$$

as shown in E108.

With these estimates of capital proportions and costs of debt and equity, the WACC is calculated in E109 as

$$k_A = \left(\frac{\text{Debt}}{\text{Total Capital}}\right) \times k_D \times (1 - T) + \left(\frac{\text{Equity}}{\text{Total Capital}}\right) \times k_E$$

$$E105 = E100 \times E102 \times (1 - \text{"Data Input"!F42}) + E101 \times E108$$

$$7.67\% = (0.7407) \times 7\% \times (1 - 0.40) + (0.2593) \times 17.57\%.$$

We now have calculated all of the input required to estimate and evaluate the cash flows to invested capital and to equity. We do this in the Evaluation sheet, considered next.

10.3.2.6 Evaluation Sheet

Exhibit 10.7 shows the Evaluation sheet that contains the Invested Capital and Equity evaluation methods. We consider the Invested Capital method first.

10.3.2.7 The Invested Capital Method

Evaluation of the project using the WACC as the discount rate is also referred to as the Invested Capital method. This method is shown in lines

	A	B	C	D	E	F	G	H	I	J	K
1					Stiliko Corporation						
2					Investment Evaluation						
3					Invested Capital and Equity Evaluation Modules						
4											
5	Invested Capital Perspective										
6					Period	0	1	2	3	4	5	6
7		Initial Costs										
8		Capital Invested				5,500	3,000	-	-	-	-	-
9		Net increase in working capital						1,500				
10		Cash Flows from operations										
11		EBIT(1 - Tax)					-	1,314	990	1,800	2,016	2,205
12		Depreciation					-	1,560	2,100	750	390	75
13		Net cash flows from operations					-	2,874	3,090	2,550	2,406	2,280
14		Terminal Value					-	-	-	-	-	6,332
15		Net cash flows to invested capital				(5,500)	(3,000)	1,374	3,090	2,550	2,406	8,612
16												
17		WACC	7.67%		PV(Operating cash flow)			$ 9,979				
18		NPV	$ 4,464		PV(Terminal value)			$ 4,065				
19		IRR	19.3%		PV(Operating flow + terminal flow)			$ 14,044				
20					PV(Initial cost)			$ 9,580				
21					NPV(Project)			$ 4,464				
22												
23												
24												
25	Equity Perspective										
26					Period	0	1	2	3	4	5	6
27		Initial Costs										
28		Capital Invested				1,100	600					
29		Net increase in working capital						300				
30		Cash Flows from operations										
31		Net income					(185)	1,054	725	1,595	1,875	2,132
32		+ Depreciation					-	1,560	2,100	750	390	75
33		- Payment to debt principal					615	1,075	1,421	1,521	1,627	1,741
34		Net Cash Flows from operations					(800)	1,539	1,404	824	638	466
35		Terminal Value					-	-	-	-	-	6,332
36		Net cash flows to invested capital				(1,100)	(1,400)	1,239	1,404	824	638	6,798
37												
38		Ke	17.57%		PV(Operating Flow)			$ 2,188				
39		NPV	$ 2,758		PV(Terminal value)			$ 2,397				
40		IRR	45.0%		PV(Operating flow + terminal flow)			$ 4,586				
41					PV(Initial cost)			$ 1,827				
42					NPV(Project)			$ 2,758				
43												

EXHIBIT 10.7   Investment evaluation. Evaluation sheet.

5–21. It begins by calculating the costs and cash flows to all investors, then calculating the NPV and IRR for these flows. To calculate the appropriate cash flows, this section pulls the following data from the Data Processing sheet: capital invested and the increase in NWC (line 15), depreciation expense and EBIT (lines 82 and 83), terminal cash flow (cell K51), and the WACC (cell E109). The tax rate is pulled from the Data Input sheet (cell F40). Using this data, the net cash flows to all investors is calculated in line 15.

All that remains is to calculate the NPV and IRR for these cash flows. The NPV for the project is calculated in cell C18, as follows:

$$\text{NPV}_{\text{Project}} = \text{Net Cash Flows}_0 + \sum_{t=1}^{6} \frac{\text{Net Cash Flows}_t}{(1 + k_A)^t}$$

(10.20)

C18 = E15+NPV(C17,F15:K15) = $4464.

The NPV function is of the form $=NPV(rate, CF_1:CF_h)$, where rate in this case refers to the discount rate, k_A (cell C17) and F15:K15 is the cell range that contains the cash flows.

It is very important to note that the Excel function $=NPV(\ )$ is inconsistent with the way we normally define NPV in the finance field. The proper definition of NPV includes the cash flows at time $t = 0$ (usually outflows), and does not discount the initial flows at $t = 0$. However, the Excel NPV function assumes that the first cash flow in its series is to be discounted by one period. Because of this assumption, we must treat the capital expenditure at $t = 0$ separately so that it is not discounted. That is why we show E15, the investment at $t = 0$, outside the NPV function. The first capital expenditure to be discounted occurs at $t = 1$, so it is the first in the series F15:K15. The project has a positive NPV, meaning that it generates enough cash to return the investors' initial capital and exceed their required rates of return.

There are three components to the project's NPV: the present values of the operating cash flows, terminal value, and initial costs. These are summarized in cells E17:H21. The present value of the operating flows from the project is calculated in cell H17 as

$$PV_{\text{Operating Flow}} = \sum_{t=1}^{6} \frac{\text{Earnings Flow to Invested Capital}_t}{(1 + k_A)^t}$$

$$H17 = NPV(C17, F15:K15).$$

(10.21)

The present value of the terminal flows is calculated in cell H18 as

$$PV_{\text{Terminal Flow}} = \frac{\text{Terminal Flow Net of Taxes}_6}{(1 + k_A)^6}$$

$$H18 = \frac{K14}{(1 + C_{17})^{k6}},$$

(10.22)

where K6 is the number of period to the end of the project, and K14 is the Terminal Flow.

Cell H19 is the present value of the project's cash flows excluding its initial costs. It represents the value of the project before subtracting the present value of the costs. That is, the value of the project is

$$\text{Project Value} = PV(\text{Operating Flow} + \text{Terminal Flow})$$
$$= PV(\text{Operating Flow}) + PV(\text{Terminal Flow})$$

(10.23)

$$H19 = H17 + H18$$
$$\$14,044 = 9979 + 4065.$$

Cell H20 calculates the present value of the initial outlays for the project as

$$PV_{\text{Initial Cost}} = \text{Capital Expenditure}_0 + \sum_{t=1}^{6} \frac{\text{Capital Expenditure}_t}{(1 + k_A)^t} \tag{10.24}$$

H20 = E8+NPV(C17,F8:K9).

The present value of the project costs is \$9580. Subtracting these from the project's cash flows yields the NPV of the project, as follows:

$$NPV = PV(\text{Op Flow} + \text{Terminal Flow}) - PV(\text{Initial Cost})$$

$$H21 = H19 - H20 \tag{10.25}$$

$$\$4464 = 14{,}044 - 9580.$$

The project value exceeds its cost by \$4464, so it is worthwhile. The NPV represents the addition to the value of the firm that should result from the adoption of the project. Thus, the NPV of \$4464 is how much higher the value of Stiliko Plastics should be with the project as compared to not adopting the project.

The other measure of the acceptability of the project is the IRR of the project. The IRR is solved as the discount rate that will equate the present value of the project's cash inflows with the present value of the pro-ject's costs. In other words, IRR is the discount rate (r) that makes the present value of all the project cash flows (inflows and outflows) equal to zero, thus,

$$\sum_{t=0}^{6} \frac{\text{Cash Flow to Invested Capital}_t}{(1 + r)^t} = 0. \tag{10.26}$$

The Total Net Flow to Invested Capital in row 15 (E15:K15) is the summary flow that is relevant for calculating IRR, so the IRR is calculated in C19 as

$$C19 = IRR(E15:K15),$$

where IRR(.) is the function that solves Equation 10.26. We see that the IRR for the project is 19.3%. This figure is to be compared with the firm's cost of capital, $k_A = 7.67\%$ in cell C17. The IRR exceeds the cost of capital, so the project is acceptable.

10.3.2.8 Equity Method

The structure of the Equity Evaluation model is the same as that for the WACC method. However, we calculate the cash flows differently and discount them using a different discount rate. The equity method evaluates the cash flows to equity as shown in Exhibit 10.3. These cash flows are discounted by the cost of equity capital, the equity investors' required rate of return. The resulting NPV represents the value of the project to the equity investor.

In the case of Stiliko Plastics, the equity method evaluates the project from the viewpoint of Flavio Stiliko and the other equity investors. The equity investors will have to contribute $2 million to help finance the purchase of project assets. In addition, equity investors will have to cover the negative equity flow of about $800,000,000 in year 1. For committing their capital and bearing much of the project's risk, the equity investors will want a return on their investment of about 17.6%. This is the rate that will be used to discount the equity cash flows to determine the NPV of the project.

The Equity Evaluation section is shown by lines 25–42 in Exhibit 10.7. It is identical to the Invested Capital section in almost all respects, differing only in that it calculates equity cash flow and discounts at the cost of equity. Otherwise, it uses the same equations as the other section and it draws its basic data from the Data Processing sheet. Lines 30–34 show the calculation of equity cash flow, following the format of Exhibit 10.3. As before, terminal value is pulled from cell K51 on the Data Processing sheet. It matches the Invested Capital section because the lenders will have been repaid when the project is ended, so the stockholders will receive the entire terminal payoff. Lines 38–42 calculate the NPV as the present value of the cash flows to equity, using the cost of equity as the discount rate. The NPV for the equity investors is $2758 (thousand); their IRR is 45.0%.

The equity NPV of $2.758 million means that the value of the project exceeds its cost by $2.758 million. The value of the project, $4.586 million, is the present value of the cash flows that will be earned by the stockholders. However, to earn these cash flows, the stockholders will have to invest their capital. They will invest a total of about $2 million consisting of $1.1 million at $t = 0$, $0.6 million at $t = 1$, $0.3 million at $t = 2$, and cover the $0.8 million outflows at $t = 1$. The NPV of their capital investment, $1827, is the equity cost of the project. Thus, the amount that the value exceeds the cost, in present value terms, is

Equity value of project	$4.586 million
Equity cost of project	1.827 million
Equity NPV	2.759 million.

This analysis indicates that the project will increase equity value by $2.759 million and will be attractive to the investors whose required rate of return is 17.57%.

10.3.3 Evaluating the Investment's Impact on the Firm

In addition to evaluating the NPV of the investment by itself, it may be important to assess the effect of the investment on the financial position of the firm as a whole, and to see how it fits into the broader set of policies and strategies of the firm. In some cases an investment project that is attractive when evaluated in isolation with the NPV criterion will have undesirable effects on other aspects of the firm. Management will want to know how the investment will affect the whole firm's sales, operating costs, earnings, ability to service debt, liquidity, and equity position. In addition, the financing for the project will affect the whole firm's financial position, and, in turn, the firm's financial mix will affect the cost of capital for the project. Thus, the effect of the investment on the firm's financial position should be explored by integrating the investment decision model with the broader model of the firm.

We can explore the impact of the proposed investment project on the firm as whole by linking the project investment model with the various modules in the firm planning model. Most of the links that we need are straightforward—we add the project's sales and expenses to those the firm would have without the project, and we add the acquired assets and issued liabilities to those of the whole firm. In evaluating a capital investment project, both the cost of equity and the WACC should reflect the capital structure mix for the whole firm that results from adding the project financing to the firm's existing capital structure. For example, suppose that O&R Corp. is an ongoing firm that is considering investing in the same plastic project that Stiliko has been evaluating. O&R has an existing capital structure consisting of the components shown in the abbreviated balance sheet (Exhibit 10.8) that shows both book values and market values of the capital structure prior to adopting the proposed project. For purposes of calculating costs of capital, we focus on the firm's invested capital—its interest-bearing debt and the equity. The bottom part of the table shows the invested capital components. The interest-bearing debt (both short-term and long-term), which has a book value of $40.2 million, has a market value of $45.9 million because the long-term bonds are trading at premium. The book value of equity is $1493.9 million, but with O&R's common stock trading at a price of $9.885 for each

Book and Market Values (in millions of dollars)

	Book Value	Market Value
Current Assets	710.0	
Net Fixed Assets	1424.1	
Total Assets	2,134.1	1634.4
Non-Interest Bearing Liabilities	600.0	600.0
Short-term Debt	0	0
Long-term Debt	40.2	45.9
Equity	1493.9	988.5
Total Liabilities & Equity	2134.1	1634.4
Invested Capital		
Interest Bearing Debt (ST + LT)	40.2	45.9
Equity	1493.9	988.5
Invested Capital	1534.1	1034.4

EXHIBIT 10.8   O&R assets and capital structure.

of its 100 million outstanding shares, the aggregate value of the stock is $988.5 million.

With the present capital structure O&R's cost of debt is $k_d = 7\%$, the cost of equity is $k_E = 8.9\%$, and the WACC is $k_A = 8.69\%$. This WACC is based on the following calculations. The capital proportions in the WACC are based on the market values of the invested capital components. Even though, in book value terms, the debt-to-total invested capital ratio is

$$\frac{\text{Debt}}{\text{Debt} + \text{Equity}} = \frac{40.2}{40.2 + 1493.9} = \frac{40.2}{1534.1} = 2.6\%,$$

in market value terms, we have

$$\frac{\text{Debt}}{\text{Debt} + \text{Equity}} = \frac{45.9}{45.9 + 988.5} = \frac{45.9}{1034.4} = 4.4\%.$$

With the capital proportions based on market values, O&R's WACC is

$$k_A = \left(\frac{\text{Debt}}{\text{Total Invested Capital}}\right) k_d (1 - T) + \left(\frac{\text{Equity}}{\text{Total Invested Capital}}\right) k_E$$

$$= \left(\frac{45.9}{1034.4}\right) 7\%(1 - 0.40) + \left(\frac{988.5}{1034.4}\right) 8.9\%$$

$$= (0.044)4.2\% + (0.956)8.9\% = 8.7\%.$$

The cost of equity is estimated using the CAPM based on a risk-free rate of 4%, a market risk premium of 5%, and O&Rs' stock beta of 0.98:

$$k_E = k_f + (k_m - k_f)\beta = 4\% + (5\%)0.98 = 8.9\%.$$

If O&R adopts the proposed project that costs $10 million, assume that it will finance it with the same 80% debt and 20% equity mix that we discussed for Stiliko. This would require that O&R borrow $8 million and provide $2 million equity to get the project started. Since O&R's resources are fully committed to their existing operations, the equity component will require that O&R issue new common stock to raise the equity portion of project financing. Raising this new capital will change the capital structure mix for the company and possibly change its cost of capital.

Assume that O&R can borrow the required $8 million at 8%. The $2 million of new equity capital will be raised by selling new common shares. The current stock price is $9.885 per share, but the flotation costs of selling new shares will reduce the proceeds from the issue to $9.00 per share. This will require that 222,222 new shares be issued ($2,000,000/9.00 = 222,222 shares).

After raising the new capital, the capital structure will be as shown in Exhibit 10.9. Note that when we add the new financing to the existing financial structure, the debt-to-capital ratio changes from 2.6% to 3.1% in book value terms, and from 4.4% to 5.2 % in market value terms.

When we calculate the WACC for evaluating the proposed investment project, we need to use the new market value capital proportions for the

	Book Value		Market Value	
		Proportion		Proportion
Debt				
Short-term	0		0	
Long-term – existing	40.2		45.9	
Long-term – new	8.0		8.0	
Total Debt	48.2	3.1%	53.9	5.2%
Equity – existing	1493.9		988.5	
Equity – new	2.0		2.0	
Total Equity	1495.9	96.9%	990.5	94.8%
Total Invested Capital	1544.1	100%	1044.4	100%

EXHIBIT 10.9 O&R capital structure with new project financing (in thousands of dollars).

firm as a whole. That is, we use the 5.2%/94.8% mix shown in Exhibit 10.9. In this way, we are treating the project's financing as if we obtain some new capital, mix it with the old capital, and use this combined mix to purchase the new project.

Even though we include both the existing and new capital in the calculation of the proportions, our costs of capital should be the marginal costs for the project by itself. The marginal cost of borrowing is the current rate, $k_d = 8\%$ for the additional debt, regardless of what the historical rate that the old debt carries. For example, O&R's existing long-term debt has a coupon rate of 9%, and prior to this project, was trading at a yield to maturity of 7%. However, both the 9% embedded coupon cost and the prior market yield of 7% are irrelevant for calculating the cost of capital. To issue new debt, the cost will be 8%, and this marginal cost is the relevant cost for calculating the cost of capital.

We also use the current marginal cost of equity in WACC. This marginal cost should not only reflect current market conditions, but it also should be based on the equity risk of the project being evaluated. Prior to adopting the new project, the O&R's stock had an unleveraged beta of 0.958, a leveraged beta of 0.98, and a cost of equity of 8.9%. However, this cost of capital is not used to evaluate the new project. The cost of equity for the project should be based on the risk of the project itself. The project, by itself, with no debt financing is expected to generate equity returns that have volatility that is equal to that of the market in general. That is, the project, with *no debt financing*, has a beta of 1.0. However, O&R will use some debt to finance the project so that debt must be considered in the calculation of the marginal cost of equity. The way we do this is to calculate a leveraged beta that reflects O&R's capital structure mix for the firm as whole. When the new financing is combined with O&R's existing financing, the market value debt-to-equity ratio will be

$$\frac{\text{Debt}}{\text{Equity}} = \frac{53.9}{990.5} = 0.054 \ (5.4\%).$$

We use this new, total firm debt-to-equity to calculate the leveraged beta for the project:

$$\beta_L = \beta_U \left[1 + \left(\frac{\text{Debt}}{\text{Equity}} \right) (1 - T) \right]$$

$$\beta_L = 1.0[1 + (0.054)(1 - 0.40)] = 1.032,$$

and the leveraged beta is used to calculate O&R's marginal cost of equity capital for the project:

$$k_E = k_f + (k_m - k_f)1.032 = 4\% + (5\%)1.032 = 9.16\%.$$

Using the marginal costs of debt and equity and the whole firm capital structure proportions from the table above, O&R's WACC for evaluating the project is

$$k_A = \left(\frac{\text{Debt}}{\text{Invested Capital}}\right)k_d(1 - T) + \left(\frac{\text{Equity}}{\text{Invested Capital}}\right)k_E$$

$$= \left(\frac{53.9}{1044.48\%}\right)(1 - 0.40) + \left(\frac{990.5}{1044.4}\right) \times 9.16\%$$

$$= (0.052) \times 4.8\% + (0.948) \times 9.16\% = 8.69\%.$$

Note that the WACC is really a weighted average of marginal costs.

For evaluating the project from O&R's viewpoint, the cash flows are calculated the same way as for Stiliko, but we use O&R's WACC of 8.69%. At this cost of capital, the NPV of the project is $3.95 million, as compared to the NPV for Stiliko of $4.464 million, which was based on Stiliko's WACC of 7.67%. The project is attractive for both firms, but is slightly more valuable to Stiliko because of their lower cost of capital. Stiliko's WACC is lower because it is using proportionally more debt to finance the project. While both firms are borrowing the same amount for the project, and the direct equity is the same in both cases, O&R is implicitly committing more equity to the project because its existing equity is backing up the project.

It is ironic that by committing more equity to the project and seeming to have less risk, O&R has a higher cost of capital. The reason for this is that Stiliko is using proportionally more of the lower cost source of funds—debt. Because of the greater financial leverage, Stiliko's marginal cost of equity is considerably higher than O&R's (17.57% versus 9.16%). Nevertheless, Stiliko's cost of equity is not so high that it offsets the advantage of using a greater proportion of debt.

The estimation of NPV of an investment project is the most widely advocated method for evaluating the acceptability of investment projects. While we have used it to evaluate a proposal to invest in new plant and equipment, the method applies to evaluating any type of long-term investment. The investment would not have to involve fixed assets. It could be a softer type investment such as a new marketing program or personnel training. These are investments in less tangible assets. The marketing program generates sales demand and brand identity, which is a type of asset; and the personnel training program increases the productivity of the employees. Both would have long-term benefits and potentially increase the firm's cash flow. Thus, both would have costs and benefits that could be evaluated in terms of NPV.

10.4 MODELING MERGERS AND ACQUISITIONS

10.4.1 Overview

Mergers and acquisitions (M&A) have become an important part of managing a business in recent years. A staple of the American business scene, they have become increasingly common in Europe, both within and across borders. M&A presents interesting, complex, and challenging modeling problems. This chapter provides a brief overview of the M&A process and then suggests several ways to model the acquisition decision taking uncertainty into account.

A primary attraction of M&A is that they offer many potential ways to increase firm value. Commonly cited sources of enhanced value that are amenable to modeling include the operating synergies that result from combining the buyer's business with the target's business in a way that improves the performance of both operations, the transfer of proprietary technology or unique skills, the use of complementary resources, and the makeover of an acquired operation. However, M&A can also destroy value. Sources of value destruction that are amenable to modeling include simply paying too much and the failure to identify and maintain the value drivers of the transaction.

The track record of M&A is not encouraging. Though M&A activity creates value on average, most acquisitions do not create value for the acquiring company's shareholders. Indeed, many acquisitions lose money and become significant drains on management resources. Financial modeling can help, especially when it includes Monte Carlo simulation. Many assumptions and projections are made when analyzing a potential acquisition. A financial model can identify which of the assumptions and projections must go right, and which cannot go wrong, if the acquisition is to succeed.

The M&A process is remarkably complex. Much time and expense goes into just identifying potential targets worthy of further analysis.*

---

* Hooke (1997) and Bruner (2004) offer excellent practical overviews of the merger and acquisition process.

Our focus in this section is limited to building a financial statement simulation model of the acquisition process. We assume that the stand-alone projections for the future sales and costs of both target and acquirer have been done and the non-operational and non-recurring cash flows have been identified and quantified. Given this information, we show how to combine the financial statements and adjust them to reflect the incremental changes stemming from the merger. This leads to complete *pro forma* financial statement projections for the target, acquirer, and prospective post-acquisition firm and estimates for the stand-alone value of the target, the acquirer, the combined entity, and the value created by the merger.

The discounted cash flow valuation method we use provides many useful insights into the M&A decision. It clearly identifies the value drivers and value destroyers of the deal. Expected synergies are the motivating factor underlying many, if not most, acquisitions. They are the increase in cash flows over and above those the two companies can generate independently. Synergies derive from increases in operational efficiency, financial efficiency, or both.* The estimated values for these cash flows are highly dependent on the opinions of experienced operating personnel. Because they are projections of future value, synergies introduce considerable uncertainty into the valuation for an acquisition or merger. Buying a company also introduces potentially large unanticipated risks. Transition plans have a hard time gauging operating risk—the risk that melding product lines, rearranging assets, and reallocating acquired personnel goes badly. More general risks include the effects of a potential recession on cash flows, the reaction of competitors (e.g., price wars), and the possibility that the expected synergies will be less beneficial than expected. This uncertainty is well suited to Monte Carlo simulation.

10.4.2 Modeling the Merger and Acquisition Problem

Given its inherent uncertainty and considerable sources for error, the merger and acquisition problem is an excellent candidate for financial modeling. The potential benefits and pitfalls from M&A are easily modeled. While modeling cannot address every uncertainty inherent in the M&A

---

* A short list of synergies includes economies of scale that stem from increased operating efficiency in both the acquiring and target companies, elimination of duplicate costs and unnecessary overhead, increased pricing power, and the exchange of complementary skills or resources.

process, it can illuminate many issues related to valuation of the combined firm. In particular, Monte Carlo simulation can identify the primary value drivers and destroyers in a deal.

The remainder of this chapter proceeds as follows. We use a simple example to illustrate one approach to modeling the M&A process based on the dynamic financial statement simulation models introduced in Chapter 4. We then convert this model to a Monte Carlo simulation model. Both models are driven by the Investment Module and take financing into account.

10.4.2.1 The Merger

Marks Pharma is considering approaching another pharmaceutical company, BAM Biotech, with a friendly takeover proposal. Marks' management believes the acquisition makes sense. Marks generates substantial free cash flow, but its drug pipeline is almost empty and its R&D department is losing its creative edge. BAM Biotech, on the other hand, has an almost full drug pipeline and an innovative and successful approach to R&D, but it lacks the money needed to pursue its many promising drugs. The two companies complement each other in other ways as well. Marks has an efficient management structure, a low-cost approach to R&D, and is much better at managing its working capital. BAM Biotech, in contrast, has developed a proprietary and remarkably efficient system of buying and managing its raw materials.

Marks managers believe that the two firms could share their expertise and make the combined entity considerably more efficient and profitable than either company on its own. They have identified the following areas for potential synergies (value drivers):

- Cost savings, including reduced CGS and SGA

- More efficient use of working capital, including lower balances in cash, receivables and inventory, and increased balances in payables

- Faster growth

These relative advantages can be seen in the current cost structures and estimated growth rates for the two companies shown in Exhibit 10.10. For the purpose of this model, the expected synergies resulting from the merger are captured by changes in the relationships between the affected

Parameters (% of sales)	Marks	BAM
COGS	0.640	0.310
SG&A	0.160	0.300
R&D	0.070	0.120
Depreciation	0.020	0.030
Cash and Equivalents	0.090	0.210
Receivables	0.105	0.200
Inventory	0.086	0.100
Gross PPE	0.405	0.460
Goodwill & Intangibles	0.140	0.150
Accounts Payable	0.230	0.200
Economy		
Riskfree rate	0.045	
Market risk premium	0.070	
Unleveraged beta	0.800	0.600
Tax rate	0.400	0.400
Interest expense	0.060	
Decision/Sensitivity		
5-year growth rate	0.10	0.10
Payout ratio	0.30	0.55
Debt portion	0.25	0.25
Debt retired	1.00	0.50
Excess cash returned	0.00	0.00
Price paid for BAM Biotech	28.00	
Terminal rates		
Terminal growth rate	0.050	0.045
Terminal payout ratio	0.70	0.60
Terminal cash returned	1	1

EXHIBIT 10.10   Cost structure for Marks pharma and BAM Biotech.

income statement and balance sheet items and forecast sales. We discuss each potential source and the ratio by which it is realized next.*

All *Parameters* are percents of sales. Variables under *Economy* reflect their names. Under *Decision/Sensitivity*, five-year growth rate is the forecast sales growth rate over the next five years; Payout ratio is the portion of the firm's net income paid out as dividends; Debt portion is the portion of new financing raised by issuing debt; Debt retired is the portion of

* Coming up with these estimates is the most difficult part of the M&A process. It is, however, not within the scope of this text. Interested readers are referred to the texts by Bruner (2004) and Hooke (1997).

excess cash used to retire outstanding debt; and Excess cash returned is the portion of excess cash returned to shareholders. Under *Terminal rates*, Terminal growth rate is used to calculate terminal value, Terminal payout ratio is the portion of net income to be paid out as dividends in perpetuity, and Terminal cash returned is a failsafe variable to make sure all excess cash flow is returned to shareholders in perpetuity.

10.4.2.2 Synergies

10.4.2.2.1 Costs.   The synergies from cost savings are expected to come from reduced operating costs (CGS) and SGAs. BAM Biotech has a much more effective cost control system than Marks. BAM's CGS averages only 31% of sales versus 64% for Marks. Marks managers expect to employ BAM Biotech's methods to increase the efficiency of the combined firm's purchasing system. They also expect to negotiate better prices from their suppliers due to their increased size and financial clout. As a result, they expect the combined firm's CGS-to-sales ratio will be 55%.

The merger team also expects to be able to reduce overall SGAs. Sales forces will be integrated and redundancies eliminated. They anticipate lower per unit overhead and advertising costs from increased economies of scale. In addition, Marks' acquisition team expects to institute their own cost control methods to further reduce BAM's costs. As a result of these changes, they expect the post-acquisition SGA-to-sales ratio to be 15%.

Finally, the merger team believes it is cheaper to acquire BAM's drugs than to develop competing drugs internally. They are interested especially in BAM's full and innovative pipeline and the R&D methods used to develop it. Even though they plan to impose some of their cost discipline on BAM's R&D department, the overall post-merger R&D budget will increase. Marks expects the post-merger ratio of R&D expense-to-sales to be 9%.

10.4.2.2.2 Working Capital Synergies.   Marks' merger team believes they can free up much of the cash currently locked in BAM's working capital via Marks' greater managerial resources. Gains are expected to come from shorter product cycles, lower inventories and receivables, larger payables, and more efficient use of cash. Marks is more efficient at managing its inventories and receivables than is BAM. On average, Marks' inventory spends 48 days in CGS versus BAM's 88 days (8.6% of sales versus BAM's 10%). The merger team expects to quickly gain control of BAM's inventory, and for this reason they expect post-merger inventory as a percent of sales to remain at 8.6%. On average, credit sales spend 38 days in Marks receivables versus nearly 60 days in BAM's receivables (10.5% of sales versus BAM's 20% of sales). The merger team anticipates post-merger receivable turnover to slow slightly as they plan to aggressively market BAM's pipeline, so receivables are expected to be 11.5% of sales. Marks is marginally better at extracting favorable payment terms from its suppliers. Its payables-to-sales ratio is 23% versus BAM's 20%. Given the increased size of the combined entity, the merger team expects even more favorable terms are possible. Their forecast for post-merger payables is 24% of sales. Finally, BAM currently carries excess cash on its books. Its ratio of cash to sales is 21% versus Marks' 9%. Post-merger, Marks expects this ratio to be 9%, in line with its current cash management policies. This will free up some cash to help with acquisition costs.

10.4.2.2.3 Growth Synergies.   The merger team expects a significant boost to the combined entity's sales. By cross-selling each other's products and integrating the sales force, they plan to move more product through the existing distribution system. By increasing R&D, they plan to develop an innovative new line of drugs. They anticipate that sales growth will be 15% over the next five years, settling into a sustainable growth rate of 5.5% thereafter. The increase in sustainable growth from 5% to 5.5% reflects the long-term value of BAM's drug pipeline. These scenarios are summarized in Exhibit 10.11.

Variable	Post-merger Expected value
COGS	0.550
SG&A	0.150
R&D	0.090
Receivables	0.115
Inventory	0.086
PPE	0.410
Accounts Payable	0.240
5-year growth	0.150
Terminal growth	0.055
Price paid	30.00

EXHIBIT 10.11   Parameter scenarios after the Marks–BAM merger.

Integration Costs by Year				
1	2	3	4	5
8,000	10,000	4,000	2,000	0

EXHIBIT 10.12 Expected costs of integration.

10.4.2.2.4 Costs of Integration. Finally, the merger team is planning for substantial post-acquisition costs. The workforce will be trimmed or relocated, plants and warehouses will be shuttered, product lines merged, and computer and information systems will be integrated. All of this takes time and costs a great deal. They expect high costs in the first two years, decreasing to zero by the fifth year. Exhibit 10.12 reports the merger team's estimates for these costs.

We now have the inputs necessary to build our base model of Marks acquisition of BAM. We assume that the beta of the combined firm will be a book value-based weighted average of the two firm's betas. This yields an unleveraged beta of 0.74 for the post-merger entity. We also assume book value-based discount rates based on the CAPM, with a risk-free rate of 4.5% and a market risk premium of 7%.

The first step in the valuation process is to estimate the stand-alone intrinsic values for the acquiring and target firms. These values are estimated by percent-of-sales models that use the ratios reported in Exhibit 10.10 and the current financial statements for each firm shown in Exhibit 10.13. Marks' stand-alone intrinsic value is $51,956 million; BAM Biotech's is $45,117 million.* We assume that the acquisition occurs by the end of year 1, but that there will have been insufficient time for any expected synergies to be realized. Therefore, year 1 revenues will be the sum of Marks' and BAM's expected year 1 revenues, year 1 CGS will be the sum of the expected year 1 CGS for each firm, and so forth.

To integrate BAM's financial statements with Marks', account balances from BAM's year 1 income statement and balance sheet are entered into the Investment Module for Marks. The expanded Investment Module is shown in Exhibit 10.14. The Investment Module then feeds them into the Accounting Module. In this way, year 1 account values in Marks' *pro forma*

---

* These stand-alone models are not shown because they are simple percent-of-sales models similar to the ones previously described in this text. The assumed five-year growth rate for both firms is 10%. Little or no external financing is required by either firm to support this rate.

Income statement ($millions)	Marks 0	BM Biotech 0
Revenue	51,790	18,106
CGS	33,054	6,532
Gross profit	18,737	11,574
SGA	6,187	5,267
R&D	2,677	2,206
Depreciation	205	239
Interest Expense	391	374
Merger Costs		
Net Income Before Taxes	10,214	3,488
Income Taxes	3,064	1,395
Net Income After Taxes	7,150	2,093
Dividends paid	3,211	2,164
Retained earnings	3,939	(71)

Balance Sheet ($millions)	0	0
Cash and Equiv	4,971	4,572
Receivables	5,423	2,968
Inventory	4,440	1,608
Total Current Assets	14,835	9,148
Gross PPE	20,984	8,706
Accumulated Depreciation	6,788	3,931
Net PPE	14,196	4,775
Goodwill & Intangibles	7,241	2,677
Goodwill from merger		
Total LT Assets	21,437	7,452
Total Assets	36,271	16,600
Accounts payable	11,567	3,561
Total Current Liabilities	11,567	3,561
Long-term Debt	5,695	7,261
Total Liabilities	17,262	10,822
Common Stock	6,974	2,711
Retained Earnings	36,145	15,795
Treasury Stock	(24,109)	(12,728)
Total Shareholders Equity	19,009	5,778
Total Liabilities & Equity	36,271	16,600

EXHIBIT 10.13   Current income statements and balance sheets for Marks and BAM.

statements are linked to BAM's stand-alone *pro forma* financial statements. The most obvious example of how this accounting adjustment is done can be seen by looking at Revenues in cell D42 of Exhibit 10.15B. Marks' revenues in year 1 are $56,969. BAM's year 1 revenues of $19,917 are added to these revenues for total revenues in year 1 of $76,886. Less obvious is how

	B	C	D	E	F	G	H
88	**INVESTMENT MODULE**	**0**	**1**	**2**	**3**	**4**	**5**
89	Current Asset Investment		11,331	-436	3,859	4,438	5,104
90	Investment in Net Fixed Assets		6,378	3,212	4,531	5,211	5,993
91	Replacement of Dep. Assets		681	806	906	1,042	1,199
92	Capital Expenditures		7,058	4,018	5,438	6,253	7,191
93	Increase in Goodwill		57,374	1,415	1,857	2,135	2,456
94	Total Investment		75,764	4,997	11,154	12,827	14,751
95							
96	**BM Biotech Contribution**	**0**	**1**				
97	*Income statement*						
98	Revenue	18,106	19,917				
99	COGS	6,532	6,174				
100	SG&A	5,267	5,975				
101	Research & Development	2,206	2,390				
102	Depreciation/Amortization	239	261				
103	Interest Expense	374	436				
104	*Balance sheet*						
105	Cash & Equivalents	4,572	4,182				
106	Accounts Receivable	2,968	3,983				
107	Inventory	1,608	1,992				
108	Gross PP&E	8,706	9,162				
109	Accumulated Depreciation	3,931	4,192				
110	Goodwill & Intangibles	2,677	2,987				
111	Accounts Payable	3,561	3,983				
112	Debt (assumed)	7,261	7,175				
113	*Merger-associated costs*		8,000	10,000	4,000	2,000	0
114	*Contribution to Goodwill*						
115	Book value of liabilities		11,158				
116	Book value of equity	5,778					
117	Number of shares	1,981					
118	Price per share paid	$ 30.00					
119	Total paid		59,430				
120	Excess over book value to Goodwill		53,652				

EXHIBIT 10.14   Expanded investment module for Marks acquisition of BAM showing account balances from BAM's *pro forma* financial statements.

	B	C	D	E
1		**Merger of Marks and BAM Biotech**		
2				**Merged**
3	**Parameters**	**Marks**	**BAM**	**Firm**
4	COGS	0.640	0.310	0.550
5	SG&A	0.160	0.300	0.150
6	R&D	0.070	0.120	0.090
7	Depreciation	0.020	0.030	0.025
8	Cash and Equivalents	0.090	0.210	0.090
9	Receivables	0.105	0.200	0.115
10	Inventory	0.086	0.100	0.086
11	Gross PPE	0.405	0.460	0.410
12	Goodwill & Intangibles	0.140	0.150	0.140
13	Accounts Payable	0.230	0.200	0.240
14	*Economy*			
15	Riskfree rate	0.045		
16	Market risk premium	0.070		
17	Unleveraged beta	0.800	0.600	0.740
18	Tax rate	0.400	0.400	
19	Interest expense	0.060		
20	*Decision/ Sensitivity*			
21	5-year growth	0.10	0.100	0.150
22	Payout	0.30	0.550	
23	Debt portion	0.25	0.250	
24	Debt retired	1	0.500	
25	Excess cash	0	0.000	
26	Price paid for BM Biotech	30.00		
27	*Terminal rates*			
28	Terminal growth	0.050	0.045	0.055
29	Terminal payout	0.7	0.600	
30	Terminal cash	1	1	

EXHIBIT 10.15A   Input sector for Marks acquisition of BAM Biotech.

B	C	D	E	F	G	H
	0	1	2	3	4	5
41 Income statement (millions)						
42 Total revenue		76,886	88,419	101,682	116,934	134,474
43 Marks	51,790	56,969				
44 BAM Biotech		19,917				
45 COGS	33,054	42,635	48,630	55,925	64,314	73,961
46 Gross profit	18,737	34,251	39,788	45,757	52,620	60,513
47 SG&A	6,187	15,090	13,263	15,252	17,540	20,171
48 R&D	2,677	6,378	7,958	9,151	10,524	12,103
49 Depreciation	205	681	806	906	1,042	1,199
50 Interest Expense	391	777	1,364	1,206	1,216	1,192
51 Merger Costs		8,000	10,000	4,000	2,000	0
52 Net Income Before Taxes	10,214	3,325	6,398	15,240	20,298	25,849
53 Income Taxes	3,064	1,330	2,559	6,096	8,119	10,339
54 Net Income After Taxes	7,150	1,995	3,839	9,144	12,179	15,509
55						
56 Dividends paid	3,211	599	1,152	2,743	3,654	4,653
57 Retained earnings	3,939	1,397	2,687	6,401	8,525	10,856
58						
59 Balance sheet ($millions)	0	1	2	3	4	5
60 Cash and Equiv	4,971	9,310	7,958	9,151	10,524	12,103
61 Receivables	5,423	9,965	10,168	11,693	13,447	15,465
62 Inventory	4,440	6,891	7,604	8,745	10,056	11,565
63 Total Current Assets	14,835	26,166	25,730	29,589	34,028	39,132
64						
65 Gross PPE	20,984	32,234	36,252	41,689	47,943	55,134
66 Accumulated Depreciation	6,788	11,661	12,467	13,373	14,415	15,614
67 Net PPE	14,196	20,573	23,785	28,316	33,527	39,520
68 Goodwill & Intangibles	7,241	10,963	12,379	14,235	16,371	18,826
69 Goodwill from merger		53,652	53,652	53,652	53,652	53,652
70 Total LT Assets	21,437	85,188	89,815	96,204	103,550	111,999
71 Total Assets	36,271	111,354	115,545	125,793	137,578	151,131
72						
73 Accounts payable	11,567	17,086	21,221	24,404	28,064	32,274
74 Total Current Liabilities	11,567	17,086	21,221	24,404	28,064	32,274
75						
76 Long-Term Debt	5,695	22,737	20,107	20,272	19,872	18,358
77 Total Liabilities	17,262	39,823	41,327	44,676	47,936	50,632
78						
79 Common Stock	6,974	58,099	58,099	58,597	58,597	58,597
80 Retained Earnings	36,145	37,541	40,228	46,629	55,154	66,011
81 Treasury Stock	-24,109	-24,109	-24,109	-24,109	-24,109	-24,109
82 Total Shareholders Equity	19,009	71,531	74,218	81,117	89,642	100,499
83						
84 Total Liabilities & Equity	36,271	111,354	115,545	125,793	137,578	151,131

EXHIBIT 10.15B   Income statement and balance sheet after Marks' acquisition of BAM Biotech.

the combined CGS is calculated. BAM's CGS for year 1 is $6174; Marks' is $36,460. The merged firm's CGS for year 1 is $42,635. The Investment Module also includes estimates for the post-acquisition integration costs associated with the merger as well as the calculation of goodwill that will go onto Marks' balance sheet.* Marks' post-acquisition input sector, income statement, and balance sheet are shown in Exhibits 10.15A and B. The formulas for the income statement and balance sheet are shown in Exhibit 10.16.

---

* To maintain simplicity, we assume that the goodwill associated with the acquisition will not be impaired in future and we ignore any associated tax consequences. A word about goodwill: an asset's value reflects the cash flows it throws off, not the cost of putting the asset into service. When a company is acquired, the price paid reflects the operating value of that company's assets, not their cost or book value. For these reasons, the price paid for a firm is typically greater than the book value of that firm. Something must make the total assets brought onto the books of the acquiring firm equal the price paid. That something is call goodwill.

	B	C	D
41	**Income statement (millions)**	0	1
42	Total revenue		=D44+D43
43	Marks	51790.3	=C43*(1+MRK_5_year_growth)
44	BAM Biotech		=D98
45	COGS	33053.6	=MRK_COGS*D43+D99
46	Gross profit	18736.7	=D42-D45
47	SG&A	6186.8	=MRK_SG_A*D43+D100
48	R&D	2677.2	=MRK_R_D*D43+D101
49	Depreciation	204.9	=IF(MRK_Depreciation*C65>C67,MAX(C67,0),MRK_Depreciation*C65)+D102
50	Interest Expense	390.8	=MRK_Interest_expense*C76+D103
51	Merger Costs		=D113
52	Net Income Before Taxes	10213.6	=D46-D47-D48-D49-D50-D51
53	Income Taxes	3064.1	=MRK_Tax_rate*D52
54	Net Income After Taxes	7149.50000000001	=D52-D53
55			
56	Dividends paid	3210.57	=D151
57	Retained earnings	3938.93000000001	=D54-D56
58			
59	**Balance sheet ($millions)**	0	1
60	Cash and Equiv	4971.2	=MRK_Cash_and_Equiv*D43+D105
61	Receivables	5423.4	=MRK_Receivables*D43+D106
62	Inventory	4440	=MRK_Inventory*D43+D107
63	Total Current Assets	=SUM(C60:C62)	=SUM(D60:D62)
64			
65	Gross PPE	20983.6	=MRK_Gross_PPE*D43+D108
66	Accumulated Depreciation	6788	=MIN(C66+D49,D65)+D109
67	Net PPE	14195.6	=D65-D66
68	Goodwill & Intangibles	7241	=MRK_Goodwill___Intangibles*D43+D110
69	Goodwill from merger		=D120
70	Total LT Assets	=C67+C68	=D67+D68+D69
71	Total Assets	=C63+C70	=D70+D63
72			
73	Accounts payable	11566.9	=MRK_Accounts_Payable*D43+D111
74	Total Current Liabilities	=C73	=D73
75			
76	Long-Term Debt	=4879+816	=C76+D134-D135
77	Total Liabilities	=C74+C76	=D74+D76
78			
79	Common Stock	6974	=C79+D136
80	Retained Earnings	36144.5	=C80+D57
81	Treasury Stock	-24109.1	=C81-D137
82	Total Shareholders Equity	19008.9	=SUM(D79:D81)
83			
84	Total Liabilities & Equity	=C77+C82	=D82+D77
85		Check =C84-C71	=D84-D71

EXHIBIT 10.16   Formulas for the post merger income statement and balance sheet.

Beginning in year 2 we assume that the expected synergies have been realized. This is captured by estimating all the usual income statement and balance sheet items as percentages of the combined revenues. The percentages used are those reported under the Merged Firm column of the Input Sector. Once you have finished modifying the Investment and Accounting modules, your model is finished. Nothing else needs to be modified. The Financing, Free Cash Flow, and Investment modules automatically adjust to the changes made in the Accounting module.

The intrinsic values for Marks and BAM as stand-alone companies and the intrinsic value after the merger are reported in Exhibit 10.17. On their own, Marks and BAM have a combined intrinsic value of $97,073; when merged, their intrinsic value is $172,420, an increase of $75,346. On a fully diluted basis, Marks' value per share was $22.82 before the merger and $55.23 afterwards.

This model, while encouraging, gives us an incomplete assessment of the proposed merger. It assumes that all expected synergies will be realized.

Tracking Intrinsic Value	
Panel a. Aggregate Equity Intrinsic Value	
Pre-Acquisiton	
Marks	51,956
BAM Biotech	45,117
Total	97,073
Post-Acquisition	**Expected Value**
	172,420
Change	75,346
Panel b. Per Share (diluted) Intrinsic Value	
Pre-Acquisition	
	$ 22.82
Post-Acquisition	
	$ 55.23
Change	$ 32.41

EXHIBIT 10.17   Summary of value added by Marks' acquisition of BAM Biotech.

This is unlikely to be the case. Some synergies may never be realized, and some costs may be grossly under-estimated. Nor does this model tell us which of the potential synergies are responsible for Marks' dramatic increase in value. While some expectations likely will be met, others will exceed or fall short of their expected values. Hence, there will be some canceling of good and bad draws. It is more realistic to assume that the input parameters can assume one of many different values within a given range. Monte Carlo simulation addresses all these issues. In the next section, we adapt the current model to Monte Carlo simulation.

10.4.3 Monte Carlo Simulation

Converting the base model to a Monte Carlo simulation model is simple. All that needs to be done is to define the appropriate input parameters as random variables and define the output variable that we wish to track. The model's structure remains the same. The conversion involves replacing the Merged Firm parameters with @RISK random variable functions. The inputs that will be defined as random variables, their distributions, and their distributions' parameters are shown in Exhibit 10.18. We now briefly discuss the rationale behind the distributions for each of these random variables.

Random Variables and Their Distributions				
Variable	Distribution	Minimum	Expected	Maximum
COGS	PERT	0.52	0.55	0.61
SG&A	Triangle	0.15	0.15	0.22
R&D	Uniform	0.07		0.11
Recievables	PERT	0.105	0.115	0.25
Inventory	PERT	0.08	0.086	0.19
PPE	Uniform	0.405	0.41	0.45
Accounts Pay.	Triangle	0.19	0.24	0.25
5-year growth	PERT	0.07	0.15	0.18
Terminal growth	Uniform	0.035		0.055

EXHIBIT 10.18 Random input variables with their distributions and parameters.

10.4.3.1 Cost of Goods Sold

Marks anticipates being able to reduce their operating costs between 15% and 20% by adapting BAM Biotech's cost control system. They most likely will be able to reduce operating costs to 55% of sales, and have a reasonable likelihood of being able to reduce them even more, to 52%. There is, however, a remote possibility that the integration will not go well. If so, the operating costs as a percent of sales would be 61%.

The PERT distribution is ideally suited to model this type of scenario.* Frame (2003) calls it "the Murphy's Law distribution." The reason can be seen in Exhibit 10.19. Observe that 95% of the draws from this distribution will cluster around 55%, while only a thin tail extends out to the worst-case draw of 61%. This says that Marks believes it has talented managers who are doing their best and that the outcome of these efforts should be significantly lower operating costs. However, if something goes wrong when adapting BAM's cost control methods, it will go very wrong.

10.4.3.2 Selling, General, and Administrative Costs

Similar logic applies to the post-merger ratio of selling, general, and administrative costs to revenues. In this case, however, Marks' merger team is less certain of the likelihood of the worst-case scenario. Because of the difficulty of integrating different cultures, they believe an adverse outcome is more likely for SGA than for CGS. Therefore, we use the triangular distribution to model this parameter.

---

* PERT stands for Program Evaluation Review Technique. It was developed by the Navy in the 1950s to plan complex projects. However, it is well suited for modeling asymmetric distributions with long thin tails.

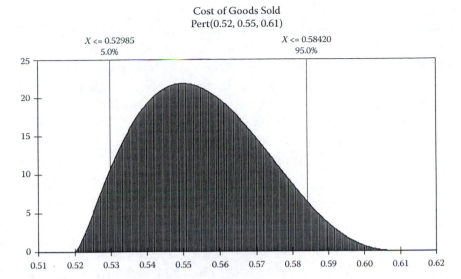

EXHIBIT 10.19   Pert distribution for CGS.

10.4.3.3 Research and Development

Marks' only expectations for post-merger R&D expenses is that they will fall somewhere between 7% and 11% of sales, with any value considered equally likely. They have selected the uniform distribution to model this expense.

10.4.3.4 Receivables and Inventory

Marks management believes that these ratios will come very close to their expected values. However, should the economy suffer a severe downturn, both would balloon. We use the Pert distribution to model these items.

10.4.3.5 Gross Plant and Equipment

Investment in fixed assets is expected to fall somewhere between 40.5% and 45% of sales, with all possible values viewed as equally likely. We use the uniform distribution to model this investment.

10.4.3.6 Accounts Payable

Marks is less secure about the value AP will assume. Similar to SGA expense, they have a fairly clear picture for the best, most likely, and worst-case values this ratio may assume, but little more. We therefore model this ratio using the triangular distribution.

10.4.3.7 Growth Rates

Marks is fairly certain that their growth over the next five years will fall somewhere between 15% and 18% as the synergies from the acquisition of BAM Biotech are realized. However, there is a remote possibility of an economic downturn, in which case Marks' short-term growth rate could fall to as little as 7%. We model this variable using the Pert distribution. Terminal growth is expected to fall somewhere between 3.5% and 5.5%, so we model it using the uniform distribution.

The converted input section, after making the above changes, is shown in Exhibit 10.20A; the formula view is shown in Exhibit 10.20B. Random variables are highlighted.

10.4.3.8 Integration Costs

Marks has no strong beliefs about the distributions for the post-merger acquisition costs, other than their worst, most likely, and best cases. We therefore model these costs using the triangle distribution based on the parameters given for the scenario simulation model.

10.4.3.9 Output

Marks plans to use this simulation to track the increase in the intrinsic fully diluted value per fully diluted share.

This model allows the user to determine the relative portions of debt and equity when issuing new financing. At an expected price paid of

	B	C	D	E	F	G	H	I	J	K
1				Merger of Marks and BAM Biotech						
2	random variables									
3				Merged			Random Variables and Their Distributions			
4	Parameters	Marks	BAM	Firm		Variable	Distribution	Minimum	Expected	Maximum
5	COGS	0.640	0.31	0.555		COGS	PERT	0.52	0.55	0.61
6	SG&A	0.160	0.30	0.173		SG&A	Triangle	0.15	0.15	0.22
7	R&D	0.070	0.12	0.090		R&D	Uniform	0.07		0.11
8	Depreciation	0.020	0.03	0.025		Receivables	PERT	0.105	0.115	0.25
9	Cash and Equivalents	0.090	0.21	0.100		Inventory	PERT	0.08	0.086	0.19
10	Receivables	0.105	0.20	0.136		PPE	Uniform	0.405	0.41	0.45
11	Inventory	0.086	0.10	0.102		Accounts Pay.	Triangle	0.19	0.24	0.25
12	Gross PPE	0.405	0.46	0.422		5-year growth	PERT	0.07	0.15	0.18
13	Goodwill & Intangibles	0.140	0.00	0.140		Terminal growth	Uniform	0.035		0.055
14	Accounts Payable	0.230	0.20	0.227						
15	Economy									
16	Riskfree rate	0.045								
17	Market risk premium	0.070								
18	Unleveraged beta	0.800	0.600	0.74						
19	Tax rate	0.400	0.40							
20	Interest expense	0.060								
21										
22	Decision/Sensitivity									
23	5-year growth	0.25	0.10	0.142						
24	Payout	0.30	0.55							
25	Debt portion	0.25	0.250							
26	Debt retired	1	0.500							
27	Excess cash	0	0.000							
28	Price paid for BM Biotech	$ 30.00								
29	Terminal rates									
30	Terminal growth	0.050	0.045	0.045						
31	Terminal payout	0.7	0.600							
32	Terminal cash	1	1							

EXHIBIT 10.20A Input sector after conversion to Monte Carlo simulation.

	B	C	D	E
1				**Merger of Marks and BAM Biotech**
2	random variables			
3				**Merged**
4	*Parameters*	**Marks**	**BAM**	**Firm**
5	COGS	0.64	0.31	=RiskPert(I5,J5,K5)
6	SG&A	0.16	0.3	=RiskTriang(I6,J6,K6)
7	R&D	0.07	0.12	=RiskUniform(I7,K7)
8	Depreciation	0.02	0.03	0.025
9	Cash and Equivalents	0.09	0.21	0.1
10	Receivables	0.105	0.2	=RiskPert(I8, J8, K8, RiskCorrmat(NewMatrix,1))
11	Inventory	0.086	0.1	=RiskPert(I9, J9, K9, RiskCorrmat(NewMatrix,2))
12	Gross PPE	0.405	0.46	=RiskTriang(I10,J10,K10)
13	Goodwill & Intangibles	0.14	0	0.14
14	Accounts Payable	0.23	0.2	=RiskTriang(I11,J11,K11)
15	*Economy*			
16	Riskfree rate	0.045		
17	Market risk premium	0.07		
18	Unleveraged beta	0.8	=0.6	0.74
19	Tax rate	0.4	0.4	
20	Interest expense	0.06		
21				
22	*Decision/Sensitivity*			
23	5-year growth	0.25	0.1	=RiskPert(I12, J12, K12, RiskCorrmat(NewMatrix,3))
24	Payout	0.3	0.55	
25	Debt portion	0.25	0.25	
26	Debt retired	1	0.5	
27	Excess cash	0	0	
28	Price paid for BM Biotech	30		
29	*Terminal rates*			
30	Terminal growth	0.05	0.045	=RiskUniform(I13,K13)
31	Terminal payout	0.7	0.6	
32	Terminal cash	1	1	

EXHIBIT 10.20B   Formula view of input sector after conversion to Monte Carlo simulation.

$30.00 per share, Marks will need to raise a total of $71,475 million by the end of year 1. Of this, $59,430 million will be paid to BAM Biotech's current shareholders. We assume the new financing will be 25% debt and 75% equity. While Marks would prefer to finance the acquisition with 100% debt, doing so would put its debt into the junk bond category and raise Marks cost of equity to 20%. Financing entirely with new equity would be too dilutive.*

All that remains is to make the above changes, run a simulation, and analyze the results. We first consider the increase in the fully diluted intrinsic value per share resulting from the merger. Since Marks is currently selling near its intrinsic value per share, $22.82, Marks' managers believe that any increase in intrinsic value per share will be reflected in the price of their stock. The simulated frequency distribution is shown in Exhibit 10.21. The mean increase is $8.02 per share or 35%. However, the range of the distribution (−$40.44 to $59.26) suggests there is considerable

---

* These values were read directly from the model. The $71,475 is the REF in year 1; the 20% cost of equity is found in the valuation module.

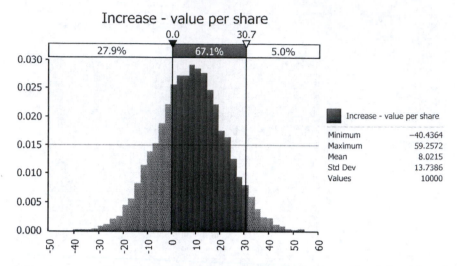

EXHIBIT 10.21 Simulated distribution of the increase in the fully diluted intrinsic value per share resulting from Marks' acquisition of BAM Biotech.

risk in this acquisition, a conclusion supported by the almost 30% likelihood of the acquisition destroying current shareholder's value.

Regression sensitivity analysis (Exhibit 10.22) reveals the key value drivers of this deal. Not surprisingly, they are cost synergies—CGS, SGAs, and R&D expenses. Marks will need to closely monitor the synergies related to these costs if it is to benefit from the deal. The terminal growth rate also has a significant effect on intrinsic value added. Marks should therefore try to position itself for stable, long-term growth at the high end of their projections.

Numbers in the bars reflect the relative influence of the variables listed. For example, the reported number for CGS is −0.60. This means that 1 standard deviation increase in CGS will decrease the intrinsic value per share added by 1 standard deviation.

Should Marks move forward with the acquisition? Clearly, it has the potential to richly reward their current shareholders, but it is not without significant downside risk. Even though this analysis does not reveal a clear answer, it does provide a much richer information set on which management can base their decision.

This example of applying Monte Carlo simulation to an acquisition is highly simplistic. Other than integration costs, it assumes that values for

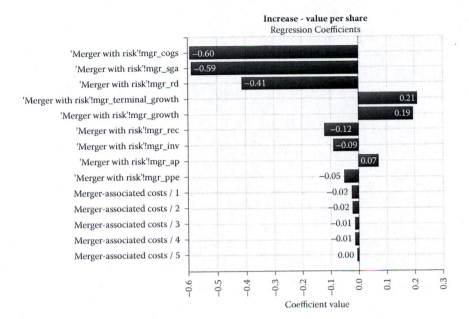

EXHIBIT 10.22   Regression sensitivity tornado graph for simulated increase in intrinsic value per share.

all parameters apply across all sub-periods in each iteration. A more realistic model would make separate but correlated draws for each parameter for each sub-period. Additional complexity could be added by assuming that different scenarios apply for many of the random parameters, each with its own distribution. While these enhancements add significant complexity, they are not particularly difficult to build. The interested reader is referred to Winston (2000).

10.5 SUMMARY

This chapter has discussed three main topics relating to modeling long-term assets. The first modeling approach discussed how to include long-term assets in a sales-driven planning model. The second focused on evaluating an investment in long-term assets. The third showed how to model the merger and acquisition decision.

Modeling long-term assets in the sales-driven model is fairly straight-forward. We simply relate the asset level with the sales level. The modeling problems that arise are how to adjust the asset level to both positive and negative changes in assets.

The capital budgeting decision is one of the most important decisions for a firm. We showed how to model the capital budgeting decision. It was noted that the crucial modeling issue is to correctly calculate the cash flows and costs of capital, and match them together. That is, when the WACC approach is used, it is important to match the project cost, cash flows, and cost of capital.

In the last section of the chapter we showed how to model a merger decision. Owing to the uncertainties of the benefits of the merger, it is important to explore the drivers of value and analyze the sensitivity of the decision to changes in the assumptions. Monte Carlo simulation was used to explore these aspects.

PROBLEMS

1 RMM

RMM manufactures and markets a line of sports cars. At the end of the current year, selected revenue, expense, and asset accounts are as follows (expressed in thousands of dollars):

Sales revenue	50,000
Depreciation expense	17,860
GFA	125,000
Accumulated depreciation	63,500
NFA	61,500

Construct a model of Rocky Mountain's fixed assets. Assume that sales drives GFA, so that for each $100 of car sales, it has to have GFA of $250. Assume that annual depreciation expense is equal to 15% of the previous year's GFA. Your model should show 10 years of results.

Based on your model, answer the following questions:

(a) If sales grow at 5% per year, how much will the gross capital investment and net capital investment have to be next year (year 1)?

(b) At 5% sales growth, what will be the annual rate of growth in NFA over the next five years?

(c) At 5% sales growth, what will be the GFA and NFA in year 6?

(d) At what rate would sales have to grow so that NFA remain unchanged?

2 RMM (2)

Assume the same initial data and ratio relationships for RMM as in problem 1. RMM's sales tend to be cyclical. That is, sales tend to go up when the economy does, and sales decline during recessions. While RMM cannot operate with less than $250 worth of GFA for each $100 dollar of sales, there is no reason it cannot produce less than $100 of output with $250 of GFA. That is, if sales decline in a given period it does not have to sell assets. It can maintain its asset level at excess capacity until sales turn up again.

Construct a 10-year model of Rocky Mountain's GFA so that assets are added when necessary, but GFA do not decline with decreases in sales.

(a) Test your model by varying the rate at which sales grow. Include sales that decline at high rates (e.g., −20% annually), low rates of sales growth (2%), and high rates of sales growth (+30% annually).

(b) Now show the results when the company faces a business cycle with sales growth rates as follows:

Year	1	2	3	4	5	6	7	8	9	10
Sales growth (from prior year) (%)	+20	+18	−35	−10	+30	+25	+60	−10	−15	+20

What are the gross and fixed net assets at the end of the tenth year?

3 Xatos Corporation

The Xatos Corporation is considering the purchase of a new Xat machine that will enable it to expand its annual rate of production and increase its expected sales. The machine will have a direct cost of $14,000 that will be paid right now ($t = 0$). In addition, the machine must be installed and tested. The installation and testing will occur over the next year, and the costs of installation and testing, which are assumed to be paid one year from now ($t = 1$) will amount to $6000. The depreciable cost of the machine includes the acquisition cost, installation cost, and testing costs. It will be depreciated according to the MACRS three-year schedule (rates are 33%, 45%, 15%, and 7%). The machine will be placed in service at the end of the first year ($t = 1$), with depreciation taken for the first time at the end of its first year of service ($t = 2$).

The equipment will be operated for 10 years, with the first year of operation beginning at time 1, with sales for the first year of operations assumed to be received at time $t = 2$. The sales in this first year of operation are expected to be $350,000. Then, sales will grow by 8% per year until $t = 11$ (the tenth year in operation). The annual operating costs (before depreciation) will consist of fixed cash operating costs of $25,000 and variable operating costs equal to 75% of sales.

To support the increased level of production, the inventory of raw materials will have to be increased. The inventory will have to be available at the beginning of the year in which the sales are made. With the first sales occurring at $t = 2$, the initial investment in inventory occurs at $t = 1$. The value of the inventory at each date will be a fixed amount equal to $5000, plus a variable amount that is equal to 5% of the sales expected at the end of the coming year. The additional inventory will be carried until the machine is scrapped, which is assumed to be at time 11. At that time the investment in inventory will be recovered. Assume there is no taxable gain or loss on the recovery of investment in inventory.

At the end of the 10-year operating life of the project (time $t = 11$), it is assumed that the equipment will be sold for $6000. Any gain over book value will be taxed at the firm's ordinary income tax rate of 40%.

If the new Xat machine is purchased, it will be financed by borrowing the required funds at an interest rate of 8%. The loan will be repaid in eight equal annual installments, with the first principal installment occurring at $t = 4$. Interest will be paid annually as it accrues. The funds will be borrowed at the dates that they will be needing to purchase the project assets.

If this project were considered by itself, with no debt financing, its beta would be 1.3. Prior to undertaking this new investment, the company's balance sheet is as follows:

Current assets	$25,000
NFA	50,000
Total assets	$75,000
Debenture bond, 10% annual coupon, due in 15 years	$20,000
Common stock, $1 par value, 1000 shares	1000
Paid in capital in excess of par value	16,000
Earned surplus	38,000
Total capital	$75,000

The debentures are currently trading at a price of 117.29 (percent of par value), to yield 8%. The common stock is trading at a price of

$46.92 per share. The beta coefficient for the stock (before the new project) is 1.56, and recent estimates of the risk premium on the market portfolio are $(k_m - k_f) = 0.05$, and the current risk-free rate is $k_f = 0.04$.

Assignment

Set this decision problem up as a spreadsheet to evaluate the acceptability of the Xat project. Your model should calculate NPV and IRR for the project. The cost of capital should take account of both the risk of the project itself, and the capital structure mix of the firm as it would be when the project is undertaken.

4 Investment Evaluation with Monte Carlo Simulation: JN Pharmaceutical Corporation Acquisition of Bri-Mark Pharmaceutical Company

Part I: Drug Development under Certainty

JN Pharmaceuticals is considering expanding its drug pipeline by purchasing Bri-Mark Pharmaceutical Company. Bri-Mark is a smaller biotech firm with one drug under development, an anti-depressant they call Woebegon. Woebegon just finished phase II testing and begins phase III testing immediately. If phase III testing is successful, Bri-Mark will seek FDA approval and, subsequent to approval, begin a marketing campaign. Table 10.1 below lists the remaining phases in the drug's development, the average number of years for completion of each phase, and the average annual cost of completing that phase. Also shown are the minimum and maximum annual costs of completing each remaining phase.

If Woebegon reaches the marketing stage and is successful, the annual cash flow generated by the product's sales would be about $60 million, beginning the year after the marketing campaign. These cash flows would be earned for eight years, at which point Woebegon would lose its patent protection and all meaningful cash flows from it would cease.

TABLE 10.1   Remaining Development Phases

Stage	Number of Years in Phase	Average Annual Cost ($ millions)	Minimum Cost ($ millions)	Maximum Cost ($ millions)
a. Phase III testing	3	8	6	10
b. Regulatory filing	2	5	4	8
c. Marketing	1	18		

To evaluate this type of investment, the cash flows should be discounted at the cost of capital adjusted for the riskiness of the project. This type of investment is considered to be very risky: its cash flows are about twice as volatile as the average stock market investment. Consequently, in the context of the CAPM, Bri-Mark's beta is about 2. The 20-year Treasury bonds are currently yielding about 4.5%. Use this as the risk-free rate.* In the current moderately risk-tolerant environment, the risk premium (returns in excess of the risk-free rate) on the market portfolio (or the average stock market investment) is about 4%.

Phase III will begin in year 1 and its first costs will be incurred at the end of year 1. Develop a spreadsheet model based on the most likely costs and cash flows with which you can evaluate the fair value of Bri-Mark Pharmaceuticals.

- What is Bri-Mark's NPV and the IRR?
- What is the maximum price JN should pay for Bri-Mark?

Part II: Modeling Uncertainty

There is considerable uncertainty about the ultimate success of Woebegon. This uncertainty should be considered in the evaluation. We can only guess about the possible success for each of the remaining phases of the research and marketing, and there is considerable variation in the possible cash flows.

In addition, there is a remote possibility that the U.S. Department of Justice (DoJ) will disallow the acquisition due to anti-trust concerns. JN plans to file a petition with the DoJ for a letter of approval as soon a purchase agreement is reached. The Justice Department then will seek comment from JN's competitors before making a decision. This alternative action will take one year, after which JN estimates there is a 10% chance that the Justice Department will nullify the merger. In this case, JN would be out the first full year of phase III testing expenses.

There are also sources of uncertainty related to each of the three remaining phases of development. Assume that each remaining phase has a probability of failure or success according to the following table (Table 10.2). Model these probabilities using the binomial distribution with n equal to 1 and the probability set appropriately.

---

* Ibbotson and Associates use the 20-year T-bond yield as their risk-free rate of return, as they believe this is the investment horizon for most long-term investors.

TABLE 10.2   Probability of Success at Each Phase

	Phase		
	Phase III	Regulatory Approval	Marketing
Probability of success (%)	70	85	80 if aggressive 50 if less aggressive

If Woebegon is unsuccessful in the phase III testing or in seeking FDA approval, the project will be terminated and no additional costs will be incurred. If it is successful at the end of a given stage, the next stage will be carried to completion. Note also that failure is recognized only at the end of each stage. For example, if a stage lasts three years, costs for that stage are incurred for the full three years, and then, if Woebegon fails, no future costs or revenues will be realized.

The costs of development at each stage are random variables drawn from a probability distribution. Assume that the development costs are from a triangular distribution with minimum, most likely, and maximum value as shown in Table 10.1. For example, the annual costs for phase 3 are drawn from a triangular distribution with a minimum of $6 million, a most likely value of $8 million, and a maximum value of $10 million. You need to make a separate draw for each year of phase III testing and seeking regulatory approval. Marketing costs are a decision variable, not a random variable.

Marketing and Sales

If Woebegon makes it through the phase III testing and receives FDA approval, then it can be marketed to the medical industry and the public. Stage c is a marketing program that will begin subsequent to FDA approval and last one year. After that year, Woebegon will be sold to the public, with the first cash flow being received one year after the end of the marketing effort. These flows will continue for eight years.

Bri-Mark is considering two marketing programs, a less aggressive one that costs $4 million and a very aggressive one that costs $18 million. The success of the marketing program depends on the amount spent. The $4 million program has a probability of success of 50%; the $18 million program has a probability of success of 80%.

If the marketing effort is successful, Woebegon will have a substantially different market than if it is less successful. Projections and

418 ■ Introduction to Financial Models for Management and Planning

TABLE 10.3   Expected Revenues from Marketing Campaign

Population Who Use the Woebegon	Successful	Less Successful
Mean	1.0%	0.5%
Standard deviation	0.15%	0.10%
Days per year taken	365	365
Wholesale price per dose	$0.125	$0.110
Profit margin	40%	35%

data relating to the markets for successful and less successful marketing programs are reported in Table 10.3. The percentage of the total U.S. population who will use the treatment will be a normally distributed random variable with the parameters reported in Table 10.3. The profit margin represents the rate of profit earned from the wholesale price, and represents the incremental cash flow to the company that will be generated from the sales.

The population in the United States prior to beginning phase III testing was approximately 300 million. It is expected that the population will grow at an annual rate of 0.8% for the next 30 years.

Expand your model to take account of the uncertainty of the Justice Department actions, phase III testing, FDA approval, and the marketing processes. Your model should permit you to conveniently switch between marketing plans. Once you have done so, run a test simulation of 1000 iterations under the aggressive marketing campaign. Your distribution should look something like that shown in Exhibit P.1. If it does, copy the worksheet and change the marketing scenario in the copy to the less aggressive program. Name the NPV cells in each worksheet according to the marketing plan in effect. Now run a simulation of 10,000 iterations and address the following managerial concerns.

- Why is the simulated expected NPV so much smaller than the NPV reported by either the model under certainty or by the NPV cell in your models under uncertainty?
- The simulated distribution has three modes. Explain them. (You might find it useful to look at the probabilities of an outcome falling into one of the three modes for both proposed marketing campaigns.)
- Now, overlay the simulated NPV distribution for the less aggressive marketing campaign on the simulated distribution for the

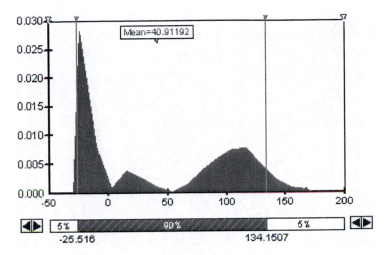

EXHIBIT P.1   Distribution for NPV aggressive/C91.

aggressive campaign. Does the less aggressive marketing campaign offer any advantages over its more expensive competitor? If Woebegon is approved by the FDA, should JN market it aggressively (i.e., spend $18 million) or less aggressively (i.e., spend only $4 million)?

- JN believes it can acquire Bri-Mark for $33.5 million. At that price, what is the likelihood that doing so will decrease the value of JN's current shareholders. (Put another way, what is the likelihood that Bri-Mark's NPV is less than $33.5 million?)
- Assume that if JN acquires Bri-Mark, it will market Woebegon aggressively.
- JN has a company policy of not investing in projects or acquisitions that have a likelihood of destroying shareholder value greater than 40%. Given this policy, what is the maximum price JN should pay for Bri-Mark?
- Given your analysis, would you recommend that JN acquire Bri-Mark? Defend your position.

Finally, suppose that JN could increase the likelihood of successfully bringing Woebegon to market by 5% points if it invests an additional $1 million into managing either the DoJ review, phase III testing, or the FDA application. Assume that the $1 million would be paid in the last year of the stage involved. (HINT: You do not need to test each of the three phases to answer this question.)

- Into which of these stages should JN invest?
- Which marketing campaign would realize the greatest increase in NPV?
- By how much would it increase Woebegon's NPV?

In a memo, briefly introduce the decisions faced by JN, describe your findings (the answers to the questions in Parts I and II), and give your recommendations (acquire Bri-Mark or not; and market aggressively or not, acquire a portfolio of companies or not). Your memo should contain the @RISK graphs of the probability distributions that support your findings.

CHAPTER 11

Debt Financing

T HIS CHAPTER EXPLAINS HOW to model the details of the firm's debt financing. In the first section we expand the basic planning model to handle greater detail in the debt financing sector of the model. In the second section, we use Monte Carlo simulation to determine an optimal capital structure. In the last section, we show how to model some of the more esoteric issues of debt financing, including duration and swaps.

11.1 DEBT IN THE LONG-TERM PLANNING MODEL

In earlier chapters we distinguished between spontaneous and discretionary sources of funds, and between internal and external sources of funds. Here we are concerned mainly with the discretionary, or external sources of debt financing. Borrowing is considered a discretionary, as opposed to a spontaneous source of funds, because an explicit decision and action must occur to tap these sources, and most borrowing does not necessarily increase spontaneously with revenues.

One of the first tasks is to determine the amount of discretionary financing that will be required. In our sales-driven model of the O&R Corp. in Chapter 4, the discretionary financing was based on the required

E.1	Required Financing$_t$	= Total Investment$_t$
E.2	Earnings Cash Flow$_t$	= Net Income$_t$ + Depreciation Expense$_t$
E.3	Minimum Dividend$_t$	= Net Income$_t$(Payout Ratio)
E.4	Internal Financing$_t$	= Earnings Cash Flow$_t$ − Minimum Dividend$_t$
E.5	Spontaneous Financing$_t$	= (Current Liabilities/Sales)Sales$_t$ − Current Liabilities$_{t-1}$
E.6	Required External Financing$_t$	= Required Financing$_t$ − Internal Financing$_t$ − Spontaneous Financing$_t$
E.7	Debt Issued$_t$	= Required External Financing$_t$(Debt Portion)
E.8	Stock Issued$_t$	= Required External Financing$_t$ − Debt Issued$_t$
E.9	Total Financing$_t$	= Internal Financing$_t$ + Spontaneous Financing$_t$ + Debt Issued$_t$ + Stock Issued$_t$

EXHIBIT 11.1   Financing module.

external financing (REF), that was shown in Exhibit 4.2E, and is repeated here as Exhibit 11.1.

The REF is the amount necessary to fund the firm's planned investment in new assets and pay the minimum dividend. The task then, is to decide how to split this financing requirement into debt and equity. In our O&R model, this decision is specified by the user in the decision variable, "Debt Portion," that splits the REF into debt and equity as shown in Equations E.7 and E.8.

In the model in Chapter 4, we did not delve very deeply into the question of how we decide on the split between debt and equity financing. The purpose of this chapter is to understand how to model the firm's borrowing decisions in a way that is consistent with the theory of capital structure so that our models can help us to make the best financing decisions.

Naturally, there are many different approaches to specifying the discretionary debt financing in the simulation model. The approach to be discussed first is aggregative and would be appropriate for long-term planning. The O&R model had only one kind of debt. We may want to provide for the issue of different varieties of debt. The most basic division would be short- and long-term debt. Exhibit 11.2 extends the financing module of Exhibit 11.1 by splitting the debt issue into short term and long term, and then tracking the debt service and balances from period to period. Short-term Portion is the decision variable input by the user to specify the proportion of the total Debt Issued that will be short-term debt, with the remainder assumed to be long-term debt.

This model assumes the short-term debt matures in one period, as reflected in Equation 11.4. Long-term debt is repaid according to a schedule

Debt Transactions		
Debt Issued$_t$	= Required External Financing$_t$(Debt Portion)	(11.1)
Short-term Debt Issued$_t$	= Debt Issued$_t$(Short-term Portion$_t$)	(11.2)
Long-term Debt Issued$_t$	= Debt Issued$_t$ − Short-term Debt Issued$_t$	(11.3)
Short-term Debt Repaid$_t$	= Short-term Debt$_{t-1}$	(11.4)
Long-term Debt Repaid$_t$	= Long-term Debt$_{t-1}$(Long-term Repayment Rate$_t$)	(11.5)
Interest and Debt Service		
Short-term Interest$_t$	= Short-term Debt$_{t-1}$(Short-term Rate$_t$)	(11.6)
Long-term Interest$_t$	= Long-term Debt$_{t-1}$(Long-term Rate)	(11.7)
Interest Expense$_t$	= Short-term Interest$_t$ + Long-term Interest$_t$	(11.8)
Debt Principal Payment$_t$	= Short-term Debt Repaid$_t$ + Long-term Debt Repaid$_t$	(11.9)
Total Debt Service$_t$	= Interest Expense$_t$ + Debt Principal Payment$_t$	(11.10)
Debt Balance		
Short-term Debt$_t$	= Short-term Debt$_{t-1}$ − Short-term Debt Repaid$_t$ + Short-term Debt Issued$_t$	(11.11)
Long-term Debt$_t$	= Long-term Debt$_{t-1}$ − Long-term Debt Repaid$_t$ + Long-term Debt Issued$_t$	(11.12)

EXHIBIT 11.2   Debt sector equations.

specified by the user with the parameter Long-term Repayment Rate, which is the proportion of the outstanding long-term debt that is repaid each period. The interest expense would be calculated using Short-term Rate and Long-term Rate parameters that are the rates on short- and long-term debts, respectively.

11.1.1 Constraining the Debt Ratios

The financing model in Exhibit 11.2 determines the borrowing based on the Debt Portion parameter specified by the user, and then splits the debt into short term and long term based on the user input. Instead of having the user specify these parameters, we may want to allow the model to determine the financing mix based on targets or limits on the debt ratios. For example, suppose our policy is to maintain our ratios within the following limits:

Current Ratio ≥ 2.0

Total Debt-to-Total Assets ≤ 45%

Short-term Issued$_t$ = 30% of Debt Issued$_t$

Long-term Borrowing$_t$ = 70% of Debt Issued$_t$.

An example of modifications to the Equations of Exhibit 11.2 to handle these constraints is shown in Exhibit 11.3, where we assume that the only current liabilities are AP and short-term debt, and long-term debt is the only long-term liability.

In Exhibit 11.3, Equation 11.13 uses the total debt-to-total asset ratio to set the limit on total debt, consisting of all liabilities. We will subtract the non-interest-bearing debt from this to find our capacity for short- and long-term interest-bearing debt. The first short-term debt limit (Equations 11.14

Input: Policy Limits		
Debt/Asset Limit	= 45%	
Current Ratio Limit	= 2.0	
Short-term Portion	= 30%	
Debt Financing		
Maximum Total Debt$_t$	= (Debt/Asset Limit)Total Assets$_t$	(11.13)
Maximum Current Liabilities$_t$	= Current Assets$_t$/(Current Ratio Limit$_t$)	(11.14)
Short-term Debt Capacity$_t$	= MAX[Maximum Current Liabilities$_t$ − Accounts Payable$_t$, 0]	(11.15)
Maximum Short-term Debt$_t$	= MIN[ST Debt Capacity$_t$, Maximum Total Debt$_t$ − (Long-term Debt$_{t-1}$ − Long-term Debt Repaid$_t$) − Accounts Payable$_t$]	(11.16)
Short-term Financing Capacity$_t$	= MAX[Maximum Short-term Debt$_t$ − (Short Debt$_{t-1}$ − Short-term Debt Repaid$_t$), 0]	(11.17)
Long-term Debt Capacity$_t$	= MAX[Maximum Total Debt$_t$ − Maximum Short-term Debt$_t$ − Accounts Payable$_t$, 0]	(11.18)
Long-term Financing Capacity$_t$	= MAX[Long-term Debt Capacity$_t$ − (Long-term Debt$_{t-1}$ − Long-term Debt Repaid$_t$), 0]	(11.19)
Short-term Debt Issued$_t$	= MIN[(Short Term Portion)Required External Financing$_t$, Short-term Financing Capacity$_t$]	(11.20)
Long-term Debt Issued$_t$	= MIN[Required External Financing − Short-term Debt Issued$_t$, Long-term Financing Capacity$_t$]	(11.21)

EXHIBIT 11.3 Ratio constraints on debt financing.

and 11.15) is set by the requirement that the current ratio must be at least 2.0. Note that when we have a subtraction such as Equation 11.15, we usually put it inside a MAX function to prevent the difference from being negative. A second short-term debt limit is set by a constraint on total debt, so Equation 11.16 says that short-term debt cannot exceed the smaller of: (a) the limit set by the current ratio; or (b) a limit set by the maximum total debt. Then, in Equation 11.17, the allowable increase in short-term debt is the difference between the maximum total short-term debt and the amount that is currently outstanding. The long-term debt capacity (Equation 11.18) is set as the residual of total debt capacity less maximum short-term debt, and long-term financing capacity is the difference between the capacity for long-term debt and the current level of debt. Equations 11.20 and 11.21 issue short- and long-term debt based on these capacity limits coupled with the policy of meeting the debt needs with 30% short term and 70% long term. So long as none of the limits are binding, the 30–70% mix will prevail. But, if the firm has hit its debt limits, it will issue debt within these limits instead of using the 30–70% mix. Naturally, there are other ways to model borrowing subject to constraints; in addition, other constraints such as interest coverage may apply. But, this simple model will give you a start.

The treatment of debt financing can be expanded or contracted according to the modeling needs of the user. For longer planning horizons, it may not be productive to distinguish between long- and short-term debt, or between different issues or tranches of debt. For shorter term planning or more detailed models, this distinction may be much more important. As the need for greater detail increases, the detail can be expanded to show various types of debt financing having different features. We have already looked at a more detailed accounting of different debt issues in the Stiliko example in Chapter 10. In the Stiliko capital investment example, long-term debt was issued at times 0, 1, and 2, with all issues amortized to mature at time 6. For each of the three different issues, we assumed constant annual debt service (principal and interest), and we noted that we could track the details of the debt with Excel® PMT(.) functions:

$$\text{Debt Service Payment}_t = \text{PMT}(\text{Interest Rate, Term to Maturity,}$$
$$\text{Amount of Loan}),$$

$$\text{Interest Payment}_t = \text{IPMT}(\text{Interest Rate, Period in Life of Loan,}$$
$$\text{Term to Maturity, Amount of Loan}),$$

$$\text{Principal Payment}_t = \text{PPMT(Interest Rate, Period in Life of Loan,}$$
$$\text{Term to Maturity, Amount of Loan).}$$

In these functions, Amount of Loan is the original principal, Term to Maturity is the length of the loan at the date it is funded, Period in Life of Loan refers to the age of the loan, that is, period 1 means the first period in the life of a loan that might be a five-year loan. The PMT(.) functions assume equal periodic payments including principal and interest, so we could not use these functions for loans with other payment terms.

The size and complexity of our model will increase when we have to deal with several different types or issues of debt. As Exhibit 11.2 showed, we need to calculate the amount issued, principal repaid, amount outstanding, and interest payment for each different type of debt and each period. So for each issue, we need at least four lines in our model. Adding to the modeling task, we usually need to keep track of which period it is in the life of the loan. If we track specific issues of debt it may be difficult to maintain a generalized model. That is, as we add different types of debt, we need to make the model specific to the types of debt. This means that the model is less flexible and applicable to other modeling problems and tasks. This is one reason it is helpful to build separate modules so that you can increase or decrease the complexity of one module or sector without having to change all the other sectors of the model. The following example shows how we can develop a more detailed sector of our model to handle various types of debt.

11.1.2 Example: A Borrowing Sector for O&R

O&R needs to finance the new production facility. O&R will need to borrow $4400 immediately (all amounts are expressed in thousands), $2400 in one year, and $1200 at the end of two years. There are several different types of debt that O&R can issue, and O&R is considering combinations of a fixed rate mortgage loan, a floating rate loan, and a fixed coupon bond. The basic terms for the different loans are as follows.

The mortgage loan would be for five years at an interest rate of 8%. It will not be fully amortizing, so it will have a balloon payment equal to 20% of the face amount of the loan at the end of the five years. The periodic payments will amortize the other 80% of the loan, with equal annual payments (principal + interest). The fees and closing costs will be 0.5% of the face amount of the loan plus $10 based on the assumption that the amount of the loan will be between $2000 and $5000.

The floating rate loan must be repaid in three equal annual principal installments, and the interest rate on the loan will be equal to the 90-day London Inter-Bank Offer Rate (LIBOR) plus 3.5%, adjusted annually. There will be a 1% loan fee paid up front.

O&R can also issue fixed coupon bonds that will be repaid in four years when the bonds mature. Bonds can be issued at 8.5%, and will involve flotation and transaction costs of 2% plus a fixed cost of $50.

O&R would like to analyze different combinations of loans to meet its needs. As a first pass, O&R is considering funding the first $4400 with a mortgage loan, then using the floating rate loan for the $2400 needed at time 1, and the bond for the funds needed at time 2. The debt sector model to help analyze these borrowing alternatives is shown in Exhibit 11.4A,

	A	B	C	D	E	F	G	H	I	J
1										
2						Odd & Rich Corporation				
3						Loan Sector				
4						Loan Tracking for Different Loan Types				
5										
6										
7	Loan Details									
8			Enter the Loan Features in this Section:							
9										
10		Loan	Amount	Interest	Borrowing	Years to	Repayment	Balloon	Fees %	Fees $
11		Type	Borrowed	Rate	Date	Maturity	Rate	Payment		
12							% per year	% of loan		
13										
14		Short Term	0	6.0%	0	1	100.0%	0.0%	0.0%	0
15										
16		Mortgage	4,400	8.0%	0	5.0	Annuity	20%	0.5%	10
17										
18		Floating	2,400	Libor +	1	3.0	33%		1.0%	
19		Rate		3.5%						
20										
21		Bond	1,200	8.5%	2	4.0	0%	100%	2.0%	50
22										
23										
24	Economic Data									
25			Enter Interest Rate forecasts in this Section:							
26										
27					Time:	0	1	2	3	4
28										
29		90 Day T-Bill Rate				3.00%				
30		90 Day Libor Rate				3.50%				
31		Libor Change from Initial Value (basis Points)				0	20	25	-15	20
32		10 Year T-Note Rate				6.50%				
33		Corp Bond, AA, 10 yr				7.00%				
34										
35	Loan Tracking									
36										
37		Mortgage								
38					Time	0	1	2	3	4
39										
40			Amount Borrowed			4,400				
41			Loan Fees			32				
42			Funds Available			4,368				
43			Total Debt Service Payment				952	952	952	952
44			Interest Payment				352	304	252	196
45			Principal Payment				600	648	700	756
46			Balloon Payment							
47			Ending Balance			4,400	3,800	3,152	2,452	1,696
48										
49			Period in Life of Loan			0	1	2	3	4
50			Remaining Term to Maturity			5	4	3	2	1
51										
52			Net Loan Cash Flow			4,368	-952	-952	-952	-952
53			Net Cost of Debt			8.25%				

EXHIBIT 11.4A   Loan sector module.

	A	B	C	D	E	F	G	H	I	J
56		Floating Rate Loan								
57					Time	0	1	2	3	4
58										
59			Libor			3.50%	3.700%	3.750%	3.350%	3.700%
60			Rate Change (basis points)			0	20	25	-15	20
61			Loan Rate			7.00%	7.20%	7.25%	6.85%	7.20%
62										
63			Amount Borrowed				2,400			
64			Loan Fees				24			
65			Funds Available				2,376			
66			Total Debt Service Payment					973	916	855
67			Interest Payment					173	116	55
68			Principal Payment					800	800	800
69			Ending Balance				2,400	1,600	800	
70										
71			Period in Life of Loan				0	1	2	3
72			Remaining Term to Maturity				3	2	1	0
73										
74			Net Loan Cash Flow			0	2,376	-973	-916	-855
75			Net Cost of Debt			7.73%				
76										
77		Bond								
78					Time	0	1	2	3	4
79			Coupon Rate			8.50%				
80			Amount Borrowed				1,200			
81			Loan Fees				74			
82			Funds Available				1,126			
83			Total Debt Service Payment						102	102
84			Interest Payment						102	102
85			Interim Principal Payment							
86			Balloon Payment							
87			Ending Balance				1,200	1,200	1,200	
88										
89			Period in Life of Loan					0	1	2
90			Remaining Term to Maturity					4	3	2
91										
92			Net Loan Cash Flow			0	0	1,126	-102	-102
93			Net Cost of Debt			10.47%				
94										
95		Summary of Loans Flows								
96					Time:	0	1	2	3	4
97										
98			Amount Borrowed			4,400	2,400	1,200		
99			Loan Fees			32	24	74		
100			Funds Available			4,368	2,376	1,126		
101			Total Debt Service Payment				952	1,925	1,970	1,909
102			Interest Payment				352	477	470	353
103			Principal Payment				600	1,448	1,500	1,556
104			Ending Balance			4,400	6,200	5,952	4,452	2,896
105										
106			Net Loan Cash Flow			4,368	1,424	-799	-1,970	-1,909
107			Net Cost of Debt			8.51%				

EXHIBIT 11.4A (*continued*).

with formulas shown for selected cells shown in Exhibit 11.4B. The model also allows for a short-term loan, but for our present discussion, we will focus on the longer term loans. This module is intended to link with a broader capital budgeting model by way of the aggregated flows.

The loan sector model shown as Exhibit 11.4A has a section for each type of loan, showing columns for just the first four years. For each loan it calculates the interest and principal payments, the balance, and the net interest cost of the loan taking account of the loan fees and transaction costs. Then, the payments and balances are aggregated to feed back to the broader firm planning model. In this way, the loan model can be expanded or contracted as necessary without having to change the main planning

	C	H
37	**Mortgage**	
38	Time	=H27
39		
40	Amount Borrowed	
41	Loan Fees	
42	Funds Available	
43	Total Debt Service Payment	=IF(AND(H$49>0,H$49<=$F$16),-PMT($D$16,$F$16,$C$16,-$H$16*$C$16),)
44	Interest Payment	=IF(AND(H$49>0,H$49<=$F$16),-IPMT($D$16,H$49,$F$16,$C$16,-$H$16*$C$16),)
45	Principal Payment	=IF(AND(H$49>0,H$49<=$F$16),-PPMT($D$16,H$49,$F$16,$C$16,-$H$16*$C$16),)
46	Balloon Payment	
47	Ending Balance	=IF(H40>0,H40,MAX(G47-H45-H46,0))
48		
49	Period in Life of Loan	=IF(AND(H38>=$E$16,H38<=$E$16+$F$16),MAX(H38-$E$16,0)," ")
50	Remaining Term to Maturity	=IF(ISNUMBER(H49),MAX($F$16-H49,0)," ")
51		
52	Net Loan Cash Flow	=H42-H43-H46
53	Net Cost of Debt	
54		
55		
56	**Floating Rate Loan**	
57	Time	=H27
58		
59	Libor	=$F$59+(H60/10000)
60	Rate Change (basis points)	=H31
61	Loan Rate	=H59+$D$19
62		
63	Amount Borrowed	
64	Loan Fees	
65	Funds Available	
66	Total Debt Service Payment	=H67+H68
67	Interest Payment	=IF(AND(H$71>0,H$71<=$F$18),G61*G69,)
68	Principal Payment	=IF(AND(H$71>0,H$71<=$F$18),MIN($G$18*$C$18,G69),)
69	Ending Balance	=IF(H63>0,H63,G69-H68)
70		
71	Period in Life of Loan	=IF(AND(H57>=$E$18,H57<=$E$18+$F$18),MAX(H57-$E$18,0)," ")
72	Remaining Term to Maturity	=IF(ISNUMBER(H71),MAX($F$18-H71,0)," ")
73		
74	Net Loan Cash Flow	=H65-H66
75	Net Cost of Debt	
76		
77	**Bond**	
78	Time	=H27
79	Coupon Rate	
80	Amount Borrowed	=IF(H$78=$E$21,$C$21,)
81	Loan Fees	=IF(H80>0,$I$21*H80+$J$21,)
82	Funds Available	=H80-H81
83	Total Debt Service Payment	
84	Interest Payment	
85	Interim Principal Payment	
86	Balloon Payment	
87	Ending Balance	=G87+H80-H85-H86
88		
89	Period in Life of Loan	=IF(AND(H78>=$E$21,H78<=$E$21+$F$21),MAX(H78-$E$21,0)," ")
90	Remaining Term to Maturity	=IF(ISNUMBER(H89),MAX($F$21-H89,0)," ")
91		
92	Net Loan Cash Flow	=H82-H83
93	Net Cost of Debt	
94		
95	**Summary of Loans Flows**	
96	Time:	=H27
97		
98	Amount Borrowed	=H40+H63+H80
99	Loan Fees	=H41+H64+H81
100	Funds Available	=H42+H65+H82
101	Total Debt Service Payment	=H43+H66+H83
102	Interest Payment	=H44+H67+H84
103	Principal Payment	=H45+H46+H68+H85+H86
104	Ending Balance	=H47+H69+H87
105		
106	Net Loan Cash Flow	=H52+H74+H92
107	Net Cost of Debt	

EXHIBIT 11.4B   Loan sector module. Formulas for selected cells.

model, which only needs to know the total payments, but not the details for each loan. Of course, the number and types of loans could be expanded without changing the broader model.

This three-loan model is structured to give the user flexibility to explore different loan features, amounts, and even borrowing dates. The loan details are entered by the user in lines 14–21, and general interest rate data are entered in lines 29–34. For each loan, the user enters the details in the loan detail section as follows:

Amount borrowed	Column B
Interest rate	Column C
Borrowing date	Column D (time 0, 1, 2, etc.)
Years to maturity	Column E
Repayment rate per year	Column F (% of loan repaid)
Balloon payment	Column G (% of loan repaid in addition to the normal repayment schedule)
Fees %	Column H (% fees based on size of loan)
Fees $	Column I (fixed $ fees independent of size)

Now, we look at the mortgage loan section to understand the model structure. For any loan, and particularly for a mortgage amortized loan, it is important to track the age of the loan and the remaining time to maturity because this information is used to calculate the interest and principal payments. Lines 66 and 67 of the mortgage section account for the age of the loan using the borrowing date and the original years to maturity as input. Ignoring for the moment the flexibility that we will discuss shortly, the period in the life of the loan would be

Period in Life of Loan = Current Period − Borrowing Date.

For example, if the current period is $t = 3$ and the borrowing date was $t = 1$, then time $t = 3$ is the second period in the loan life. However, we need to make sure that the period is not negative, so we do that with the function = Max(Current Period − Borrowing Date,0).

Our model allows the user to specify different possible borrowing dates for each type of loan (for the mortgage loan, cell D16 of the loan detail input section). But this flexibility complicates the calculation of the period in the life of the loan. We handle this in cell G66 (time $t = 2$) with the formula

Period in Life of Loan =IF(AND(G55>=D16,G55<=D16+E16),
MAX(G55−D16,0),""),

where G55 is the number of the current period, D16 the borrowing period, and E16 the years to maturity. This formula says that if the current date is

between the borrowing date and the maturity date, the period in the life of
the loan is Current Period − Borrowing Date. Otherwise, a blank is entered
in the cell.

Cell G67 calculates the time remaining in the life of the loan in
period 2 as

$$\text{Remaining Time to Maturity} = \text{IF(ISNUMBER(G66)},$$
$$\text{MAX(E16−G66,0),"")},$$

where E16 is the original years to maturity and G66 the period in the
life of the loan from the previous calculation. The =ISNUMBER(G66)
function returns a "True" value if the entry in the cell G66 is a number as
opposed to a non-number, so this formula calculates time to maturity as
Original Maturity − Period in Life of Loan if there is a number in cell G66.
There will be a number in G66 only if the current period is between the
borrowing date and the maturity date of the loan.

The effect of these loan life formulas is to enable the user to specify any
starting date for the loan and the formulas will track the loan life. For
example, the user could input the borrowing date as any time within the
model planning horizon, and the model will track the life of the loan until
maturity. The loan life figures are particularly important for tracking the
mortgage amortized loans when we use the Excel PMT functions. Turning
now to those functions, for the mortgage amortized loan, the periodic loan
payment is calculated in line 60 with the payment function of the form:

=PMT(Interest Rate, Years to Maturity, Amount Borrowed,
Balloon Payment).

However, we want the payment to be calculated only for periods between
the borrowing date and the maturity date, both of which we want to be
flexible within the model. This is done with the formula (e.g., for period 2,
column G)

$$\text{Total Debt Service Payment} = \text{IF(AND(G66>0,G66<=E16)},$$
$$\text{−PMT(C16,E16,B16,−G16*B16),)},$$

which calculates the debt service if the period in the life of the loan (G66)
is between 0 and the time to maturity (E16), and where C16 is the interest
rate, E16 the time to maturity, B16 the amount of the loan, and G16*B16

the balloon payment at the end, with G16 being the balloon payment as a percent of the amount of the loan (B16). The effect of the blank space at the end of the IF function is to return zero if the AND function is not TRUE. That is, if the current period is not between the borrowing date and the maturity date, the cell content will be zero.

The total debt service payment of the mortgage amortized loan is split into interest (line 61) and principal (line 62) with the IPMT and PPMT functions. Both functions follow the same format for the parameters to be entered in the function:

> =IPMT(Interest Rate,Period in Life of Loan,Years to Maturity, Amount Borrowed, Balloon Payment),

and

> =PPMT(Interest Rate,Period in Life of Loan,Years to Maturity, Amount Borrowed,Balloon Payment).

These functions differ from the total payment function because splitting the payment into interest and principal depends on the period in the life of the loan, which is the second parameter, G66, in each of the following statements where the loan payment functions are included in an IF statement of the same form as the earlier total payment calculation:

> Interest Payment =IF(AND(G66>0,G66<=E16,–IPMT
> (C16,G66,E16,–G16∗B16),),

> Principal Payment =IF(AND(G66>0,G66<=E16,–PPMT
> (C16,G66,E16,–G16∗B16),).

Our O&R example calls for a balloon payment at the end of the five-year loan life, amounting to 20% of the amount of the loan. Line 63 of the mortgage loan section accounts for this balloon payment. The payment formula for period 1, for example, is

> Balloon Payment =IF(G67=0,G16∗B16,0),

where G67 is the remaining term to maturity, G16 is the balloon payment as a percent of the face amount of the loan, and B16 the face amount of the loan.

Line 64 is the balance on the loan at each date, which is calculated as

> Ending Balance =IF(G57>0,G57,MAX(G64–G62–G63,0)),

which says the ending balance will be equal to the amount borrowed (G57), if the funds were borrowed that period, and otherwise the balance will be

Previous balance	(G64)
− Principal payment	(G62)
− Balloon payment	(G63)
= Ending balance	(G64)

However, the MAX function is used to assure that the balance is not negative.

The last part of the mortgage section is a calculation of the net interest cost of the loan using an IRR function. The cash flows for the loan are calculated in line 69 for each period as

Amount borrowed	(G57)
− Loan fees	(G58)
= Funds available	(G59)
− Total debt service payment	(G60)
− Balloon payment	(G63)
= Net loan cash flow	(G69)

In the period the funds are borrowed, there will be no payments, so the cash flow is positive for the borrower. In subsequent periods, payments will be made, but the amount borrowed will be zero that period, and the cash flow will be negative.

The interest cost in cell E70 is

$$\text{Net Cost of Debt} = \text{IRR(E69:O69)},$$

so the interest cost is net of loan fees and takes account of all payments made by the borrower. In this example, the stated rate on the five-year mortgage loan is 8%, and the loan fees are sufficiently small that the effective rate is still at 8.00%.

Each of the other types of loans has a section that follows the same general format. However, they are not mortgage amortized loans, so they do not use the PMT functions. In the case of the floating rate loan, the interest rate is calculated each period based on the LIBOR + Spread for that period, and the principal payment is simply a percent of the face amount of the loan. In the case of the bond, the interest is based on the coupon rate, and the entire principal is paid at the end of the loan. In each case, we should track the life of the loan to show when the loan starts and ends just as with the mortgage loan. Of course, the net cost of each loan is calculated with the IRR function.

The loan flows and costs are combined and summarized in the Summary section (lines 115–128). Each of the lines in the summary section corresponds to the individual loan sections, with each summing the amounts borrowed, payments, and balances for the individual loans. The net cost in the summed section combines all the flows, so the cost is the consolidated cost for the package of loan. For our O&R example, with three loans of differing borrowing dates, amounts, interest rates, and repayment rates, the net cost is about 8.3% as shown in cell E128.

In the context of a larger firm planning model, these combined figures are the only ones that need to be passed to the broader model. That is, the broader corporate planning model only needs to know the total debt issued, total debt payments, total interest expenses, and so on. It does not need to know the specific details for each of the firm's loans. This way, the loan section can be expanded to include any number of different loans with different features without having to change the firm-level planning model.

The original O&R model was set up to determine the firm's required financing based on a sales forecast. If we have a separate loan sector as we have been discussing, we are inserting our own specific financing decisions in the model, and we need to link these user specified decisions with the basic model. This is done in the financing module of the basic model. In the financing module of the O&R model, the debt financing was originally calculated as

$$\text{Debt Financing}_t = (\text{Debt Portion}) \text{ REF}_t.$$

If part of the firm's financing is set by the user in a loan sector, we can feed this specific financing into the model and still allow the model to solve for any additional financing with the following modification:

$$\text{Debt Financing}_t = \text{Max}[(\text{Debt Portion}) \text{ REF}_t - \text{Total Borrowing from} \\ \text{Loan Sector}_t, 0] + \text{Total Borrowing from Loan Sector}_t.$$

This allows the model to determine the financing necessary for the firm's growth, and also include the loan sector borrowing in the total.

11.1.3 Analyzing the Financing Decision

Our discussion has focused on ways to include debt in our planning model. The next step is to use the results of the model to evaluate the financing decision. Questions we might want our model to answer include:

(a) Will the firm's ratios be within the bounds acceptable to management, lenders, and stockholders?

(b) Will the company be able to make the required debt payments?

(c) What happens if sales decline?

(d) How will the financial structure affect the value of the stock?

(e) Is there a better or best financial structure?

A planning model similar to our O&R is well suited to help answer most of these questions. The financial ratios are part of the output so we can look at the liquidity, leverage, and coverage ratios that are part of the usual analysis of debt financing. Sensitivity analysis that explores different scenarios will easily help us to see whether we will be able to service the debt under a range of likely scenarios. And, with cost of capital and valuation included as part of the model, we can explore the interaction between financing, capital costs, and equity valuation. These tools give us some guidance regarding the effects of our financing plan. However, it is still difficult to find the best financial structure with this model. But, with the theory of capital structure to guide us, in the next section we will see how we can develop a model to find an optimal financial structure.

11.2 USING MONTE CARLO SIMULATION TO FIND THE OPTIMAL CAPITAL STRUCTURE

One of the most important and enduring questions in finance is what is the best financing mix for the firm? The issue has been debated for over 40 years and is still not fully resolved. There is still wide disagreement as to whether an optimal capital structure even exists, and we will not restate the debate here.* Despite a diversity of opinions, many experts now agree that the best capital structure is the one where the tax benefits of debt and

* The capital structure literature is so extensive that we cannot begin to list all the relevant articles here. For basic summaries of the debates about capital structure, see Chapter 18 in Brealey, Myers, and Allen (2008), and Chapters 15 and 16 in Ross, Westerfield, and Jaffe (2008).

the costs of financial distress are in balance. In this section, we will show how we can use Monte Carlo simulation to model this balance.

11.2.1 Default Risk, Tax Savings, and Bankruptcy Costs

Stewart Myers (1984) characterized the widely accepted view of optimal capital structure as the "static trade-off hypothesis," which he concisely summarized by stating

> "A firm's optimal debt ratio is usually viewed as determined by a tradeoff of the costs and benefits of borrowing,.... The firm is portrayed as balancing the value of interest tax shields against various costs of bankruptcy or financial embarrassment." (p. 577)

According to this view, the firm's optimal capital structure occurs at the mix of debt and equity where the "net benefit of debt" is maximized. The net benefit of debt can be expressed as

Expected Present Value of Tax Savings from Interest − Expected Present Value of Bankruptcy Costs = Net Benefit of Debt

or, abbreviated as

$$NB(D) = E[PV(TS(D))] - E[PV(BC(D))],$$

where NB(D) denotes the net benefits from debt level D, and E[PV(TS(D))] and E[PV(BC(D))] the expectation of the present values of tax savings and bankruptcy costs, respectively. The optimal capital structure would be the amount of debt where NB(D) is maximized. We need to be more specific to be able to calculate E[PV(TS(D))] and E[PV(BC(D))].

11.2.1.1 Default Risk

To introduce the notation, we will not be specific at this stage about the events that bring about default, but we will say simply that if the firm cannot meet the terms required by its debt covenants, it will default, and we will assume that it enters the bankruptcy process. The probability of defaulting in period t, having survived for $t-1$ periods will be denoted by q_t. The probability of not defaulting in period t is $P_t = (1 - q_t)$. The probability of surviving for t periods and being able to service the debt in period t is

$$P_t = \prod_{j=1,t} P_j = \prod_{j=1,t} (1-q_j),$$

and the probability of surviving $t-1$ periods and then failing in period t is

$$Q_t = P_{t-1}\, q_t.$$

11.2.1.2 Tax Savings

In period t the taxes saved from paying interest I_t is TI_t, where T is the tax rate. The present value of this tax saving is $TI_t(1 + k_d)^{-t}$, and if the tax savings were certain in each period t, the present value of tax savings over the n period life of the debt would be

$$\sum_{t=1}^{n} TI_t \left(\frac{1}{1+k_d}\right)^t.$$

However, the tax savings would not be certain. Assume that if bankruptcy occurs, the firm will not pay interest or taxes, and no tax savings will be realized subsequent to the bankruptcy event. Since P_t is the probability of surviving through t periods, it represents the probability of paying interest for t periods and realizing tax savings in each period through time t. The expected value of tax savings over the n period life of the debt is

$$E[PV(TS)] = \sum_{t=1}^{n} P_t T I_t \left(\frac{1}{1+k_d}\right)^t.$$

11.2.1.3 Bankruptcy Costs

Assume that bankruptcy cost of BC_t would be paid if bankruptcy occurs in period t. The expected present value of bankruptcy cost for period t would be

$$P_{t-1}\, q_t\, BC_t \left(\frac{1}{1+k_d}\right)^t$$

and the expected value of bankruptcy costs over the n period life of the debt would be

$$E[PV(BC)] = \sum_{t=1}^{n} P_{t-1}\, q_t\, BC_t \left(\frac{1}{1+k_d}\right)^t.$$

11.2.1.4 Net Benefits from Debt

Combining these terms, the net benefit of using amount of debt D would be

$$NB(D) = \sum_{t=1}^{n} P_t T I_t \left(\frac{1}{1+k_d}\right)^t - \sum_{t=1}^{n} P_{t-1} q_t BC_t \left(\frac{1}{1+k_d}\right)^t.$$

The Net Benefit would be maximized at the point where the first derivative of NB with respect to the amount of debt, D, is zero. That is, where the marginal E[PV(TS)] is equal to the marginal E[PV(BC)]. Conceptually, this optimization is a simple calculus problem.* However, to take derivatives and solve this problem is difficult in a realistic situation because of the complexity of the E[PV(TS)] and E[PV(BC)] terms. Even though the problem is difficult to solve analytically, it is relatively easy to solve with simulation. We will show how the problem can be solved with Monte Carlo simulation.

11.2.2 The Monte Carlo Model

With Monte Carlo simulation, for a given level of debt, D, we can simulate the frequency distribution of the present value of tax savings. The simulated mean of that frequency distribution is our estimate of the expected value of the present value of tax savings, E[PVTS(D)]. Similarly, we can simulate the frequency distribution of the present value of bankruptcy costs. The mean of that simulated frequency distribution is our estimate of the expectation of the present value of bankruptcy costs. The difference of the simulated means is our estimate of the net benefit of the given level of debt. Then, by varying the amount of debt, we can find the debt level, D*, that maximizes the net benefit of debt. This is the optimal amount of debt based on the trade-off model of capital structure.

We will explain this approach in stages. First, we will use Monte Carlo simulation to explore the risk of bankruptcy. Second, armed with our ability to determine the risk of bankruptcy, we can estimate the expected present value of tax savings and bankruptcy costs. Third, we vary the debt level until we find the D* that maximizes the net benefit.

11.2.2.1 Simulating the Risk of Default and Bankruptcy

Default and bankruptcy are complex occurrences, and the line between corporate survival and corporate death is sometimes subtle and vague.

---

* Morris (1982) has shown that under reasonable conditions, NB(D) is concave, and reaches a maximum at less 100% debt.

Corporate death is usually the result of a complicated sequence of events of which default and bankruptcy are just the unfortunate denouement. Firms seldom go bankrupt suddenly from a single event. A typical sequence of events leading to bankruptcy is a story of decline: declining profits, declining liquidity, declining asset and equity values, and declining ability to meet debt payments and covenants. Because of this complexity, two different firms could face substantially the same situation, and one would go bankrupt and the other would not. Consequently, it is often difficult to say when bankruptcy will occur; and, it is even more difficult to accurately model default and bankruptcy. Part of our difficulty is because we have numerous terms dealing with financial failure, and some may not be very precise. For example, consider the following definitions of financial distress, failure, default, insolvency, and bankruptcy.

Financial distress is an imprecise term referring to the difficulty a firm faces when it struggles to meet its financial obligations, and when creditors and the financial markets become concerned that the firm may not be able to survive.

A firm is a failure when its costs exceed its revenue, or when it cannot earn a rate of return on its invested capital at least as high as its cost of capital. The firm has failed when it ceases to operate due to its inability to earn a profit.

Default is the failure to make payments required by creditors, and/or failing to meet conditions required by creditors.

Insolvency is generally regarded as having two definitions. A firm is insolvent in the equity sense when it is unable to meet its obligations as they fall due. A firm is insolvent in the bankrupt sense when the value of its assets is less than its debt obligations.

Bankruptcy also has two meanings. A firm is bankrupt when the value of its assets is less than the value of its debt obligations (same as insolvency). The second use of bankruptcy refers to the legal process. A firm is bankrupt when it applies for court protection for bankruptcy reorganization, or for court supervised liquidation according to the bankruptcy laws.

Despite the numerous concepts of financial distress, we will try to develop a model that will simulate financial distress, default, and bankruptcy. Our models will necessarily simplify the process and will not reflect all of its complexity. With this caveat, we will proceed with our first simple model of financial failure.

To develop our initial model of financial failure, we will focus on default as the crucial event, and make the following assumptions.

When the firm's operating earnings (EBITDA) are insufficient to cover the required debt service, it will default. That is, default is assumed to occur when

$$EBITDA_t < Debt\ Service_t,$$

where Debt Service includes required interest and principal payments. Note that this definition of default is simplified in that it does not make a distinction between principal and tax deductible interest payments in causing default, and it does not take account of other resources the firm might have to help service the debt and stave off default. Upon default, it immediately fails, and is bankrupt. From the standpoint of current security holders, all payments cease, interest and principal payments to creditors will end, and stockholders will get no further payments.

With the termination of interest payments, the tax savings from interest payments end.

Bankruptcy costs are incurred at the time of default and bankruptcy.

Let us apply these concepts with an example. Assume that the firm's EBITDA is normally distributed with mean, μ_{EBITDA}, of $1000, and standard deviation, σ_{EBITDA}, of $100, with a probability density as shown in Exhibit 11.5. The distribution is the same for each period t.

Assume the firm has $3673 of debt at an interest rate of 6% that requires equal annual payments of $872 (principal + interest) for five years. The probability of default in period t is

$$q_t = Pr(EBITDA_t < 872)$$

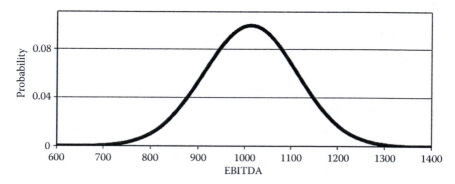

EXHIBIT 11.5   Probability distribution of EBITDA.

The standardized value of the debt service of 872 is calculated as

$$Z = \frac{872 - \mu_{EBITDA}}{\sigma_{EBITDA}}$$

$$= \frac{872 - 1000}{100} = -1.28.$$

The standardized value, Z, represents the distance, measured in standard deviation units, that 872 is from the mean, $\mu_{EBITDA} = \$1000$. That is, Debt Service of $872 is $128 less than the mean EBITDA of $1000. With a standard deviation of $100, this is 1.28 standard deviation units below the mean. If we look at a standard normal probability table, the probability of a standard normal variable being less than 1.28σ below the mean is 10%, as is shown in Exhibit 11.6. That is, the probability of default has been found to be $q_t = 10\%$.

With $q_t = 10\%$ for all periods t, we can calculate other probabilities. The probability of not failing in period t is $P_t = (1 - q_t) = (1 - 0.10) = 0.90$, and the probabilities of surviving through t periods, P_t, and the probability surviving for $t - 1$ periods and then failing in period t, Q_t, are calculated as shown below.

$P_1 = (1 - q_1) = 0.90$ $\qquad$ $Q_1 = q_1 = 0.10$

$P_2 = P_1(1 - q_2) = 0.90(0.90) = 0.81$ $\qquad$ $Q_2 = P_1q_2 = 0.90(0.10) = 0.09$

$P_3 = P_2(1 - q_3) = 0.81(0.90) = 0.729$ $\qquad$ $Q_3 = P_2q_3 = 0.81(0.10) = 0.081$

$P_4 = P_3(1 - q_4) = 0.729(0.90) = 0.656$ $\qquad$ $Q_4 = P_3q_4 = 0.729(0.10) = 0.073$

$P_5 = P_4(1 - q_5) = 0.656(0.90) = 0.59$ $\qquad$ $Q_5 = P_4q_5 = 0.656(0.10) = 0.066$

For example, the probability of surviving for three periods and then failing in period 4 is $P_3q_4 = 7.3\%$. The probability of surviving through the full

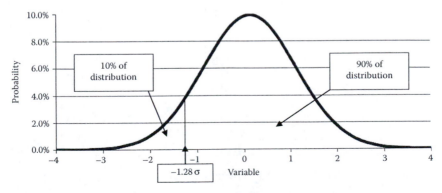

EXHIBIT 11.6   Standardized normal probability distribution.

five period life of the debt is $P_5 = 59\%$, and the probability of failing some-time during the five periods is $\sum_{t=1,5} Q_t = 0.10 + 0.09 + 0.081 + 0.073 + 0.066 = 41\%$, or, more simply, $(1 - P_5) = 1 - 0.59$.

With the probability distribution for EBITDA given, what are the probabilities of failure for different amounts of debt? Using the same five-year, 6% mortgage amortized loan, and varying the amount of debt, we get the probabilities as shown in Exhibit 11.7 and are graphed in Exhibit 11.8.

The analysis thus far provides us with sufficient information to apply the concept of debt capacity popularized by Donaldson (1961, 1962). Donaldson's chance constrained approach to debt capacity is the idea that the firm's management will determine a level of default risk that it is willing to tolerate, and issue debt up to the point where it reaches that risk

Calculation of the Probability of Default as a Function of the Amount of Debt				
Loan Terms				
Rate		6%		
Years to Maturity		5		
EBITDA Parameters				
Mean		$1,000		
Std Dev		$100		
Amount of Debt	Debt Service	Z		Probability q
$1,000	$237	-7.63		0.00%
2,000	475	-5.25		0.00%
3,000	712	-2.88		0.20%
3,200	760	-2.40		0.81%
3,400	807	-1.93		2.69%
3,600	855	-1.45		7.30%
3,673	872	-1.28		10.02%
3,800	902	-0.98		16.38%
4,000	950	-0.50		30.71%
4,200	997	-0.03		48.83%
4,400	1,045	0.45		67.20%
4,600	1,092	0.92		82.13%
4,800	1,140	1.40		91.85%
5,000	1,187	1.87		96.92%
6,000	1,424	4.24		100.00%

EXHIBIT 11.7   Calculation of the probability of default as a function of the amount of debt.

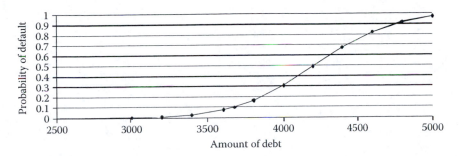

EXHIBIT 11.8 Probability of default as a function of the amount of debt.

level, but will not exceed it. In the above example, if the firm did not want to exceed a 10% probability of risk, with five-year, 6% debt, it could issue up to $3673 of debt. If they could tolerate a 15% chance of bankruptcy, they could issue a little more and push the debt to $3776.

This chance-constrained approach to debt capacity can also help us to compare different maturities in terms of default risk. In our example, if we consider a 10-year, mortgage amortized loan, at 7% (a higher interest rate because of the longer maturity), we can borrow $6125 (instead of $3673 for five years at 6%) and still be within our 10% risk limit.

11.2.2.2 Simulating Default Risk

The equations and notation introduced above aid our discussion of default risk but they may look intimidating for the less mathematically inclined. The good news is that with the aid of Monte Carlo simulation we can simplify these calculations. This section will explain how to analyze default risk and debt capacity with simulation run with @Risk.

Exhibit 11.9A shows the output of an Excel model with @Risk to determine the probability of default. Line 9 is EBITDA, representing the amount available to service the debt. It is a normally distributed random variable with mean of $1000 and standard deviation of $100.

The Excel model equations for the first few columns are shown in Exhibit 11.9B, which we will now explain. EBITDA is entered in line 9 as the @Risk probability function =RiskNormal(mean,standard deviation), so that EBITDA is generated randomly from this normal distribution. Debt service (line 10) is from the PMT function in G23. Line 11 is Funds in Excess of Debt Service = EBITDA − Debt Service, and line 12, Funds Surplus, shows a positive amount when the Funds in Excess is positive and zero otherwise.

	A	C	D	E	F	G	H	I	J
1									
2									
3				Model to Simulate Default					
4									
5	Debtor Corporation								
6				Period					
7					1	2	3	4	5
8									
9	Funds Available to Service Debt:		EBITDA		$1,006	$1,059	$843	$0	$0
10	Debt Service				$872	$872	$872	$872	$872
11	Funds In Excess of Debt Service				$134	$187	($29)	($872)	($872)
12	Funds Surplus				134	187	0	0	0
13	Funds Deficit				0	0	29	872	872
14									
15	Failure Event				1	1	0	0	0
16	Failure Product				1	1	0	0	0
17									
18									
19	EBITDA Parameters			Loan Terms					
20	Mean	$1,000		Amount of Debt		$3,673	Input		
21	Std Dev	100		Rate		6%			
22				Years to Maturity		5			
23				Debt Service		$872			
24									

EXHIBIT 11.9A Simulating default with @Risk. Default occurs when EBITDA < debt service.

	A	C	D	E	F	G
1						
2						
3				Model to Simulate Default		
4						
5	Debtor Corporation					
6					Period	
7					1	2
8						
9	Funds Available to Service Debt:		EBITDA		=RiskNormal($C$20,$C$21)	=RiskNormal($C$20,$C$21)*F16
10	Debt Service				=$G$23	=$G$23
11	Funds In Excess of Debt Service				=F9-F10	=G9-G10
12	Funds Surplus				=IF(F9>F10,F9-F10,0)	=IF(G9>G10,G9-G10,0)
13	Funds Deficit				=IF(F10>F9,F10-F9,0)	=IF(G10>G9,G10-G9,0)
14						
15	Failure Event				=IF(F13=0,1,0)	=IF(G13=0,1,0)
16	Failure Product				=F15	=F16*G15
17						
18						
19	EBITDA Parameters			Loan Terms		
20	Mean	1000		Amount of Debt		3673
21	Std Dev	100		Rate		0.06
22				Years to Maturity		5
23				Debt Service		=-PMT(G21,G22,G20,0,0)
24						

EXHIBIT 11.9B Formulas for default model.

Our model is programmed so that once the firm fails, it remains failed, and it will receive no income subsequent to failure. The way this is done is to use 0,1 variables to denote failure and success. The formula in line 13

Funds Deficit =IF(Debt Service>Funds Available,
Debt Service − Funds Available,0)
=IF(F10>F9,F10−F9,0)

tests whether there is a deficit. If there is a deficit, the Funds Deficit is F10-F9, the positive amount by which the Debt Service exceeds EBITDA. Line 15, Failure Event, is a 0,1 variable that tests whether there is a funds deficit in line 13. If there is a deficit, the Failure Event variable is 0. Line 16 tracks success or failure through time by multiplying the values in line 15 in subsequent columns. Thus, if failure occurs in any period (column), the values in line 16 will be zero for all subsequent periods. Note in cell G9 the formula =RiskNormal($C$20,$C$21)*F16. By multiplying the RiskNormal function by F16, we are setting the contents of G9 to zero if failure occurred in any previous period. Thus, if failure occurs in period 1, then there can be no EBITDA in period 2. To check the probability of failure in each period, independently of previous periods, we would remove the 0,1 variable, F16, from the formula in G9, and subsequent columns. Now return to the results of the model shown in Exhibit 11.9A. Debt Service is $872 (G23), and we see in line 9 that EBITDA exceeds the debt service in periods 1 and 2, but in period 3 (column H), the random draw of EBITDA is only $843, so it is insufficient to pay the debt service of $872. Line 12 shows a zero surplus, and line 13 shows a deficit of $29. The firm is assumed to fail, so the failure event cell for period 3 (H15) is zero. The zero in H15 is carried forward so we show zero EBITDA in periods 4 and 5. That is, once the company is dead, it remains dead in all subsequent periods.

If EBITDA is insufficient to service the debt, the firm is assumed to fail, and all subsequent cash flows cease. In the Monte Carlo simulation, the proportional frequency of having insufficient cash flow to service the debt is our estimate of the probability of bankruptcy. To obtain this estimate, we designate the Funds in Excess of Debt Service as the @Risk Output variables. To do this, select cell F11:J11 and click the Add Output icon on the @Risk menu. When you run the simulation, the frequency distribution of Funds in Excess of Debt Service is the frequency distribution of survival/bankruptcy. The estimate of the probability of default is the relative frequency of Funds in Excess of Debt Service falling below zero.

Exhibit 11.10 shows the frequency distribution from a simulation run of our default model in @Risk. The assumed amount of debt was $3673. With that amount of debt and 5000 iteration in the simulation, our estimated risk of default is 10%. Note the slider (delimiter) at $0 along the top. This is placed approximately zero, and represents the cut-off between failure and survival. The proportion of observations to the right of the arrow represents the probability (90% shown in the top band) that Funds in Excess of Debt Service will be positive—that is, the firm will survive that

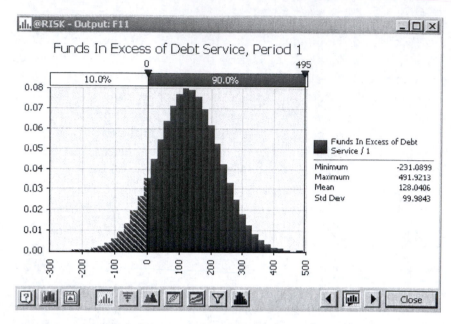

EXHIBIT 11.10   @Risk frequency distribution showing probability of default as the point where funds in excess of debt service equals zero.

period. The probability of default is the proportion of observations where Funds in Excess of Debt Service are negative—to the left of the arrow, or 10% =[1 − Pr(survival)], shown on the left in the top band. In the probability notation introduced above, our probability estimates for the first period are p_1 = 90% and q_1 = 10%. To estimate p_t and q_t for each period, we would designate Funds in Excess of Debt Service in each period t as the output variable and check the frequency distribution for each period. The model shown as Exhibit 11.9 would be modified by removing the 0,1 variables that multiply EBITDA in line 9 to obtain these estimates for each period.

11.2.2.3 Balancing Taxes and Bankruptcy Costs

We have built a simple model to analyze the probability of default. Now we want to extend that model to determine the expected values of tax savings and bankruptcy costs, and to determine the financing mix that maximizes the net benefit of debt.

As stated earlier, the net benefit of debt would be calculated as

$$\text{NB(D)} = \sum_{t=1}^{n} P_t T I_t \left(\frac{1}{1+k_d} \right)^t - \sum_{t=1}^{n} P_{t-1} q_t \text{BC}_t \left(\frac{1}{1+k_d} \right)^t.$$

If we know the probabilities and the tax and bankruptcy parameters, we can easily apply this equation to determine the net benefit for an amount of debt, D. Then, we can incrementally change D to find the debt level, D*, that maximizes the net benefit.

To see how to calculate the net benefit of debt, we will continue the earlier example with EBITDA normally distributed with mean $1000 and standard deviation $100. Assume the debt amount is $3673 for five years at 6% interest, so the probability of bankruptcy each year is $q_t = 10\%$. Assume the income tax rate is $T = 40\%$ and bankruptcy costs are equal to 20% of the outstanding debt principal at the date of bankruptcy.

Exhibit 11.11 shows the calculation of the net benefit. This is the "analytical calculation" rather than the simulation. The first section shows the probabilities of default and survival. The probabilities of default in period t, $Q_t = P_{t-1}q_t$ sum to 40.95%, which is the probability of defaulting in any one of the five periods, and is equal to $1 - P_5 = 1 - 0.5905$, the probability of surviving for five periods. The next section shows the calculation of the present value of tax savings from interest. The last line of the section, the expected value of tax savings in each period t is calculated by multiplying

Tax Savings & Bankruptcy Costs									
Loan Terms				**Cost Parameters**					
Amount of Debt	$3,673			Tax Rate		40%			
Rate	6%			Bankruptcy Cost		20%			
Years to Maturity	5								
Debt Service	872.0								
Probabilities of Default & Survival					Year				
					1	2	3	4	5
Probability of Default		q_t			10.00%	10.00%	10.00%	10.00%	10.00%
Probability of Survival		P_t			90.00%	90.00%	90.00%	90.00%	90.00%
Probability of Surviving t periods		P_t			90.00%	81.00%	72.90%	65.61%	59.05%
Probability of Default in period t		Q_t			10.00%	9.00%	8.10%	7.29%	6.56%
Value of Tax Savings from Interest									
Interest					$220.38	$181.29	$139.85	$95.92	$49.36
Tax Savings T*I					88.15	72.51	55.94	38.37	19.74
PVIF @ k d					0.9434	0.8900	0.8396	0.7921	0.7473
PV(TS)					83.16	64.54	46.97	30.39	14.75
E[PV(TS)]		P_tPV(TS)			74.85	52.28	34.24	19.94	8.71
					Expected PV Tax Savings (Sum)				190.01
Value of Bankruptcy Costs									
Debt Balance at Start of Period					3673.00	3021.42	2330.75	1598.64	822.60
Bankruptcy Cost					734.60	604.28	466.15	319.73	164.52
PVIF @ k d					0.9434	0.8900	0.8396	0.7921	0.7473
PV(BC t)					693.02	537.81	391.39	253.25	122.94
E[PV(BC t)]		Q_tPV(BC)			69.30	48.40	31.70	18.46	8.07
					Expected PV Bankruptcy Costs (Sum)				175.94
Net Benefit of Debt =		Expected Value of Tax Savings - Bankruptcy Costs							14.07

EXHIBIT 11.11 Calculating the net benefit of debt.

the discounted tax savings by the probability of surviving through period t, P_t. The sum of $190.01 is the expectation of the value of tax savings over the five period life of the debt. The last section shows the calculation of the expected present value of bankruptcy costs. Each period's bankruptcy cost is calculated as the amount of the debt at the beginning of the period multiplied by the 20% bankruptcy cost. That amount is discounted, and then multiplied by the probability of defaulting that period, Q_t, to obtain the period t expected bankruptcy cost. The summation over all periods, amounting to $175.94, is the expected present value of bankruptcy costs for the five-year debt. The difference of $14.07 is the net benefit of borrowing $3673 over five years at 6%.

11.2.2.4 Simulating the Net Benefit of Debt

Exhibit 11.11 showed the analytical calculation of the net benefit. A simulation can make the calculation easier. Exhibit 11.12A shows the tax

	A	C	D	E	F	G	H	I	J
1									
2									
3				Model to Simulate Default					
4									
5	Debtor Corporation								
6				Period					
7					1	2	3	4	5
8									
9	Funds Available to Service Debt:		EBITDA		$1,044	$1,045	$861	$0	$0
10	Debt Service				$872	$872	$872	$872	$872
11	Funds In Excess of Debt Service				$172	$173	($11)	($872)	($872)
12	Funds Surplus				172	173	0	0	0
13	Funds Deficit				0	0	11	872	872
14									
15	Success or Failure				1	1	0	0	0
16	Failure Product				1	1	0	0	0
17									
18	Interest				$220.38	$181.29	$139.85	$0.00	$0.00
19	Tx Sv				88.2	72.5	55.9	0.0	0.0
20	Tax Savings from Interest				88.2	72.5	55.9	0.0	0.0
21	PV(Tax Savings)			$194.67					
22									
23									
24	Bnkrpt Cst				734.6	604.3	466.2	322.0	322.0
25	Bankruptcy Cost				0.0	0.0	466.2	0.0	0.0
26	PV(Bankruptcy Cost)			$391.39					
27									
28	Net Benefit			($196.72)					
29									
30									
31	Debt Service Accounting								
32	Debt Service				$871.96	$871.96	$871.96	$871.96	$871.96
33	Interest (contract)				$220.38	$181.29	$139.85	$95.92	$49.36
34	Principal Payment (contract)				$651.58	$690.67	$732.11	$776.04	$822.60
35	Principal Balance (contract)			$3,673	$3,021.42	$2,330.75	$1,598.64	$822.60	$0.00
36									
37	Debt Service (actual)				$871.96	$871.96	$860.84	$0.00	$0.00
38	Interest (actual)				$220.38	$181.29	$139.85	$0.00	$0.00
39	Principal (actual)				$651.58	$690.67	$721.00	$0.00	$0.00
40	Principal Balance (actual)			$3,673	$3,021.42	$2,330.75	$1,609.76	$1,609.76	$1,609.76
41									
42									
43	EBITDA Parameters			Loan Terms			Cost Parameters		
44	Mean	$1,000		Amount of Debt		$3,673	Tax Rate		40%
45	Std Dev	100		Rate		6%	Bankruptcy Cost		20%
46				Years to Maturity		5			
47				Debt Service		$872			

EXHIBIT 11.12A Monte Carlo model for simulating tax savings and bankruptcy costs.

	A	C	E	F	G
3			Model to Simulate Default		
4					
5	Debtor Corporation				
6				Period	
7				1	2
8					
9	Funds Available to Service Debt:			=RiskNormal($C$44,$C$45)	=RiskNormal($C$44,$C$45)*F16
10	Debt Service			=$G$47	=$G$47
11	Funds In Excess of Debt Service			=RiskOutput($A$11,1)+F9-F10	=RiskOutput($A$11,2)+G9-G10
12	Funds Surplus			=IF(F9>F10,F9-F10,0)	=IF(G9>G10,G9-G10,0)
13	Funds Deficit			=IF(F10>F9,F10-F9,0)	=IF(G10>G9,G10-G9,0)
14					
15	Success or Failure			=IF(F13=0,1,0)	=IF(G13=0,1,0)
16	Failure Product			=F15	=F16*G15
17					
18	Interest			=MIN(F33,F9)	=MIN(G33,G9)
19	Tx Sv			=$J$44*F38	=$J$44*G38
20	Tax Savings from Interest			=F19	=G19*F16
21	PV(Tax Savings)	=NPV(G45,F20:J20)			
22					
23					
24	Bnkrpt Cst			=$J$45*E40	=$J$45*F40
25	Bankruptcy Cost			=IF(F16=0,F24,0)	=IF(G16=0,G24,0)*F16
26	PV(Bankruptcy Cost)	=NPV(G45,F25:J25)			
27					
28	Net Benefit	=E21-E26			
29					
30					
31	Debt Service Accounting				
32	Debt Service			=-PMT($G$45,$G$46,$G$44,0,0)	=-PMT($G$45,$G$46,$G$44,0,0)
33	Interest (contract)			=-IPMT($G$45,F7,$G$46,$G$44,0)	=-IPMT($G$45,G7,$G$46,$G$44,0)
34	Principal Payment (contract)			=-PPMT($G$45,F7,$G$46,$G$44,0)	=-PPMT($G$45,G7,$G$46,$G$44,0)
35	Principal Balance (contract)	=G44		=E35-F34	=F35-G34
36					
37	Debt Service (actual)			=MIN(F32,F9)	=MIN(G32,G9)
38	Interest (actual)			=MIN(F33,F37)	=MIN(G33,G37)
39	Principal (actual)			=F37-F38	=G37-G38
40	Principal Balance (actual)	=E35		=E40-F39	=F40-G39
41					
42					
43	EBITDA Parameters		Loan Terms		
44	Mean	1000	Amount of Debt		='Default Risk'!G20
45	Std Dev	100	Rate		='Default Risk'!G21
46			Years to Maturity		='Default Risk'!G22
47			Debt Service		=-PMT(G45,G46,G44,0,0)
48					

EXHIBIT 11.12B   Formulas for tax savings and bankruptcy costs.

savings—bankruptcy cost model in Excel with @Risk, and Exhibit 11.12B shows the first two columns of Excel code for the model. The tax savings—bankruptcy cost simulation model extends the default model, Exhibit 11.11, repeating the top 16 lines of the default model. The tax savings and bankruptcy costs calculations are on lines 18–28, debt service is in lines 31–40, and the input data is in the bottom line 43–47. The amount of the contractual interest and principal payments are in lines 32–35. However, if EBITDA is insufficient to make the contractual payment, lines 37–40 calculate the amount actually paid assuming that the entire EBITDA will be paid to cover the debt service, with interest being paid first. In the example shown, period 3's EBITDA of $861 is insufficient to cover the required debt service of $872. The whole $861 (line 37) is paid to the lenders, consisting of interest of $139.85 (line 38), and the remaining amount of $721 (line 39) going toward the principal. Since this is insufficient, the firm is in default, and is assumed to fail. Failure in period 3 results in zero EBITDA in periods 4 and 5. Zero EBITDA in periods 4 and 5 (I9 and J9) is modeled by the multiplication of

EBITDA by the zero Failure Product variable in line 16. The tax savings (line 20) are $88.20, $72.50, and $55.90 in periods 1, 2, and 3, and, because of failure in period 3, zero in periods 4 and 5. The present value of the three periods of tax savings amounts to $194.677 (E21), where the interest rate on the debt is used as the discount rate.

The bankruptcy cost (line 24) is assumed to be equal to 20% of the debt principal balance at the end of the previous period. However, this cost is incurred only at the time default occurs, so the bankruptcy cost in line 25 is calculated as

$$\text{Bankruptcy Cost}_t = (\% \text{ Bankruptcy Cost})(\text{Principal Balance}_{t-1})$$
$$\text{if Default in period } t,$$
$$= 0 \text{ otherwise.}$$

For period 2 (column G) this is

$$\text{G25} = \text{IF(G16=0,G24,0)}*\text{F16,}$$

where G24 is 20% of the previous principal balance, G16 the failure product (0 if failure occurs), and F16 the previous period's failure product. In this way, if default occurs in period 2, Bankruptcy Cost is a positive amount, but if default occurred in a previous period, it is zero. In the example, default occurs in period 3, so Bankruptcy Cost is $466 (20% × 2330) at time 3, and zero in all other periods. The present value of bankruptcy cost is $391.39 (E26). The Net Benefit (E28) of the debt is the difference between the present value of the tax savings ($194.67) and the present value of bankruptcy costs ($391.39), which amounts to −$196.72. Thus, in this case, the benefits of debt were outweighed by the costs of financial distress. However, before the fact, we do not know which sequence of events will occur and we do not know if the firm will fail. That is why we run the simulation.

The results shown in Exhibit 11.12A are for one random draw in the simulation. Monte Carlo simulation makes many draws from the specified probability distributions, so we can see the results of numerous scenarios instead of just one of the possible outcomes. To simulate the expected Net Benefit, we use @Risk to run a Monte Carlo simulation of the model with Net Benefit, cell E28, designated as the @Risk output variable. The mean of the simulated distribution of Net Benefit will be our estimate of the expected Net Benefit.

Exhibit 11.13 shows the frequency distribution from a simulation using 1000 random draws. The mean of $35.25 is our estimate of the expected net

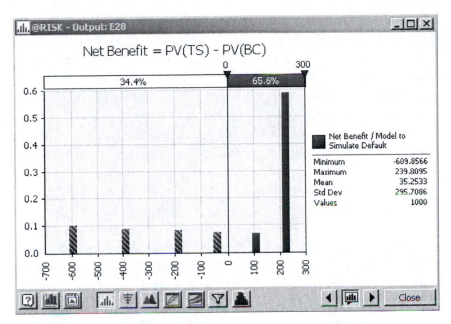

EXHIBIT 11.13 Frequency distribution of net benefit of debt.

benefit associated with using $3673 of debt. The frequency distribution in Exhibit 11.13 consists of six spikes, with each spike representing the relative frequency of each of the six discrete outcomes. The six outcomes are survival and failure at times 1, 2, 3, 4, and 5, respectively. If the firm survives the full five periods, the present value of tax savings is $239.81, which, with no bankruptcy costs, is the net benefit. If it survives for four periods and fails in the fifth period, the net benefit is $116.87, and failure in periods 4, 3, 2, and 1 generate net benefits of −$28.20, −$196.72, −$390.11, and −$609.86, respectively. The spikes in the frequency diagram are at these points.

While $35.25 is a valid estimate of the expected benefit of the debt, a look at the frequency chart of net benefits will suggest that we need to be careful about interpreting this number. Note that the frequency distribution in Exhibit 11.13 does not resemble a symmetric, continuous distribution such as the normal that we are familiar with. We will get one of the six possible outcomes, and not one of them is $35.25. Still, we evaluate the benefit of the debt as being $35.25 because it is the expected value of the six possible outcomes, and we will use this number in comparison to other debt amounts and strategies.

We have calculated the net benefit from using $3673 of debt financing. Now we have to determine the optimal amount of debt. The question is whether we can obtain a higher net benefit with a different amount of debt. For example, would we obtain higher net benefits if we borrowed $4000 instead of $3673? To answer this question, we increase the debt to $4000, and run the simulation to estimate the expected net benefit. With $4000 of debt, the simulated mean of the Net Benefit turns out to be negative −$280.86, so borrowing $4000 is worse than $3673.

To find the amount of debt that maximizes the net benefit, we vary the amount of debt in increments so that we can trace out the values of tax savings and bankruptcy costs for each level of debt. Different levels of debt were put into the model and the simulation was run to obtain the distribution means of tax savings, bankruptcy costs, and net benefits, respectively. The results for various levels of debt are shown in tabular form in Exhibit 11.14, and as a graph in Exhibit 11.15. As can be seen on the graph, the net benefit reaches its maximum at about $3100 of debt, where the probability of bankruptcy is still quite small at 0.25%. Beyond that level of debt, the probability of default starts to rise very rapidly, as does the expected present value of bankruptcy costs. This risk quickly overwhelms the tax savings. This simulation shows us that for this firm, with fairly large proportional bankruptcy costs, there is not much capacity to increase debt before it becomes too risky.

Debt Amount	Default Probability	Expected Tax Savings	Expected Bankruptcy Cost	Expected Net Benefit
500	0.00%	32.64	0.00	32.64
1,000	0.00%	65.3	0.0	65.3
1,500	0.00%	97.9	0.0	97.9
2,000	0.00%	130.6	0.0	130.6
2,500	0.00%	163.2	0.0	163.2
2,750	0.55%	179.4	0.7	178.8
2,900	0.35%	189.2	1.2	188.0
3,000	0.45%	195.2	4.1	191.1
3,100	0.25%	201.6	5.9	195.7
3,250	1.25%	209.1	20.6	188.5
3,500	4.20%	216.4	75.9	140.6
3,673	10.85%	209.9	174.7	35.3
3,800	14.15%	204.1	267.9	-63.8
4,000	30.20%	180.2	449.8	-269.7
4,500	76.75%	124.3	790.3	-666.0
5000	96.70%	116.2	936.1	-819.9

EXHIBIT 11.14   Results from Monte Carlo simulations of expected net benefit, tax savings, and bankruptcy costs at various levels of debt.

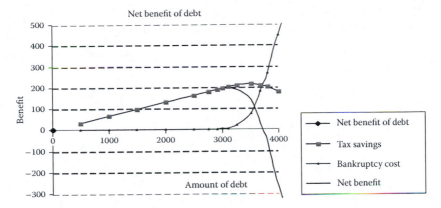

EXHIBIT 11.15 Net benefit, tax savings, bankruptcy costs, and probability of default.

11.3 ADDITIONAL DEBT MODELING CONCEPTS

In this section we introduce two additional topics relating to modeling debt. First, we briefly explain the concept of duration, and then show how to model duration in the spreadsheet. Second, we introduce swaps and show how to construct a model to analyze simple swap opportunities.

11.3.1 Duration

The duration of a security measures two related aspects of the security: the length of its life, and its sensitivity to changes in interest rates. These aspects are potentially important for both the issuer of the security and the investor who buys the security.

11.3.1.1 Duration as Average Life of Cash Flows

Duration is usually applied to debt securities, particularly bonds. For bonds we think of the life of the bond as the time until it matures. For example, a bond that pays the principal at the end of five years is considered to be a bond with five years to maturity. But because we receive several coupon interest payments in the intervening periods, the time to maturity may not tell us much about the timing of the cash flows we get from the bond. Duration originated as a measure of the life of the security that takes account of the timing of all the payments, not just the last one at the maturity date.

To understand duration, consider the following example. Suppose you loan me $1000. The interest rate on the loan is 10%, and I promise to make the following annual payments over four years to repay the loan and pay you 10% interest:

Period	1	2	3	4
Payment	100	900	150	78

At a 10% discount rate, these payments have a present value of $1000, calculated as

$$100\left(\frac{1}{1.10}\right)^{1} + 900\left(\frac{1}{1.10}\right)^{2} + 150\left(\frac{1}{1.10}\right)^{3} + 78\left(\frac{1}{1.10}\right)^{4} = 1000,$$

confirming that the rate on the loan is 10%.

Ask yourself, "What is the time to maturity of the loan?" Surely, you will answer, "Four years." However, when do you actually get your payments? Do you have to wait four years to get your payments? The answer is no. By the end of the second period, you have received $1000, the amount you originally loaned me. In terms of the length of time until you receive your money, the time is something less than four years because you receive most of your money sooner than four years. Therefore, you are not really waiting a full four years to receive your money back.

Duration is a measure of the average time it takes to collect all the payments. Duration is a weighted average of the dates that you receive the payments, where the weights are the *proportion of the value* of the investment that you receive at each date.

In this example, the present value of all the payments is $1000. Of that $1000, the first payment of $100 has a present value of $100(0.9091) = $90.91, which is $90.91/$1000 = 0.0991 = 9.91% of the value that you will receive. The second payment of $900 has a present value of 900 (0.8264) = $743.80, which constitutes 74.38% of the value of all the payments. Similar calculations show that the third payment of $150 has a present value that is 11.27% of the total present value ($150(0.7513) = $112.70), and the last payment of $78 is 5.328% of the total value ($78(0.687) = $53.28). Of the total present value of $1000 that you receive, you will receive 9.91% in period 1, 74.38% in period 2, 11.27% in period 3, and 5.33% in period 4.

While it takes four periods to receive all the funds, most of the money is received earlier than four periods. Duration takes account of this pattern. Duration is calculated as the weighted average:

$$\text{Duration} = \sum_{t=1}^{m} w_t t = 0.0909(1) + 0.7438(2) + 0.1127(3) + 0.0533(4) = 2.13,$$

(11.22)

where the weights

$$w_t = \frac{\text{Cash Flow}_t \left[1/(1+r)^t \right]}{\sum_{j=1}^{m} \text{Cash Flow}_t \left[1/(1+r)^{jt} \right]}$$

(11.23)

are the proportions of the present value received at each date, and t is the date at which you receive the payment. In this case, you receive payments at times $t = 1, 2, 3,$ and 4, but the duration is 2.13 years. This indicates that the average time to receive the value of the payments is 2.13 years, instead of four years.

Exhibit 11.16 shows the calculation of duration for three different bonds. Bond A is the one we just explained, bond B is an 8% annual coupon bond that matures in 10 years, and bond C is a 10% annual coupon bond that matures in 16 years. The cash flows from each bond are shown in columns C, H, and L, respectively. The numbers in column P are the present values of $1, [1/(1 + r)^t]$, commonly abbreviated PVIF. The amounts in columns

	A	B	C	D	E	F	G	H	I	J	K	L	M	N	O	P
1			Bond A					Bond B				Bond C				
2	Duration					2.13			7.04					9.81		
3																
4				Value	% of											
5	Period		Cash	Value		t·(%of Value)		Cash				Cash				PVIF
6	0		Flow	1000				Flow	877.11			Flow	1000			1/(1+r)
7																
8	1		100	90.91	0.091	0.0909		80	72.73	0.083		100	90.91	0.104		0.9091
9	2		900	743.72	0.744	1.4874		80	66.12	0.151		100	82.64	0.188		0.8264
10	3		150	112.40	0.112	0.3372		80	60.11	0.206		100	75.13	0.257		0.7513
11	4		78	52.97	0.053	0.2119		80	54.64	0.249		100	68.30	0.311		0.6830
12	5							80	49.67	0.283		100	62.09	0.354		0.6209
13	6	=C11*P11						80	45.16	0.309		100	56.45	0.386		0.5645
14	7							80	41.05	0.328		100	51.32	0.410		0.5132
15	8					=SUM(F8:F11)		80	37.32	0.340		100	46.65	0.425		0.4665
16	9	=D11/$D$6						80	33.93	0.348		100	42.41	0.435		0.4241
17	10							1080	416.39	4.747		100	38.55	0.440		0.3855
18	11											100	35.05	0.440		0.3505
19	12	=A11*E11								7.04		100	31.86	0.436		0.3186
20	13					=((I17/$I$6))*$A17						100	28.97	0.429		0.2897
21	14											100	26.33	0.420		0.2633
22	15				=DURATION(DATE(2011,1,1),DATE(2020,12,31), 0.08,0.1,1)							100	23.94	0.409		0.2394
23	16											1100	239.39	4.367		0.2176
24																

EXHIBIT 11.16 Duration calculations for three bonds.

D, I, and M are the present values of each of the annual payments, with the sum of these present values being the value of the bonds in D6, I6, and M6. For bond A, cells E8:E11 are the weights calculated according to Equation 11.22, and F8:F11 are the product tw_t. The sum Σtw_t is the duration in F2, which as shown earlier is 2.13 years for bond A. For bonds B and C, J8:J17 and N8:N23 are the calculations of tw_t, and these figures are summed to obtain the durations in J2 and N2. The durations are 7.04 years for the 10-year, 8% bond, and 9.81 years for the 16-year, 10% bond. Note that with a duration of 9.81 years for the 16-year bond, the duration is just a little greater than half of the time to maturity.

Excel has a duration formula that is in cell J19 of the exhibit, and shows the duration of the 10-year bond. The format for the duration function is

$$=\text{Duration(settlement date,maturity date,coupon,}$$
$$\text{yield,frequency,basis),}$$

where settlement date is the date you pay for the bond, maturity date is the last payment (principal plus interest), coupon is the bond's coupon rate, yield is the yield to maturity based on the current market price of the bond, frequency is the number of coupon payments per year, and basis is the number of days per year. As you can see, the duration formula returns the same answer as our model for this standard bond. However, it only applies to standard bonds, and is not suited to more unusual cash flow patterns such as bond A. For securities other than standard fixed coupon bonds, the Excel duration formula does not work well, and you are better off making your own model similar to Exhibit 11.16.

11.3.1.2 Duration as Interest Rate Sensitivity

Many discussions of duration emphasize that interpreting duration as a measure of time is misleading. They say that the more relevant interpretation is that duration is a measure of the sensitivity of the value of a security to changes in the discount rate. This is because in response to small changes in the discount rate, the (approximate) percentage change in the value of the security can be written as

$$\frac{d\text{Value}}{\text{Value}} = -\text{Dur}\left(\frac{dr}{1+r}\right), \qquad (11.24)$$

where dValue denotes the change in the value, Dur the duration, and dr the change in the interest rate, r. This says that the greater the duration, the

	A	B	C	D	E	F	G	H	I	J	K	L	M	N	O	P
1			Bond A					Bond B				Bond C				
2	Duration					2.12				6.94				9.38		
3																
4				Value	% of											
5	Period		Cash		Value	t·(%of Value)		Cash				Cash				PVIF
6	0		Flow	981				Flow	823.32			Flow	926.2084			1/(1+r)
7																
8	1		100	90.09	0.092	0.0918		80	72.07	0.088		100	90.09	0.109		0.9009
9	2		900	730.38	0.745	1.4891		80	64.93	0.158		100	81.16	0.197		0.8116
10	3		150	109.39	0.112	0.3345		80	58.50	0.213		100	73.12	0.266		0.7312
11	4		78	51.08	0.052	0.2083		80	52.70	0.256		100	65.87	0.320		0.6587
12	5							80	47.48	0.288		100	59.35	0.360		0.5935
13	6							80	42.77	0.312		100	53.46	0.390		0.5346
14	7							80	38.53	0.328		100	48.17	0.410		0.4817
15	8							80	34.71	0.337		100	43.39	0.422		0.4339
16	9							80	31.27	0.342		100	39.09	0.427		0.3909
17	10							1080	380.36	4.620		100	35.22	0.428		0.3522
18	11											100	31.73	0.424		0.3173
19	12									6.94		100	28.58	0.417		0.2858
20	13											100	25.75	0.407		0.2575
21	14											100	23.20	0.394		0.2320
22	15											100	20.90	0.381		0.2090
23	16											1100	207.12	4.025		0.1883
24																
25	Discount Rate		11%													

EXHIBIT 11.17   Bond values and duration at 11% interest.

more the value will change in response to changes in the interest rate. Longer term bonds that have greater duration will be more sensitive to changes in interest rates than bonds with short duration.

To see the link between interest rates, value, and duration, return to the example of the three bonds of Exhibit 11.16. Keep in mind that at an interest rate of 10%, the values of the bonds were $1000, $877.11, and $1000 for A, B, and C, and their durations were 2.13, 7.04, and 9.81 years, respectively. Exhibit 11.17 is the same spreadsheet, except we changed the interest rate from 10% to 11%. With the discount rate at 11%, the bond values have declined to $981, $823, and $926, which are percentage changes of −1.9%, −6.13%, and −7.38%. Using Equation 11.24, the calculation of the percentage change in value for each of the three bonds would be

$$\frac{dValue}{Value} = -Dur\left(\frac{dr}{1+r}\right) = -2.13\left(\frac{0.01}{1.10}\right) = -0.0194 = -1.94\% \text{ for A,}$$

$$= -7.04\left(\frac{0.01}{1.10}\right) = -0.064 = -6.4\% \text{ for B,}$$

and

$$= -9.81\left(\frac{0.01}{1.10}\right) = -0.0892 = -8.9\% \text{ for C.}$$

As you see, these figures approximate the change in values that we computed. However, they are just approximations because the formula is accurate only for very small changes in the interest rate. A rate change from 10% to 11% is relatively large in this context, and there is some error. The error is because the relationships are non-linear, and the duration formula does not pick that up. The non-linear relation is referred to as convexity, and for a more accurate calculation we need to take the convexity into account. However, like so many topics, that is beyond the scope of this chapter, and you are referred to more detailed discussions such as Hull (2009), Ho and Lee (2004), and Reilly and Brown (2006).

11.3.1.3 Duration and Immunization

Duration is useful for understanding the interest rate risk of a security. But in the context of financial planning for corporate financial management, one of the most interesting applications of the concept is immunization. Immunization has two different meanings in this context. The first refers to balancing the risk of changes in *value* due to interest rate changes against *reinvestment* risk.* The second meaning refers to insulating the value of the equity of the firm from changes in interest rates. We will confine the remainder of this section to the second interpretation.

From our duration discussion we see that the longer the duration of an investment, the more sensitive the value of the security is to changes in interest rates. If rates go up, investments with longer durations will decline more in value than those with shorter durations. And, if interest rates decline, longer duration investments will increase in value more than shorter duration investments.

Note that we used the term investment in reference to the sensitivity of the value. The reason is that the sensitivity to interest rates does not just apply to debt securities such as bonds. It applies to all investments whose values are based on the present value of future cash flows. With that in mind, we can think of a firm consisting of different claims on the cash flows. The assets generate a stream of future cash flows that come into the firm and are subsequently paid out to the creditors and the equity owners. With the equity receiving the residual flow after payments to the creditors, the value of equity is the value of the assets minus the value of the liabilities:

$$\text{Equity Value} = \text{Asset Value} - \text{Liability Value}.$$

---

* Reinvestment risk is an important concept for managing investment portfolios—particularly those such as retirement funds that have a target payout in the future. Fabozzi (2004) and Granito (1984) provide greater detail about immunization in the presence of reinvestment risk.

Each of the components in this expression has a duration of its cash flows, and each has a sensitivity to interest rate changes.

With Equity being the residual claimant, consider the effects of interest rate changes on the value of the equity. Suppose the asset cash flow is of very long duration, and the liabilities are of short duration, and the interest rates increase. The increase in rates will make the longer duration component decline in value more than the shorter duration component. In this case, the value of the long duration assets will decline more than the short duration liabilities. The net effect is that the value of the equity will decline simply because the asset value declines more than the liability value. On the other hand, if rates decline, the long duration assets will increase more in value than the shorter duration liabilities, and the value of equity will increase.

The point is that the duration of the assets and liabilities impacts the interest rate risk of the equity. The interest rate risk of the equity can be managed by adjusting the maturities and duration of the assets and liabilities. In the example, if management wants to decrease equity's exposure to interest rate risk, they could restructure the debt to increase the duration of their liabilities. With appropriate balancing of the duration of assets and liabilities, the value of equity should be immunized against changes in interest rates. We will not go more deeply into this interesting issue than to alert you to one of the applications of the duration concept for financial management.*

11.3.2 Swaps

In the last section we discussed the effects of interest rate changes on the value of securities. In this section we explain a technique for dealing with interest rate variability: interest rate swaps. Interest rate swaps are a mechanism for borrowers and lenders to shift from fixed to variable rate, or variable rate to fixed rate securities.† For example, suppose a borrower has a variable rate loan, but a fixed rate loan would fit its needs better. The variable rate borrower could arrange a swap with another borrower who

---

* The following sources provide detailed explanations of the effect of duration and immunization on the value of equity: Fabozzi (2004), Bierwag (1987), Haugen and Wichern (1974, 1975), Boquist, Racette, and Schlarbaum (1975), and Grove (1966, 1974).

† The fixed for variable rate swaps are the most common and the easiest to understand, and are termed "plain vanilla" swaps. The name suggests that there are other flavors. There are many kinds of swaps including currency swaps, and various kinds of swaps involving risk shifting. See Hull (2009) and Dubofsky and Miller (2003) for more detailed discussions.

has a fixed rate loan and wants a variable rate. These two borrowers make the swap in a way that results in the formerly variable rate borrower now making fixed rate interest payments, and the formerly fixed rate borrower now making variable rate payments. The effect is as if the two borrowers each agree to make the payments for the other one.

The reason for interest rate swaps is that some firms are better able to handle the uncertainty of interest rate variability than others. Yet, the best deal for a loan that they can get may be a floating rate loan. They can accept the floating rate loan and then make an arrangement to swap it with another borrower whose best borrowing opportunity might have been a fixed rate loan and who may prefer to have a floating rate loan. An example will help to show how this works and show how we can model it.

11.3.2.1 Example: Swapping Plain Vanilla

The Varbic Company has a loan for $100 million. The loan matures in five years, with the entire principal paid at the end of the five years. The interest rate on the loan is variable, being equal to the six-month LIBOR rate plus 1.2% (120 basis points). Interest is paid at the end of each six-month interval based on the rate at the beginning of the period.

Varbic's income available to pay interest (EBIT) tends to be relatively stable over the economic cycle. However, for several years they have been meeting their financing needs with variable rate borrowing. The result is that their income after paying interest (EBIT - Interest) is quite variable. Exhibit 11.18 diagrams the pattern of the company's EBIT, interest cost, and income after interest. EBIT has been slowly trending upward with minor variations. But the variability of their interest costs translates directly into their income stream. The managers have concluded that if they could stabilize their interest costs, their income would be as smooth as their EBIT.

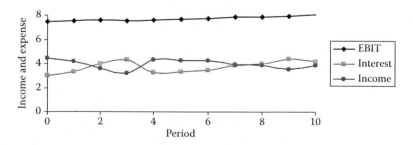

EXHIBIT 11.18 Varbic income with variable rate debt.

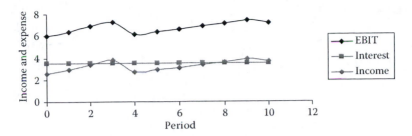

EXHIBIT 11.19 Varbic income with fixed rate debt.

Varbic management examined their income after interest costs assuming stable interest costs. Exhibit 11.19 shows the company's past income assuming that their debt had been a fixed rate loan over the last five years, with interest at an annual rate of 7.5%, which they estimate would have been the rate they could have got at the start of the five-year period. They currently have a five-year variable rate loan that at the moment is carrying a rate of 7.2%. At current rates, they could obtain a new fixed rate loan at 8.2%, but they think they can make a better deal by engaging in swap with another company that has a fixed rate loan.

The Fixion Company has a $100 million fixed rate loan that matures in five years. The rate on the loan is 7%, with interest paid every six months. Principal is paid at the end of the five years. The income stream generated by Fixion's assets is variable and tends to follow the cycle of the economy and interest rates. Fixion's management feels that they could stabilize their income by swapping to a variable rate loan. As shown in Exhibit 11.20, their income over that last five years shows considerable cyclical variability. The dotted line in the middle represents interest expense, and the bottom line is the income after interest. The variability of income has been increased by the fact that their interest costs remain constant when their EBIT is fluctuating.

Exhibit 11.21 shows Fixion's income based on the assumption that they used floating rate debt with the same pattern as Varbic. The interest cost follows the same pattern as their EBIT, so the difference, EBIT − Interest, is quite stable as shown by the income line at the bottom.

Using a bank as the broker to make the arrangements, Varbic and Fixion agree to a swap. The swap has the effect of the two companies trading payments. It is as if Varbic makes Fixion's payments, and Fixion makes Varbic's payments. The way this is actually implemented is that

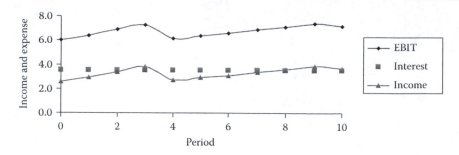

EXHIBIT 11.20   Fixion income and expense with fixed rate debt.

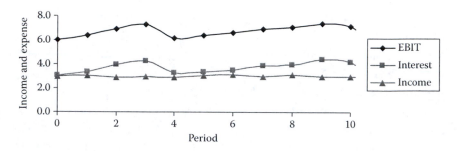

EXHIBIT 11.21   Fixion income and expense with variable rate debt.

the payments are netted against each other and one counterparty pays the other the difference. Exhibit 11.22 models how this would work. The upper left section (A5:F14) shows the input. Cells D9:F9 indicate the rate each of the companies would pay for a fixed rate loan, and D11:F12 show the rates each would have for a variable rate loan. Fixion's loan is a fixed rate at 7%, and Varbic's is a floating rate at LIBOR plus 1.20%. The current LIBOR rate is 6%, and hypothetical future LIBOR rates are shown in B19:B27 to demonstrate what could happen over the five-year remaining life of the loans.

   With the fixed rate loan at 7% on $100 million principal, Fixion's interest payments would be $3.5 million every six months for the next five years as shown in D19:F28. Given the hypothetical LIBOR rates shown in B18:B27, Varbic's interest payments would be as shown in F19:F28, varying every six months based on the variation of the LIBOR over the period. The figures in column H (H19:H28) show the difference between the fixed payments by Fixion and the floating rate payments by

	A	B	C	D	E	F	G	H	I	J	K	L	M	N	O	P
1	Swap															
2																
3	CounterParties			Fixion Co		Varbic Co										
4																
5	Principal ($ million)			100		100										
6	Term			5		5										
7	Number of Payments			10		10										
8																
9	Fixed Interest Rate			7.0%		8.2%										
10	Interest Rate - Floating															
11	Base	LIBOR		6.00%		6.00%										
12	Margin above Base			0.30%		1.20%										
13																
14	Loan - Fixed or Float?			Fixed		Floating										
15											Fixion Co				Varbic Co	
16		Period						Difference		Payment		Net		Payment		Net
17								Fx - Fl		Received	Made	Payment		Received	Made	Payment
18	0	6.00%														
19	1	7.00%		3.50		3.60		-0.10		0.00	0.10	3.60		0.10	0.00	3.50
20	2	8.00%		3.50		4.10		-0.60		0.00	0.60	4.10		0.60	0.00	3.50
21	3	8.50%		3.50		4.60		-1.10		0.00	1.10	4.60		1.10	0.00	3.50
22	4	7.00%		3.50		4.85		-1.35		0.00	1.35	4.85		1.35	0.00	3.50
23	5	6.00%		3.50		4.10		-0.60		0.00	0.60	4.10		0.60	0.00	3.50
24	6	4.00%		3.50		3.60		-0.10		0.00	0.10	3.60		0.10	0.00	3.50
25	7	5.00%		3.50		2.60		0.90		0.90	0.00	2.60		0.00	0.90	3.50
26	8	4.00%		3.50		3.10		0.40		0.40	0.00	3.10		0.00	0.40	3.50
27	9	3.50%		3.50		2.60		0.90		0.90	0.00	2.60		0.00	0.90	3.50
28	10			3.50		2.35		1.15		1.15	0.00	2.35		0.00	1.15	3.50

EXHIBIT 11.22   Swap payment by Varbic and Fixion.

Varbic. The negative numbers indicate that Varbic's payments exceed the fixed rate payment by Fixion. Positive numbers indicate that Fixion's payments are greater.

In a swap deal, the counterparties do not really take over each other's payments. Rather, they simply pay each other the difference between the payments. Look at the loan payments made by the two companies in the first month. Fixion pays $3.5 million (D19), and Varbic pays $3.6 million (F19), for a difference of −$0.10 million (H10) against Varbic. To make up this difference, Fixion will pay Varbic $0.10 million (K19). Varbic's receipt of the payment is shown in N19. The effect of Fixion paying this difference to Varbic is to stabilize Varbics net interest payments and to turn Fixion's payments into floating rate payments. Without this payment from Fixion, Varbic's cost would be $3.6 million (F19), but receiving a payment of $0.10 million reduces Varbic's net cost to $3.5 million as shown in P19. Without having to make the payment, Fixion's cost would be $3.5 million (D19), but net of the payment, Fixion's cost in $3.6 million (L19). Look at each payment period. The net payment by Fixion (column L) varies, so it is equal to the payment that Varbic would have made without the swap (F19:F28). Now, with the swap, Varbic's net payments in column P are fixed at $3.5 million every six months. The effect of the companies agreeing to pay the differences in the payments is as if they had traded loans—Fixion getting the variable rate loan, and Varbic getting the fixed rate loan.

Why would these companies do this? The first reason we already discussed. The variable rate loan fits better with Fixion's pattern of income and the fixed rate loan is better suited to Varbic's pattern of income. So why does not each company simply arrange the kind of loan they need? Without going into all the detail of the comparative advantage argument,[*] it may be that Varbic can make a better deal with the lender by taking the variable rate loan, and then swapping, and Fixion can get its best deal by taking the fixed rate loan and then subsequently swapping.

The variable rates on LIBOR in Exhibit 11.22 are hypothetical to demonstrate the impact of the variation in rates. Before the fact, we do not know what rates will do, and the borrower who ends up paying the floating rate will face the uncertainty of future rates. The model shown in Exhibit 11.22 can easily be adapted to use Monte Carlo simulation to explore what might happen from the random variation of future rates. We show a Monte Carlo model next. First, we need to construct a model of randomly varying interest rates.

11.3.2.2 Stochastic Interest Rates

Obviously interest rates vary considerably based on conditions in the economy and financial markets. To show the variation of interest rates, we need to model the process that interest rates follow over time. Vasicek (1977)[†] developed a model of the random variation of interest rates where rates are assumed to follow a mean reverting Weiner process, or what he calls an elastic random walk, of the form

$$dr = a(b - r)dt + \sigma z \sqrt{dt}, \qquad (11.25)$$

where dr denotes the change in the short-term interest rate, r, over a short period of time, dt; b a long-run average (mean) rate of interest; σ the instantaneous standard deviation of the interest rate, expressed as an annual rate; z the standard normal variable with mean 0 and standard deviation 1; and a the response rate governing the speed of adjustment. The mean rate b is a rate toward which interest rates tend to drift over time. The term $(b - r)$ is the difference between the long-run mean and the

---

[*] Hull (2009), Chapter 7 elaborates on the comparative advantage argument.

[†] Vasicek's (1977) interest rate model is one of many that portray the random variation of rates over time. Hull (2009, Chapter 30) provides an overview of many of these interest rate models.

current rate, r. When the current rate is below the average, $b - r > 0$, the rate will have a tendency to increase proportionally to the gap between the current rate and the long-run rate. Conversely, if the current rate is very high, $b - r < 0$, rates will have a tendency to drift down toward the long-run mean, b. The rate at which the adjustment occurs is governed by the response rate, a. If a is large, the adjustment will be rapid, and vice versa.

When interest rates follow the elastic random walk portrayed by Equation 11.25, they tend to vary randomly over short periods of time, but the further away the rate is from the average, the stronger the tendency to move back to the average, with random variation along the way. The random variation comes from the error term $\sigma z \sqrt{dt}$. In this equation z represents a random draw from a standard normal distribution (a normal distribution with mean 0 and standard deviation 1).

We use the Vasicek elastic random walk interest rate model to portray the movements of the LIBOR rate in our swap example. When we put some numbers into the interest rate equation it will be a little more understandable. Assume the long-run average interest rate is $b = 0.03$ (3%), the standard deviation of the interest rate change is $\sigma = 0.006$ (0.6%) per year, the response parameter is $a = 0.10$, and the time interval is $dt = 0.50$ (one half year).* Assume the current LIBOR rate is 6%, well above the long-run average of $b = 3\%$. The change in the rate in the next six-month interval will be

$$dr = a(b - r)dt + \sigma z \sqrt{dt}$$
$$= 0.10(0.03 - 0.06)0.50 + 0.006z\sqrt{0.50}$$
$$= 0.10(-0.03)0.50 + 0.00424z$$
$$= -0.0015 + 0.00424z,$$

where z is the random draw from the standard normal distribution. Suppose the random draw is $z = 1.40$. The change in the interest rate over the six-month interval would be

$$dr = -0.0015 + 0.00424(1.4) = 0.0044,$$

---

* In the interest rate model, the time interval, dt, is intended to be just an instant. That is, the limit as $dt \to 0$. However, for our example, we approximate the process by assuming that dt is a discrete period of time such as six months.

	A	B	C	D	E	F	G	H	I
34				Model of Stochastic Interest Rates					
35									
36									
37				dr = a(b-r)dt + σz(dt)$^{1/2}$					
38									
39									
40				Interest Rate Input Parameters					
41		Response Rate, a				0.1			
42		Long Run Average, b				0.03			
43		Rate Volatility, σ				0.006			
44		Time Interval, dt				0.5			
45									
46	Period					Rate		LIBOR	
47		Drift		Random		Change		Rate	
48	0							0.06	
49	1	=$F$41*($F$42-H48)*$F$44		=RiskNormal(0,1)*$F$43*SQRT($F$44)		=B49+D49		=H48+F49	
50	2	=$F$41*($F$42-H49)*$F$44		=RiskNormal(0,1)*$F$43*SQRT($F$44)		=B50+D50		=H49+F50	
51	3	-0.001874		-0.00472		-0.00659		0.0609	
52	4	-0.001545		-0.002429		-0.003974		0.05692	

EXHIBIT 11.23   Model to generate random interest rate changes.

and the interest rate will move from the initial rate of 6% to 6.44% (0.06 + 0.0044). In the next interval, a new random value of z will occur, and the rate will change from the base of 6.44%. Note that even though this process has a tendency to move back toward the long-run average, depending on the random draw, z, the rate can move up, as it did from 6% to 6.44%. But the further the rate goes above the average, the greater the tendency to move back toward the mean.

This elastic random walk of interest rates is easy to model. Exhibit 11.23 shows a Monte Carlo model to generate the interest rate series. The parameters from the example in the previous paragraph are input in cells F41:F44. Only the first four lines of the model are shown, with the formulas for periods 1 and 2 in lines 49 and 50, and the output for periods 3 and 4 in lines 51 and 52. The formulas in the drift column B are the $a(b-r)dt$ components of the model. The random part is in column D, which calculates the $\sigma z\sqrt{dt}$ part of process. The random variable z is drawn by @Risk with the =RiskNormal(0,1) function. The drift and the random components are added together in column F, and the resulting interest rate change is added to the previous period's rate in column H. The formulas are copied down the column for as many periods as needed.

When we run the model we get a sequence of rates starting from the initial value and varying randomly over time. Exhibit 11.24 shows the sequence of rates over 20 periods (10 years) generated with three different Monte Carlo simulation runs. Each different run produces as different sequence of rates. But with the mean reverting model, the sequences will tend to fluctuate randomly, but generally drift downward toward the

Three rate sequences

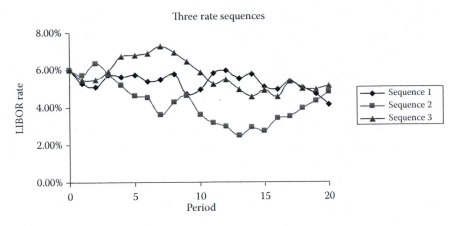

EXHIBIT 11.24　Different rate sequences from mean reverting random walk.

	A	B	C	D	E	F	G	H	I	J	K	L	M	N	O	P
1	Swap															
2																
3																
4	CounterParties			Fixion Co		Varbic Co										
5								Model of Stochastic Interest Rates								
6	Principal ($ million)			100		100										
7	Term			5		5		$dr = a(b-r)dt + \sigma dz$								
8	Number of Payments			10		10										
9																
10	Fixed Interest Rate			0.07		0.082		Response Rate, a		0.10						
11	Interest Rate - Floating							Long Run Average, b		3.00%						
12	Base	LIBOR		6.00%		6.00%		Rate Volatility, σ		0.60%						
13	Margin above Base			0.30%		1.20%		Time Interval, dt		0.50						
14																
15	Loan - Fixed or Float?			Fixed		Floating										
16											Fixion Co				Varbic Co	
17	Period							Difference	Payment		Net			Payment		Net
18								Fx - Fl	Received	Made	Payment			Received	Made	Payment
19	0	6.00%														
20	1	5.48%		3.50		3.60		-0.10	0.10	0.00	3.60			0.00	0.10	3.50
21	2	5.56%		3.50		3.34		0.16	0.00	0.16	3.34			0.16	0.00	3.50
22	3	5.27%		3.50		3.38		0.12	0.00	0.12	3.38			0.12	0.00	3.50
23	4	5.98%		3.50		3.24		0.26	0.00	0.26	3.24			0.26	0.00	3.50
24	5	5.34%		3.50		3.59		-0.09	0.09	0.00	3.59			0.00	0.09	3.50
25	6	4.85%		3.50		3.27		0.23	0.00	0.23	3.27			0.23	0.00	3.50
26	7	5.28%		3.50		3.02		0.48	0.00	0.48	3.02			0.48	0.00	3.50
27	8	5.25%		3.50		3.24		0.26	0.00	0.26	3.24			0.26	0.00	3.50
28	9	4.93%		3.50		3.23		0.27	0.00	0.27	3.23			0.27	0.00	3.50
29	10			3.50		3.06		0.44	0.00	0.44	3.06			0.44	0.00	3.50
30																
31	Annual Rate over 5 Years			7.12%		6.73%					6.73%					7.12%

EXHIBIT 11.25　Swap model with Monte Carlo model of stochastic interest rates.

long-run average, which in this example is assumed to be 3%. This example shows the Monte Carlo model of interest rates. The next step is to use the interest rate model in the swap model.

Exhibit 11.25 is the swap model shown earlier with the LIBOR rates generated as a mean reverting random walk. One particular realization of the randomly generated LIBOR rates are shown in cells B19:B28. The payments by each counterparty to the swap are in columns J–P, just as in Exhibit 11.22. Note that Varbic ends up with fixed payments of

$3.5 million every six months no matter what the sequence of LIBOR rates that might be generated. On the other hand, Fixion ends up paying the floating rate.

Since the LIBOR rate can vary over time, it is difficult ahead of time to know what the floating rate loan will cost. The Monte Carlo simulation allows us to see what the possibilities are. There are different ways to measure the cost. The way we will gauge the cost of the loan is to calculate the IRR on the loan over the five-year loan period. Treating the loan principal as a cash inflow at time $t = 0$, and the semi-annual interest payment and the principal payment at the end of the five years as outflows, the cost of the loan over the five years can be calculated with the =IRR( ) function in Excel.

A single random sequence of rates will generate a single IRR. What we want to know is what does the probability distribution of all possible IRRs look like. When we run the Monte Carlo simulation with many iterations, we generate a frequency distribution of IRRs.

Exhibit 11.26 shows the frequency distribution of Fixion's annual interest cost (IRR) over the five years with the swap. The mean cost is 6.75% with a standard deviation of 0.623%. The maximum rate was 8.72% and

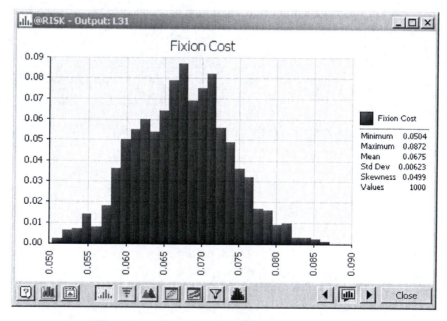

EXHIBIT 11.26　Frequency distribution of annual rate.

the minimum was 5.04%. With the swap, even though the rate is floating randomly, Fixion is able to stabilize their income stream by swapping to a variable rate loan, and in this case, the company is able to decrease the expected interest cost. Without the swap, their cost would have been fixed at 7.12%, but with the swap, the expected (mean) cost has been reduced to 6.75%. The reason for the reduction in cost is that with the mean reverting process, rates tend to drift downward if they are currently on the high end of their range. Since rates were high at the time of the swap, the counter-party with the floating rate is able to ride the rates downward. On the other hand, if rates had been low at the time of the swap, it is more likely that the floating rate borrower would end up with higher rates over the five years.

In this section we have introduced you to the plain vanilla swap where a fixed rate loan is swapped for a variable rate loan. The model we presented enables us to analyze the effects of the swap and see the distribution of possible interest costs for a borrower who undertakes the variable rate debt. This Monte Carlo model of borrowing rates and costs is helpful for understanding the costs and benefits of the swap. There is a wide variety of swaps and similar arrangements. This model is just scratching the surface, but the model shows you how to get started to develop a model that will analyze more complex situations.

11.4 SUMMARY

The purpose of this chapter was to show different ways to model the firm's debt. The first section developed a detailed model for tracking the debt in the context of a long-term planning model such as our ubiquitous O&R model. With this elaboration, the planning model can track the details of different debt issues, so the user can explore the implications of alternative borrowing strategies.

The second section presented a Monte Carlo model for analyzing default risk and for determining the optimal capital structure. The Monte Carlo model allowed us to balance the tax savings and bankruptcy costs associated with the debt so as to determine the optimal amount of debt.

The last section of the chapter discussed two aspects of debt: duration and swaps. We only touched on these topics so that you would know what they were, and have a starting point should you have to delve more deeply into the topics and build models to deal with these issues. Like so many topics in this text, we are only able to get you started, and leave it up to you to extend the models as required for your applications.

PROBLEMS

1 Evaluating the Risk of Default for Laric Corporation

You are the financial manager of the Laric Corporation. The chief executive officer (CEO), Al Laric has stated that he would like you to determine the maximum amount of debt that the company could issue and not exceed certain limits on the risk of default on the debt. Specifically, he wants you to determine the maximum amount of long-term debt that could be issued and not exceed respectively: (a) 5% chance of bankruptcy; (b) 10% chance of bankruptcy; and (c) 20% chance of bankruptcy in 2011.

The current date is year end 2010. Laric's financial statements through the end of 2010 are shown in Table 11.1. Dollar amounts are stated in thousands ($000). For the purpose of making forecasts of Laric's future financial condition, the following assumptions are made.

Laric's annual sales revenue is a normally distributed random variable that will be drawn *independently* from normal distributions with expected values and standard deviations as follows:

Year	2011	2012	2013	2014	2015
Expected sales	33,000	33,000	33,000	33,000	33,000
Standard deviation	3000	3000	3000	3000	3000

CGS will consist of fixed and variable components. The variable cost is a random variable from a triangular distribution with a minimum value of 55% of sales, a most likely value of 60%, and a maximum value of 65%. The fixed portion of CGS will be $3000 per year.

SGA consists of a fixed component equal to $5000 per year plus a variable component that is a random variable from a uniform distribution that ranges from 8% to 12% of sales.

> The Depreciation expense (a non-cash expense) is equal to 8.5% of GFA at the end of the previous year.
> Interest expense is equal to 9% of the previous year's debt balance for both new and old debt.
> The income tax rate is 40%.
> The target Cash balance is 8% of *expected* Sales.
> Receivables will be 2.2% of *expected* Sales.
> Inventories will be 15.6% of *expected* Sales.
> GFA will be 25.5% of *expected* Sales.
> Other Long-term Assets will be equal to a fixed amount of $1400 plus a variable amount equal to 10% of *expected* Sales.

TABLE 11.1   Financial Statements for Laric Corporation

	A	B	C	D	E	F	G
1					**Laric Corp**		
2					**Income Statement**		
3					($ millions)		
4		Year Ended December 31:					
5				2006	2007	2008	2009	2010
6							
7	Sales		$31,380	$32,510	$31,846	$32,004	$33,000
8	Cost of Goods Sold		22,466	23,665	22,458	22,193	22,800
9	Selling, General & Administrative		6,759	7,164	8,637	8,459	8,300
10	Other Inc/Exp		100	233	-465	-6	0
11	EBITDA		2,255	1,914	286	1,346	1,900
12	Dep		637	710	695	717	667
13	EBIT		1,618	1,204	-409	629	1,233
14	Interest		663	673	477	426	0
15	EBT		955	531	-886	203	1,233
16	Tax		361	195	-318	89	493
17	Net Income		594	336	-568	114	371
18	Extraordinary Items		0	0	-137	-16	0
19	Adjusted Net Income		594	336	-705	98	371
20							
21	Dividend	(Pfd + Cmn)	590	536	248	161	440
22	Pfd Dividend		38	36	33	29	0
23	Common Dividend		552	500	215	132	440
24							
25	Retained Earnings		4	-200	-953	-63	300
26							
27					**Laric Corp**		
28					**Balance Sheet**		
29					($ millions)		
30		December 31					
31				2006	2007	2008	2009	2010
32	Assets						
33	Cash		$511	$1,233	$944	$2,840	$2,474
34	Accounts Receivable		4,415	1,138	893	698	705
35	Inventory		6,031	5,947	5,269	4,930	4,945
36	Other CA		168	154	151	209	229
37	Current Assets		11,125	8,472	7,257	8,677	8,353
38							
39	Gross Fixed Assets		8,333	8,195	8,062	8,317	8,154
40	Accum Depreciation		2,875	2,883	2,948	3,328	3,253
41	Net Fixed Assets		5,458	5,312	5,114	4,989	4,901
42	Other Assets		7,055	7,104	7,371	4,382	4,613
43	Total Assets		23,638	20,888	19,742	18,048	17,867
44							
45	Liabilities						
46	Accounts Payable		1,496	1,480	1,948	1,551	1,792
47	Notes Payable		2,362	955	250	935	0
48	Other CL		2,112	2,030	2,037	2,013	4,477
49	Current Liabilities		5,970	4,465	4,235	4,499	6,269
50							
51	Long Term Debt		7,143	5,844	5,448	5,179	0
52	Other LT Liabilities		3,356	3,351	3,800	2,241	0
53	Preferred Stock		475	446	399	363	0
54	Common Stock		2,850	3,266	3,294	3,330	3,423
55	Earned Surplus		3,844	3,516	2,566	2,436	8,175
56	Total Equity		6,694	6,782	5,860	5,766	6,037
57							
58	Total Liabilities & Equity		$23,638	$20,888	$19,742	$18,048	$17,867

AP will be equal to 5.3% of *expected* Sales.

Other Current Liabilities will be equal to 14% of *expected* Sales.

All earnings and cash flow in excess of the amount necessary to maintain assets at their required levels and meet the current period's financial obligations are paid out as dividends to the stockholders.

At the end of the planning horizon (end of 2015), the firm will be liquidated. Liabilities will be paid from the proceeds from the sale of the assets, which will be sold for their book values. Any value in excess of the liabilities will be paid to the stockholders as a liquidating dividend. If the firm defaults and fails prior to the horizon, the stockholders receive nothing from that point forward, and the creditors receive the subsequent cash flow and proceeds from the asset sale at the end.

For purposes of testing different financial structures, assume that a specified amount of debt is issued at the end of 2010. (i.e., you simply insert this amount of debt into the 2010 capital structure). The cash generated by this borrowing will be used, first to meet the firm's needs for external financing, and second, what is left will be paid to stockholders as a dividend. That is, if issuing new debt at the end of 2010 results in excess cash, the excess cash will be used to pay an immediate dividend to the common stockholders at the end of 2010, or any other year when debt financing generates excess cash.

At the end of 2010, prior to issuing any new debt, there is no short-term or long-term debt outstanding. Any new long-term debt that is issued will be issued at the end of 2010. The debt issued at the end of 2010 will have a term to maturity of five years (i.e., it will mature at the end of 2015), and it will require five equal annual payments (principal + interest) at an interest rate of 9%. If any additional external financing is necessary after the end of 2010, it will be long-term debt at 9%, with equal annual payments (principal + interest) so as to be completely repaid by the end of 2015.

For purposes of this analysis, it is assumed that financial distress (bankruptcy) would occur if the EBITDA falls below the required debt service. Debt service is the sum of interest and principal (sinking fund) payments. Default is the event that EBITDA < Debt Service. The probability of default in a given year, t, having survived for $t-1$ years, is

$$q_t = \Pr(\text{EBITDA}_t < \text{Debt Service}_t).$$

The probability of surviving in year t is

$$p_t = (1 - q_t) = \Pr(\text{EBITDA}_t > \text{Debt Service}_t).$$

The probability of surviving for t periods is

$$P_t = \prod_{j=1,t} p_j = \prod_{j=1,t} (1 - q_j),$$

and the probability of surviving for $t - 1$ periods and then failing in period t is

$$Q_t = P_{t-1} q_t.$$

With no debt outstanding at the end of 2010, the first date that bankruptcy can occur is when the first debt service is due at the end of 2011.

The current five-year Risk-free interest rate is 3.5%, and this rate is expected to continue to prevail for the remainder of the planning horizon.

The debt issued for financial restructuring will be issued at the end of 2010. If there is new investment projected in any year, assume it occurs at the end of the year; and new financing necessary to meet growth needs in subsequent years will occur at the end of the year.

Assignment

Develop a spreadsheet model for Laric Corporation that will enable you to determine the probability of default associated with different levels of long-term debt. Find the largest amount of long-term debt that could be issued at the end of 2010 so as to yield probabilities of default of (a) 5%; (b) 10%; and (c) 20%, respectively, in year 2011. For each of these debt amounts, find the probability of defaulting in each subsequent year, 2012–2015. For each case, what is the probability that you will default at any time over the planning horizon? Note: Your estimate of probability of default should be based on running the Monte Carlo simulation and evaluating $\Pr(\text{EBITDA}_t < \text{Debt Service}_t)$ in each period.

2 Laric Corporation: Optimal Capital Structure

In Problem 1 you were asked to determine how much debt the company could incur and not exceed specified levels of default Risk. Now you are asked to use that same model to determine the company's optimal capital structure.

According to the widely accepted "trade-off model" of capital structure, the firm's optimal capital structure occurs at the mix of debt and equity where the "net benefits of debt" are maximized. In this context, the net benefits of debt are defined as

$$\text{Net Benefits} = E[\text{PV(Tax Savings)}] - E[\text{PV(Bankruptcy Costs)}]$$
$$NB = E[\text{PV(TS)}] - E[\text{PV(BC)}].$$

You can determine the optimal debt level by calculating the total Net Benefits for different levels of debt, and find the level of debt where the maximum is reached. That is, you assume a level of debt and calculate NB. Then increase the assumed debt and recalculate NB. Keep doing this until you can graph the functional link between the amount of debt and NB.

To do this, make distribution assumptions just as you did in Problem 1, and assume

- Dividends are paid only if there are funds remaining after servicing the debt.
- Financial distress (bankruptcy) occurs if EBITDA falls below the required debt service. Debt service is defined as the sum of the interest payment and the principal payment on the debt.
- If bankruptcy occurs, bankruptcy costs will be incurred that will equal 15% of the face value of the outstanding debt.
- If bankruptcy occurs, the firm stops operating. However, in the period of bankruptcy, if the firm has any cash flow available, it pays what it has to the lenders. Interest is paid before principal. Assume that there are tax savings in that period equal to the tax rate x interest actually paid.
- Subsequent to bankruptcy, no further tax savings or bankruptcy costs will occur.

In the simulation, the value of tax savings is calculated as

$$\text{PV(TS)} = \sum_{t=1,n} TI_t (1+k_d)^{-t},$$

and the value of bankruptcy costs is calculated as

$$\text{PV(BC)} = \sum_{t=1,n} BC_t (1+k_d)^{-t}$$

for each single draw in the simulation. These are designated as the Output variables in @Risk.

The *means* of these values in the simulation are estimates of E[PV(TS)] and E[PV(BC)], respectively. If your model has a variable defined as

$$\text{NB} = \text{PV(TS)} - \text{PV(BC)}$$

as these items are calculated above, when you run the simulation, you get a frequency distribution of NB. The *mean* of this distribution is the estimate of the expected net benefit.

3 Swapping by Ying and Shuang

Ying Corporation and the Shuang Company each have loans that they are interested in swapping. Ying has a loan for $200,000 that pays interest only for five years, with the principal due at the end of five years. The interest rate on this loan is fixed at 8% over the five years. Shuang has a variable rate loan that pays interest only for five years, with the principal due at the end of the five years. Shuang's variable rate is reset every six months, equal to the LIBOR plus 2% per year. The current LIBOR rate is 6.5%. If the swap is made, the interest paid at the end of each six-month period is based on the LIBOR at the start of the six-month period. The future LIBOR rate is unknown, but assume that the changes in the LIBOR follow an "elastic random walk" function of the form: $dr = 0.15(0.04 - r)dt + 0.005z(dt)^{1/2}$.

Assignment

Construct a spreadsheet model similar to Exhibit 11.22 that will show the results of a potential swap. Your model should calculate the IRR for each loan. Using the random LIBOR function simulate with 5000 draws, the interest rates over the next five years. Compute and show the frequency distribution of the IRR for the borrower who pays the variable rate in the swap.

CHAPTER 12

Modeling Working Capital Accounts

IN THIS CHAPTER WE explain different approaches to modeling the working capital accounts—that is, the categories of accounts in current assets and current liabilities. The accounts we will consider include cash, marketable securities, receivables, inventories, and AP.

If the purpose of the simulation is to analyze the effects of broad strategic decisions over a long planning horizon, then the details of working capital are unimportant. Only the general details of working capital would have to be modeled to pursue the implications of long-run policies. With a long planning horizon comprised of long sub-periods, any seasonal variation in current assets and liabilities disappears. Generally, the longer the horizon, the less the need to distinguish between the component accounts of working capital. Thus, for longer run planning purposes it is usually satisfactory to let the balances in the current asset and current liability accounts be directly proportional to sales, and even to lump all these accounts together as we did in the O&R model.

However, as the planning horizon and the planning periods are shortened, greater detail may be needed in modeling current asset and liability accounts. Indeed, one of the most useful applications of financial simulation models is to explore short-run decisions that affect seasonal current balances, so it is necessary to build much more detailed accounting for current accounts by increasing the number of accounts and variables, and more carefully linking current assets and liabilities to the sales forecast.

A sales-driven model such as the percent-of-sales approach is particularly appropriate for working capital because the primary motive for most of the current accounts is to meet the needs of sales. Inventory is a good example because it is normally held solely to meet the requirements of future sales. However, this statement itself points out one of the difficulties of modeling working capital investment: should working capital balances in the model be related to current sales, past sales, or future sales? Other questions include whether working capital investment should be strictly proportional to sales, and what degree of detail should be built into the working capital sector. We will consider these questions as we discuss the various working capital categories.

12.1 CASH

Our choice of the way to model the cash account would depend on the type of model, the length of the sub-periods and planning horizon, and, of course, the purpose of the modeling effort. If the sub-periods and the planning horizon are long, and the purpose of the planning model is to address very broad planning issues, then one would not be as concerned with the specific detail of the balances of liquid assets. In such a case it would usually be appropriate to combine cash and marketable securities in a single account, and use the percent-of-sales approach for modeling the desired balance of liquid assets. Cash normally is held to meet the transaction requirements that are generated by the firm's production and sales, so that the percent-of-sales approach fits the situation, and the cash balance held at the end of period t could be modeled as

$$\text{Cash}_t = C \times \text{Sales}_t, \tag{12.1}$$

where parameter C is obtained from historical data or is based on the anticipated policy for the planning period.

An alternative approach that would be appropriate under the same circumstances would be to relate the balance of liquid assets to the stock of Total Assets, that is, Total Assets would depend on sales, and cash would be a constant proportion of Total Assets:

$$\text{Cash}_t = C \times \text{Total Assets}_t. \qquad (12.2)$$

However, one must be aware of the way that Total Assets are computed in the model to avoid the problem of simultaneity. That is, if the balance in Cash must be known to calculate Total Assets, then the model has simultaneous equations if the amount of Total Assets must also be known to calculate the Cash balance.

The approach to modeling the cash balance would also depend on whether the planning model is a "stock balancing" or a "flow balancing" model. In the standard stock balancing model that is driven by the sales forecast, the sales forecast leads to the required asset balances, and the liability balances are adjusted to equal total assets. In this case, it is permissible to have the cash balance adjust to the forecast of sales or total assets independently from the balancing mechanism for the system that forces liabilities to equal assets.

However, if the model is of the flow balancing variety, it cannot be assumed that the liquid assets would automatically be a constant proportion of either Sales or Total Assets. This is because in the typical flow balancing model the mechanism for maintaining equilibrium is the adjustment of the cash and liquid asset account. The equation that balances the system is of the general form

$$\text{Net Change in Cash Balances}_t = \text{Net Cash Inflow}_t.$$

One advantage of the flow balancing approach is that it helps us to avoid the problem of cash disappearing into a financial black hole. When we discussed balancing in reference to the O&R example, we noted that one problem we can encounter is that the model can generate cash that does not seem to flow either to the cash balance or to stockholders in the form of dividends. When this happens, we have cash flowing into the firm, but it does not show up in any asset account. This is what we call a financial black hole. If our model clearly accounts for all the cash flow, we avoid the black hole. The flow balancing approach is better for accounting for all the cash flow. Moreover, a flow balancing model with cash as the balancing account is well suited to a planning problem where the planning horizon and the sub-periods are short and greater detail about cash and marketable security accounts is needed.

For the cash flow balancing model to function properly, the model must be equipped with some decision rules that tell it how to balance the system in a way that realistically reflects management's policies. In the absence of such instructions, there is nothing to prevent the cash balance getting excessively large or small, or even negative. The kinds of decision rules necessary are instructions stating the desired balance of cash and stating what should be done when the actual balance departs from the desired balance. Exhibit 12.1 shows an example of a cash balance section for a short-term planning model that is based on the cash flow equations of Exhibit 4.2C of Chapter 4. In developing the equations in Exhibit 12.1, the assumptions and policies are

(a) The firm will maintain its cash balance at a level assigned to the policy variable Desired Cash.

(b) Any cash in excess of Desired Cash that would build up from the firm's operations would be invested in marketable securities.

(c) If operations generate a cash outflow so that the cash balance would be less than Desired Cash, the cash balance is restored to the desired level by first selling marketable securities, and second by borrowing.

Cash Flow from Operations$_t$	=	Earnings Cash Flow$_t$ − NWC Investment$_t$
Source of Cash$_t$	=	Cash Flow from Operations$_t$ − Capital Expenditures$_t$ + Capital Financing$_t$ − Dividend$_t$ + Securities Matured$_t$
Trial Cash Balance$_t$	=	Cash$_{t-1}$ + Source of Cash$_t$
Trial Excess Cash$_t$	=	Trial Cash Balance$_t$ − Desired Cash$_t$
Excess Cash$_t$	=	Maximum(0,Trial Excess Cash$_t$)
Cash Deficit$_t$	=	0 − Minimum(0,Trial Excess Cash$_t$)
Securities Purchased$_t$	=	Excess Cash$_t$
Securities Sold$_t$	=	Minimum(Cash Deficit$_t$,Marketable Securities$_{t-1}$ − Securities Matured$_t$)
Short-term Borrowing$_t$	=	Maximum(0,Cash Deficit$_t$ − Securities Sold$_t$)
Net Cash Flow$_t$	=	Trial Source of Cash$_t$ − Securities Purchased$_t$ + Securities Sold$_t$ + Short term Borrowing$_t$
Cash$_t$	=	Cash$_{t-1}$ + Net Cash Flow$_t$

EXHIBIT 12.1 Cash balance adjustment sector.

(d) If security sales are necessary, the amount of Securities Sold is limited to the amount of securities in the portfolio at the beginning of the period, Marketable Securities$_{t-1}$.

(e) If Cash Deficit exceeds the amount that can be obtained by selling securities, then the remainder of Cash Deficit is financed by short-term borrowing.

(f) In reference to NWC Investment, Other CA Investment excludes cash and marketable securities, and Spontaneous Financing refers to current liabilities other than short-term borrowing.

(g) Cash and security investments are the residual in the model, and dividends are based on a target payout ratio.

The equations in Exhibit 12.1 provide the model with a mechanism for managing the cash account with the target Desired Cash. This target level of cash is not generated in the model, rather, it is specified by the user as a policy parameter. The target cash balance would be influenced by needs such as compensating balances required by the bank, anticipated cash payments, and emergency requirements. There are several ways that modeling can help us to determine the minimum or target level for cash including simulation to find a comfortable cash target, or even optimization models that solve for the best cash target.

A priori, we may not know the appropriate level of the Desired Cash Balance. We can get a better idea by specifying a Desired Cash balance and then exploring different scenarios for the inflows and outflows, allowing us to trace out the implications of using different cash target levels. If too little cash is maintained, the firm will have to liquidate securities or borrow too frequently and its transaction costs will be excessive. If too much cash is maintained, the transactions are avoided, but the firm foregoes the income it could have earned if the funds were invested.*

There are two ways we can simulate: we can use scenario simulation, or we can use Monte Carlo simulation. With scenario simulation, for a given policy we simply try out different cash flow scenarios to trace out the results for each scenario. Then, we change the policy and iterate over the different cash flow scenarios again. We compare the results from the different policies and determine which one yields the best results. With

---

* Cash management is explained in greater detail in Kallberg and Parkinson (1996), Masson and Wikoff (1995), Smith (1979), and Orgler (1970).

Cash balance

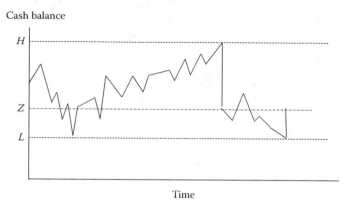

Time

EXHIBIT 12.2   Cash balances using the M&O model.

Monte Carlo simulation, we draw the cash flow inputs as random variables from a probability distributions, and see the results as a probability, or frequency, distribution of outcomes. We evaluate the resulting probability distributions to see which policy yields the most desirable result.

Another approach to determining the appropriate cash balance has been to borrow from the techniques of inventory theory to derive decision rules regarding the inventory of cash. For example, Miller and Orr (1966) presented a model for managing cash balances that allows the cash balance to wander randomly between an upper limit, H, and a lower limit, L. When the cash balance hits either the upper or lower limit, actions are taken to restore the balance to a restoration level, Z. The limits and the restoration point are determined by a formula that considers the parameters of the probability distribution that describes the random daily cash flows, the transaction cost of restoring the cash balance, and the interest rate that represents the opportunity cost of holding funds as cash.* With the Miller and Orr (M&O) (H, Z, L) policy, the cash balance would follow the pattern shown in Exhibit 12.2, where securities are purchased when the cash balance hits the upper limit H, to restore the balance to Z, and securities are sold (or funds borrowed) when the lower limit, L, is breached. The M&O cash management decision rules could be built into a short-run planning model, such as Exhibit 12.1, with the equations shown in

* The M&O formula is $Z = (3b\sigma^2/4r)1/3$, and, $H = 3z$, $L = 0$, where b is the fixed dollar cost of a transaction, r the interest rate per day, and σ^2 the variance of the random, normally distributed daily cash flows.

Cash Flow from Operations$_t$	=	Earnings Cash Flow$_t$ − NWC Investment$_t$
Source of Cash$_t$	=	Cash Flow from Operations$_t$ − Capital Expenditures$_t$ + Capital Financing$_t$ − Dividend$_t$ + Securities Matured$_t$
Trial Cash Balance$_t$	=	Cash$_{t-1}$ + Source of Cash$_t$
Z	=	$3 \times B \times$ Cash Flow Variance$/4 \times$ Interest Rate
H	=	$3 \times Z$
L	=	0
Excess Cash$_t$	=	Maximum(0,Trial Cash Balance$_t$ − H)
Cash Deficit$_t$	=	Maximum(0,L − Trial Cash Balance$_t$)
Securities Purchased$_t$	=	Trial Cash Balance$_t$ − Z, IF(Excess Cash$_t$ ≥ 0)
	=	0 otherwise.
Securities Sold$_t$	=	Minimum(Z − Trial Cash Balance$_t$,Marketable Securities$_{t-1}$),IF(Cash Deficit$_t$ > 0)
	=	0 otherwise.
Borrowing$_t$	=	Minimum(Z − Trial Cash Balance$_t$ − Securities Sold$_t$, 0),IF(Cash Deficit$_t$ > 0)
Net Cash Flow$_t$	=	Source of Cash$_t$ − Securities Purchased$_t$ + Securities Sold$_t$ + Borrowing$_t$
Cash$_t$	=	Cash$_{t-1}$ + Net Cash Flow$_t$

EXHIBIT 12.3   Equations for a cash management model using the M&O policy.

Exhibit 12.3. The parameters for the limit formulas would be estimated from a study of the firm's operating cash flows. In the model described by Exhibit 12.3, it is assumed that the costs of borrowing are identical to the rate earned on marketable securities.

12.2   MARKETABLE SECURITIES

If the planning horizon and the sub-periods making up the horizon are long, there is little need to distinguish between cash and marketable securities; they can be combined in a liquid asset category for purposes of analyzing a long-term plan, and a straightforward percent-of-sales approach could be used. On the other hand, if the planning horizon is short and the purpose of the model is to consider the effects of short-run investment strategies, then more detailed modeling of marketable securities would be appropriate.

For the longer run model, the percent-of-sales approach might specify marketable securities as

$$\text{Marketable Securities}_t = M \times \text{Sales}_t, \qquad (12.3)$$

where M is a parameter based either on historical relationships or on anticipated policy. Alternatively, securities could be specified as a percent of assets as in

$$\text{Marketable Securities}_t = M \times \text{Total Assets}_t, \qquad (12.4)$$

but care must be taken to avoid simultaneity.

For the shorter run model focusing on working capital management, the details of the securities accounts would be of concern, and more detailed modeling would be required. A good starting point for modeling securities is simply to track the balance with an equation such as

$$\text{Marketable Securities}_t = \text{Marketable Securities}_{t-1} - \text{Securities Matured}_t$$
$$- \text{Securities Sold}_t + \text{Securities Purchased}_t, \quad (12.5)$$

where the balances are stated in terms of face value of the securities. We only need to expand on the components of this equation to have a complete model of marketable securities.*

Exhibit 12.4 shows equations that would be developed to model security transactions over a short planning horizon. We will introduce some notation so that we can show the relationships concisely—with apologies to the less mathematically inclined readers. Two subscripts will be used: the first subscript will denote the period in which a transaction occurs, and the second subscript will denote the period in which a security matures. For example, $SP_{t,j}$ is the amount of securities purchases in period t that will mature in period j. This is stated in terms of the face, or maturity, value of the security. $SS_{t,j}$ denotes the face value of securities sold in period t that mature in period j. The index 0 refers to securities held at the beginning of the planning horizon, and h denotes securities that mature at or beyond the end of the horizon. The equation at the bottom of the exhibit, for $t = 1, \ldots, h - 1$, means that for each period in the planning horizon, there is an equation corresponding to 1 through 7.

---

* A model of short-term security transactions in the context of a linear programming (LP) model for working capital management is presented in Orgler (1970), and a simplified example of Orgler's model will be presented in the optimization chapter.

(1) Securities Purchased$_t$	$= \sum_{j=t+1,h} SP_{tj}$
(2) Securities Sold$_t$	$= \sum_{j=t+1,h} SS_{tj}$
(3) Securities Matured$_t$	$= \sum_{i=0,t-1} SP_{it} - \sum_{i=1,t-1} SS_{it}$
(4) Security Balance$_t$	$=$ Security Balance$_{t-1}$ + Securities Purchased$_t$ − Securities Sold$_t$ − Securities Matured$_t$
(5) Marketable Securities$_t$	$= \sum_{i=0,t-1} \sum_{j=t+1,h} a_{j-t} SP_{ij} - \sum_{i=1,t-1} \sum_{j=t+1,h} a_{j-t} SS_{ij}$
(6) Cash Flow from Securities$_t$	$=$ Securities Matured$_t$ − $\sum_{j=t+1,h} a_{j-t} SP_{tj}$ + $\sum_{j=t+1,h} a_{j-t} SS_{tj}$
(7) Security Income$_t$	$=$ Securities Matured$_t$ − $\sum_{i=0,t-1} a_{t-i}$ $(SP_{it} - SS_{it}) + \sum_{j=t-1,h} (1 - a_{j-t})SS_{ti}$ for $t = 1, \ldots, h - 1$

Note: Decision variables SP$_{tj}$, SS$_{tj}$, and the a summary balances are stated in terms of face value for discount securities. The coefficients, a_{t-j}, represent the discounted value per dollar of FV for a security maturing in $t - j$ periods.

EXHIBIT 12.4   A model for short-term marketable securities.

For this model it is assumed that the securities are short-term discount securities, such as Treasury bills, that are bought at a discount from face value, and the increase in value from purchase until sale or maturity is interest income, and is shown as Security Income$_t$. The total values of Securities Purchased$_t$, Securities Sold$_t$, Securities Matured$_t$, and the portfolio balance, Security Balance$_t$ are all stated in terms of face, or maturity, value. To convert from face value to market value, we define the coefficient $a_{t-j} = 1/(1 + r_{t-j})$ as the present value factor that converts the face (maturity) value of securities to the discounted value at time t, with r_{t-j} representing the interest rate for $t - j$ periods. For example, $a_{5-3} = a_2$ is the present value for the two periods from periods 3 to 5. Suppose securities are purchased at time 3 to mature at time 5 with a maturity value of $1000. With an interest rate of 1% per period, the amount spent for the purchase at time 3 is

$$a_{35}SP_{35} = \left(\frac{1}{1.01}\right)^2 SP_{35} = (0.9803) \times 1000 = 980.30.$$

Equations 1 through 4 of the exhibit total the face values of securities purchased, sold, matured, and held. Equation 1 says that the face value of securities purchased in period t is the sum of all securities purchased that period that mature in subsequent periods. Equation 2 totals the sales of securities that mature in future periods that are sold in period t. Equation 3 says that the face value of securities matured in period t is the sum of all past purchases that would mature in t, less the amount of securities sold in past periods that mature in period t. Equation 5 gives the market value of the security portfolio at the end of each period t based on all past purchases and sales; and Equation 6 shows the cash flowing in and out of the portfolio each period. The income Equation 7 calculates income in terms of the gain from the discounted purchase price to the amount realized on the sale or maturity of the security.

Now we will see how to implement the equations shown in the exhibit in a spreadsheet.

12.2.1 Example: Managing Securities for the Mogul Corporation

The Mogul Corporation has a portfolio of cash and short-term securities. In managing the cash and securities, the objective is to maximize value of cash and securities held at the end of the three-month planning horizon. The planning horizon is divided into one-month time intervals. At the start of the planning period, the firm has an initial balance of cash and an endowment of securities as follows:

Cash balance	$5000		
Marketable securities			
Maturity date	**Month 1**	**Month 2**	**Month 3**
Face value ($)	1000.00	2000.00	500.00
Market value ($)	990.10	1960.59	485.30

In addition to holding the initial balance of cash and securities, the following cash flows are expected in each month:

	Month 1	**Month 2**	**Month 3**
Expected cash flow ($)	1000	−3000	500

The cash inflow will be available to be invested, and the cash outflow of $3000 must be funded from the cash balance or from the security portfolio.

Securities are traded at prices that are discounted from face, or maturity value. The current and expected future short-term interest rates that we use to discount the face value to the market value of the securities are shown in the table below. The number in the first column indicates the month in which the security is purchased, and the numbers in the top row indicate the month the security matures. Thus, a security purchased in month 1 that matures in month 3 is expected to have a monthly interest rate of 1.2%. Securities purchased in period 0 refers to securities already held at the start of the planning horizon. Period 4 represents securities maturing after the end of the three-month planning horizon. We will assume period 4 is one month beyond the end of the planning horizon.

Monthly Interest Rates for Marketable Securities

Period Security is Purchased	Period Security Matures			
	1	2	3	4
0	1.0%	1.0%	1.0%	1.0%
1		1.1%	1.2%	1.3%
2			1.3%	1.4%
3				1.4%

Our task is to construct a spreadsheet model to track our security portfolio for three months so that we can see the results of alternative strategies for buying and selling securities. The decisions we can make are how much to invest in securities each month, which month the securities will mature, and, when, how many securities, and of which maturity should we sell? We can purchase securities at times 1, 2, and 3 to mature at any subsequent date up to time 4. The purchase price of the security is the present value of the face value of the security. Thus, if, in month 1 we purchase a security that matures in month 3 at face value of $1000, the purchase price (market value) will be

$$\text{Purchase Price} = \text{Market Value} = 1000\left(\frac{1}{(1+1.012)^2}\right)$$
$$= 1000(0.97643)$$
$$= \$976.43.$$

If we hold the security for the two months until it matures, we will receive the face value of $1000 at the end of month 3. The income will be the appreciation in price amounting to

$$\text{Income} = \$1000 - 976.43 = \$23.57.$$

The security can be sold prior to maturity. If this security is sold at time 2, the sale price will be

$$\text{Sale Price} = \$1000\left(\frac{1}{1.013}\right) = \$1000(0.98717) = \$987.17.$$

The income earned over the one-month holding period would be

$$\text{Income} = \$987.17 - 976.43 = \$10.74.$$

The discount rate used above is the 1.3% that is shown in the table as the rate for securities purchased at $t = 2$ to mature at $t = 3$.

In this example, the portfolio is given at $t = 0$, and we can make purchase decisions at times 1, 2, and 3, with time 4 being beyond the horizon. The six purchase decision variables are enumerated as follows:

Security Purchase Decision Variables

	Period that Security Matures		
Period of Purchase	2	3	4
1	SP_{12}	SP_{13}	SP_{14}
2		SP_{23}	SP_{24}
3			SP_{34}

The six security sales decision variables are

Security Sale Decision Variables

	Period that Security Matures		
Period of Sale	2	3	4
1	SS_{12}	SS_{13}	SS_{14}
2		SS_{23}	SS_{24}
3			SS_{34}

To analyze the implications of different security transactions, we have constructed a spreadsheet model that tracks the security transactions, keeping track of purchases, sales, security balances, and earnings from the portfolio. Exhibit 12.5A shows the model with the decisions inserted that will be explained shortly. Exhibit 12.5B shows selected key formulas for the model. Now, we take a brief tour of the model.

Mogul Corporation
Marketable Security Portfolio

Input Data

Enter initial data in this section

	Period	0	1	2	3	4

Cash	$5,000.00
Marketable Securities (MV)	3,435.99 <-------- Do not enter data here
Cash & Securities	8,435.99

	1	2	3	4
Cash Inflow	1,000.00	-3,000.00	500.00	0.00
Securities Maturing (face value)	1,000.00	2,000.00	500.00	0.00

Interest Rate

Purchase Period	Maturity Period 1	2	3	4
0	0.010	0.010	0.010	0.010
1		0.011	0.012	0.013
2			0.013	0.014
3				0.014

Objective

Cash & Security Value at Horizon: 7,130.11

Present Value Factors

Purchase Period	Maturity Period 1	2	3	4
0	0.99010	0.98030	0.97059	0.96098
1		0.98912	0.97643	0.96199
2			0.98717	0.97258
3				0.98619

User Input: Decisions

Enter purchases & sales in this section
except the initial portfolio (time 0)

Security Purchases

Face Value

Purchase Period	Maturity Period 1	2	3	4	Total FV
0	1,000.00	2,000.00	500.00	0.00	0.00
1		1,000.00	4,000.00	500.00	500.00
2			1,500.00	0.00	0.00
3				500.00	0.00

Market Value

Purchase Period	Maturity Period 1	2	3	4	Total MV
0	990.10	1,960.59	485.30	0.00	3,435.99
1		989.12	3,905.70	481.00	5,375.82
2			1,480.75	0.00	1,480.75
3				493.10	493.10

Security Sales

Face Value

Sale Period	Maturity Period 1	2	3	4	Total FV
1	0	0	0	0	0.00
2			500		500.00
3			0		0.00

Market Value

Sale Period	Maturity Period 1	2	3	4	Total MV
1	0.00	0.00	0.00	0.00	0.00
2			493.58		493.58
3			0.00		0.00

EXHIBIT 12.5A Spreadsheet model for marketable securities.

#	A	B	C	D	E	F	G	H	I	J	K	L	M	N	O	P
47	Securities Maturing															
48																
49			Maturity Period	1	2	3	4									
50				1,000.00	3,000.00	5,500.00	1,000.00									
51	Security Balance								Security Balance							
52										Market Value						
53			Maturity Period							Maturity Period						
54		Period Held		1	2	3	4	Total FV			Period Held	1	2	3	4	Total MV
55		0		1,000.00	2,000.00	500.00	0.00	3,500.00			0	990.10	1,960.59	485.30	0.00	3,435.99
56		1		0.00	3,000.00	4,500.00	500.00	8,000.00			1		2,967.36	4,393.91	481.00	7,842.27
57		2			0.00	5,500.00	500.00	6,000.00			2			5,429.42	486.29	5,915.71
58		3				0.00	1,000.00	1,000.00			3				986.19	986.19
59																
60	Cash Flow			1	2	3			Cash & Security Balance			1	2	3		
61				-3,375.82	-987.17	5,506.90				Cash		1,624.18	637.02	6,143.92		
62										Securities (MV)		7,842.27	5,915.71	986.19		
63	Security Income									Cash & Securities		9,466.45	6,552.72	7,130.11		
64				1	2	3										
65				9.90	-1,949.71	128.31										
66	Security Accounting															
67																
68	Sales of Securities Purchased in Period 0								Security Balance After Sale: Purchased @ 0							
69			Maturity Period	1	2	3	4			Maturity Period		1	2	3	4	
70		Sale Period							End of Period							
71		1			0.00	0.00	0.00				0	1,000.00	500.00	0.00		
72		2				0.00	0.00				1	2,000.00	500.00	0.00		
73		3				500.00	0.00				2		0.00	0.00		
74											3			0.00		
75																
76	Sales of Securities Purchased in Period 1								Security Balance After Sale: Purchased @ 1							
77			Maturity Period	1	2	3	4			Maturity Period		1	2	3	4	
78		Sale Period							End of Period							
79		2					0.00				1	1,000.00		500.00		
80		3				0.00	0.00				2	4,000.00	4,000.00	500.00		
81											3			500.00		
82																
83	Sales of Securities Purchased in Period 2								Security Balance After Sale: Purchased @ 2							
84			Maturity Period	1	2	3	4			Maturity Period		1	2	3	4	
85		Sale Period							End of Period							
86		3					0.00				2			0.00		
87											3		1,500.00	0.00		
88																
89																

EXHIBIT 12.5A (continued).

Table 1 (columns B–F)

	B	C	D	E	F
30	Security Purchases				
31	Face Value				
32	Maturity Period		1	2	3
33	Purchase Period				
34	0		=D16	=E16	=F16
35	1	=IF(P35>C10+D		1000	4000
36	2	=IF(P36>L61+E			1500
37	3	=IF(P37>M61+F			
38					
39	Security Sales				
40	Face Value				
41	Maturity Period		1	2	3
42	Sale Period				
43	1	=IF(OR(E43>E5		0	0
44	2	=IF(OR(F44>F5			500
45	3	=IF(G45>G57,"			
46					
47	Securities Maturing				
48					
49	Maturity Period		1	2	3
50			=D16	=E16+E35-E43	=F16+F35+F36-F43-F44
51	Security Balance				
52					
53	Maturity Period		1	2	3
54	Period Held				
55	0		=D16	=E16	=F16
56	1		=D55-D50	=E55+E35-E43	=F55+F35-F43
57	2			=E56-E50	=F56+F36-F44
58	3				=F57-F50

Callout (pointing to C35): =IF(P35>C10+D50+P43,"infeasible"," ")

Callout (pointing to C43): =IF(OR(E43>E55,F43>F55,G43>G55),"infeasible"," ")

Table 2 (columns B–G)

	B	C	D	E	F	G
60	Cash Flow		1	2	3	
61			=D14+D50-P35+P43	=E14+E50-P36+P44	=F14+F50-P37+P45	
62						
63	Security Income		1	2	3	
64			=(D34-L34)+E72*(M22-N	=(E34-M34)+(E35-M35)	=(F34-N34)+(F35-N35)	
65						
66	Security Accounting					
67						
68	Sales of Securities Purchased in Perio					
69	Maturity Period		1	2	3	4
70						
71	Sale Period					
72	1			=MIN(E43,E55)	=MIN(F43,F55)	=MIN(G43,G55)
73	2				=MIN(F44,M72)	=MIN(G44,N72)
74	3					=MIN(G45,N73)
75						
76	Sales of Securities Purchased in Perio					
77	Maturity Period		1	2	3	4
78						
79	Sale Period					
80	2				=MIN(F44-F73,M79)	=MIN(G44-G73,N79)
81	3					=MIN(G45-G74,N80)
82						
83	Sales of Securities Purchased in Perio					
84	Maturity Period		1	2	3	4
85						
86	Sale Period					
87	3					=MIN(W74-G81,N86)

Callout (Security Accounting, C66): =(D34-L34)+E72*(M22-M21)+F72*(N22-N21)+G72*(O22-O21)

Table 3 (columns K–P)

	K	L	M	N	O	P
30	Security Purchases					
31	Market Value					
32	Maturity Period	1	2	3	4	Total MV
33	Purchase Period					0
34	0	=(1/(1+D21))*D34	=(1/(1+E21)^2)*E34	=(1/(1+F21)^3)*F34	=(1/(1+G21)^4)*G34	=SUM(L34:O34)
35	1		=E35*M22	=F35*N22	=G35*O22	=SUM(M35:O35)
36	2			=F36*N23	=G36*O23	=SUM(N36:O36)
37	3				=G37*O24	=SUM(O37)
38						
39	Security Sales					
40	Market Value					
41	Maturity Period	1	2	3	4	Total MV
42	Sale Period					
43	1		=E43*M22	=F43*N22	=G43*O22	=SUM(M43:O43)
44	2			=F44*N23	=G44*O23	=SUM(M44:O44)
45	3				=G45*O24	=SUM(M45:O45)
46						
47						
48						
49						
50						
51	Security Balance					
52	Market Value					
53	Maturity Period	1	2	3	4	Total MV
54	Period Held					
55	0	=L21*D55	=M21*E55	=N21*F55	=O21*G55	=SUM(L55:O55)
56	1		=E56*M22	=F56*N22	=G56*O22	=SUM(L56:O56)
57	2			=F57*N23	=G57*O23	=SUM(L57:O57)
58	3				=G58*O24	=SUM(L58:O58)
59						
60	Cash & Security Balance	1	2	3		
61	Cash	=C10+D14+D50-P35+P43	=L61+E14+E50-P36+P44	=M61+F14+F50-P37+P45		
62	Securities (MV)	=P56	=P57	=P58		
63	Cash & Securities	=L61+L62	=M61+M62	=N61+N62		

EXHIBIT 12.5B Formulas for selected parts of the Mogul model.

In Exhibit 12.5A the input data describing the endowment of cash and securities, the expected exogenous cash flow, and expected future interest rates are in block A5:G25. The data input by the model user includes:

Input Data	Cells
Initial cash balance	C10
Expected cash inflow	D14:G14
Beginning security balances (FV)	D16:G16
Expected interest rates	D21:G24

The user's security purchase and sale decisions are input in block A27:G45. Specifically, the decision variables are entered in

Input Decision Variables	Cells
Security purchases, SP_{tj}	D34:G37
Security sales, SS_{tj}	E43:G45

where the amounts inserted are the face values of the securities to be purchased or sold at each date. The market values of the security purchase and sales amounts are calculated in the block, J30:P45, to the right of the input fields. The market values are calculated using the present value factors shown in J18:O25 that are based on the monthly interest rates input by the user in A18:G24. Generally, the values on the left side of the model (columns A–H) are in face value terms (FV), and the right side of the model (columns J–P) shows the market values (MV).

The face values of the securities that mature at the end of each period are shown in D50:G50. The balances (FV) of securities held at the end of each period are shown in D55:H58. For example, cell E56 is the amount of securities held at the end of period 1 that mature in period 2, and F56 is the amount held at $t = 1$ that will mature at $t = 3$. H55:H58 shows the total amount of securities held at the end of each period. The corresponding market values are shown on the right side in J51:P58.

The cash flow generated each period is calculated in line 61. The cash flow is calculated according to Equation 6 in Exhibit 12.4, taking account of the exogenous cash flow, maturing securities, purchases, and sales.

The total balance of cash and securities (MV) is tracked in J61:N63. The objective is to maximize the value of the cash and security balance at the end of the horizon, time 3. Cell N63 is the value to be maximized. It is reflected at the top right of the model in N14 as the objective.

The income earned on the security portfolio in each period is shown in D64:F64. Income is calculated according to Equation 7 in Exhibit 12.4 as to total increase in security values from the initial purchase price to maturity or sale price. For example, the income earned in period 2 is calculated as

$$\text{Income}_2 = (\text{SP}_{0\,2} - a_{0\,2}\text{SP}_{0\,2}) + (\text{SP}_{1\,2} - a_{1\,2}\text{SP}_{1\,2}) - (a_{1\,2} - a_{0\,2})\text{SS}_{1\,2}^0$$
$$+ (a_{2\,3} - a_{0\,3})\text{SS}_{2\,3}^0 + (a_{2\,3} - a_{1\,3})\text{SS}_{2\,3}^1 + (a_{2\,4} - a_{0\,4})\text{SS}_{2\,4}^0$$
$$+ (a_{2\,4} - a_{1\,4})\text{SS}_{2\,4}^1,$$

where SS_{2j}^n denotes sales in period 2 of securities maturing in period j that were purchased in period n. In spreadsheet formula in cell E64 is

$$E64 = (E34 - M34) + (E35 - M35) - (M22 - M21) \times E72$$
$$+ (N23 - N21) \times F73 + (N23 - N22) \times F80$$
$$+ (O23 - O21) \times G73 + (O23 - O22) \times G80.$$

This calculates income in period 2, taking account of (a) the gain on securities purchase in periods 0 and 1 that mature in 2:

$$(\text{SP}_{0\,2} - a_{0\,2}\text{SP}_{0\,2}) + (\text{SP}_{1\,2} - a_{1\,2}\text{SP}_{1\,2}) = (E34 - M34) + (E35 - M35),$$

offset by (b) securities that should have matured in 2, but were sold in 1:

$$-(a_{1\,2} - a_{0\,2})\text{SS}_{1\,2}^0 = -(M22 - M21) \times E72,$$

plus (c) the gain on securities sold in 2 that would mature in later periods:

$$(a_{2\,3} - a_{0\,3})\text{SS}_{2\,3}^0 + (a_{2\,3} - a_{1\,3})\text{SS}_{2\,3}^1 + (a_{2\,4} - a_{0\,4})\text{SS}_{2\,4}^0 + (a_{2\,4} - a_{1\,4})\text{SS}_{2\,4}^1$$
$$= (N23 - N21) \times F73 + (N23 - N22) \times F80 + (O23 - O21) \times G73$$
$$+ (O23 - O22) \times G80.$$

Similar equations apply to periods 1 and 3.

The security accounting block, B66:N87 tracks securities purchased in each period until they are sold or mature. It is assumed they are sold on a first in–first out basis. The reason this tracking is necessary is that we need to know when and at what price a security was purchased so we can calculate the gain on its sales. To see how this works, consider cell F73, the sales in period 2 of securities maturing in 3, which were originally purchased in period 0, $\text{SS}_{2\,3}^0$. The formula in F73 is

$$\text{SS}_{2\,3}^0 = \text{MIN}(\text{Total SS}_{2\,3}, \text{Balance at } t = 1 \text{ of securities}$$
$$\text{maturing at 3 that were purchased at } t = 0).$$

In cell terms:

$$F73 = MIN(F44, M72),$$

where F44 is the total purchases in at $t = 2$ to mature in 3, $SS_{2\,3}$, and M72 the balance at the end of the last period, $t = 1$, of securities maturing at $t = 3$, that were purchased at $t = 0$. This balance is tracked for each period's originating portfolio until it either matures or is completely sold.

Clearly, we do not want to buy securities if we do not have the money, nor do we want to sell securities that we do not own. Consequently, we need to include constraints on our decisions or put some warnings in the model so the user will know that an infeasible decision is being made.

This model has warnings in cells C35:C37 and C43:C45 that test for feasibility and provide a warning when the security purchase and security sale decisions are infeasible. For the security purchase decisions, the budget constraint in each period t is

$$\text{Total Security Purchase}_t \leq \text{Cash}_{t-1} + \text{Cash Inflow}_t$$
$$+ \text{Securities Matured}_t + \text{Securities Sold}_t.$$

The cells in C35:C37 have a test of this constraint that gives the user a message if the purchase decision will violate the budget. For example, the formula in C35 is

$$=IF(P35>C10+D50+P43, \text{"infeasible"}, \text{" "}).$$

This way, if the user inputs a purchase decision that violates the budget, a message appears that it is an infeasible decision, and the model does not compute. Nothing shows in the cell if the constraint is not violated. That is why the cells are empty in the exhibit.

There are similar constraints in the security sales section, E43:G45, where the budget test is

$$\text{Security Sales}_t \leq \text{Securities Held}_{t-1}.$$

For example, in the security sale decision input section (cells E43:G45), there are constraint tests that inform the user when a decision would violate the constraint. For period 2, the sale is infeasible if

$$\text{Security Sales}_{2\,3} > \text{Securities Held}_{1\,3}$$

or

$$\text{Security Sales}_{2\,4} > \text{Securities Held}_{1\,4},$$

so the formula for second period sales in cell C44 is

$$=\text{IF(OR(F44>F56,G44>G56),"infeasible", " ")},$$

which informs the user when she tries to sell more securities that are left in the portfolio from last period.

At this stage we do not know what the best decisions are. But we will explore the consequences of some alternative choices with our spreadsheet model. The first decisions we test are to purchase securities in period 1 totaling $5500, consisting of $1000 to mature in period 2, $4000 to mature in period 3, and $500 to mature in period 4. In period 2 we will purchase $1500 to mature in period 3, and in period 3, purchase $500 to mature in period 4. That is,

$$\text{SP}_{12} = \$1000,\ \text{SP}_{13} = \$4000,\ \text{SP}_{14} = \$500,\ \text{SP}_{23} = \$1500,\ \text{and}\ \text{SP}_{34} = \$500.$$

In addition, in period 2 we will sell $500 of securities to mature in period 3:

$$\text{SS}_{23} = \$500.$$

With these decision variables input to our spreadsheet model, the results are shown in Exhibit 12.5A. We see in cell N14 that the value of the objective function has a cash and security balance at time 3 of $7130.11. We can change the investment decisions to find if there is a better set of decisions. Note that selling securities at time 2 does not seem like a good decision because we give up interest income that would be earned on these securities. We change the security sale decision to $\text{SS}_{23} = 0$, and the objective function increases to $7136.53. With a little more adjustment of the decision variables to make sure our money is fully invested, we increase SP_{13} to $4600, increase SP_{23} to $1530, set SP_{34} at 0, to make the ending cash and security balance reach $7138.96. At this stage we do not know if this is really the best we can do. To find the best, we need to turn to an optimization routine to optimize our investment decisions. We will leave that to Chapter 17.

This section has presented a simple model for analyzing decisions to invest in short-term securities. This model allowed us to track the security portfolio to see the results of purchase and sale decisions over a three-period horizon. This is the type of model where we would have to carefully consider the costs and benefits of developing it. With just three investment periods, we see that the model is relatively large and unwieldy. Even though the model is not complex, if we were to extend the horizon, the model would increase exponentially in size. The effort to develop such a model can be worthwhile if it enables us to analyze a wider range of decision alternatives than we could without a model. As the number of periods and securities expands, we could find it almost impossible to evaluate the exponentially expanding number of decision combinations. That is when we may come to appreciate the ease with which we can easily explore the consequences of different decision combinations—once the model is built.

Our model just opens the door to many different types of security investment decisions. There are many types of short-term security problems we have not considered in this section, but this model gives you some ideas on how to get started. In the next section we will discuss ways to model receivables.

12.3 MODELING RECEIVABLES AND CREDIT

This section presents different methods for modeling the firm's accounts receivable (AR) or credit portfolio. The importance of credit to the firm's planning will depend on the role of credit in the business. For many firms, receivables arise in conjunction with selling their products on credit. For this type of firm, receivables would be secondary to the main line of business—selling the product—and receivables would be a spontaneous asset that would respond to sales. In this case, credit policy probably would be less crucial to the future of the business, and there would be fewer policy decisions to be explored in the model. At the other end of the spectrum would be a business where credit is the focus. Banks are the best example, where their raison d'etre is to make loans. In addition, there are other firms whose main line nominally may be production or sales, but credit is a major part of their business. Examples include a retailer like Sears Roebuck, whose consumer credit division frequently was an important source of its income, or an auto company whose auto credit division is a major contributor to its profitability.

Clearly, there is a wide range of problems that could be the subject of a model concerned with receivables, and at least as wide a range of models that could be used. For long-range planning for a firm where credit is not central, we may not have to delve in the fine detail of the AR, and it would be sufficient to model receivables in broad, non-detailed terms. For short-term planning, and for firms where credit is central to their business, we may want a model that allows us to see the detailed implications of our actions and policies. In this section, we will present examples for longer range planning, and we will only scratch the surface of the modeling possibilities for the firm where credit is central and a finely tuned, detailed credit model is necessary. For the more specialized credit modeling we will direct the reader to sources that specialize in that field.

ARs are normally generated by credit sales, so we can link receivables directly to sales. For long-range planning when it is not important to follow the short-run details of the AR, it may be sufficient to use a percent-of-sales approach for receivables. For a model with a long horizon and planning periods of a year or more, receivables at the end of the year would depend on sales that year, and only in unusual cases would they depend at all on sales prior to that year. Consequently, receivables can be expressed as a simple proportion of sales:

$$\text{Receivables}_t = a \times \text{Sales}_t, \tag{12.6}$$

where parameter a is the anticipated ratio of receivables-to-sales. The receivables-to-sales ratio (AR/Sales) can be estimated directly from past observations of receivables and sales, or can be based on the anticipated credit policy of the firm. One way we can link receivables to sales is via the anticipated collection period, or days sales outstanding (DSO). The DSO is calculated as

$$\text{DSO} = \frac{\text{AR}}{(\text{Sales}/360)}. \tag{12.7}$$

So the ratio of receivables-to-sales is

$$\frac{\text{AR}}{\text{Sales}} = \frac{\text{DSO}}{360}. \tag{12.8}$$

For example, if annual sales are $1,000,000 (sales per day of $2778) and ARs are outstanding for an average of 30 days, the receivables balance is 30 (2778) = $83,340, and the ratio of receivables-to-sales is

$$\frac{\text{AR}}{\text{Sales}} = \frac{\$83,340}{\$1,000,000} = 8.33\%,$$

or

$$\frac{\text{AR}}{\text{Sales}} = \frac{\text{DSO}}{360} = \frac{30}{360} = 8.33\%.$$

In this case, the receivables in the planning model would be expressed as

$$\text{AR}_t = (0.0833)\,\text{Sales}_t.$$

By linking DSO to AR/Sales, we have a way to link receivables in our percent-of-sales model to an easily and widely understood measure of collection efficiency.

As the length of the periods is shortened, the amount of detail can be increased with the AR shown as dependent on the sequence of past observed sales. A distributed lag relationship of a form such as

$$\text{Receivables}_t = a_0 \times \text{Sales}_t + a_1 \times \text{Sales}_{-1} + a_2 \times \text{Sales}_{t-2} \qquad (12.9)$$

may be appropriate, where parameters a_0, a_1, and a_2 are estimated from past data or, perhaps, based on credit policy and anticipated payment practices. For example, assume the model has a planning horizon of one year and sub-periods of one month, that credit sales are 80% of total sales, and 60% of the credit sales are collected in the first month after the sales occur, 30% are collected in the second month, and 10% are collected in the third month. Assuming there is no uncertainty in the process, the receivables collected would be

$$\text{Receivables Collected}_t = 0.60(0.80)\text{Sales}_{t-1} + 0.30(0.80)\text{Sales}_{t-2}$$
$$+ 0.10(0.80)\text{Sales}_{t-3}$$
$$= 0.48\text{Sales}_{t-1} + 0.24\text{Sales}_{t-2} + 0.08\text{Sales}_{t-3},$$

and the balance at the end of period t would be

$$\text{Receivables}_t = 0.80\text{Sales}_t + 0.40(0.80)\text{Sales}_{t-1} + 0.10(0.80)\text{Sales}_{t-2}$$
$$= 0.80\text{Sales}_t + 0.32\text{Sales}_{t-1} + 0.08\text{Sales}_{t-2}, \qquad (12.10)$$

where the parameter estimates are $a_0 = 0.80$, $a_1 = 0.32$, and $a_2 = 0.08$. With this percent-of-sales model, the primary input is the sales forecast. For the distributed lag example, the Receivables forecast at any *future* date t would be generated by the forecast of Sales at dates $t - 1$ and t. Suppose the sales forecasts for the next 12 months are as shown in Exhibit 12.6; with the parameters shown in Equation 12.10, the Receivables balance in each month will be as shown in the second line. For example, Receivables in the eighth month would be

$$\text{Receivables}_8 = 0.8\text{Sales}_8 + 0.32\text{Sales}_7 + 0.08\text{Sales}_6$$
$$= 0.8(350) + 0.32(400) + 0.08(350) = \$436.$$

To the extent that receivables and/or collections are affected by economic conditions, estimates may be improved by adding appropriate economic

Month (t)	1	2	3	4	5	6	7	8	9	10	11	12
Sales ($000)	100	150	200	250	300	350	400	350	300	250	200	150
Receivables ($000)	144	164	216	276	336	396	456	436	384	324	264	204

EXHIBIT 12.6   Sales and receivables forecast with equation: Receivables$_t$ = 0.80 Sales$_t$ + 0.32 Sales$_{t-1}$ + 0.08 Sales$_{t-2}$.

variables to the estimated predictive equation. Just remember that while such elaborations may generate better estimates using past data, to use them in the simulation of future periods, the future values of the economic variables must also be forecasted. The economic forecast may be more elusive than forecasting receivables.

Of course, there are numerous other ways in which receivables can be modeled. For example, the accounting relationship,

$$\text{Receivables}_t = \text{Receivables}_{t-1} - \text{Receivables Collected}_t$$
$$+ \text{Credit Extended}_t \tag{12.11}$$

can serve as the basis for our prediction, where we estimate the variables Receivables Collected$_t$ and Credit Extended$_t$ with regression equations that might be based on a mix of firm data and economic data. Credit collections in month t will be influenced by the past receivables balances and past and current economic conditions. Credit extended might be driven by firm sales and interest rates. This would suggest that we could estimate credit and collections with regression equations such as

$$\text{Receivables Collected}_t = A_1 \text{Receivables}_{t-1} + A_2 \text{Interest Rate}_{t-1}$$
$$+ A_3 \text{Unemployment}_{t-1}$$
$$\text{Credit Extended}_t = B_1 \text{Sales}_t + B_2 \text{Interest Rate}_{t-1}.$$

With this econometric approach, we would estimate the regression coefficients (A_1, A_2, A_3, B_1, and B_2) based on past economic and firm data. Of course, we would have to investigate several alternative predictive models to find the one that best described and predicted our firm's variables.

12.4 INVENTORIES

As with other current asset categories, there are numerous ways in which to model inventories, ranging from the simplest with inventories as a proportion of sales, to more complex models that link inventories to a production sector of the planning model.

For the longer horizon model, the simple proportional relation

$$\text{Inventories}_t = b \times \text{Sales}_t \tag{12.13}$$

may be sufficiently accurate because inventories at the end of each year may bear a stable relation to sales over a period of years, and the relation could be estimated easily from past data. For long-run strategic planning purposes where rough approximations prove sufficient, and where it is unnecessary to track the seasonal details of inventories, there is little reason to burden the model with greater complexity.

The relation between Inventories and Sales, such as Equation 12.13, can be based on past relationships and estimated from past data using a linear regression approach, or it may be included in the model so as to reflect an inventory policy decision. As an example of the latter approach, the policy might be to maintain an inventory balance equivalent to 50 days' sales (evaluated at CGS). Assume CGS is 70% of sales, with sales per day of

$$\text{Sales per Day} = \frac{\text{Annual Sales}}{365}.$$

Evaluated in terms of cost, the cost of goods sold per day would be

$$0.70 \times \text{Sales per Day,}$$

or

$$0.70 \times \frac{\text{Annual Sales}}{365},$$

and to have inventory to meet 50 days' sales, the firm would have

$$\text{Inventory} = 50 \times 0.70 \times \frac{\text{Annual Sales}}{365}$$
$$= 0.096 \times \text{Annual Sales,}$$

so this inventory policy would call for inventories being maintained at 9.6% of sales to meet the 50-day requirement.

Greater detail and accuracy may be attainable by showing inventories in terms of the variables that most directly influence them—specifically, both production and sales. For example, the equation

$$\text{Inventories}_t = \text{Inventories}_{t-1} + (\text{Units Produced}_t - \text{Units Sold}_t) \times \text{Cost per Unit}_t \tag{12.14}$$

is true by definition and could accurately track inventories if production, sales, and costs are accurately forecasted. This would be especially suited to a short horizon where seasonal variations in production and sales are important, and where there is a production sector that deals in physical units. However, the relation need not be expressed in terms of physical units. It can be expressed directly in terms of dollar value so as to bypass the difficulties of working with physical units and developing data from which to evaluate the cost per unit.

Since production is usually based on current and future expected sales, the inventory Equation 12.14 may be rederived in a form so that inventories are driven solely by sales. For example, a simple adaptive model of production could be of the form

$$\text{Units Produced}_t = \text{Units Sold}_t + a(\text{Units Sold}_t - \text{Units Sold}_{t-1}), \quad (12.15)$$

where current production is equal to current sales plus some proportion, a, of the current increase in sales so as to increase inventory to meet an anticipated sales increase next period. Substituting Equation 12.15 for Units Produced$_t$ in Equation 12.14 yields

$$\text{Inventories}_t = \text{Inventories}_{t-1} + a \times (\text{Units Sold}_t - \text{Units Sold}_{t-1}) \\ \times \text{Cost per Unit}. \quad (12.16)$$

There are numerous decision models that are intended to specify an optimal inventory policy based on various factors of demand, interest rates, and ordering costs for goods held in inventory.* These optimization models can provide the basis for forecasting inventories in a simulation model when it is assumed the firm follows the optimal policy. For example, the simplest inventory model is the economic order quantity (EOQ) model that prescribes the optimal amount of inventory to order and implies that the average inventory will be

$$\text{Inventory} = \text{Sales}^{\frac{1}{2}} \times \frac{1}{2}\left(\frac{\text{Ordering Cost}}{\text{Cost per Unit} \times \text{Interest Rate}}\right)^{\frac{1}{2}},$$

where Ordering Cost represents the fixed cost of placing one new order for the goods carried in inventory, and Cost per Unit × Interest Rate is the opportunity cost of carrying one unit in inventory for a year.

* Inventory policy models are explained in Fogarty, Blackstone and Hoffman (1991), or Chapter 12 in Mathur and Solow (1994).

For the firm following the EOQ policy, the average inventory would increase with sales at a decreasing rate. Thus, inventories would not be strictly proportional to sales, and the inventory as a proportion of sales would decrease as forecasted sales increase. Note that this model relates the average inventory to the Ordering Cost and the Carrying Cost, so that to utilize it in the financial planning context would necessitate forecasting not just sales, but these costs (including interest rates) as well.

There are numerous other more sophisticated inventory models that potentially could be useful in modeling inventories in a simulation model. However, the sophistication may be self-defeating. The more sophisticated models may impose an impractical burden of forecasting on the planner if it requires that too many variables be forecast in excessive detail. Generally, the simpler the model, the more useful it is.

12.5 SPONTANEOUS FINANCING

This section will deal with the sources of short-term financing that are spontaneous. Spontaneous financing refers to financing sources that grow more or less automatically with sales and the firm's assets, whereas discretionary means that a decision must be made to utilize a particular source of external financing. The spontaneous source we consider here is financing in the form of accounts payable (AP). The other source that we would include in this spontaneous short-term category would be accruals, but our focus will be on payables.

AP is a good example of spontaneous financing from current liabilities. First, note that APs for one firm are the ARs for another firm, so that the various methods for modeling receivables that we discussed earlier apply to modeling the payables. AP represents credit purchases of goods and materials used in producing the output. Thus, sales drives purchases, which, in turn, determines the balance in AP. In a very aggregative model with long periods comprising the planning horizon, these relationships would be summarized in a percent-of-sales model for payables of the form

$$AP_t = a \times Sales_t, \tag{12.17}$$

where a is the parameter that represents the payable as a percentage of sales. The parameter a would be determined from observations of past data, modified by judgments regarding the future relationship.

With respect to financing growth of sales and assets, it is the increase in current liabilities that is the relevant source of spontaneous financing. For

AP, the spontaneous increase, denoted by ΔAP, corresponding to Equation 12.17, is expressed as

$$\Delta AP_t = a(\text{Sales}_t - \text{Sales}_{t-1}). \tag{12.18}$$

As with other account balances, as the planning horizon and the periods comprising the horizon are shortened, it may be necessary to model payables in greater detail. There are many different approaches, and a good starting point is the definitional equation

$$AP_t = AP_{t-1} - \text{Payments on Account}_t + \text{Credit Purchases}_t. \tag{12.19}$$

Each of the components, Payments on Account and Credit Purchases, would be modeled separately. For example, if the length of the interval is one week and accounts are always paid in four weeks, we would have

$$\text{Payments on Account}_t = AP_{t-4},$$

and if purchases are equal to 60% of sales forecasted for two weeks in the future, we have

$$\text{Credit Purchases}_t = 0.60 \times \text{Sales}_{t+2}.$$

This example is based on a specific policy that is assumed to apply over the planning horizon. In other cases, it may be more appropriate to use observed past data to make econometric estimates of the relationships between payments or purchases and sales. An example of a linear regression relation that would be estimated using lagged variables is

$$\text{Payments on Account}_t = A + B \times AP_{t-1} + C \times AP_{t-2}$$

where the data are used to estimate parameters A, B, and C; and purchases are estimated from a relation such as

$$\text{Credit Purchases}_t = A + B \times \text{Sales}_{t-1} + C \times \text{Sales}_t + D \times \text{Sales}_{t+1}.$$

Other types of current liabilities, such as accrued expenses, are usually spontaneous and would be modeled with a percent-of-sales approach.

There is seldom much benefit from modeling such accounts in great detail in a planning context. The other major source of short-term financing would be short-term borrowing. We will consider that when we consider debt financing in a later chapter.

12.6 SUMMARY

This chapter has discussed ways in which we can model selected working capital accounts. We continued to assume that our task is to develop a sales-driven planning model for the firm, and presented suggestions for modeling the working capital accounts. In each case, you need to keep in mind that there are as many different ways to model particular problems or accounts as there are model builders. Each planning problem will be different and require a different approach, and our models and methods will only provide a starting point. The challenge and the fun of financial modeling is in devising new approaches and solutions to planning problems.

PROBLEMS

1 M&O Corporation Cash Balance

The M&O Corporation has asked you do examine their cash management processes and recommend a policy for managing their cash balances. As part of this assignment, you have decided to test the Miller & Orr cash balance model with Monte Carlo simulation.

The Miller & Orr model assumes the daily cash inflow is a normally distributed random variable with an average (mean) daily inflow of 0 (called zero drift), and a standard deviation of daily flow of σ (variance of σ^2). The cost associated with this policy consists of the interest foregone from holding the cost of making transfers when securities are bought and sold. The policy allows the cash balance to drift randomly between an upper limit H, and a lower limit, L. If the cash balance hits either boundary, securities are bought or sold so as to restore the cash balance to a restoration point, Z. The formulas for these policy limits are

$$Z = (3b \ \sigma^2 \ /4 \ r)^{1/3},$$
$$H = 3z, \text{ and}$$

L is a set by management based on their minimum desired cash balance, and where b is the fixed dollar cost of a transaction, and r is the interest rate per day.

You have examined the company's cash processes and estimated the following parameters and data:

Current cash balance	$10,000
Minimum allowable cash balance, set by management	$6000
Current balance of marketable securities	$30,000
Interest rate earned annually on marketable securities	12%
Interest rate per day, based on a 360-day year	0.033%

The daily cash flow is a normally distributed random variable with a mean of 0 and standard deviation of $500.

If cash is transferred into or out of marketable securities there is a fixed cost per transaction of $50.

Modeling Assignment

Develop a model of the company's daily cash flow and cash balance over a 30-day period, where the cash account is managed according to the Miller & Orr model. The model should calculate the Miller & Orr transaction policy boundaries (H, Z, L), and the total cost of the policy consisting of the total interest foregone from carrying a cash balance, and the total transactions costs over the 30 days. Graph the daily cash balance. Use @Risk to test the model. Make random draws from the daily cash flows to apply the model and follow the cash balances, security transactions, and costs. For example, for 1 run of the model over 30 days, each day's cash flow would be a single draw from the @Risk function, RiskNormal(0, 500), which could lead to securities being bought or sold that day, and result in a cash balance at the end of the day, and a cost for the day consisting of the interest foregone and perhaps a transaction cost. Graph the time pattern of the cash balance over the 30 days for a single draw of the simulation.

Run the simulation model with 5000 draws for the 30-day period. This will yield frequency distributions for the transactions cost, interest cost, and total cost over the 30 days. Provide a summary of the distributions of these costs.

2 Managing Marketable Securities: Placidia Corporation

The Placidia Corporation has a portfolio of cash and short-term securities. In managing the cash and securities, the objective is to maximize value of cash and securities held at the end of the three-month planning horizon. The planning horizon is divided into intervals of one month. The first date that decisions can be implemented is time 1, the end of the first month, which is when the first cash flow will occur, and securities will mature. The cash from maturing securities and securities sold will be available immediately to be used for other investments or other use.

At the start of the planning period, the firm has an initial balance of cash and an endowment of securities as follows:

Cash balance	$2500		
Marketable securities			
Maturity Date	**Month 1**	**Month 2**	**Month 3**
Face value ($)	400	1000	500

In addition to holding the initial balance cash and securities, the firm expects a cash flow in each month. The cash flows expected in each of the three months of the planning horizon are

	Month 1	**Month 2**	**Month 3**
Expected cash flow ($)	−3000	700	800

The cash inflow will be available to be invested, and the cash outflow of $3000 must be funded from the cash balance or from the security portfolio.

Securities are traded at prices that are discounted from face, or maturity value. The current and expected future short-term interest rates that we use to discount the face value to the market value of the securities are shown in the table below. The number in the first column indicates the month in which the security is purchased, and the numbers in the top row indicate the month the security matures. Therefore, a security purchased in month 1 that matures in month 3 is expected to have a monthly interest rate of 0.9%. Securities purchased in period 0 refers to securities already held at the start of the planning horizon. Period 4 represents securities maturing after the end of the three-month planning horizon. Assume period 4 is 1 month beyond the end of the planning horizon.

Monthly Interest Rates for Marketable Securities

	Period Security Matures			
Period Security is Purchased	**1**	**2**	**3**	**4**
0	0.5%	0.6%	0.7%	1.0%
1		0.8%	0.9%	1.2%
2			1.2%	0.8%
3				0.5%

Your task is to construct a spreadsheet model to track our security portfolio for three months so that we can see the results of alternative strategies for buying and selling securities.

You can purchase securities at times 1, 2, and 3 to mature at any subsequent date up to time 4. The purchase price of the security is the present value of the face value of the security. Thus, if, in month 1 you purchase a security that matures in month 3 at face value of $1000, the purchase price (market value) will be

$$\text{Purchase Price} = \text{Market Value} = 1000 \left(\frac{1}{(1+0.009)^2} \right)$$

$$= 1000(0.98224)$$

$$= \$982.24.$$

If you hold the security for the two months until it matures, we will receive the face value of $1000 at the end of month 3. The income will be the appreciation in price amounting to

$$\text{Income} = \$1000 - 982.24 = \$17.76.$$

A security can be sold prior to maturity. If this security is sold prior to maturity, it will be sold at a discounted price based on the discount rate prevailing at the date it is sold, with the rate reflecting the remaining time to maturity at the date it is sold.

The decisions you have to make are

- When to buy securities
- Which securities to buy
- When to sell securities
- Which securities to sell

You want to make these decisions with the objective of maximizing the value at time 3 of the cash balance and the market value of securities held at time 3. Of course, you may not allow your cash balance to be negative at any date. You cannot buy securities unless you have the cash to pay for them. You cannot sell securities that you do not have, and you cannot sell securities short.

PART V

Modeling Security Prices and
Investment Portfolios

CHAPTER 13

Modeling Security Prices

THIS AND THE FOLLOWING two chapters are concerned with modeling investment opportunities. This chapter starts by developing models of security prices that will be used in subsequent chapters. The objective of the price model is to portray mathematically and graphically the movement of security prices as time passes. This can be useful for understanding the distribution of possible future prices, and it will also be useful for devising investment strategies, constructing security portfolios, and understanding the pricing of derivative securities. We will look at two different ways to model the time path of security prices: the binomial model in discrete time and the continuous time model.

13.1 THE BINOMIAL MODEL OF STOCK PRICE MOVEMENT

The first price model we develop is the binomial model which assumes that the price of the security will move to either of two prices by the next point in time. Although this model may seem unrealistic, it can be made sufficiently accurate to portray most price patterns, and it is particularly helpful for understanding and modeling the pricing of derivative securities such as call and put options.

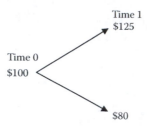

EXHIBIT 13.1   Binomial stock price movement.

As an introduction to the binomial model, assume the length of the period is one year, the current stock price is S_0, and the price at the end of the year will be S_1. Letting u denote the wealth relative return (1 + rate of return) when the stock moves up, and d the wealth relative return when the stock moves down, the year-end stock price will be either $S_1 = S_0 u$ or $S_1 = S_0 d$. Assume the initial stock price is $100 and the wealth relative returns over the year are either $u = 1.25$ or $d = 0.80$ (rates of return of 25% or −20%). Then the year-end price will be either

$$S_1 = S_0 u = 100(1.25) = 125,$$

or

$$S_1 = S_0 d = 100(0.80) = 80,$$

as shown in Exhibit 13.1. This simple single period model with just two outcomes has been shown to be particularly useful for understanding option pricing. Next, we extend the model to several periods.

Let us increase the number of periods to four and consider the prices that could result from this binomial process. Exhibit 13.2 shows the prices over four periods, starting with $S_0 = 100$ at $t = 0$, and assuming that at each node (date) the price can go up by proportion $u = 1.25$ or down by $d = 0.80$. Even though there are 16 different possible price paths to the fourth node (labeled a–p on the right), only five different prices can occur at $t = 4$: $41, $64, $100, $156.3, and $244.1.* The reason there are relatively few possible prices is that there are numerous ways to arrive at the same

---

* With two outcomes at each node, the number of price paths over four periods is $2^4 = 16$, and in general, over n periods, there would be 2^n different paths. However, with u and d the same each period, the number of different prices is $n + 1$, or with four periods, five different prices.

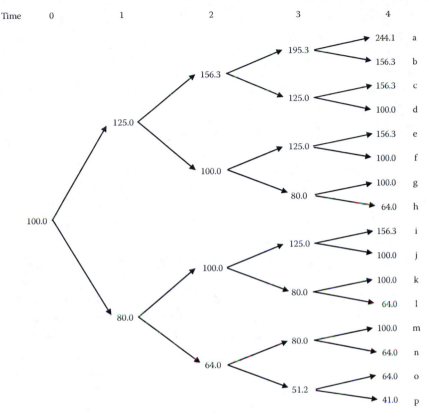

| Time | 0 | 1 | 2 | 3 | 4 | |

EXHIBIT 13.2   Price paths over four periods with binomial model.

price. For example, out of 16 price paths, $100 occurs six times (nodes d, f, g, j, k, and m). This results from the fact that u and d are the same amounts in each period. This produces six sequences of us (ups) and ds (downs) that lead from $100 to $100 in four steps. For $u = 1.25$ and $d = 0.80$ at every step, we have

$$S_4 = S_0 \, uudd = S_0 \, udud = S_0 \, uddu = S_0 \, duud = S_0 \, dduu.$$

The order of sequence of ups and downs does not matter—they all lead to the same terminal price of $100.

One of our concerns is the probability distribution of prices at a given date. Monte Carlo simulation is helpful for this purpose. If we have information about the probability of each of the moves, u and d, it is easy to see what the probability of each stock price is when there are only a few steps. However, determining the distribution of stock prices can be more difficult when there are many steps in the process. We can simulate the process to get a better estimate of the distribution of possible outcomes.

Suppose that we estimate the probability of an up move to be 78% and the probability of a down move is 22%, and the moves are independent in different periods. What is the probability of each possible price at time 4 give the initial price of $100 at time 0?

Exhibit 13.3 is a four-period model of stock prices with the up and down movement at each stage chosen randomly with an @Risk Monte Carlo simulation. The input data are described in cells A4:C16, and the output consisting of stock moves and prices at each of the four steps is shown in cells E8:H12. The initial stock price of $100 is in cell H8. The price then moves up or down at times 1–4 based on a random draw of u and d, until it reaches time 4 in cell H12. The up move is 1.25 each period with a probability of 78% (specified in cell B16), and the down move is 0.80 with a probability of 22% (cell C16). The random draws are modeled in column F. For example, cell F9 has the @Risk uniform distribution function =RiskUniform(0,1) that generates a random variable between 0 and 1. G9 then translates this draw into the up (u) or down (d) by choosing $d = 0.80$ if the fraction drawn with the uniform distribution is <0.22, and choosing $u = 1.25$ if the fraction is >0.22. By using the uniform distribution with all fractions from 0 to 1 having an equal likelihood of occurring, $d = 0.80$ will occur 22% of the time, and $u = 1.25$ will appear 78% of the time.

	A	B	C	D	E	F	G	H	I
1					Binomial Stock Price Movement				
2					Four Period Model				
3									
4	Data Input					Model of Prices			
5									
6	Initial Stock Price		100			Step	Random	Price	Stock
7							Draw	Move	Price
8			Stock			0			100.00
9			Returns & Movement			1	0.67	1.2500	125.00
10						2	0.92	1.2500	156.25
11			Outcome			3	0.84	1.2500	195.31
12			#1	#2		4	0.82	1.2500	244.14
13			Up	Down	=RiskUniform(0,1)				=H8*G9
14	Rate of Return per Year	25%	-20%						
15	Wealth Relative Return (u & d)	1.25	0.80		=IF(F9<$C$16,$C$15,$B$15)				
16	Probability	78%	22%						
17									

EXHIBIT 13.3   Excel model of binomial price movement over four periods.

The output of interest is the frequency distribution of the price at time 4. Exhibit 13.4 shows an @Risk graph of the frequency distribution, with the relative frequencies of each of the five prices turning out to be as follows:

Price ($)	41	64	100	156.3	244.1
Frequency (%)	0.2	3.2	17.8	42	36.8

This section introduced the binomial model and showed how to construct a simple Excel® model of binomial price movements. As noted earlier, you may be inclined to think such a model is unrealistic because stock prices seem to move in far more complex patterns than just up or down. While it is true that a simple up or down pattern is an oversimplification, we can adjust the model to provide more realistic movements by dividing time into smaller units with prices going up or down in very small increments. In fact, we can make the model mimic very closely the stock movements that would result from a continuous process. We show how to use the binomial model to approximate a continuous process in a subsequent section. But first we will explain and model the continuous stock price process.

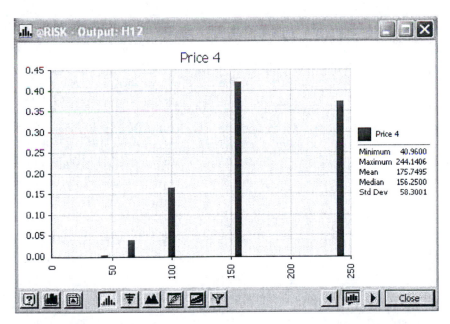

EXHIBIT 13.4   Distribution of price at time 4 with @Risk Monte Carlo simulation.

13.2 STOCK PRICES AS A RANDOM WALK IN CONTINUOUS TIME

The continuous time model of stock returns and prices was developed by Merton (1973) and underlies the Black and Scholes (1973) option pricing model. This model hypothesizes that stock prices follow a process where, over short periods of time of length dt,* the return is generated according to the equation

$$\text{Return} = \frac{dP}{P} = \mu\,dt + \sigma z\sqrt{dt}, \tag{13.1}$$

where P is the price, dP the change in the price, dP/P the rate of return, μ the expected return per period, σ the standard deviation of return per period, and z the standard normal variable with mean 0 and standard deviation 1. Equation 13.1 says that the return on a given stock during a period of length dt will be the expected return plus an error that is a random variable from a normal distribution with mean 0 and standard deviation σ. For example, if the time is defined as years, with μ and σ expressed as annual rates, the return over a quarter of a year (d$t = 0.25$) would be

$$\text{Return} = \frac{dP}{P} = \mu(0.25) + \sigma z\sqrt{0.25} \tag{13.2}$$

and if the period was one year (d$t = 1$), the return over the year would be $\mu + \sigma z$.

When the return is generated by the process described in Equation 13.1, we say the return follows a Weiner process. This is similar to the Brownian motion in chemistry that refers to the random motion of a particle suspended in a liquid, except in this case it refers to the random movement of stock prices. In this case, the stock price follows what is called a geometric Brownian motion, where the random movement of the stock price is proportional to the level of the price; it is the return that follows the Weiner process, or Brownian motion. This process is what has come to be known as a random walk of stock prices.

---

* In the continuous time model, dt denotes the limit as the time increment approaches zero, and in most discussions of the model, a distinction is made between dt and a discrete time change denoted by, for example, δt. Our price model will be a discrete time approximation of the continuous process, but we will not make the distinction between the limit dt and the discrete time version, and we use dt to denote the change in time over a small discrete increment.

Using Equation 13.1 with our Monte Carlo software, we can simulate the distribution of returns on a stock by making random draws from a standardized normal 0, 1 distribution that is represented by z in the formula. Suppose the expected return per year on a stock is μ = 12%, with a standard deviation of return per year of σ = 20%, and we wish to simulate the distribution of returns over quarter year. Exhibit 13.5 shows a model of stock returns that follow the random walk in a discrete time version of the Merton model. Cells B10, B11, and B15 are the inputs (μ, σ, and dt), B22 is the random variable, z, that is drawn from a standard normal probability distribution, and C22 is the simulated return. The return in C22 is generated by the formula = B10 × B15 + B11 × C22 × (B15^0.5). The display shows one particular random draw of the standard normal variable z = 1.369. This produces the random error of

$$\sigma z \sqrt{dt} = 0.20(1.369)\sqrt{0.25} = 0.1369$$

	A	B	C	D
4		P		
5				
6	Data Input			
7				
8	Initial Stock Price	100		
9				
10	Mean Return	12.0%	per year	
11	Standard Deviation	20.0%	per year	
12				
13	Length of Sub-Period	3 months		
14	Sub-Periods Per Year	4		
15	Length of Period	0.25	fraction of year	
16				
17		Standard	Return	
18		Period Normal	For	Price
19		Variate, z	Period	
20	= RiskNormal(0,1)			
21		0		100
22		1	0.637	9.37%	109.37
23				
24		= B10*B15+B11*B22*(B15^0.5)		

EXHIBIT 13.5   Model of stock returns that follow a Weiner process (random walk).

and the quarter year return for this particular draw of the simulation is

$$\text{Return} = r = 0.12(0.25) + 0.1369 = 0.1669,$$

as shown in cell C22. Given the initial stock price of $100, the stock price at the end of the first time interval is $116.69 in cell D22.

By making repeated draws for the random variable, z, and tracking the resulting return, we can estimate the probability distribution of the return at a single point in time. Suppose we wish to simulate the distribution of quarter year returns. We set the period at 0.25 year (cell B15) and make 5000 draws. Exhibit 13.6 shows the results for simulated distribution of stock returns over a three-month period. The simulated three-month return is normally distributed with a mean of 3% and a standard deviation of 10%, and minimum and maximum returns of −37.32% and 42.63%, respectively. These figures describe the probability distribution for a three-month return generated by the Weiner process as specified by Equation 13.2.

Now we change the perspective of the simulation to see what might happen to the price of a share of stock over time. Assume that the stock

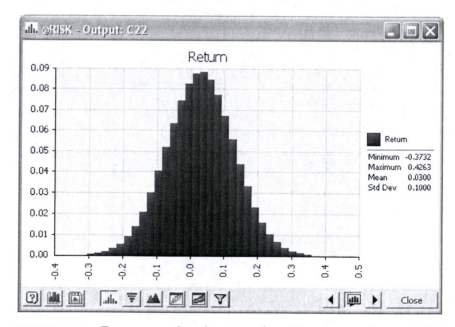

EXHIBIT 13.6 Frequency distribution of stock returns from @Risk simulation.

return follows the process described by Equation 13.2, and we use one week as the length of each subinterval. The return would be

$$\text{Return per Week} = 0.12(0.0192) + 0.20\, z\, \sqrt{0.0192}, \qquad (13.3)$$

where the 0.12 and 0.20 are the expected return and standard deviation of return per year, and the length of the period is $dt = 1/52 = 0.0192$ of a year.

With this process, the price of the stock at date t will be

$$\text{Price}_t = \text{Price}_{t-1}\,(1 + \text{Return}_t). \qquad (13.4)$$

Assume the initial price of the stock is $100. The @Risk model of the return and stock price is shown in Exhibit 13.7.

This is a model of weekly prices with a beginning price of $100. The returns based on Equation 13.3 are drawn randomly from their distributions that are independent from week to week. The end of week price is calculated according to Equation 13.4, which is shown as a cell formula in D22 for the

	A	B	C	D	E	F	G	H	I
1	Stock Returns With Continuous Process								
2									
3	$\dfrac{dP}{P}$	$= \mu\, dt + \sigma\, z\, \sqrt{dt}$							
4									
5									
6	Data Input								
7									
8	Intitial Stock Price	100							
9									
10	Mean Return	12.0%	per year						
11	Standard Deviation	20.0%	per year						
12									
13	Length of Sub-Period	1 Week				=$B$10*$B$15+$B$11*B22*($B$15^0.5)			
14	Sub-Periods Per Year	52							
15	Length of Period	0.019231	fraction of year						
16					=D21*(1+C22)				
17		Standard	Return				Standard	Return	
18	Period	Normal	Per	Price		Period	Normal	Per	Price
19		Variate, z	Period				Variate, z	Period	
20	=RiskNormal(0,1)								
21	0			100					
22	1	0.528	1.69%	101.69		14	-1.992	-5.29%	91.48
23	2	-0.314	-0.64%	101.04		15	-0.557	-1.31%	90.28
24	3	-0.673	-1.63%	99.39		16	-0.996	-2.53%	87.99
25	4	-0.153	-0.19%	99.20		17	-0.710	-1.74%	86.46
26	5	1.455	4.27%	103.43		18	1.125	3.35%	89.36
27	6	0.767	2.36%	105.87		19	-0.103	-0.05%	89.31
28	7	0.230	0.87%	106.79		20	-1.019	-2.59%	86.99
29	8	-0.465	-1.06%	105.66		21	0.105	0.52%	87.45
30	9	-1.898	-5.03%	100.34		22	-0.046	0.10%	87.54
31	10	-1.141	-2.93%	97.40		23	-0.408	-0.90%	86.75
32	11	-0.108	-0.07%	97.33		24	0.844	2.57%	88.98
33	12	0.000	0.23%	97.55		25	1.260	3.72%	92.30
34	13	-0.438	-0.99%	96.59		26	-0.854	-2.14%	90.32

EXHIBIT 13.7 Model of random stock prices.

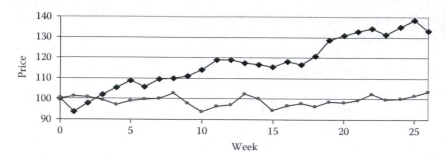

EXHIBIT 13.8   Two paths of stock prices from the same probability distribution of returns.

price at the end of week 1. The random results of one particular run of the simulation are shown in the return and price cells of Exhibit 13.7. Each run would result in a different series of prices. Exhibit 13.8 shows two different price series resulting from two different runs of the simulation. The two price series were drawn from the same probability distribution of returns, but by the chance of the draw, the prices follow different paths.

13.3 BINOMIAL APPROXIMATION OF THE CONTINUOUS PRICE PROCESS

In the last section, we modeled stock prices as a continuous process, although we used discrete time increments of one week as an approximation. For some purposes such as option pricing, it is easier to use a binomial model to approximate the continuous Weiner process. That is, instead of assuming that the return will be generated according to the Equation 13.1, we will assume that the price follows a binomial process over very short increments in time. If we use sufficiently short time increments, the process can be made to approximate very closely the continuous process.

   For the binomial model to approximate the continuous time model, we use formulas developed by Cox, Ross, and Rubinstein (1979) to relate the binomial up and down moves to the length of the step periods and the parameters of the continuous time model. Assuming the annual expected return in continuous time is μ, and the standard deviation is σ, the equivalent "up" and "down" moves over a short period of time, dt, are

$$u = e^{\sigma\sqrt{dt}} \tag{13.5}$$

$$d = e^{-\sigma\sqrt{dt}} = \frac{1}{u} \tag{13.6}$$

	A	B	C	D	E	F	G	H	I	J	K	L
1					Binomial Stock Price Movement							
2					Partial Year Model							
3												
4												
5												
6	Data Input			Step	Random	Price	Stock		Step	Random	Price	Stock
7			=RiskUniform(0,1)	Draw	Draw	Move	Price			Draw	Move	Price
8	Initial Stock Price	100		0			100.00					
9			=IF(E9<$B$25,$B$22,$B$21)	1	0.08	0.9726	97.26		14	1.00	1.0281	111.73
10	Annual Data			2	0.55	1.0281	100.00		15	0.24	0.9726	108.68
11	Risk Free Interest Rate	5%		3	0.81	1.0281	102.81		16	0.57	1.0281	111.73
12	Mean Return	12%	=G8*F9	4	0.54	1.0281	105.70		17	0.55	1.0281	114.88
13	Standard Deviation Per Year	20%		5	0.71	1.0281	108.68		18	0.76	1.0281	118.11
14				6	0.96	1.0281	111.73		19	0.57	1.0281	121.43
15	Length of Sub-Period	1 week		7	0.28	0.9726	108.68		20	0.66	1.0281	124.84
16	Sub-Periods Per Year	52		8	0.45	0.9726	105.70		21	0.33	0.9726	121.43
17				9	0.96	1.0281	108.68		22	0.13	0.9726	118.11
18	Step Length	0.0192		10	0.52	1.0281	111.73		23	0.72	1.0281	121.43
19	(Step Length)$^{1/2}$	0.1387		11	0.44	0.9726	108.68		24	0.88	1.0281	124.84
20				12	0.58	1.0281	111.73		25	0.87	1.0281	128.35
21	Up Wealth Relative Return	1.0281		13	0.38	0.9726	108.68		26	0.96	1.0281	131.96
22	Down Wealth Relative Return	0.9726										
23												
24	Up Probability	0.535										
25	Down Probability	0.465										

EXHIBIT 13.9 Model of binomial stock price movement. Weekly sub-periods.

and

$$p = \frac{e^{\mu dt} - d}{u - d} = \text{probability of an up move, } u. \qquad (13.7)$$

For our example, as with the continuous time example, assume the mean return per year is $\mu = 0.12$, and the standard deviation of annual return is $\sigma = 0.20$. Based on Equations 13.5 and 13.6, with a step length of one week (0.0192) this gives us $u = 1.0281$ and $d = 0.9726$, with the probability of prices moving either up by u with probability p, or down by d with probability $(1 - p)$, where

$$p = \frac{e^{\mu dt} - d}{u - d} = \frac{e^{0.12(0.0192)} - 0.9726}{1.0281 - 0.9726} = 0.535 \qquad (13.8)$$

Exhibit 13.9 shows the binomial model with the parameters calculated as shown above, with return and price calculations for weekly sub-periods over a horizon of 26 weeks. In the model, cell B21 contains the formula of Equation 13.5, B22 has the formula of Equation 13.6, and B24 has that of Equation (13.7).

The returns and prices shown in Exhibit 13.9 are for one particular run of the Monte Carlo simulation. Each iteration of the simulation would result in a different sequence of prices over the 26 weeks. If you run 5000 iterations, you will potentially generate 5000 different price paths.* Exhibit

---

* For the binomial process with 26 steps, the number of different price paths would be 2^{26} (about 67 million paths), leading to 27 different ending prices.

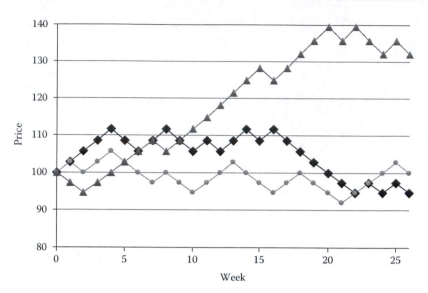

EXHIBIT 13.10 Price paths for three different simulations with weekly binomial price moves.

13.10 shows three of those possible price paths which lead to ending prices of $94.60, $100, and $131.90.

Exhibit 13.11 shows the frequency distribution of prices at the end of 26 weeks when the simulation model was run with 5000 iterations. The mean price at the end of week 26 was $106.16, with standard deviation of $14.76. On close examination you will note that the distribution is skewed to the right, which is confirmed by the positive skewness parameter of 0.3326. The positive skew of the distribution of stock prices means that the distribution is not normal. In fact, when the returns are normally distributed, the price distribution will be log-normally distributed with the positive skew we see here.

In the binomial model, at each step the price moves either up or down by a discrete percentage, so in a weekly model with a 26-week horizon the binomial model will end at any one of 27 different prices. In contrast, in the continuous model the price can move to any value within the domain of the probability distribution so that there are an infinite number of different prices possible at the end of 26 weeks. Nevertheless, the binomial model can be made to approximate the distribution of continuous outcomes by using smaller time intervals. For example, we can easily adjust the binomial model in Exhibit 13.9 by using daily price moves over the same half year. To do this, change the length of the sub-period in B15 in

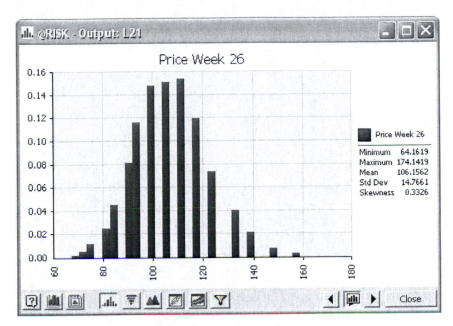

EXHIBIT 13.11   Monte Carlo simulation output. Frequency distribution of stock price at the end of 26 weeks.

the spreadsheet to read "1 day" and the sub-periods per year in B16 to read 250 (number of trading days in a half year). Increase the number of price outputs starting in D9 from 26 steps to 125 steps (i.e., Copy the Random Draw (columns E and J), Price Move (columns F and K), and Stock Price (columns G and L) to a total of 125 steps). Specify the cell that has the last price as the output and run the simulation. Exhibit 13.12 shows the frequency distribution of prices at the end of the half year (day 125). The mean and standard deviation of the price are $106.22 and $15.35, respectively, as compared to $106.16 and $14.76 that we got with weekly steps, so the results are about the same. However, as you can see, the frequency distribution is filled in almost completely because there are many more possible terminal prices ($n + 1 = 126$ in this case), approximating what would occur with a continuous process.

13.4 RETURNS FOR A PORTFOLIO OF SECURITIES

We have seen that it is easy to model the process that a stock price might follow over time. In this section we will see how to simulate the results when we combine two or more securities in a portfolio. This will set the stage for the next chapter that delves more deeply into modeling portfolios.

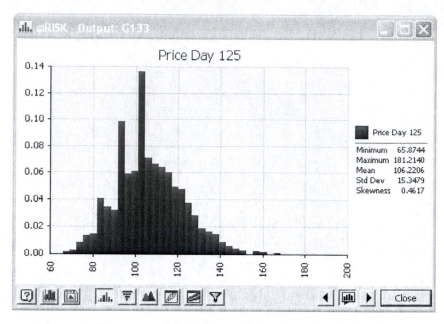

EXHIBIT 13.12 Frequency distribution of stock price at the end of a half year with daily steps.

In our portfolio model we will evaluate an investment in two securities, each of which follows the continuous time Weiner process that we introduced earlier. Exhibit 13.13 is similar to the model of stock returns for one stock shown in Exhibit 13.7. However, in this example there are two different investments, Stock A and Stock B, each of whose returns follow a Weiner process described by the equation

$$\text{Return} = \frac{dP}{P} = \mu dt + \sigma z \sqrt{dt}, \tag{13.8}$$

where each security has a different mean return μ, and standard deviation of return σ. The mean returns are in cells C15 and D15 as 12% and 18%, respectively, and the standard deviation of annual returns are in cells C16 and D16 as 20% and 35%. The current market prices of the two stocks are $100 and $60 per share, as shown in cells C12 and D12. The other data input is the correlation coefficient between the returns on the two stocks, which has been estimated to be 0.30 as shown in cell C18, and which will be explained shortly.

For the investment portfolio, the decision variables are the total amount we will invest and the proportions of our total investment that we will put

	A	B	C	D	E	F	G	H	I	J	K	L	M
1		Stock Returns With Continuous Process											
2													
3		$\dfrac{dP}{P}$	$= \mu\,dt + \sigma z \sqrt{dt}$										
4													
5													
6	Data Input												
7		Portfolio	Stock A	Stock B									
8													
9	Total Amount Invested	100,000	70,000	30,000									
10	Proportion Invested		70%	30%									
11													
12	Initial Stock Price		100.00	60.00									
13	Shares Purchased		700	500		=RiskNormal(0,1,RiskCorrmat(ReturnCorrelations,1))							
14													
15	Mean Return		12.0%	18.0%		=$C$15*$C$22+$C$16*B32*($C$22^0.5)							
16	Standard Deviation		20.0%	35.0%									
17						=D31*(1+C32)							
18	Correlation		0.30										
19						=D32*$C$13							
20	Length of Sub-Period		1 Week										
21	Sub-Periods Per Year		52			=(E32/$E$31)-1							
22	Length of Period		0.0192	fraction of year									
23													
24													
25													
26			Stock A						Stock B				
27			Standard	Return					Standard	Return			
28		Period	Normal	Per	Price	Stock	Return	Period	Normal	Per	Price	Stock	Return
29			Variate, z	Period		Value	to Date		Variate, z	Period		Value	to Date
30													
31		0			100	70,000		0			60	30,000	
32		1	-1.165	-3.00%	97.00	67,900	-3.00%	1	1.259	6.46%	63.87	31,937	6.46%
33		2	1.317	3.88%	100.77	70,537	0.77%	2	0.823	4.34%	66.65	33,323	11.08%

EXHIBIT 13.13 Model of portfolio of two stocks with returns following a Weiner process.

in stocks A and B. The total investment for our example is $100,000 in cell B9, and we have decided to invest 70% in stock A and 30% in stock B as shown in cells C10 and D10.

The returns will be simulated for weekly sub-periods over 26 weeks (a half year). The number of sub-periods per year is input as 52 in C21, and C22 translates that into the fraction of the year as 1/52.

The Monte Carlo simulation of the returns on stock A start in line 31, and proceed by weeks in the same way as shown earlier in Exhibit 13.7. Cell B32 is the probability distribution function =RiskNormal(0,1), and this and other return formulas are copied down column B for 26 periods to row 57. C32 has the return, calculated according to the Weiner process formula (13.8). The stock price in D32 is calculated as

$$Price_t = Price_{t-1}(1 + Return_t) = D31 \times (1 + C32).$$

The stock values in column E are simply the price times the number of shares (C13) in the initial portfolio. "Return to Date" in column F calculates the cumulative return at each date as

$$Cumulative\ Return_t = (Price_t/Price_0) - 1.$$

The returns and prices for stock B are calculated in the same way in columns I–M.

	H	I	J	K	L	M	N	O	P	Q
22										
23							=(O32/O31)-1	=(O32/$O$31)-1		
24						=(O32/O31)-1				
25										
26		Stock B				=E32+L32		Portfolio		
27		Standard	Return						Portfolio Return	
28	Period	Normal	Per	Price	Stock	Return		Portfolio	for	to
29		Variate, z	Period		Value	to Date		Value	Period	Date
30										
31	0			60	30,000			100,000		
32	1	-1.617	-7.50%	55.50	27,750	-7.50%		101,663	1.66%	1.66%
33	2	-1.783	-8.31%	50.89	25,445	-15.18%		98,800	-2.82%	-1.20%
34	3	-0.131	-0.29%	50.74	25,370	-15.43%		99,464	0.67%	-0.54%
35	4	0.193	1.28%	51.39	25,696	-14.35%		99,234	-0.23%	-0.77%
36	5	0.023	0.46%	51.63	25,813	-13.96%		98,178	-1.06%	-1.82%
37	6	-0.768	-3.38%	49.88	24,941	-16.86%		95,428	-2.80%	-4.57%

EXHIBIT 13.14   Model of portfolio of two stocks. Portion of model showing calculations for second stock and for portfolio.

Armed with total value of each of the two stocks at each date, we need to combine them in the portfolio. Exhibit 13.14 shows the part of the model that does the return and value calculations for stock B and for the portfolio as a whole. The portfolio calculations are in columns O, P, and Q. Portfolio value in column O simply adds together the total values of stocks A and B to get the total of the two taking account of the price of each stock and the number of shares. Column P is the return for each sub-period calculated as

$$\text{Portfolio Return for Period}_t = (\text{Portfolio Value}_t/\text{Portfolio Value}_{t-1}) - 1,$$

and column Q calculates the return to date based on the original investment as

$$\text{Portfolio Return to Date}_t = (\text{Portfolio Value}_t/\text{Portfolio Value}_0) - 1.$$

All the formulas are copied down to row 57, so we have returns and values through week 26. Our primary concern is the value of the portfolio and the cumulative portfolio return at the end of the 26-week horizon. But before discussing the end results, we need to explain one of the most important aspects of the portfolio analysis—the correlation of returns.

13.4.1 Correlating the Returns

When we combine two or more securities in a portfolio, we need to take account of the correlations in their returns. The correlation refers to the way in which the security prices and returns move with or against each other. If stocks tend to go up together, they are positively correlated, and if they tend to move in opposite directions, they are negatively correlated. As we observe from the daily news reports, most stocks tend to move in the same general direction. That is, share price of individual companies tend to move in the same direction as other companies in the same industries and as the market as a whole—after all, the market is simply the sum total of all the individual stocks, and movements in stock prices tend to be positively correlated.

Correlation is important in constructing a portfolio because it affects the overall risk of the portfolio. Holding stocks that are highly positively correlated does little to reduce portfolio risk. However, holding stocks that have low or negative correlation can reduce the portfolio risk. Hence, we must take the correlation into account when we model the portfolio value and returns.

In @Risk, variables are correlated by specifying the correlation between the probability distributions. In this model we will take account of the correlation between the returns by specifying a correlation between the draws of the standard normal distributions for stocks A and B that are in columns B and I. That is, cells B32:B57 each contain the @Risk distribution function =RiskNormal(0,1) that generates the random draw of the standard normal variate, z, for stock A for each of the 26 periods; and cells I32:I57 contain the same distribution formulas for stock B. We want to make the random draws for stock B in column I correlated with the draws for stock A in column B. We do this with the correlation dialog in @Risk. The steps for our model are as follows:

1. Select the cells containing probability distributions that you wish to correlate. Select cells B32:B57 that contain the distribution formulas, =RiskNormal(0,1), for stock A for periods 1–26. Then, holding the Ctrl key, also select cells I32:I57 that relate to stock B.

2. As shown on the left half of Exhibit 13.15, click the Define Correlations icon on the @Risk menu. This brings up the Define Correlations dialog box shown on the right that shows a few cells of the matrix of correlation coefficients for the selected random variable cells.

   (a) Matrix Name: In the Matrix Name box, type in a name for our correlation matrix. We name the matrix "ReturnCorrelations" and type it into the box as shown.

   (b) Location Box indicates the location on the spreadsheet where the correlation matrix will be placed. In this case, the correlation

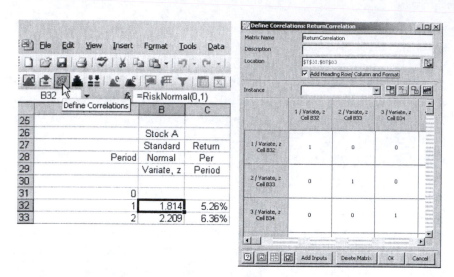

EXHIBIT 13.15   Define correlations in @Risk.

matrix will be placed in cells T31:BT83, which is indicated by entering the cell address for the upper left corner of the matrix. Click OK. @Risk places the correlation matrix in this location. Exhibit 13.16 shows the part of the initial matrix.

3. Enter Correlations: The correlation coefficients could have been entered in the matrix shown as part of the dialog box in step 2. However, in this case we want the correlations to be set equal to the contents of the correlation cell C18 that is part of the model input. One way to do this is to enter =$C$18 in all of the appropriate cells in the correlation matrix that was inserted in cells T31:BT83. The appropriate cells are those representing the intersection of each period's stock A and stock B =RiskNormal(0,1) functions. Note in Exhibit 13.16 that the rows and columns of the correlation matrix are labeled with the cell addresses.

	T	U	V	W	X	Y
		1 / Variate, z in $B$32	2 / Variate, z in $B$33	3 / Variate, z in $B$34	4 / Variate, z in $B$35	5 / Variate, z in $B$36
31	@RISK Correlations					
32	1 / Variate, z in $B$32	1				
33	2 / Variate, z in $B$33	0	1			
34	3 / Variate, z in $B$34	0	0	1		
35	4 / Variate, z in $B$35	0	0	0	1	
36	5 / Variate, z in $B$36	0	0	0	0	1
37	6 / Variate, z in $B$37	0	0	0	0	0

EXHIBIT 13.16   Upper left portion of the matrix of correlation coefficients.

For example, the first correlation cell in U32 shows the row heading as "1/Variate, z in $B$32," and the column heading is the same. This is a diagonal cell in the matrix, so the correlation is shown as 1, indicating that the B32 is perfectly correlated with itself. Cell U33 is off diagonal, with row heading "2/Variate, z in $B$33" and column heading "1/Variate, z in $B$32." This cell refers to the correlation between the random variable in B32 and B33—that is, the random draw for stock A in period 1 and the draw for stock A in period 2. A zero correlation in this case means that the returns in period 1 are uncorrelated with the returns in period 2 for stock A, which is what we want because we do not want the returns to be serially correlated over time. The cells where we want to enter the correlations are cells linking stocks A and B returns in the same period. Exhibit 13.17 shows a few of these cells of the correlation matrix. Cell U58 has row heading "1/Variate, z in $I$32" and column heading "1/Variate, z in $B$32." This intersection refers to cells B32 and I32 in the model. These are the random draws for stocks A and B in period 1. First, note that the number in the cell is 0.30, and second, note in the formula line at the top that the formula in the cell is =$C$18. This specifies that the correlation between the draws is 0.30, but that this number is drawn from the input cell C18. That way, if we change the input correlation in C18, it will be changed in the correlation matrix.

	File	Edit	View	Insert	Format	Tools	Data	Window	@RISK	Help	S&P	Adol

U58		fx =$C$18	

	T	U	V	W
30				
31	@RISK Correlations	1 / Variate, z in $B$32	2 / Variate, z in $B$33	3 / Variate, z in $B$34
32	1 / Variate, z in $B$32	1		
33	2 / Variate, z in $B$33	0	1	
34	3 / Variate, z in $B$34	0	0	1
35	4 / Variate, z in $B$35	0	0	0
57				
58	1 / Variate, z in $I$32	0.30	0	0
59	2 / Variate, z in $I$33	0	0.30	0
60	3 / Variate, z in $I$34	0	0	0.30
61	4 / Variate, z in $I$35	0	0	0

EXHIBIT 13.17   Correlations between stock A and stock B.

Now look at the other cells along a diagonal from U58, where each shows 0.30 as the correlation. Each one refers to the correlation between stock A and stock B in periods 1, 2, 3, and so on. The formula =$C$18 has been copied into each cell along this diagonal down to cell AT83, which relates returns in the period 26, so the correlation between returns is 0.30 in every period. If we want to change the input correlation, we have only to change the input cell C18, and all the relevant cells in the correlation matrix will be changed.

13.4.2 Running the Model

Now that we have a model that simulates returns for stocks A and B, with the correlations specified, we can run it to see the result. We will look at two types of output: the simulated pattern of stock returns and the values of the stocks and the portfolio over the 26 weeks. If we run the simulation, there are an infinite number of different returns and values that could be generated. One particular random draw of the simulation produced the pattern of returns that is shown in Exhibit 13.18. The solid line shows the return for stock A, which is slightly less volatile than stock B. The more

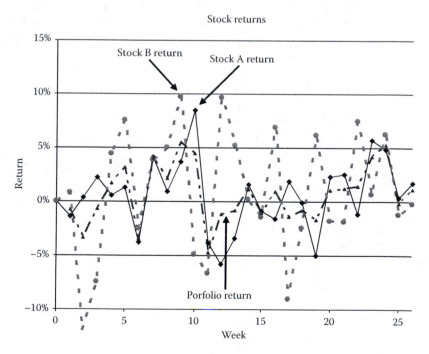

EXHIBIT 13.18 Simulated returns for stocks A and B, and portfolio with correlation of 0.30.

volatile dashed line represents the returns for stock B, which has a higher standard deviation of returns. The darker dashed line between A and B represents the return on the portfolio composed of 70% in stock A and 30% in stock B. Note that the returns on stocks A and B tend to move together some times and other times they move in opposite directions. For example, in week 3 the return on both stocks are negative. This reflects the positive correlation between the two stocks. However, they do not perform exactly in tandem. Some periods A is doing poorly when B is doing well as in week 12. This demonstrates the fact that the returns are imperfectly correlated.

Contrast that pattern with the one we might get for securities that are negatively correlated. We change the correlation by changing input cell C18 from 0.30 to −0.90 and run the simulation. Exhibit 13.19 shows the results from one random draw of the simulation. Note that when one stock has high returns, the other tends to have low returns. For example, in week 5 stock A has a negative return and stock B has a positive return. However, even with highly negative correlation, the stocks do not always move in the opposite direction. So long as the returns are not perfectly negatively

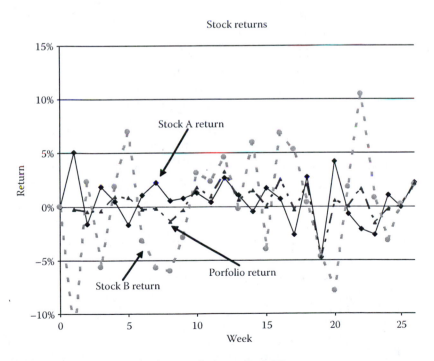

EXHIBIT 13.19 Returns with correlation of −0.90.

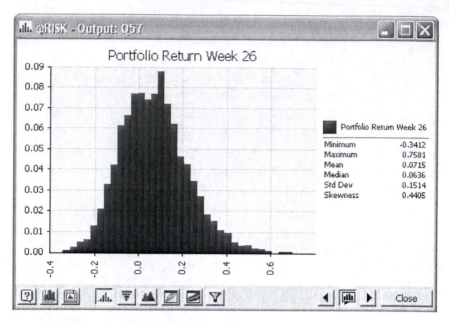

EXHIBIT 13.20   Frequency distribution of returns for two security portfolio at the end of 26 weeks.

correlated (correlation = −1.0), there is randomness in the relationship. For example, in week 12 both A and B are generating positive returns.

The simulation of returns on the two stocks did not really tell us what happened to the portfolio as whole. We now look at the distribution of returns for the portfolio that is a mix of the two securities. Exhibit 13.14 showed the part of the model with the portfolio value and return. To see the distribution of portfolio returns over the whole investment period, we designate the cumulative return at the end of the 26-week period as the output for the simulation, and then run the simulation. Exhibit 13.20 shows the distribution of portfolio returns with 5000 iterations. The mean and standard deviation of the distribution were 7.15% and 15.14% as shown to the right of the histogram. These are the 26 week, or semi-annual rates. Converted to annual rates, the annual mean and standard deviation are 13.63% and 21.41%, respectively.*

When you consider these results, the beneficial effects of diversification may not be immediately obvious. But consider that if we just took a

---

* We are using continuous compounding in the return model, where (1+ the compound return) over period dt is $e^{r\,dt}$, where r is the annual return, and the standard deviation is $\sigma(dt)^{1/2}$. With $dt = 1/2$ year, we solve for the annual return as $r = \ln(1.0715)/0.5 = 0.13625$, and the annual standard deviation is $\sigma = 0.1514/(1/2)^{1/2} = 0.2141$.

weighted average of the risks (standard deviations) of the two stocks, where the weights are the portfolio proportions, we have 0.70 (20%) + 0.30 (35%) = 24.5%, which would be the portfolio standard deviation if the two stocks were perfectly correlated. In our example with a correlation of 0.30, the portfolio standard deviation is about 21.4%. While the difference is not great, it certainly helps that the range of variation of returns on the portfolio is smaller than would result with no benefit from diversification.

This distribution of returns is a function of the proportion of the total investment put in each of the two securities. In addition, it is a function of the correlation between the returns. It is left to the reader to explore these aspects by developing the model and varying these and other inputs. In the next chapter we will delve more deeply into modeling portfolios of securities.

13.5 SUMMARY

The purpose of this chapter is to show how we can simulate the pattern of stock prices and returns over time. We used two models of stock prices, the binomial model and the continuous time Weiner process. With the binomial model, stock prices are assumed to move up or down over discrete intervals of time. With the continuous time model, the return over very short intervals of time is generated as a Weiner process. The Weiner process results in stock prices following a random walk, meaning that the movement at the next moment in time is random and cannot be predicted from the past pattern of stock prices.

We showed how to construct a model of stock prices for each of these processes. Then, we showed how the binomial process can be used to approximate the continuous process. The way to do this is to use small increments of time where the up and down movements are small. When the very small movements are repeated over many short periods, the resulting pattern of stock prices closely approximates the random walk pattern we get from the continuous process. With these models, we used Monte Carlo simulation with @Risk to simulate the price path for a stock.

Having developed stock price models for a single stock, we then showed how we can combine two stocks in a portfolio. When we combine stocks in portfolio, we need to take account of the correlation between their returns. We showed how to add correlation to the @Risk model so we could simulate the returns on a portfolio. This portfolio model leads us to consider how we can mix securities in portfolios, so that we can get the best risk-return trade off. That is the topic of the next chapter.

PROBLEMS

1 A Model of Binomial Stock Prices

Construct a spreadsheet model of binomial stock prices similar to Exhibit 13.9. Use monthly intervals and assume that each month the price of a stock can make an up move of $u = 1.106$ with a probability of 55%, or a down move of $d = 0.904$ each month with a probability of 45%. Assume the initial stock price is $100 and the risk-free interest rate is 5%. Use @Risk to simulate the movements of the stock over the course of two years.

(a) Show the graph of the frequency distributions of stock values at the end of one year, and at the end of two years. What are the parameters of the distribution of stock values at the end of each of these periods?

(b) Show the simulated distribution of stock returns at the end of one year, and at the end of two years. What are the parameters of the distribution of stock returns at each period?

(c) What is the probability that you will lose money (negative cumulative return) over the intervals of one and two years?

2 A Model for Continuous Stock Returns

Construct a spreadsheet model of stock returns that follow a continuous process such as is shown in Exhibit 13.7. Assume the stock's initial price is $100 per share, and the mean annual return of the stock is 18% and standard deviation of the annual return is 34.9%. With weekly intervals, use @Risk to simulate the stock price and the return on the stock over a period of 52 weeks.

(a) Show the frequency distributions of stock prices and stock returns at the end of 52 weeks. What are the parameters of the frequency distribution?

(b) Add a second stock whose initial price is $100 per share, and the mean annual return is 8%, with a standard deviation of annual return of 40%. Assume the correlation between the returns on the first and second stocks is 0.50.

Assume you invest 40% in the first stock and 60% in the second stock.

Simulate the returns on the portfolio over one year using weekly intervals.

Show the frequency distribution of returns for the portfolio at the end of the year. What are the parameters of the distribution of returns?

CHAPTER 14

Constructing Optimal Security Portfolios

14.1 INTRODUCTION

When Harry Markowitz (1952) published his paper on portfolio selection and mean-variance efficiency, he ushered in the new discipline of financial economics. His work earned him the 1990 Nobel Prize in Economics (along with William Sharpe and Merton Miller) and led to the development of the CAPM by William Sharpe (1964), Lintner (1965), and Mossin (1966). In the decades since, this seminal work continues to be used to determine asset allocation recommendations as well as to structure and create benchmarks for professionally managed portfolios.

Markowitz's contribution was to show how to measure the probabilistic effects of diversification on security returns. He showed how to calculate the mean and standard deviation of a portfolio of securities based on the parameters of the probability distributions of the component stocks and the proportion of the total investment committed to each security, and

536 ■ Introduction to Financial Models for Management and Planning

originated the concept of an efficient frontier and showed how to construct the efficient portfolios comprising it.

In this chapter we use Markowitz's theory to model the risk and return of portfolios of securities and show how to use Solver to find efficient portfolios. We assume the reader has been exposed to the basic ideas of portfolio analysis in previous financial management courses, and so will spend little time reviewing that material.* We show how to build models that combine a few securities into efficient portfolios, but can be easily expanded to include many more.† The models we build require the use of some rudimentary linear algebra, Excel®'s array functions, and Excel's linear and quadratic program called Solver.

This chapter begins by reviewing the mean-variance problem in a two-asset world, the one typically introduced in introductory financial management and investments courses. We then review the elementary linear algebra needed to set up the problem when the opportunity set includes more than two assets. Next, we use Excel's array functions and optimization add-in, Solver, to find the efficient frontier. We end the chapter with a brief discussion of the potential pitfalls of naïve mean-variance optimization and an overview of portfolio value-at-risk (VaR).

14.2 THE MEAN-VARIANCE PROBLEM FOR TWO ASSETS

Modern portfolio theory shows that holding two or more assets in a portfolio is often better than holding a single asset. This is because two securities do not always move in the same direction at the same time. When one security moves up and the other moves down, the value of a portfolio containing both may change very little. Indeed, it is often possible to create portfolios of assets with risk and return attributes superior to any single asset. For example, consider Exhibit 14.1. This graph plots the returns for portfolios comprised only of a bond fund and an S&P 500 index fund against their respective risks as measured by the standard deviation of return. Though the S&P index is considerably riskier than the bond fund,

---

* For a review of portfolio analysis see any standard investments text such as Elton, Gruber, Brown, and Goetzman (2006) or Reilly and Brown (2006), or a financial management text, such as Brigham and Ehrhardt (2008), Brealey, Myers, and Allen (2008), Ross, Westerfield, and Jaffe (2008), or Berk and DeMarzo (2007).

† When a large number of securities are involved (e.g., considerably more than 50), the computational challenges are beyond the scope of this text and the capacity of Excel. For greater detail and references relating to efficient portfolio computation, see "Techniques for Calculating the Efficient Frontier," Chapter 6, in Elton et al. (2006).

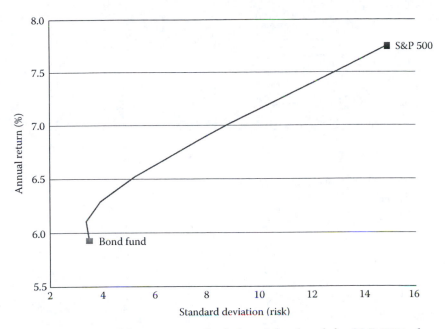

EXHIBIT 14.1   Portfolios comprised of a bond fund and the S&P 500 index.

adding a little of the S&P index to the bond fund creates a portfolio with higher returns and less risk than the bond fund alone. For example, when 93% of a portfolio is allocated to the bond fund with the remaining 7% allocated to the S&P index fund, the expected return on that fund is 6.06% (versus the bond fund's 5.93%) whereas its standard deviation is 3.3% (versus the bond fund's 3.48%).

The mathematics of two-asset portfolio return and risk are necessary to proceed. Expected portfolio return is a simple weighted average of the expected returns of the assets that comprise the portfolio. The formula is shown in Equation 14.1

$$E[r_p] = x_1 E[r_1] + x_2 E[r_2]$$

$$= \sum_{i=1}^{2} x_i E[r_i], \tag{14.1}$$

where $E[r_i]$ is the expected return on asset i, x_i the proportion of asset i in the portfolio relative to the total portfolio, and $x_1 + x_2 = 1$.

The formula for portfolio risk, defined as the standard deviation of its returns, is shown in Equation 14.2

Fund statistics

	Bond fund	S&P index
Average annual return	5.93	7.75
Standard deviation	3.48	14.92

Correlation matrix

	Bond fund	S&P index
Total Bond	1.00	
S&P500	(0.09)	1.00

EXHIBIT 14.2   Covariance matrix and annual returns for bond and S&P index funds.

$$\sigma_p = \left[x_1^2 \sigma_1^2 + x_2^2 \sigma_2^2 + 2 x_1 x_2 \rho_{12} \sigma_1 \sigma_2 \right]^{\frac{1}{2}}$$

$$= \left[\sum_{i=1}^{2} \sum_{j=1}^{2} x_i x_j \rho_{ij} \sigma_i \sigma_j \right]^{\frac{1}{2}}, \tag{14.2}$$

where x_1 and x_2 are the proportions of assets 1 and 2 relative to the total portfolio, σ_1 and σ_2 the standard deviations of assets 1 and 2, and ρ_{12} the correlation between the return on assets 1 and 2.

The source of the benefits from diversification can be seen in the last term of the first line of Equation 14.2, $2 x_1 x_2 \rho_{12} \sigma_1 \sigma_2$. The smaller the correlation, ρ_{12}, between two assets, the greater the potential benefits gained from holding both of them. Indeed, under the extreme condition of a perfect negative correlation, portfolio risk can be eliminated.

Consider the following numerical example. Suppose an investor decides to allocate her total portfolio between two funds—a bond fund and an S&P 500 index fund. The performance of these two funds, and the relationship between their returns, are summarized in Exhibit 14.2.*

Suppose, too, that our investor allocates 40% of her total investments to the bond fund and 60% to the S&P index. Then the expected return and standard deviation for this portfolio are shown in Equation 14.3.

---

* We use data for six Vanguard mutual funds from 1996 through 2007 throughout this chapter. End-of-month closing values, adjusted to reflect dividend and interest payments, were downloaded from finance.yahoo.com. Continuously compounded monthly returns and variances were calculated and then aggregated to create the annual statistics.

$$E[r_p] = x_{Bond}E[r_{Bond}] + x_{S\&P}E[r_{S\&P}]$$
$$= 0.4 \times 5.93 + 0.6 \times 7.75$$
$$= 7.02\%$$

$$\sigma_p = [x_{Bond}^2 \sigma_{Bond}^2 + x_{S\&P}^2 \sigma_{S\&P}^2 + 2x_{Bond}x_{S\&P}\rho_{Bond,S\&P}\sigma_{Bond}\sigma_{S\&P}]^{\frac{1}{2}}$$
$$= [0.4^2 \times 3.48^2 + 0.6^2 \times 14.92^2 + 2 \times 0.4 \times 0.6 \qquad (14.3)$$
$$\times(-0.09) \times 3.48 \times 14.92]^{\frac{1}{2}}$$
$$= 8.9\%$$

Before moving forward, we need to introduce the covariance matrix. While the benefits of diversification derive from correlation, mathematics of portfolio analysis takes variances and covariances as its inputs.* For example, using variances and covariances, Equation 14.3 may be rewritten as Equation 14.4

$$\sigma_p = [x_{Bond}^2 \sigma_{Bond}^2 + x_{S\&P}^2 \sigma_{S\&P}^2 + 2x_{Bond}x_{S\&P}\sigma_{Bond,S\&P}]^{\frac{1}{2}}$$
$$= [0.4^2 \times 12.1 + 0.6^2 \times 222.5 + 2 \times 0.4 \times 0.6 \times (-4.9)]^{\frac{1}{2}}$$
$$= 8.9\%, \qquad (14.4)$$

where $\sigma_{Bond}^2 = \sigma_{Bond}\sigma_{Bond} = 3.48 \times 3.48 = 12.1$, $\sigma_{S\&P}^2 = \sigma_{S\&P}\sigma_{S\&P} = 14.92 \times 14.92 = 222.5$, and $\sigma_{Bond,S\&P}$ is covariance between the bond and S&P index funds,

$$\sigma_{Bond,S\&P} = \rho_{Bond,S\&P}\sigma_{Bond,}\sigma_{S\&P} = (-0.09) \times 3.48 \times 14.92 = -4.9.$$

There is a compact way to organize variances and covariances, it is called a covariance matrix. A covariance matrix uses the same structure as the correlation matrix that was discussed in Chapter 12, except that it reports covariances instead of correlations for the off-diagonal terms (the terms not equal to 1 in the correlation matrix), and variances on the diagonal.† Equation 14.4 shows the variances of the bond and S&P index as 12.1 and

---

* The covariance between two securities, denoted by σ_{12}, is the product of the correlation between the two securities and the standard deviations of each of the securities, expressed as $\sigma_{12} = \rho_{12}\sigma_1\sigma_2$.

† The covariance matrix is also referred to as the variance–covariance matrix because the entries on the diagonal are variances. However, variances are merely the covariance of an asset with itself, just as the correlation of an asset with itself is 1. To see this, consider $\sigma_{11} = \rho_{11}\sigma_1\sigma_1 = 1 \times \sigma_1\sigma_1 = \sigma_1^2$.

Covariance matrix		
	Bond fund	**S&P index**
Total Bond	12.1	(4.9)
S&P500	(4.9)	222.5

EXHIBIT 14.3   Covariance matrix for the bond and S&P index funds. Variances (12.1 and 222.5) are reported on the diagonal. The covariance (−4.9) is reported on the off-diagonal.

222.5, respectively, and the covariance between their returns as −4.9. Exhibit 14.3 shows how these data are organized in a covariance matrix.*

It is relatively straightforward to set up this two-asset problem in Excel as a dynamic model. An example with cell formulas is shown in Exhibit 14.4. The user enters her desired percentage allocation to the bond fund in cell A14, and the model returns the weight for the S&P index fund and the resulting portfolio's expected return and risk (standard deviation).

Now suppose that our investor is particularly risk averse. She wants to bear the minimum possible risk, given a portfolio comprised of the bond and S&P index funds. What portion should she allocate to each fund? Our investor wishes to invest in the portfolio known as the minimum variance portfolio, or MVP. As its name implies, the MVP is the portfolio that has the least risk among all of the portfolios that can be formed from a given group of assets, in this case, from the bond fund and S&P index.

Mathematically, her problem is shown in Equation 14.5.

$$\underset{\{x_{\text{Bond}}, x_{\text{Index}}\}}{\text{MIN}} \sigma_p = [x_{\text{Bond}}^2 \sigma_{\text{Bond}}^2 + x_{\text{Index}}^2 \sigma_{\text{Index}}^2 + 2x_{\text{Bond}} x_{\text{Index}} \sigma_{\text{Bond,Index}}]^{\frac{1}{2}} \quad (14.5)$$

subject to

$$x_{\text{Bond}} + x_{\text{Index}} = 1$$

Equation 14.5 says that the problem is to find the weights (x_{Bond} and x_{Index}) for the bond and S&P funds that minimize the standard deviation of a portfolio with the constraint (subject to) that the weights must sum to 1.

---

* Excel's Data Analysis ToolPak has functions that will quickly calculate correlation and covariance matrices. The Data Analysis ToolPak can be found on the "Data" ribbon in Excel 2007 or on the "Tools" dropdown menu in Excel 2003. The Data Analysis ToolPak is an Excel Add-In. To add it into Excel 2007, click on the "Office button," select "Excel options," then "Add-Ins" and select Data Analysis ToolPak. To add it into Excel 2003, click on "Tools," then "Add-Ins" and check the Data Analysis box.

	A	B	C	D
2	**Expected returns**			
3		**Bond fund**	**S&P index**	
4	Annual return	5.93	7.75	
5				
6	**Covariance matrix**			
7		**Bond fund**	**S&P index**	
8	Bond fund	12.1	(4.9)	
9	S&P index fund	(4.9)	222.5	
10		=1-A14	=A14*B4+B14*C4	
11		Portfolio return and risk		
12		**Weights**	**Return**	**Sigma**
13		**Bond fund**	**S&P index**	
14		0.4	0.6	7.02	8.93
15		=SQRT(A14^2*B8+B14^2*C9+2*A14*B14*B9)		

EXHIBIT 14.4   Asset returns and covariance matrix, and the expected return and standard deviation for a two-asset portfolio.

Certainly, the solution to this simple problem can be found quickly by trial and error, or calculus, or the formulas found in many investment textbooks. However, for larger portfolios some type of program with quadratic programming capability is needed. Excel's Solver is one such program. We use Solver for this problem, even though the example and mathematics are transparent.

Before having Solver find the weights to the MVP, we need to add a cell to the model shown in Exhibit 14.4 that calculates the sum of the portfolio weights (the condition under "subject to" in Equation 14.5). The portfolio model with this additional cell (B15) and the Solver dialog window are shown in Exhibit 14.5.

Invoke Solver* and fill in its dialog window as follows.

- Set Target Cell: D14 (the cell that calculates the portfolio's standard deviation).

- Equal to: Min (we seek to minimize the portfolio's standard deviation).

---

* Solver is invoked by clicking on Solver on the "Data" ribbon in Excel 2007 or in the "Tools" dropdown menu in Excel 2003. Solver, too, is an Excel Add-In. See footnote on the previous page, if you need to add it in.

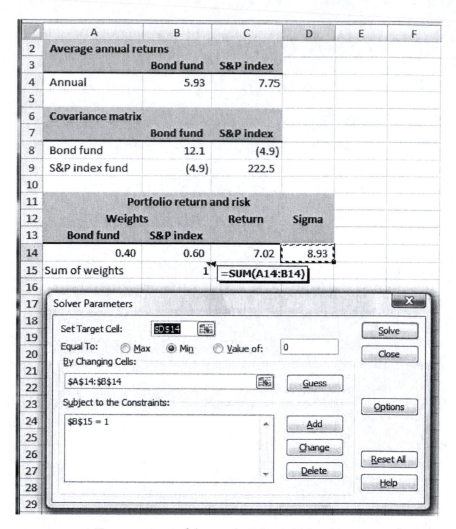

EXHIBIT 14.5   Two-asset portfolio with Solver dialog box set to find the minimum variance portfolio.

- By Changing Cells: A14:B14 (the cells that contain the portfolio weights).

- Subject to the Constraints: B15 = 1 (the weights must sum to 1).

In addition, we assume no short selling. This means all weights must be positive. Click on the "Options" link and select "Assume Non-negative." Finally, click on "Solve." Solver quickly returns the solution shown in Exhibit 14.6.

Portfolio return and risk			
Weights		Return	Sigma
Bond fund	S&P index		
0.93	0.07	6.06	3.30

EXHIBIT 14.6   Two-asset minimum variance portfolio.

Note the benefit from diversification. Combining a fund that is nearly 4 times riskier than the bond fund with the bond fund (a standard deviation of 14.9% versus the bond fund's 3.5%) creates a portfolio with less risk and a higher expected return than the bond fund alone.

14.3 A LITTLE LINEAR ALGEBRA

The two-asset model is easy to set up using the standard Excel functions. However, the simple algebraic formulas quickly become intractable as more assets are considered. For example, for our two-asset portfolio, the code for the expected return cell contained two terms and code for the standard deviation cell contained three terms. This code is easy to manage. However, the number of terms required to calculate a portfolio's standard deviation expands quickly with the number of securities in the portfolio. For example, the formula for the standard deviation of a 20-asset portfolio requires 210 terms. It would be difficult to enter a formula with 210 terms into a cell in Excel without making an error. A little linear algebra and Excel's array functions quickly resolve this problem. Though instruction in linear algebra is beyond the scope of this text, we do show how to use Excel's array functions to make the necessary portfolio calculations for portfolios of many securities.[*]

Excel array functions operate on vectors and matrices, both of which are termed arrays in Excel. An array is a rectangular block of numbers. For example, consider Exhibit 14.7. The numbers in the block B4:C4 form an array that has 1 row and 2 columns. We say it is a 1×2 array, where 1×2 refers to the number of rows by the number of columns in the array.

An array can have any number of rows and columns. When it has more than 1 row and more than 1 column, we call it a matrix. In Exhibit 14.8, the block of cells B8:C9 is a 2×2 matrix, that is, it has two rows and two columns. Covariance matrices are square and symmetric. Square means that the number of columns is the same as the number of rows and symmetric

---

[*] We refer interested scholars to Strang (2003) or any other linear algebra textbook.

	A	B	C
2	**Expected returns**		
3		BF	SP
4	Annual	5.93	7.75

EXHIBIT 14.7　A 1 × 2 array in cells B4:C4.

	A	B	C
6	**Covariance matrix**		
7		BF	SP
8	Bond fund	12.1	(4.9)
9	S&P index fund	(4.9)	222.5

EXHIBIT 14.8　A 2 × 2 array in cells B8:C9.

means that the numbers in the upper right portion of the matrix, such as (4.9) in C8, are the same as those in the lower left portion of the matrix, (4.9) in B9. This is because the covariance of the returns on asset BF with those on SP is the same as that on asset SP with those on BF. In mathematical terms, $Cov(r_{BF}, r_{SP}) = Cov(r_{SP}, r_{BF}) = -4.9$.

We use two array functions to calculate portfolio returns and standard deviations: =MMULT(array1, array2) to multiply arrays, and =TRANSPOSE (array) to change the shape of arrays. The steps for matrix multiplication are as follows:

1. Select (highlight) the block of cells into which the product of the matrix multiplication will be placed in the spreadsheet.

2. In a cell in the selected block, type the array function.

3. When you have finished typing the array function, enter it by simultaneously pressing the Ctrl-Shift-Enter keys.* This action puts { } brackets around the formula you have entered, telling Excel that it is an array formula, and completes the formula entry. Excel enters the formula into each of the cells in the selected block.

Before proceeding with the multiplication in the model, let us briefly look at matrix multiplication in the spreadsheet. Suppose we want to

---

* If you inadvertently press "Enter" you will get Excel's #VALUE! error message. If so, press F2 and then Ctrl + Shift + Enter.

multiply the 1×2 array of portfolio weights for two securities, (w_1, w_2) times the 2×2 covariance matrix

$$\begin{bmatrix} \sigma_{11} & \sigma_{12} \\ \sigma_{21} & \sigma_{22} \end{bmatrix}.$$

We write the matrix multiplication as follows:

$$\begin{bmatrix} w_1 & w_2 \end{bmatrix} \begin{bmatrix} \sigma_{11} & \sigma_{12} \\ \sigma_{21} & \sigma_{22} \end{bmatrix} = [(w_1 \sigma_{11} + w_2 \sigma_{12})(w_1 \sigma_{12} + w_2 \sigma_{22})].$$

The product is the 1×2 array $[(w_1\sigma_{11} + w_2\sigma_{12})(w_1\sigma_{12} + w_2\sigma_{22})]$, where the terms in parentheses each represents one number.

To use the array functions in Excel, we need to have enough knowledge of matrix algebra to recognize that the product of the above operation will be a 1×2 array. Assume for the moment that we know these dimensions. The following list shows how to do the multiplication in Excel.

1. Select (highlight) the 1×2 cell block B16:C16 as in Exhibit 14.9A.

2. With the two cells selected, type in =MMULT(B13:C13,B8:C9). Do not hit the "Enter" key. The result of this sequence is shown in Exhibit 14.9A.

3. Hit the key combination "Ctrl-Shift-Enter."

The "Ctrl-Shift-Enter" enters the formula as an array formula into both the highlighted cells (B16:C16) and reports the products of the matrix multiplication as shown in Exhibit 14.9B. Note that the product of the 1×2 array multiplied by the 2×2 array is the 1×2 array of numbers [1.90, 131.53]. These numbers are not meaningful for us at the moment. The example was used simply to show the method for array multiplication in Excel.

Armed with this brief introduction to matrix operations in Excel, we now show how to calculate the return and risk for a portfolio comprised of assets BF and SP.* Exhibit 14.10 shows the portfolio model for our two securities, BF and SP. The expected returns are in cells B4:C4, the

---

* The reader will recognize that these are the same calculations coded in simple algebra in Section II.

	A	B	C
2	**Expected returns**		
3		BF	SP
4	Annual	5.93	7.75
5			
6	**Covariance matrix**		
7		BF	SP
8	Bond fund	12.1	(4.9)
9	S&P index fund	(4.9)	222.5
10			
11	**Portfolio weights**		
12		BF	SP
13		0.40	0.60
14			
15	**Array Product**		
16		=MMULT(B13:C13,B8:C9)	

EXHIBIT 14.9A   Arrays of investment data.

covariance matrix in B8:C9, and the portfolio weights in B13:C13. Each of these blocks of numbers is an array.

We want to calculate the expected return and standard deviation of return for the portfolio consisting of securities BF and SP held in the

B16		f_x	{=MMULT(B13:C13,B8:C9)}	
	A	B	C	D
2	**Expected returns**			
3		BF	SP	
4	Annual	5.93	7.75	
5				
6	**Covariance matrix**			
7		BF	SP	
8	Bond fund	12.1	(4.9)	
9	S&P index fund	(4.9)	222.5	
10				
11	**Portfolio weights**			
12		BF	SP	
13		0.40	0.60	
14				
15	**Array Product**			
16		1.90	131.53	

EXHIBIT 14.9B   Resulting product of array multiplication in cells B16:C16.

▲	A	B	C	D	E	F	G
2	**Expected returns**						
3		BF	SP				
4	Annual	5.93	7.75				
5							
6	**Covariance matrix**						
7		BF	SP				
8	Bond fund	12.1	(4.9)				
9	S&P index fund	(4.9)	222.5				
10							
11	**Portfolio weights**						
12		BF	SP				
13		0.40	0.60				
14							
15	**Array Product**						
16		1.90	131.53				
17							
18	**Portfolio statistics**						
19	Expected Return	7.02	=MMULT(B13:C13,TRANSPOSE(B4:C4))				
20	Variance	79.68	=MMULT(MMULT(B13:C13,B8:C9),TRANSPOSE(B13:C13))				
21	Standard deviation	8.93	=SQRT(B20)				

EXHIBIT 14.10 Expected returns and standard deviation for the two security portfolio using Excel's array functions.

weights indicated in B13:C13. The results of these array multiplications and the code used for each calculation are shown in Exhibit 14.10. The portfolio expected return and standard deviation were computed to be 7.02% and 8.93% as shown in B19 and B21.

We now address a few remaining characteristics of array functions—dimension, multiplication order, the TRANSPOSE function, and nested MMULT statements. We have already encountered dimension—the number of rows and columns in an array. Multiplication order refers to the order the arrays are listed in the MMULT function. Dimension and multiplication orders are inextricably linked. When multiplying one array by another, the column dimension (the second number) of the first array must match the row dimension (the first number) of the second array. For example, in Exhibit 14.9B, we multiplied B13:C13 (a 1×2 array) with B8:C9 (a 2×2 array) and got the product B16:C16 (a 1×2 array). The multiplication worked because the dimensions "fit:" $(1 \times 2) \times (2 \times 2) = (1 \times 2)$.

Now consider the calculation for portfolio return. We multiply the weight array, B13:C13 (a 1×2 array) and the return array, B4:C4 (also a 1×2 array). The dimensions for these two arrays do not fit. The column dimension of the first array is 2 and the row dimension of the second array is 1, giving you $(1 \times 2) \times (1 \times 2)$. Here is where the TRANSPOSE function comes into play. If we change the dimensions of the second array (returns) to 2×1, then the dimensions do fit—$(1 \times 2) \times (2 \times 1)$—and the operation will succeed. The TRANSPOSE function does just that—it turns a row array into a column or vice versa. Exhibit 14.11 shows how this works. Line 24 shows the return array B4:C4 with dimension 1×2. Lines 26 and 27 transpose this array from a row to a column, giving it the dimensions 2×1. Using the transposed B4:C4 array, the operation =MMULT (B13:C13,TRANSPOSE(B4:C4) succeeds.

This brings us to the code for portfolio variance, =MMULT(MMULT (B13:C13,B8:C9),TRANSPOSE(B13:C13)). This is a nested MMULT statement—one MMULT function is nested inside the other. The operation proceeds as follows. Excel calculates the nested MMULT command first, multiplying the arrays B13:C13 and B8:C9. As we saw earlier, this produces the array [1.90, 131.53]. Then the outer MMULT multiplies this

	B26		f_x	{=TRANSPOSE(B4:C4)}			
	A	B	C	D	E	F	G
2	**Expected returns**						
3		BF	SP				
4	Annual	5.93	7.75				
5							
6	**Covariance matrix**						
7		BF	SP				
8	Bond fund	12.1	(4.9)				
9	S&P index fund	(4.9)	222.5				
10							
11	**Portfolio weights**						
12		BF	SP				
13		0.40	0.60				
14							
15	**Array Product**						
16		1.90	131.53				
17							
18	**Portfolio statistics**						
19	Expected Return	7.02	=MMULT(B13:C13,TRANSPOSE(B4:C4))				
20	Variance	79.68	=MMULT(MMULT(B13:C13,B8:C9),TRANSPOSE(B13:C13))				
21	Standard deviation	8.93	=SQRT(B20)				
22							
23	**Array or function**	Values		Dimension			
24	B4:C4	5.93	7.75	1x2			
25							
26	TRANSPOSE(B4:C4)	5.93		2x1			
27		7.75					

EXHIBIT 14.11 Excel's =TRANSPOSE(.) function.

new 1 × 2 array into the 2 × 1 transposed array. This second product is the portfolio's variance, cell B20 in Exhibit 14.11.

14.3.1 Naming the Arrays

When working with small arrays, it is a relatively simple and error-free process to type in the array function and enter the arrays by highlighting them or typing in their cell addresses. However, when arrays become large, this process becomes quite cumbersome. In addition, models coded using cell addresses are not easy to troubleshoot. Excel's Name function resolves both of these problems. To see how it works, we recalculate portfolio returns and variance using named arrays.

The first step in using named cells is to name them. To do this, highlight the array and type the name you wish to use for it into the name box, then press "Enter." This operation is shown in Exhibit 14.12. Choose a name that will make sense to you should you revisit your model long after building it. I chose the name "omega_2" for the covariance matrix because covariance matrices are frequently denoted by the Greek letter omega. The name "returns_2" applies to the array of expected returns, and "weights_2" names the array of portfolio weights. The extension "_2" is appended to the names to indicate

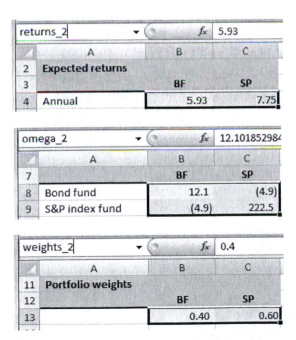

EXHIBIT 14.12 Naming the return, covariance, and weight arrays.

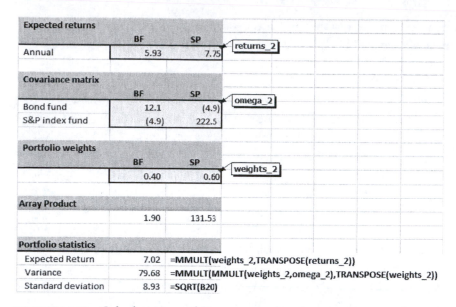

Expected returns			
	BF	SP	returns_2
Annual	5.93	7.75	

Covariance matrix			
	BF	SP	omega_2
Bond fund	12.1	(4.9)	
S&P index fund	(4.9)	222.5	

Portfolio weights			
	BF	SP	weights_2
	0.40	0.60	

Array Product		
	1.90	131.53

Portfolio statistics		
Expected Return	7.02	=MMULT(weights_2,TRANSPOSE(returns_2))
Variance	79.68	=MMULT(MMULT(weights_2,omega_2),TRANSPOSE(weights_2))
Standard deviation	8.93	=SQRT(B20)

EXHIBIT 14.13   Calculating portfolio return and variance using named arrays.

the number of securities in the portfolio to distinguish these calculations from those for the six-security examples that come later in this chapter.

Having named the arrays, you need to only type in their names as you write your code. Exhibit 14.13 repeats Exhibit 14.10, this time using names in place of cell references. We realize that using names for the first time can be daunting. However, the reward for perseverance is great. One last comment: The code shown in Exhibit 14.13 will not change regardless of the number of securities in a portfolio. Only the names of the arrays will change.

14.4 THE EFFICIENT FRONTIER WITH MULTIPLE ASSETS

We now find the efficient frontier with no short selling* using six assets— the bond and S&P index funds from the previous example plus an

---

* Throughout this section we assume that there is no short selling of securities. In the model, short selling would be indicated by portfolio weights less than zero. When solving the optimization problem with Solver, we require that the portfolio weights be non-negative, thereby prohibiting short sales in our solution. We believe that no short selling is a realistic approach while finding the efficient frontier because the mutual funds held by most individual investors cannot be sold short. In addition, the unlimited short sales assumption presumes that the short seller will have use of the funds generated by the short sale. Again, this is rarely if ever the case for individual investors. Finally, the costs involved in short selling are typically large. Granted, Exchange Traded Funds (ETFs) are rapidly replacing mutual funds and can be sold short. Even so, the latter two constraints on short selling would still obtain.

international fund, a small capitalization fund (small cap), a real estate fund (REIT), and an emerging markets fund. All are Vanguard funds. Adjusted closing values were downloaded from finance.yahoo.com. The fund ticker symbols for the bond, index, international, small cap, REIT, and emerging market funds are VBMFX, VFINX, VWIGX, NAESX, VGSIX, and VEIEX, respectively. Monthly continuously compounded returns for the years 1996–2007 were then calculated and used to calculate average monthly returns and as input to Excel's Correlation and Covariance functions in the Data Analysis ToolPak. The annualized covariance matrix and the average annual returns were found by multiplying monthly covariances and average monthly returns by 12. The covariance and correlation matrices for these funds, along with their average annual returns and standard deviations, are reported in Exhibit 14.14.

The only changes to the code for portfolio returns and standard deviation, that need to be made, are to the names of the arrays. Once again we append numbers to the names to distinguish these arrays from those used in the two-asset example. The weight and return arrays (weight_6 and return_6) are 1 × 6 row arrays; the covariance matrix (omega_6) is a 6 × 6 array.

The only way to find the efficient frontier when no short selling is allowed is by finding one efficient portfolio at a time. Solver must be used repeatedly to maximize the expected returns for a given portfolio standard deviation

Annual statistics	Bond fund	S&P index	International	Small cap	Emerging markets	REIT
Average return	5.93	7.75	8.00	8.62	10.19	12.29
Standard deviation	3.48	14.92	15.23	19.12	24.33	14.72

Correlation matrix	Bond fund	S&P index	International	Small cap	Emerging markets	REIT
Bond fund	1.00					
S&P index	(0.09)	1.00				
International	(0.15)	0.79	1.00			
Small cap	(0.14)	0.75	0.76	1.00		
Emerging markets	(0.18)	0.72	0.84	0.74	1.00	
REIT	0.03	0.32	0.33	0.46	0.34	1.00

Covariance matrix	Bond fund	S&P index	International	Small cap	Emerging markets	REIT
Bond fund	12.1	(4.9)	(8.0)	(9.4)	(15.2)	1.6
S&P index	(4.9)	222.5	179.8	212.9	260.8	71.0
International	(8.0)	179.8	231.9	222.7	309.8	74.0
Small cap	(9.4)	212.9	222.7	365.5	345.2	130.0
Emerging markets	(15.2)	260.8	309.8	345.2	592.1	122.5
REIT	1.6	71.0	74.0	130.0	122.5	216.6

EXHIBIT 14.14 Fund statistics used to calculate and graph the six-asset efficient frontier.

and its output saved to find enough efficient portfolios to clearly trace out the efficient frontier. Therefore, the first step in finding the efficient frontier is to set up the portfolio calculations with an output structure that facilitates easy copying and pasting to a table. We call these portfolio calculations the "Solver portfolio" and show our structure in Exhibit 14.15.

We draw your attention to two aspects of this structure. First, the "Output row" (C20:J20) is set up to report the portfolio's standard deviation first, followed by the return, and then the portfolio weights. This will facilitate drawing the efficient frontier in Excel once the necessary efficient portfolios have been identified by Solver. Second, all the output data are on one line, so that the entire line can easily be copied and pasted (as values) into a table. Third, note the "Target sigma" cell, C22. As mentioned earlier, Solver must be used repeatedly to map the efficient frontier. The modeler asks Solver to find the portfolio weights that maximize the return on a portfolio for a given standard deviation. The standard deviation is then increased and Solver is asked to find the portfolio weights that maximize the return on a portfolio given the new standard deviation. Then the process is repeated. This means constantly having to change one of the constraints in Solver's dialog window. One way to simplify the process is to link the standard deviation constraint in Solver to a specific cell (C22 in our model). Then one needs to only change the value in this cell before running Solver to find another efficient portfolio.

	B	C	D	E	F	G	H	I	J
3	Covariance matrix	Bond fund	S&P index	International	Small cap	Emerging markets	REIT	omega_6	
4	Total Bond	12.10	(4.90)	(8.05)	(9.43)	(15.16)	1.58		
5	S&P500	(4.90)	222.49	179.78	212.87	260.76	70.98		
6	Intl growth	(8.05)	179.78	231.92	222.70	309.79	73.96		
7	Small cap	(9.43)	212.87	222.70	365.52	345.22	129.99		
8	Emerging markets	(15.16)	260.76	309.79	345.22	592.12	122.50		
9	REIT	1.58	70.98	73.96	129.99	122.50	216.64		
10									
11							return_6		
12	Annual statistics	Bond fund	S&P index	International	Small cap	Emerging markets	REIT	10-year treasuries	
13	Return	5.93	7.75	8.00	8.62	10.19	12.29	4.05	
14	Standard deviation	3.48	14.92	15.23	19.12	24.33	14.72		
15									
16	=SQRT(MMULT(MMULT(weight_6,omega_6),TRANSPOSE(weight_6)))								
17	=MMULT(weight_6,TRANSPOSE(return_6))								
18					Portfolio weights			weight_6	
19	Solver Portfolio	Sigma	Return	Bond fund	S&P index	International	Small cap	Emerging markets	REIT
20	Output row	12.28	8.80	0.17	0.17	0.17	0.17	0.17	0.17
21	Sum of weights			1.00	=SUM(weight_6)				
22	Target sigma	3.50							
23				return_6 = 1x6 vector of average annual returns					
24				omega_6 = 6x6 covariance matrix					

EXHIBIT 14.15 Solver portfolio for six-asset efficient frontier.

B	C	D	E	F	G	H	I	J
					Portfolio weights			
	Sigma	Return	Bond fund	S&P index	International	Small cap	Emerging markets	REIT
MVP	3.24	6.18	0.91	0.01	0.06	0.00	0.00	0.02
	3.24	6.25	0.90	0.01	0.06	0.00	0.00	0.03
	3.25	6.31	0.89	0.01	0.06	0.00	0.00	0.04
	3.26	6.35	0.89	0.01	0.06	0.00	0.00	0.04
	3.28	6.41	0.88	0.01	0.06	0.00	0.01	0.05
	3.30	6.46	0.87	0.01	0.06	0.00	0.01	0.06
	3.35	6.55	0.86	0.00	0.05	0.00	0.01	0.07
	3.40	6.63	0.85	0.00	0.05	0.00	0.01	0.09
	3.50	6.76	0.83	0.00	0.05	0.00	0.01	0.11
	3.60	6.86	0.82	0.00	0.04	0.00	0.02	0.12
	3.80	7.04	0.80	0.00	0.03	0.00	0.02	0.15
	4.00	7.19	0.78	0.00	0.03	0.00	0.02	0.17
	4.50	7.53	0.73	0.00	0.01	0.00	0.03	0.23
	5.00	7.82	0.69	0.00	0.00	0.00	0.04	0.27
	6.00	8.35	0.61	0.00	0.00	0.00	0.04	0.35
	8.00	9.32	0.45	0.00	0.00	0.00	0.04	0.51
	10.00	10.23	0.31	0.00	0.00	0.00	0.04	0.65
	12.00	11.12	0.17	0.00	0.00	0.00	0.04	0.79
	14.72	12.29	0.00	0.00	0.00	0.00	0.00	1.00

EXHIBIT 14.16   Six-asset efficient portfolios with no short selling.

Next, set up the table into which Solver's output will be pasted. Exhibit 14.16 shows one way to structure this table. At a minimum, you need to copy and paste each efficient portfolio's standard deviation and return. We recommend tracking the portfolio weights as well, as simple or naïve mean-variance optimization can produce counterintuitive recommendations (more on this later).

The efficient frontier begins with the MVP. As we did for the two-asset example, we use Solver minimize the Solver portfolio's standard deviation. The only constraints are that the weights sum to 1 and the solution weights are non-negative. This dialog window is shown in Exhibit 14.17.* Run Solver and copy and paste the output to the first row of your table.

The remaining efficient portfolios are found by using Solver to maximize the portfolio return for a given standard deviation, where the standard deviation constraint is entered in the "Target sigma" cell. Then the standard deviation is increased slightly and Solver is run again. You need to "Add" this constraint to the Solver dialog box as shown in Exhibit 14.18. The incremental increases in the target sigma should be small at first, as small changes in sigma near the MVP result in relatively large increases in

* We named the portfolio standard deviation cell *sigma*, the portfolio returns cell *return*, and the sum of the weights cell *sum_weights*. Cells were named in the same way that the arrays were—by clicking on a given cell and typing the desired name in the "Name box." Although naming cells is certainly not necessary, it makes for a more transparent dialog window.

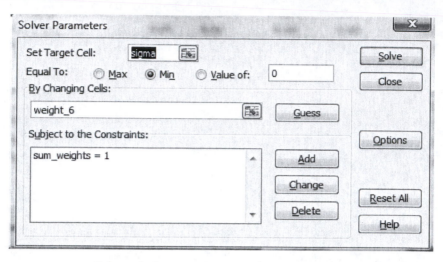

EXHIBIT 14.17 Solver dialog window set to find the minimum variance portfolio. (Not shown: Options set to "Assume non-negative" solutions so as to prohibit short sales.)

portfolio return. As the target sigma increases, the incremental changes can also increase.[*]

The table for the efficient frontier is shown in Exhibit 14.16. The table was constructed using Solver set as shown in Exhibit 14.18 and running it repeatedly. After each run, the output was copied and pasted as values into the table as another row. Then the value in the target sigma cell was increased, so that the next run of Solver would find an efficient portfolio with a little more risk. This process was repeated until the sigma constraint equaled the fund with the highest return, the end of the efficient frontier with no short selling.[†]

Once your table is complete, drawing the chart is easy. In Excel 2007, highlight the sigma and return columns in the table, go to the "Insert" ribbon, and click on "Scatter" in the "Charts" segment and select "Scatter with smooth lines and markers" (we will remove the markers shortly). The chart will now float over your current worksheet. To add the underlying assets, right-click on the chart and choose "Select data." Click on "Add" in the new

---

[*] Alternatively, one can maximize the slope of the line connecting some arbitrarily chosen y-intercept to the efficient frontier, and then increase the y-intercept.

[†] A program called SolverTable can automate this process. It can be downloaded from http://www.kelley.iu.edu/albrightbooks/Free_downloads.htm.

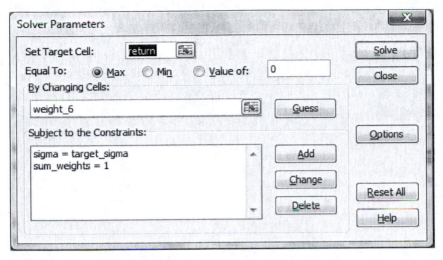

EXHIBIT 14.18   Solver dialog box for maximizing portfolio return for a target sigma. (Not shown: Options set to "Assume non-negative" solutions.)

dialog window, then add the assets, either individually (if you wish to identify them) or as a series (if you just wish to show the assets as a group). At this point, your chart will look something like that in Exhibit 14.19.

To remove the markers from the efficient frontier curve, right-click on the data series, choose "Format Data Series," then "Marker options," and then "None." To remove the lines connecting your assets series, right-click on the Assets series, select "Format Data Series," select "Line Color," and

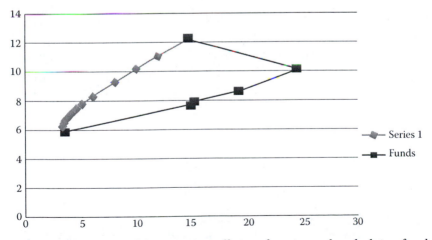

EXHIBIT 14.19   Chart of the six-asset efficient frontier and underlying funds (as it is likely to appear before adding titles and formatting the series).

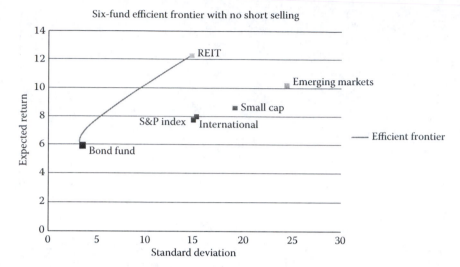

EXHIBIT 14.20   Efficient frontier.

then "No line." To change the name of Series 1, right-click on the efficient frontier series, choose "Select Data," highlight Series 1 and choose "Edit." Type in the name you desire (e.g., Efficient frontier) or point the "Name" box to the cell containing the name you wish to use. To add chart and axis titles, click on your graph, then on the "Chart Tools" tab that floats above the ribbon names. Select the Chart Layout that you want, then enter the desired titles, which finishes the process. You are finished. Our finished chart is shown in Exhibit 14.20.*

14.5 EXTENSIONS OF MODERN PORTFOLIO THEORY

The preceding pages explain how to find the efficient frontier with no short selling. We now consider some extensions to this basic model: finding the tangency portfolio (the market portfolio for the CAPM), and the problem of counterintuitive portfolio allocation recommendations and constrained optimization.

14.5.1 The Tangency Portfolio

It is common in most discussions of modern portfolio theory to assume that investors can borrow and lend at a risk-free rate of return. This

---

* In the graph shown in Exhibit 14.19, each asset was entered as a separate series. Once on the graph, each asset's point was "right-clicked" and "Add Data Labels" was selected. By default, Excel reports the y-coordinate. Each series was right-clicked again, "Format Data Labels" was selected and the "Series name" option chosen.

assumption results in a linear efficient frontier that begins at the risk-free asset (the y-intercept) and is tangent to the efficient frontier of risky assets. However, to draw this line one needs to identify the tangency portfolio. In the discussion that follows, we assume a 10-year investment horizon and use as our risk-free rate of return the recent annual return on the 10-year U.S. Treasury note, 4.05%. It should be noted that the 10-year horizon is arbitrary. Investors should match the risk-free rate of return to their investment horizon. For example, Ibbotson and associates assume an investment horizon of 20 years use the rate of return on Treasury bonds with maturities of 20 years.

Finding the tangency portfolio requires the addition of two cells to our Solver portfolio, one containing the risk-free rate of return, the other containing the slope of the line connecting the risk-free asset to the efficient frontier. The numerator of the slope is the return on the Solver portfolio less the risk-free rate; the denominator is the standard deviation of the Solver portfolio. The weights of the portfolio that Solver returns after maximizing this slope identify the composition of the tangency portfolio. The formula, using the names given the arrays in our workbook, is shown in Equation 14.6.

$$\max_{\{weight\_6\}} \ [slope] = \left[\frac{(weight\_6 \ x \ TRANSPOSE(return\_6) - riskfree\_rate)}{(weight\_6 \ x \ omega\_6 \ x \ TRANSPOSE(weight\_6))^{\frac{1}{2}}} \right]$$

$$= \frac{(return - riskfree\_rate)}{sigma} \tag{14.6}$$

These changes to our workbook, along with the tangency portfolio weights, are shown in Exhibit 14.21. The tangency portfolio weights were found by asking Solver to maximize Equation 14.6, the slope cell (C37) in Exhibit 14.21.

	B	C	D	E	F	G	H	I	J
31		sigma	return				Portfolio weights		
32	Solver Portfolio	Sigma	Return	Total Bond	S&P500	Intl growth	Small cap	Emerging markets	REIT
33	Output row	3.86	7.09	0.79	–	0.03	–	0.02	0.16
34	Sum of weights			1.00					
35	Target sigma	3.50							
36	Risk free rate	4.05	riskfree_rate						
37	Slope	0.79	=(return-riskfree_rate)/sigma						
38									

EXHIBIT 14.21   The tangency portfolio found by having Solver maximize the slope of the linear efficient set.

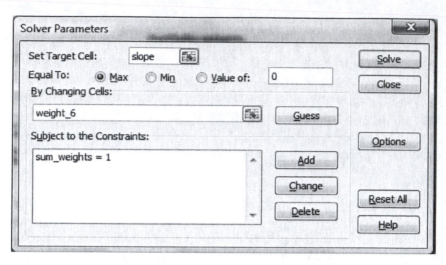

EXHIBIT 14.22   Solver dialog window to find tangency portfolio.

The Solver dialog window used to find this solution, assuming no short sales, is shown in Exhibit 14.22.

Once the tangency portfolio is found, the line can be added to the chart by constructing a table with three observations. The first point on the line is the risk-free asset. The second point is the tangency portfolio just identified. The third point is a portfolio with an arbitrarily high standard deviation that extends the linear efficient beyond the tangency portfolio. We chose a standard deviation of 20% for this portfolio, large enough to extend the line beyond the REIT fund. The return on the third portfolio is found using the formula: $r_p = r_f + [(r_{tangency} - r_f)/\sigma_{tangency}]\sigma_p$, where σ_p is the standard deviation of this third high-risk portfolio. (The reader will likely recognize this formula as the formula for the CAPM's Capital Market Line.) To add this line to your portfolio, build the table shown in Exhibit 14.23, where "Extension" refers to this third portfolio. Right-click on your chart, select "Source Data," click on "Add", and fill in the appropriate dialog boxes.

The final result will look like the chart shown in Exhibit 14.24.

14.5.1.1 Extension of the Model to the CAPM

The only differences between the linear efficient set just identified and the CAPM's ex post Capital Market Line are the data used and, of course, the assumptions upon which the CAPM is based. If one were to accept the simple CAPM's assumptions and use investors' expectations for the future

	L	M	N
31	Linear efficient set		
32		*Sigma*	*Return*
33	Riskfree	0.00	4.05
34	Tangency	3.86	7.09
35	Extension	20.00	19.80
36	=N33+(N34-N33)/M34*M35		
37			

EXHIBIT 14.23   Table used to graph the linear efficient set.

returns and covariances for all assets in the economy to construct the above graph and allow short selling, then the tangency portfolio identified would be the market portfolio and the linear efficient set would be the Capital Market Line.

14.5.2 Counterintuitive Recommendations

Simple or naïve mean-variance optimization suffers from several criticisms. By simple or naïve, we mean that the optimizer is run without restricting its choice of weights other than no short selling. First, the recommended portfolios are often unstable, in that subtle shifts in the estimated asset returns, shifts that are well within estimation error, can produce a huge shift in recommended weights. Closely related to this criticism is the problem that the recommended portfolios can be counterintuitive.

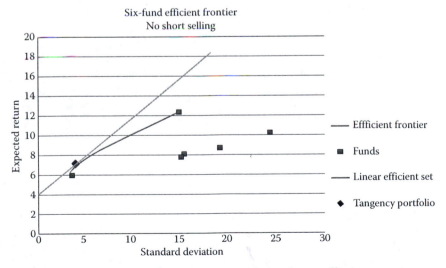

EXHIBIT 14.24   Linear efficient frontier with no short selling.

Both of these problems are present in the recommended portfolios for the efficient frontier developed earlier in this chapter. Exhibit 14.25 reports the standard deviations, returns, and asset weights for this efficient frontier. Two assets dominate the portfolio recommendations—the bond fund and the REIT fund. Two portfolios that are typically assumed to be core assets, the S&P index and the international fund, receive almost no weight. Further, the small cap index, arguably an important supporting part of any portfolio, is never included in the naïve optimizer's recommendations.

These problems stem from two problems in the data. First, the return on the REIT fund over our observation period is large relative to the other funds and its total risk (standard deviation) is small relative to its returns. Second, the returns on the emerging market fund are negatively correlated with the bond fund and larger than the returns on the S&P index, international, and small cap funds. Hence, as the desired expected return increases, naïve mean-variance optimization selects the REIT and the emerging market funds to combine with the bond fund while excluding all others in the opportunity set.

The second problem can be seen by looking at the S&P index-international-small-cap-fund cluster in Exhibit 14.26. The international fund has a slightly higher return than the S&P index, lower risk than the small cap fund, and the lowest correlation with the bond fund. These attributes lead mean-variance optimization to select the international fund over the S&P index and small cap funds. Had either the S&P index or small cap fund had higher returns or lower correlations with the bond fund (or both) than the international fund, they would have dominated the other two funds in this cluster.

		Portfolio weights					
Sigma	Return	Bond fund	S&P index	nternationa	Small cap	Emerging markets	REIT
3.24	6.18	0.91	0.01	0.06	0.00	0.00	0.02
3.50	6.76	0.83	0.00	0.05	0.00	0.01	0.11
4.00	7.19	0.78	0.00	0.03	0.00	0.02	0.17
4.50	7.53	0.73	0.00	0.01	0.00	0.03	0.23
5.00	7.82	0.69	0.00	0.00	0.00	0.04	0.27
6.00	8.35	0.61	0.00	0.00	0.00	0.04	0.35
8.00	9.32	0.45	0.00	0.00	0.00	0.04	0.51
10.00	10.23	0.31	0.00	0.00	0.00	0.04	0.65
12.00	11.12	0.17	0.00	0.00	0.00	0.04	0.79
14.72	12.29	0.00	0.00	0.00	0.00	0.00	1.00

EXHIBIT 14.25   Efficient portfolios under naïve mean-variance optimization.

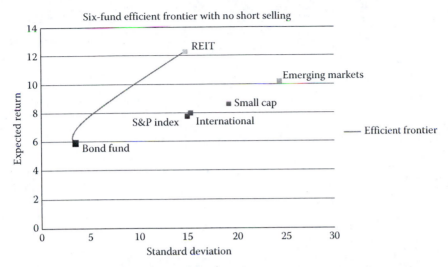

EXHIBIT 14.26   Naïve efficient frontier with underlying assets.

The solution to these problems is quite straightforward. The first step is to categorize the assets to be included in the optimal portfolio as either core or supporting assets. Suppose you have determined that you wish the core of your portfolio to be formed from the bond, S&P index, and international funds with minimum weights of 10%, 25%, and 15%, respectively. Suppose, too, that you wish the weights for each of the supporting funds to fall between 3% and 8%.*

These constraints are easy to integrate into our current Solver portfolio and Solver. One method is shown in Exhibit 14.27A. Once the weight constraints have been added to the Solver portfolio model, invoke Solver and constrain the weight vector to be greater than the minimum weights and less than the maximum weights by adding two more constraints in the Solver dialog window. The resulting dialog window is shown in Exhibit 14.27B. Calculate the constrained efficient frontier in the same manner as the unconstrained frontier. Indentify the MVP, then maximize portfolio returns subject to the additional constraints. The results are shown in Exhibits 14.28A and B.

* The values for these weight constraints are the subject of much debate among financial planning practitioners. The actual weights selected will be a function of an investor's age and level of risk aversion. The weights shown here are somewhat aggressive, given the low minimum for bonds and the relatively high maximums for the supporting funds. Still, the MVP of the constrained efficient frontier allocates 51% of the assets to the bond fund.

	B	C	D	E	F	G	H	I	J
28	Solver Portfolio			Portfolio weights					
29		Sigma	Return	Bond fund	S&P index	International	Small cap	Emerging markets	REIT
30		12.95	8.25	0.10	0.25	0.42	0.08	0.08	0.07
31	Sum of weights			1					
32	Target sigma	12.95							
33	Weight constraints								
34	Minimum			0.1	0.25	0.15	0.03	0.03	0.03
35	Maximum			1	1	1	0.08	0.08	0.08

EXHIBIT 14.27A   Solver portfolio with weight constraints.

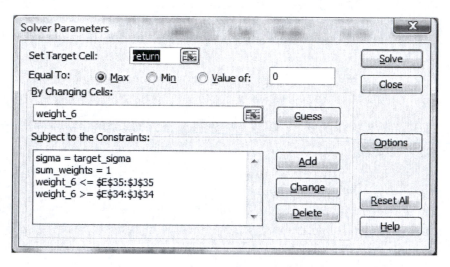

EXHIBIT 14.27B   Solver dialog window for constrained optimization.

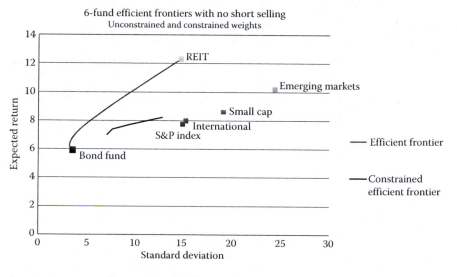

EXHIBIT 14.28A   Constrained efficient frontier.

		Portfolio weights					
Sigma	Return	Bond fund	S&P index	International	Small cap	Emerging markets	REIT
6.95	7.09	0.51	0.25	0.15	0.03	0.03	0.03
7.00	7.14	0.50	0.25	0.15	0.03	0.03	0.04
7.50	7.46	0.45	0.25	0.15	0.03	0.04	0.08
8.00	7.57	0.42	0.25	0.15	0.03	0.07	0.08
9.00	7.74	0.37	0.25	0.15	0.07	0.08	0.08
10.00	7.89	0.29	0.25	0.22	0.08	0.08	0.08
12.00	8.18	0.16	0.25	0.35	0.08	0.08	0.08
12.86	8.30	0.10	0.25	0.41	0.08	0.08	0.08

EXHIBIT 14.28B   Portfolio weights for constrained efficient frontier.

The constrained frontier differs significantly from the unconstrained frontier, beginning with the MVP. Whereas the bond fund comprised over 90% of the unconstrained MVP, it comprises just 51% of the constrained MVP. This is the most that can be allocated to the bond fund given the minimum values for the other assets in the opportunity set for this example. As a result, the constrained MVP has more risk than its unconstrained counterpart and the constrained efficient frontier starts further to the right in Exhibit 14.28A. In addition, the expected returns on the constrained efficient frontier are considerably smaller than those for the unconstrained frontier. This is a direct outcome of constraining the maximum weight on the REIT fund to 8%. Finally, the constrained frontier is considerably shorter than the unconstrained efficient frontier. The maximum return of 8.3% (the end of the constrained efficient frontier) is achieved by assuming a portfolio standard deviation of 12.86%. In this portfolio, the two funds with the lowest returns (the bond and S&P 500 funds) are allocated their minimum weights, the three supporting funds are allocated their maximum weights (all have higher returns than the core assets), and the remaining weight (41%) is allocated to the highest yielding core fund, the international fund.

The counterintuitive recommendations in this chapter's example result, at least in part, from the use of historical returns instead of forecast returns. In particular, real estate and emerging market funds experienced historically high returns over our observation period, especially relative to the more staid returns on the S&P index, international, and small cap funds. Using forecast returns partly resolved this problem. Lummer, Riepe, and Siegel (1994) suggest using CAPM betas and the security market line to forecast future returns. Haugen and Baker (1996) use as many as 40 independent variables to forecast future returns with impressive results.

Ultimately, the source of the returns used is a judgment call. Forecasting, whether it involves projecting historical returns forward, using an asset pricing model such as the CAPM or Arbitrage Pricing Theory (APT), or using an ad hoc multiple factor model, must be viewed as the process of substituting error for chaos. In the end, mean-variance optimization recommendations are simply the starting point for one's investment decisions, not the ending point.

14.6 VALUE-AT-RISK

Up to this point, this chapter has defined portfolio risk as the standard deviation of portfolio returns. For those comfortable with statistics, and assuming normally distributed returns, the mean and standard deviation of a portfolio's returns provide the information necessary to compare the risks and rewards of different risky portfolios. However, many people desire a simple and intuitive measure of risk. Value at risk, or VaR, is such a measure. VaR is a compelling measure of portfolio risk because it conveys the market risk of a portfolio in one number, and that number, either in terms of dollars or returns, is the major motivation for assessing risk—the potential loss in value of a portfolio (Hendricks, 1996).

VaR tells the investor the loss on a portfolio she can expect to experience some specified percent of the time over a given investment horizon. Typically, the critical probability (the small percent) is either 1% or 5% and the given investment horizon is either one day or two weeks. For example, if the two-week 5% VaR for a given portfolio of $100 million is a loss of $1.5 million, this tells the investor that 5% of the time she can expect to lose more than $1.5 million; that is she can expect to lose $1.5 million or more once over any 20 two-week periods. Put another way, VaR express the price risk of a portfolio of assets in terms of the frequency with which a specific loss will be exceeded some given percentage of the time (Boudoukh, Richardson, and Whitelaw, 1995).

There are many ways to calculate VaR.* We present three: computation, historical simulation, and Monte Carlo simulation. In the discussion that follows, we calculate the VaR for two portfolios—the constrained MVP and the constrained maximized return portfolio (MxRP) calculated earlier in this chapter. We use all three methods to estimate the VaR of both portfolios over one-day and two-week horizons for a critical value of 5%, leaving VaR with 1% critical value as an exercise for the reader.

---

* The interested reader is referred to Hendricks (1996) for 12 examples.

14.6.1 Computational Method

The computational method is perhaps the simplest and fastest way to estimate VaR. We begin by making the assumption that returns are normally distributed and serially uncorrelated, then combine the Z-statistic from the standard normal distribution for the critical p-value with the return and standard deviation of returns for a given portfolio to calculate that portfolio's VaR.* The application is simple. For a critical value of 5%, the return VaR is the return for the fifth percentile of the portfolio's probability distribution (the expected median return being the 50th percentile). Suppose the expected return for a risky portfolio is μ and the standard deviation of its returns is σ. The Z-statistic for the fifth percentile of a standard normal distribution (the left 5% tail) is -1.645.[†] The expression for the fifth percentile return is shown in Equation 14.7.

$$VaR_{\alpha=0.05} = \mu - 1.645_\sigma \tag{14.7}$$

Applying Equation 14.7 to real data is straightforward. We begin with the constrained MVP. The expected annual return and standard deviation of returns for this portfolio are 0.0709 and 0.0695, respectively. Assuming there are 240 trading days per year, the daily return for the MPV portfolio is calculated as

$$E_{daily}[r_{MVP}] = \frac{0.0709}{240}$$
$$= 0.0002954. \tag{14.8}$$

The daily standard deviation of returns is calculated as

$$\sigma_{MVP, daily} = \frac{\sigma_{MVP,annual}}{\sqrt{240}}$$
$$= \frac{0.0695}{\sqrt{240}}$$
$$= 0.00449. \tag{14.9}$$

Substituting these values into Equation 14.7 gives us the fifth percentile return for the MVP portfolio, thus

* Serially uncorrelated returns means that the returns for one period are not related to returns in previous periods. See Chapter 6 for a more complete explanation.
† The Z-statistic corresponding to the critical probability is returned by the Excel function =NormSInv(0.05), where the argument of the function, 0.05, is the critical probability, Z.

$$\text{VaR}_{\text{MVP}, \alpha=0.05} = \mu - 1.645_\sigma$$
$$= 0.000295 - 1.645 \times 0.00449$$
$$= -0.0071. \tag{14.10}$$

This tells us that the MPV portfolio can expect one-day losses greater (more negative) than −0.71% one out of every 20 trading days. If the portfolio is currently worth $100 million, then the investor can expect losses to exceed $710,000 once every 20 trading days, that is, once a month.

VaR is a measure of relative risk. When viewed in isolation, it tells us very little about the risk of a given portfolio. It is best used to compare the risks of different portfolios. For this reason, we calculate the VaR for the constrained MxRP from earlier in the chapter. This portfolio has an expected annual return of 0.0830 and a standard deviation of returns of 0.1286. Equation 14.11 converts these to daily values and calculates the 5% critical value VaR returns.

$$E_{\text{daily}} \left[r_{\text{MxRP}} \right] = \frac{0.0830}{240}$$
$$= 0.000346$$
$$\sigma_{\text{MxRP, daily}} = \frac{0.1286}{\sqrt{240}} \tag{14.11}$$
$$= 0.0083$$
$$\text{VaR}_{\text{MxRP}, \alpha=0.05} = 0.000346 - 1.645 \times 0.0083$$
$$= -0.0133.$$

This tells us that an investor in the MxRP portfolio can expect losses in excess of −1.33% one out of every 20 trading days. If this portfolio, too, is currently worth $100 million, then the investor can expect losses to exceed $1.33 million once every 20 trading days, or once a month. Now the relative nature of VaR is clear. The extra return earned by MxRP portfolio comes at a price. The losses expected to be exceeded 5% of the time on any given trading day are nearly 90% greater than the similarly defined losses for the MVP portfolio.

The 5% VaR over a two-week or 10-trading-day horizon is found by increasing the daily returns by a factor of 10 and the daily standard deviations by the square root of 10. These calculations are shown in Equations 14.12 and 14.13.

$$E_{\text{two week}} \left[r_{\text{MVP}} \right] = 0.000295 \times 10$$
$$= 0.00295$$

$$\sigma_{MVP,\text{two week}} = \sigma_{MxRP,\text{ daily}} \times \sqrt{10}$$
$$= 0.004499 \times \sqrt{10}$$
$$= 0.0142 \qquad\qquad (14.12)$$
$$VaR_{MVP,\text{two week}} = 0.00295 - 1.645 \times 0.0142$$
$$= -0.0204$$
$$E_{\text{two week}}[r_{MxRP}] = 0.000346 \times 10$$
$$= 0.00346$$
$$\sigma_{MxRP,\text{daily}} = \sigma_{MxRP,\text{ annual}} \times \sqrt{10} \qquad\qquad (14.13)$$
$$= 0.0083 \times \sqrt{10}$$
$$= 0.0262$$
$$VaR_{MxRP,\alpha=0.05} = 0.00346 - 1.645 \times 0.0262$$
$$= -0.0396.$$

As would be expected, the size of the loss exceeded 5% of the time over any given two-week horizon is considerably greater than the loss of a one-day horizon for both portfolios.

14.6.2 Historical Simulation

The historical simulation method is more data intensive but less mathematically intensive than the computational method. To implement it, collect a daily series of portfolio values over some given horizon, calculate the one-day and two-week dollar and percentage changes for that series, then sort the changes series by size, and lastly find the observation that lays one observation above the bottom 5% of observations. Alternatively, once the changes and returns have been calculated, you can use Excel's =PERCENTILE(*array, percentile*) function to find the VaR. Be aware, however, that this function reports an extrapolated percentile value rather than an actual observation.

To apply the first method to the MVP and MxRP portfolios, we collected the daily adjusted closing prices for the six funds in our observation set from May 1, 2006 through May 1, 2008. Next, we determine the number of shares of each fund in the opportunity set held by the MVP and MxRP portfolios. This is done using the size of the portfolio, the weights of the underlying funds in each portfolio, and the prices of the underlying funds. We assume both portfolios were worth $100 million on May 1, 2006. The weights are those determined by Solver for the constrained minimum variance and maximum return portfolios. The prices we take from the adjusted closing values on May 1, 2006 are as reported

	I	J	K	L	M	N	O
21	**Portfolio weights and shares**						
22	Assumed portfolio value	100,000,000					
23	May 1, 2006 prices	8.78	115.48	19.69	31.13	22.02	19.39
24							
25	**Weights**	**VBMFX**	**VFINX**	**VWIGX**	**NAESX**	**VEIEX**	**VGSIX**
26	MVP	0.51	0.25	0.15	0.03	0.03	0.03
27	MxRP `=J$22*J26/J23`	0.10	0.25	0.41	0.08	0.08	0.08
28	**Shares if purchased now**	**VBMFX**	**VFINX**	**VWIGX**	**NAESX**	**VEIEX**	**VGSIX**
29	MVP `=J$22*J27/J23`	5,808,656	216,488	761,808	96,370	136,240	154,719
30	MxRP	1,138,952	216,488	2,082,275	256,987	363,306	412,584

EXHIBIT 14.29   Number of shares in each fund held by the MVP and MxRP portfolios on May 1, 2006.

by finance.yahoo.com. For example, suppose we wish to calculate the number of shares of the bond fund held by the MVP portfolio. The weight allocated to the bond fund in the MVP portfolio is 0.51. The bond fund's closing price was $8.78. Hence, the number of shares of the bond fund in the MVP portfolio is

$$\text{(Number of shares)} = \frac{\text{(Portfolio size)} \times \text{weight}}{\text{Price}}$$
$$= \frac{(\$100 \text{ million}) \times 0.51}{\$8.78}$$
$$= 5,808,656. \tag{14.14}$$

Exhibit 14.29 reports prices, weights, number of shares held in each portfolio, and the code used to calculate the number of shares.

Having determined the share holdings for each fund, we constructed portfolio value series for our two target portfolios over the May 2006–2008 time period. We then calculated one-day and overlapping two-week returns for both portfolios.* The first 14 observations and calculations are shown in Exhibit 14.30.

Once the changes in values are calculated, copy the changes and paste them as values in another table. Next, sort each series from largest to smallest using Excel's "Sort" function (found on the "Data" ribbon in Excel 2007 or under the "Data" dropdown menu in Excel 2003). After sorting, determine the number of observations below the fifth percentile. For

---

* We wish to capture the greatest two-week loss. We acknowledge that this likely creates a downward bias for the two-week VaR, but non-overlapping periods would likely introduce a positive bias. Since downside risk is the focus of VaR, we chose the former bias.

	P	Q	R	S	T	U
6			MPV Portfolio			
7	Value	Daily Change	Daily return	2-week change	2-week return	Date
8	100,000,000	=P9-P8	=Q9/P8			5/1/2006
9	100,484,883	484,883	0.0048			5/2/2006
10	100,161,096	(323,787)	(0.0032)			5/3/2006
11	100,458,185	297,089	0.0030			5/4/2006
12	101,165,119	706,934	0.0070			5/5/2006
13	101,151,334	(13,785)	(0.0001)			5/8/2006
14	101,239,758	88,424	0.0009			5/9/2006
15	101,115,879	(123,880)	(0.0012)			5/10/2006
16	100,393,875	(722,003)	(0.0071)	=P18-P8	=S18/P8	5/11/2006
17	99,664,243	(729,632)	(0.0073)			5/12/2006
18	99,517,743	(146,500)	(0.0015)	(482,257)	(0.0048)	5/15/2006
19	99,436,596	(81,148)	(0.0008)	(1,048,287)	(0.0104)	5/16/2006
20	98,428,978	(1,007,617)	(0.0101)	(1,732,117)	(0.0173)	5/17/2006
21	98,189,006	(239,973)	(0.0024)	(2,269,179)	(0.0226)	5/18/2006
22	98,329,075	140,069	0.0014	(2,836,044)	(0.0280)	5/19/2006

EXHIBIT 14.30   Calculating one-day and overlapping two-week dollar and percentage changes in portfolio value for the MVP portfolio.

example, we have 504 observations in our one-day series. Five percent of 504 equals 25.2, making the 26th observation from the bottom the 5% VaR. The result is shown in Exhibit 14.31. For the MVP portfolio, the dollar loss that was exceeded 5% of the time is $998,331; the percentage loss that was exceeded 5% of the time is −0.91%. For the MxRP portfolio, these values are $2,050,494 and −1.87%, respectively.

Excel's PERCENTILE function offers an alternative to copying, pasting, and sorting the change data. After creating the dollar and percentage changes, invoke the percentile function (in the Statistics library of functions), highlight the data series for which you wish to find the VaR, then enter the percentile desired. Exhibit 14.32 reports the VaR found using this function. Note the extrapolation that has occurred for the changes. For the MVP portfolio, the historical value VaR for the dollar change is −$998,331; the PERCENTILE function reports it as −$993,491, a slight overstatement of the cost.

14.6.3 Monte Carlo Simulation

14.6.3.1 Simulated Portfolios

We begin by assuming that the returns on the six funds in our opportunity set follow a simple Weiner process as described in Chapter 13. We then model the daily price movements over two weeks (10 trading days) for each

	MPV Portfolio		MxR Portfolio		Observation Rank	
	Change	% change	Change	% change	Descending	Ascending
5% VaR	(998,331)	(0.0091)	(2,050,494)	(0.0187)	26	479
	(1,007,617)	(0.0093)	(2,065,505)	(0.0188)	25	480
	(1,022,844)	(0.0094)	(2,066,267)	(0.0188)	24	481
	(1,045,139)	(0.0095)	(2,079,663)	(0.0188)	23	482
	(1,053,069)	(0.0096)	(2,109,657)	(0.0190)	22	483
	(1,060,735)	(0.0097)	(2,143,471)	(0.0191)	21	484
	(1,091,534)	(0.0097)	(2,193,755)	(0.0192)	20	485
	(1,098,690)	(0.0101)	(2,207,737)	(0.0195)	19	486
	(1,185,807)	(0.0103)	(2,222,962)	(0.0199)	18	487
	(1,187,529)	(0.0103)	(2,372,672)	(0.0204)	17	488
	(1,213,971)	(0.0104)	(2,429,668)	(0.0216)	16	489
	(1,236,107)	(0.0110)	(2,527,226)	(0.0217)	15	490
	(1,262,837)	(0.0111)	(2,649,717)	(0.0224)	14	491
	(1,268,920)	(0.0113)	(2,667,849)	(0.0229)	13	492
	(1,301,993)	(0.0116)	(2,675,749)	(0.0235)	12	493
	(1,311,473)	(0.0116)	(2,682,048)	(0.0235)	11	494
	(1,345,635)	(0.0116)	(2,872,417)	(0.0237)	10	495
	(1,358,034)	(0.0117)	(2,913,205)	(0.0240)	9	496
	(1,429,013)	(0.0127)	(2,937,319)	(0.0256)	8	497
	(1,444,986)	(0.0127)	(2,963,026)	(0.0259)	7	498
	(1,472,531)	(0.0128)	(3,159,096)	(0.0271)	6	499
	(1,529,086)	(0.0132)	(3,170,356)	(0.0272)	5	500
	(1,570,814)	(0.0135)	(3,239,132)	(0.0279)	4	501
	(1,605,385)	(0.0137)	(3,477,286)	(0.0286)	3	502
	(1,751,725)	(0.0154)	(3,699,570)	(0.0336)	2	503
	(1,834,793)	(0.0166)	(3,887,848)	(0.0353)	1	504

EXHIBIT 14.31 One-day 5% VaR for the MVP and MxRP portfolios found using the historical simulation method.

fund and track the dollar and percentage changes over the first day and the cumulative dollar and percentage changes over the two-week horizon. Finally, we run a Monte Carlo simulation of 5000 iterations and read the 5% VaRs for each portfolio from the simulated probability distributions.

	5% VaR from Historical simulation				
	Values found using =PERCENTILE(*array, percentile*)				
	MPV Portfolio			MxR Portfolio	
Horizon	Change	% change		Change	% change
1 day	(993,491)	(0.0091)		(2,050,011)	(0.0187)
2 week	(2,708,757)	(0.0238)		(5,837,581)	(0.0501)

EXHIBIT 14.32 Five percent VaR values as determined by Excel's PERCENTILE(.) function.

	B	C	D	E	F	G	H
2	**Inputs**						
3	Historical annual performance						
4		Bond fund	S&P index	International	Small cap	Emerging markets	REIT
5	Return	5.93%	7.75%	8.00%	8.62%	10.19%	12.29%
6	Standard deviation	3.48%	14.92%	15.23%	19.12%	24.33%	14.72%
7							
8	Length of sub-period (days)	1					
9	Sub-periods per year	240					
10	Length of period	0.0042					

EXHIBIT 14.33   Input sector for Monte Carlo simulation of VaR.

The first step is to build the Input section, as shown in Exhibit 14.33. It contains the annual returns and their standard deviations for each of the six funds. Because we model one-day changes, the length of the sub-period in days is 1. Assuming 240 trading days per year, the sub-period length in years is $1/240 = 0.0042$.

Next, we model the standard normal random variables, return processes, and price series for each of the six funds. We assume that the standard normal variables are correlated for each day according to their historical correlations (the correlation matrix calculated earlier in this chapter), but are serially uncorrelated. Entering all of the correlations in one correlation matrix would require a 60×60 matrix. To simplify the process, we chose instead to create separate correlation matrices for each trading day in our model. This resulted in ten 6×6 matrices that were filled in by copying and pasting the correlation matrix, facilitating data entry and greatly reducing the likelihood of error. After creating the 10×6 array of standard normal random variables and correlating each day's draws, return and price series were built following the procedures described in Chapter 13. Initial prices were based on the adjusted closing fund prices on May 1, 2008, though any arbitrary price could have been used. In practice, the modeler will know the current prices and share holdings for the assets of their portfolios. The resulting return process model is shown in Exhibit 14.34; three of the correlation matrices are shown in Exhibit 14.35.

Once the models of the returns processes and individual fund price series have been built, all that remains is to use the simulated price series data to construct two two-week series that simulate the values for the MVP and MxRP portfolios, calculate and track the one-day and two-week dollar and percentage changes, run a Monte Carlo simulation, and read the desired VaR from the simulated probability distributions.

Once again, the first step in this final model-building phase is determining the shares of each fund in the opportunity set held by the MVP and

	B	C	D	E	F	G	H
4		Bond fund	S&P index	International	Small cap	Emerging markets	REIT
5	Return	5.93%	7.75%	8.00%	8.62%	10.19%	12.29%
6	Standard deviation	3.48%	14.92%	15.23%	19.12%	24.33%	14.72%
7							
8	Length of sub-period (days)	1					
9	Sub-periods per year	240					
10	Length of period	0.0042					
11							
12	*Return processes*	=RiskNormal(0,1,RiskCorrmat(day_1,1))					
13	Normal processes	Bond fund	S&P index	International	Small cap	Emerging markets	REIT
14	1	0	0	0	0	0	0
15	2	0	0	0	0	0	0
16	3	0	0	0	0	0	0
17	4	0	0	0	0	0	0
18	5	0	0	0	0	0	0
19	6	0	0	0	0	0	0
20	7	0	0	0	0	0	0
21	8	0	0	0	0	0	0
22	9	0	0	0	0	0	0
23	10	0	0	0	0	0	0
24							
25	Simulated daily returns	Bond fund	S&P index	International	Small cap	Emerging markets	REIT
26		=C$5*$C$10+C$6*C14*SQRT($C$10)					
27	1	0.000247	0.000323	0.000333	0.000359	0.000425	0.000512
28	2	0.000247	0.000323	0.000333	0.000359	0.000425	0.000512
29	3	0.000247	0.000323	0.000333	0.000359	0.000425	0.000512
30	4	0.000247	0.000323	0.000333	0.000359	0.000425	0.000512
31	5	0.000247	0.000323	0.000333	0.000359	0.000425	0.000512
32	6	0.000247	0.000323	0.000333	0.000359	0.000425	0.000512
33	7	0.000247	0.000323	0.000333	0.000359	0.000425	0.000512
34	8	0.000247	0.000323	0.000333	0.000359	0.000425	0.000512
35	9	0.000247	0.000323	0.000333	0.000359	0.000425	0.000512
36	10	0.000247	0.000323	0.000333	0.000359	0.000425	0.000512
37		=(1+C27)*C39					
38	Simulated daily prices	Bond fund	S&P index	International	Small cap	Emerging markets	REIT
39	0	10.12	129.27	24.11	31.77	32.31	22.21
40	1	10.12	129.31	24.12	31.78	32.32	22.22
41	2	10.12	129.35	24.13	31.79	32.34	22.23
42	3	10.13	129.40	24.13	31.80	32.35	22.24
43	4	10.13	129.44	24.14	31.82	32.36	22.26
44	5	10.13	129.48	24.15	31.83	32.38	22.27
45	6	10.14	129.52	24.16	31.84	32.39	22.28
46	7	10.14	129.56	24.17	31.85	32.41	22.29
47	8	10.14	129.60	24.17	31.86	32.42	22.30
48	9	10.14	129.65	24.18	31.87	32.43	22.31
49	10	10.15	129.69	24.19	31.88	32.45	22.32

EXHIBIT 14.34   Return process model to simulate VaR.

MxRP portfolios. The process is the same as used in the previous section on historical simulation, except in this case we use prices from May 1, 2008 as our opening prices. The results are shown in Exhibit 14.36.

After determining the number of shares held in each portfolio, we construct the two-week portfolio value series for each fund. We use Excel's SUMPRODUCT function to multiply the number of shares of each fund held by a given portfolio times their respective simulated prices for each trading day in the series. Judicious use of the dollar to make the weight array an absolute reference (meaning that it will not change when the cell is dragged to a new position) facilitates the series construction. Enter the

	J	K	L	M	N	O	P
12	Day 1						
13	@RISK Correlations	1 / Bond fund in $C$14	1 / S&P index in $D$14	1 / International in $E$14	1 / Small cap in $F$14	1 / Emerging markets in $G$14	1 / REIT in $H$14
14	1 / Bond fund in $C$14	1					
15	1 / S&P index in $D$14	-0.094415097	1				
16	1 / International in $E$14	-0.151943743	0.791424902	1			
17	1 / Small cap in $F$14	-0.141800106	0.746462021	0.764883749	1		
18	1 / Emerging markets in $G$14	-0.179064044	0.718441344	0.835974861	0.742054667	1	
19	1 / REIT in $H$14	0.030923262	0.323316812	0.329956348	0.461954351	0.342043079	1
20	Day 2						
21	@RISK Correlations	2 / Bond fund in $C$15	2 / S&P index in $D$15	2 / International in $E$15	2 / Small cap in $F$15	2 / Emerging markets in $G$15	2 / REIT in $H$15
22	2 / Bond fund in $C$15	1					
23	2 / S&P index in $D$15	-0.094415097	1				
24	2 / International in $E$15	-0.151943743	0.791424902	1			
25	2 / Small cap in $F$15	-0.141800106	0.746462021	0.764883749	1		
26	2 / Emerging markets in $G$15	-0.179064044	0.718441344	0.835974861	0.742054667	1	
27	2 / REIT in $H$15	0.030923262	0.323316812	0.329956348	0.461954351	0.342043079	1
28	Day 3						
29	@RISK Correlations	3 / Bond fund in $C$16	3 / S&P index in $D$16	3 / International in $E$16	3 / Small cap in $F$16	3 / Emerging markets in $G$16	3 / REIT in $H$16
30	3 / Bond fund in $C$16	1					
31	3 / S&P index in $D$16	-0.094415097	1				
32	3 / International in $E$16	-0.151943743	0.791424902	1			
33	3 / Small cap in $F$16	-0.141800106	0.746462021	0.764883749	1		
34	3 / Emerging markets in $G$16	-0.179064044	0.718441344	0.835974861	0.742054667	1	
35	3 / REIT in $H$16	0.030923262	0.323316812	0.329956348	0.461954351	0.342043079	1

EXHIBIT 14.35 Correlation matrices for the first three trading days.

	B	C	D	E	F	G	H
37		=(1+C27)*C39					
38	Simulated daily prices	Bond fund	S&P index	International	Small cap	Emerging markets	REIT
39	0	10.12	129.27	24.11	31.77	32.31	22.21
40	1	10.12	129.31	24.12	31.78	32.32	22.22
41	2	10.12	129.35	24.13	31.79	32.34	22.23
42	3	10.13	129.40	24.13	31.80	32.35	22.24
43	4	10.13	129.44	24.14	31.82	32.36	22.26
44	5	10.13	129.48	24.15	31.83	32.38	22.27
45	6	10.14	129.52	24.16	31.84	32.39	22.28
46	7	10.14	129.56	24.17	31.85	32.41	22.29
47	8	10.14	129.60	24.17	31.86	32.42	22.30
48	9	10.14	129.65	24.18	31.87	32.43	22.31
49	10	10.15	129.69	24.19	31.88	32.45	22.32
50							
51	Portfolio weights and shares						
52	Assumed portfolio value	100,000,000					
53							
54	Weights	VBMFX	VFINX	VWIGX	NAESX	VBEX	VGSIX
55	MVP	0.51	0.25	0.15	0.03	0.03	0.03
56	MxRP	=$C$52*C55/C39 → 0.10	0.25	0.41	0.08	0.08	0.08
57	Shares if purchased now	VBMFX	VFINX	VWIGX	NAESX	VBEX	VGSIX
58	MVP	=$C$52*C56/C39 → 5,039,526	193,394	622,148	94,429	92,851	135,074
59	MxRP	988,142	193,394	1,700,539	251,810	247,601	360,198
60		=SUMPRODUCT($C$58:$H$58,C40:H40)			=SUMPRODUCT($C$59:$H$59,C40:H40)		

	B		C	D	E		F	G	H
61	10-day Simulated Porfolio Performance								
62			MVP	Cumulative			MxRp	Cumulative	
63			Value	Change	Return		Value	Change	Return
64		0	100,000,000				100,000,000		
65		1	100,029,554	29,554	0.000296		100,034,573	34,573	0.000346
66		2	100,059,117				100,069,158		
67		3	100,088,689				100,103,756		
68		4	100,118,270				100,138,366		
69		5	100,147,860				100,172,988		
70		6	100,177,459				100,207,623		
71		7	100,207,068				100,242,270		
72		8	100,236,685				100,276,929		
73		9	100,266,312				100,311,601		
74		10	100,295,948	295,948	0.002959		100,346,285	346,285	0.003463
75	=RiskOutput("MVP: 10-day change w/ corr")+C74-$C$64						=RiskOutput("MPV: 10-day return w/ corr")+D74-$C$64		

EXHIBIT 14.36 Fund prices, share holdings, portfolio prices, and dollar and percentage changes in value for the MVP and MxRP portfolios.

initial portfolio value of $100 million for day 0. In the next row, use SUMPRODUCT to multiply the weights by their respective prices and sum the products, dollar-signing both sides of the array address for the weights. Then drag the cell down nine more rows. The value series is finished. See the code in cell C65 in Exhibit 14.36 for an example of how to do this.

Next, calculate the cumulative dollar and percentage changes for the first and tenth trading day of each of the portfolio series. Again, see the code in cells D74 and D75 of Exhibit 14.36 for examples. These are the values we track to find the one-day and two-week VaR. Now define each of the four cells for each portfolio series as @Risk output variables and run a 5000 iteration simulation. Bring up the histograms for each of the output variables and take from them the values shown at the 5th percentile (the 90% confidence interval is @Risk's default setting). These are the simulated 5% values-at-risk. Exhibit 14.37 shows the histograms for the one-day and 10-day (two-week) MVP portfolios. The 5th percentile one-day return is reported as −7.16 thousandths, or −0.72%; the two-week 5% return VaR as −20.5 thousandths or −2.05%. Exhibit 14.38 shows the histograms and values for the MxRP portfolio, −1.34% and 3.95%, respectively.

14.6.4 Discussion of Results

Exhibit 14.39 summarizes the one-day and two-week return VaR for the MVP and MxRP portfolios determined by the three methods considered in this chapter. Two important findings stand out. First, the computed VaR are statistically insignificantly different from those found using Monte Carlo simulation. This makes sense. They are based on the same underlying return process assumptions. Second, both the computational method and the Monte Carlo simulation method understate the actual VaR found using historical simulation by nearly 20%.

The cause of this difference likely stems from the assumption of normally distributed returns. Many returns distributions have fatter tails than predicted by the normal distribution (Duffie and Pan, 1997). VaR measures the tail risk for the returns on a given asset. If the tails of the actual return distribution are fatter than those predicted by the normal distribution, then VaR based on the normal distribution will understate the tail risk. While this problem more typically shows up when the critical value is 1%, it is nonetheless present for the 5% VaR in our limited data set.

There are several solutions. The first is to use historical simulation to estimate VaR. This is the method recommended by Hendricks. Alternatively, when the normal distribution is used, the analyst should recognize that

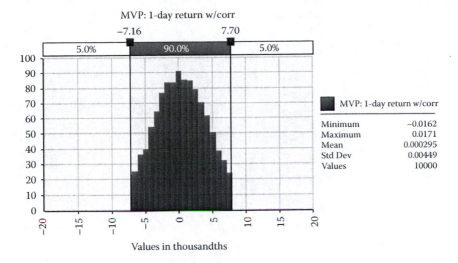

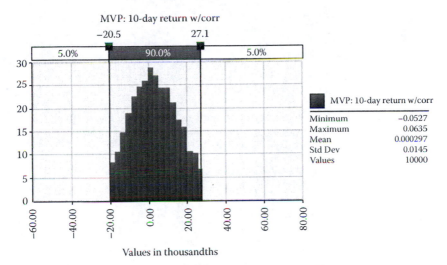

EXHIBIT 14.37 Simulated probability distributions of the one-day and two-week percentage changes in value for the MVP portfolio.

VaR likely understates the true downside risk. Finally, the return process can be adjusted to produce fatter tailed return distributions. Interested readers are referred to Duffie and Pan (1997) and Zangari (1996).

14.7 SUMMARY

This chapter considered the risk and return for portfolios of risky assets. We began by showing how to use Excel's array functions to calculate the

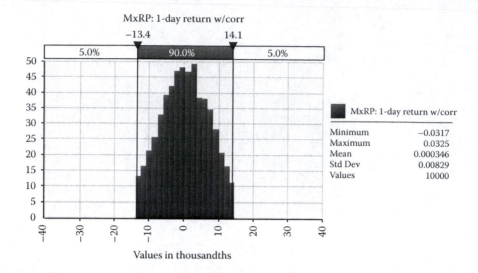

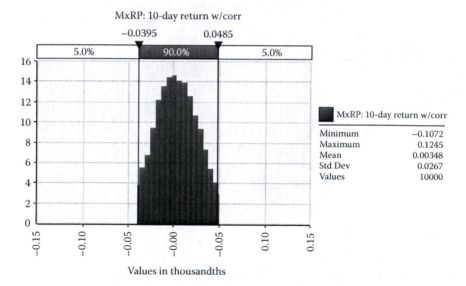

EXHIBIT 14.38 Simulated probability distributions of the one-day and two-week percentage changes in value for the MxRP portfolio.

returns and the standard deviation of returns for portfolios with many securities. We first calculated the expected return and standard deviation of returns for a portfolio of six risky assets, each of which comprised 1/6th of the total portfolio's value. We then considered the benefits of efficient portfolios. We used Solver to find the portfolio that offered the least of

Horizon	5% VaR from computations		
	MVP % change		MxRP % change
1 day	-0.71%		-1.33%
2 week	-2.04%		-3.96%
	5% VaR from Historical simulation		
1 day	-0.91%		-1.87%
2 week	-2.38%		-5.01%
	5% VaR from Monte Carlo simulation		
1 day	-0.71%		-1.34%
2 week	-2.03%		-3.94%

EXHIBIT 14.39   Summary return VaR findings.

amount of risk (the MVP) and portfolios that promised the highest returns for a given level of risk. The resulting portfolios formed the efficient frontier of risky assets.

Next, we showed how and why these portfolios are not necessarily the ones investors would want to hold. We then offered one solution to this problem—constraining Solver to find more intuitive portfolio allocation recommendations.

Lastly, we looked at an alternative measure of portfolio risk called VaR. VaR tells the investor the loss on a portfolio she can expect to experience some small percent of the time over a given investment horizon. It is a more intuitive way to measure portfolio risk than the standard deviation of returns, because it conveys the market risk of a portfolio in one number that is likely most important to an investor—the potential for significant losses.

PROBLEMS

1 Security Returns and Portfolio Composition

Use the following data to answer questions (a)–(j). You will also find these data on the CD that accompanies this text.

(a) Calculate the average quarterly returns and the standard deviations of those returns for each asset in Table 14.1.

(b) Use Excel's Correlation function (in DataAnalysis) to calculate the correlation matrix for these assets. Which assets promise the greatest benefits from diversification? Why?

(c) Use Excel's Covariance function (in DataAnalysis) to calculate the covariance matrix for these assets. Fill in the upper right-hand portion to create a square, symmetric covariance matrix.

TABLE 14.1   Return Data

							Emerging Market Stock	Emerging Market Bond	High-yield Bond
		Quarterly Index Returns							
		All returns are expressed as percentages							
Year	Quarter	Total Market	Commodities	Total Bond	Internationl	Small cap	Emerging Market Stock	Emerging Market Bond	High-yield Bond
1	I	0.447	2.486	-0.775	2.306	-5.108	7.037	2.067	0.316
	II	15.024	-4.959	3.63	13.08	15.981	8.549	10.967	4.924
	III	8.439	2.907	3.321	-1.489	14.526	-2.07	7.265	3.83
	IV	1.919	-9.032	2.862	-9.92	-3.422	-12.219	-4.652	2.286
2	I	11.914	-5.001	1.401	12.103	9.642	4.31	5.595	3.06
	II	1.438	-9.822	2.406	0.774	-5.105	-21.878	-8.004	1.361
	III	-12.311	-6.529	4.157	-13.66	-22.499	-29.067	-36.075	-1.361
	IV	19.108	-15.073	0.325	16.365	15.283	12.687	12.327	2.436
3	I	3.041	4.058	-0.488	-1.006	-5.794	8.94	2.796	1.854
	II	7.212	2.029	-0.984	4.771	15.745	19.715	4.891	-0.791
	III	-7.209	11.152	0.82	-0.08	-6.182	-2.407	0	-1.064
	IV	14.709	-0.173	-0.164	19.774	17.131	36.603	13.174	2.639
4	I	4.042	6.547	2.425	5.829	6.652	8.48	8.229	-1.575
	II	-3.648	6.144	1.427	-3.289	-3.115	-10.84	-0.58	2.094
	III	0.406	2.042	3.101	-10.429	0.775	-13.001	4.916	1.287
	IV	-9.701	6.949	3.893	-1.221	-7.589	-15.274	1.647	-2.591
5	I	-13.253	-7.211	3.179	-12.695	-6.658	-9.03	3.215	4.114
	II	6.304	-4.873	0.85	-2.492	13.834	6.484	3.963	-2.293
	III	-17.29	-6.575	4.144	-17.411	-23.093	-28.172	-3.963	-2.611
	IV	10.749	-6.598	-0.135	11.7	18.978	24.867	5.804	3.637
6	I	0.627	11.063	0.135	0.444	3.908	10.548	6.267	1.267
	II	-14.385	0.07	2.802	-4.534	-8.199	-6.572	-7.267	-1.523
	III	-19.353	6.584	3.745	-23.293	-23.958	-17.081	-0.504	-3.117
	IV	7.217	3.687	1.385	8.094	5.894	10.046	8.547	4.892
7	I	-3.556	2.588	1.366	-7.86	-4.617	-7.053	5.561	2.726
	II	14.617	2.281	2.556	16.444	20.061	20.555	11.973	5.933
	III	2.941	4.321	-0.12	7.446	8.248	13	0.775	1.373
	IV	11.254	11.236	0.24	13.571	13.916	16.755	3.788	4.445
8	I	1.816	10.893	2.607	6.001	6.358	9.178	3.65	1.938
	II	0.909	-4.622	-2.487	-2.218	0.993	-9.411	-4.521	-0.857
	III	-2.345	6.162	3.07	0.155	-2.259	6.614	8.503	4.211
	IV	9.221	-5.077	0.926	13.421	13.001	17.502	6.231	3.046
9	I	-2.656	10.728	-0.462	-0.542	-3.814	1.695	1.18	-1.816
	II	1.784	-5.846	2.965	-1.092	4.736	5.886	7.098	2.613
	III	3.515	15.349	-0.676	10.974	5.068	18.731	4.751	0.593
	IV	1.547	-4.061	0.564	4.627	1.156	6.451	2.985	1.371
10	I	4.744	-3.546	-0.677	9.462	11.491	10.909	2.451	1.544
	II	-2.444	4.76	-0.227	-0.107	-4.821	-8.047	-2.635	-0.576
	III	4.094	-7.623	3.787	3.726	-0.269	6.896	6.246	3.409
	IV	6.412	3.707	1.411	10.04	8.118	17.959	4.895	3.477

(d) Use Solver to find the allocation weights, expected quarterly return, and standard deviation of returns for the MVP. Assume no short selling.

(e) Identify 10 efficient portfolios by having Solver maximize the quarterly returns for 10 portfolios, where each portfolio has a different standard deviation. Begin with three standard deviations only incrementally larger than the standard deviation of the MVP you identified in question (d). Assume no short selling. Plot the efficient frontier and the underlying assets on a chart with expected quarterly returns on the y-axis and the standard deviation or returns on the x-axis. The efficient frontier begins at the MVP and ends at the asset that promises the highest return.

(f) Now assume that you can borrow and lend at the risk-free rate of 1% per quarter. What are the quarterly returns and the standard deviation of returns for the tangency portfolio?

(g) Compute the 5% VaR for the MVP over one quarter, expressed as a percentage. Assume that the parameters you estimated in question a will hold in the future.

(h) Use historical simulation to identify the 5% VaR for the MVP over one quarter, expressed as a percentage. Combine the weights for the MVP with the assets returns for each quarter to generate the time series of portfolio returns. (Hint: Excel's SUMPRODUCT function is useful for this type of calculation.)

(i) What are the 1% VaR (dollars and percentages) over a one-day horizon for the MVP and the MxRP discussed in Section 14.7 of this chapter? Use the computational and historical simulation methods. (Hint: You will find the VaRs for the latter method in Exhibit 14.31.)

(j) How might you apply the methods in this chapter to manage your retirement savings?

CHAPTER 15

Options

15.1 INTRODUCTION

In this chapter we review the basics of options and learn how to build some basic models of options. A thorough treatment of options is beyond the scope of this book, and it is assumed that the reader has already had some exposure to options and option valuation models.* The emphasis here will be on how to build basic models of options so you have a foundation for undertaking more complex modeling tasks.

The first order of business is to introduce options and the terminology of options. Most of the discussion is confined to the standard options: calls and puts. If you learn about these basic derivatives and learn how to model them, it should not be difficult to extend your modeling to more complex derivatives at a later date. Here are some definitions and details.

---

* See the sources listed in the references at the end of the text for more detailed explanations of options. The more understandable introductions include Hull (2009), Dubofsky and Miller (2003), Ross, Westerfield, and Jaffe (2008), and Berk and DeMarzo (2007).

15.1.1 Introducing Calls and Puts

The two most common options are calls and puts. A call is an instrument that gives the owner of the call the option to *buy* a share of stock at a specified price by a specified date. The price at which you can by the stock is called the *exercise price*, and will be denoted by X. The last date by which you must exercise the option to purchase is called the *expiration date*, denoted by T. We refer to this instrument as an *option* because it gives the holder of the option the right, but not the obligation to buy the stock. That is, the option holder is free not to exercise the option, and if he/she does not exercise the option, it expires.

Most options traded in the U.S. markets are referred to as American options. An American option can be exercised anytime until the expiration date. In contrast to the American option is the European option that can be exercised only on the date of expiration. Much of the development of the theory of option pricing was based on the European option simply because it is easier to analyze. But most of the analysis can be extended to American options.

Whereas a call is the option to buy the stock, a put is just the opposite, it is an option to *sell* the stock at a specified price by a specified date.

If you buy an option, there must be someone on the other side of the transaction who is the seller. That is, if you buy a call giving you the right to buy the stock by a future date, there is someone else who is selling you this right. The person who is selling the call is termed the *writer* of the call. This person is undertaking an obligation to sell the stock should you choose to exercise your option. That is, the call writer must sell you the shares, and does not have a choice. So the call *writer* does not have an option; he/she has the *obligation* and must perform if the option is exercised.

As with the call, for each buyer of a put option, there is seller on the other side of the transaction—the put writer. The writer of the put has the obligation to buy the stock should the buyer of the put option choose to exercise the right to sell the stock.

Acquiring an option involves a cost. The option writer who undertakes the obligation to sell the stock in the case of the call, or to buy the stock in the case of a put, would not normally be willing to undertake the obligation for free. The price the option writer earns for undertaking the obligation is referred to as the option *premium*, or simply the option price. We denote the premium paid for the call as C, and the premium paid for a put as P.

How much an investor should be willing to pay to acquire an option, and the factors that determine that price, or premium, has been one of the most interesting questions in finance. Black and Scholes (1973) and Merton (1973) developed models of option prices that have revolutionized the field of finance and earned Nobel Prizes for both Scholes* and Merton. One of the purposes of this chapter is to explain the famous Black–Scholes model. We also explain the more easily understood binomial model of option prices developed by Cox, Ross, and Rubinstein (1979).

15.2 PAYOFFS FROM OPTIONS

15.2.1 Buying Options

The starting point for modeling options is to understand the payoffs from the options. Most of the discussion focuses on the call. If we understand the call and can model its payoffs, the put easily follows.

The reason the call buyer wants to buy the call is to profit from an increase in the price of the underlying stock. At the expiration date, T, if the market price of the stock, S_T, is higher than the exercise price of the call, the call buyer can exercise the call and buy at the exercise price, X. The stock can be sold immediately for S_T, and the payoff is the difference, $S_T - X$. If the stock price is below the exercise price at expiration, the call buyer can refuse to exercise the call and get nothing. Thus, the call buyer will either make a profit on exercise, or nothing. In notation, before considering the premium originally paid for the call, the payoff to the call buyer at expiration, which is termed the *intrinsic value*, is

$$\text{Max}(S_T - X, 0). \tag{15.1}$$

So long as $S_T > X$, the payoff at expiration is positive, and the call buyer will profit (prior to taking account of the call premium paid earlier). The higher the stock price at expiration, the greater the profit, so the call buyer is betting that the stock price will increase. In fact, as we will see later, buying a call is much like a leveraged, or margined, position in the stock.

The put is just the opposite of the call in that the put buyer will profit from a decline in the price of the stock. At expiration, the put buyer has

---

* Black did not share the Nobel Prize simply because he died prior to its being awarded, and the prize is not given posthumously.

the option to sell the stock at exercise price, X, when the stock price is S_T. If the market price is below the exercise price, the payoff is positive and the lower the stock price the greater the payoff from exercising the put option. If the stock price is higher than the exercise price, the payoff from exercising the put is negative, so the put buyer will refuse to exercise the option and get nothing. The payoff to the put buyer at expiration (the intrinsic value of the put) can be written as

$$\text{Max}(X - S_T, 0) = \text{Max}(-(S_T - X), 0). \tag{15.2}$$

The payoffs for the options are easy to model. Exhibit 15.1 shows a model of the payoffs from the call and the put. The input data are in B6:D11, indicating that the initial (current) stock price is $100, and the exercise price for both the call and the put is $100 (cells C9:D9); there is one year to expiration, and the cost premium to buy either option is $10. Stock prices at expiration ranging from 0 to $140 are shown in column A. The intrinsic

	A	B	C	D	E	F	G	H
1			Option Payoff					
2			at Expiration					
3								
4	Data Input							
5								
6	Initial Stock Price	100						
7	Option Features							
8	Type of Option		Call	Put				
9	Exercise Price	X	100	100		=MAX(A19-$C$9,0)		
10	Time to Expiration	T	1	1				
11	Premium to Buy Option		10.00	10.00				
12						=MAX($D$9-A19,0)		
13								
14		Stock Price	Call Payoff			Put Payoff		
15		at Expiration	Intrinsic	Net of		Intrinsic	Net of	
16			Value	Premium		Value	Premium	
17		0	0	-10.00		100	90.00	
18		10	0	-10.00		90	80.00	
19		20	0	-10.00		80	70.00	
20		30	0	-10.00		70	60.00	
21		40	0	-10.00		60	50.00	
22		50	0	-10.00		50	40.00	
23		60	0	-10.00		40	30.00	
24		70	0	-10.00		30	20.00	
25		80	0	-10.00		20	10.00	
26		90	0	-10.00		10	0.00	
27		100	0	-10.00		0	-10.00	
28		110	10	0.00		0	-10.00	
29		120	20	10.00		0	-10.00	
30		130	30	20.00		0	-10.00	
31		140	40	30.00		0	-10.00	

EXHIBIT 15.1 Model of payoffs for call and put.

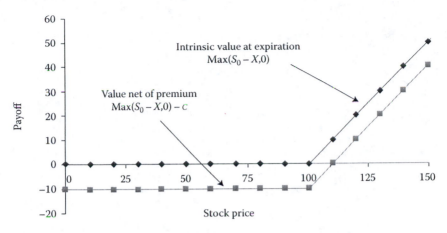

EXHIBIT 15.2   Payoffs at expiration from the call option.

value of the call at expiration [Max $(S_0 - X,0)$] is shown in C17:C31, and the payoff net of the cost of buying the call for $10 is shown in D17:D31. The same calculations for the put are in F17:G31.

At expiration, if the stock price is below the call exercise price of $100, the call buyer will not exercise the option, and the intrinsic value is zero as shown in C17:C27. However, the call buyer paid $10 to buy the option, so the payoff net of the $10 premium is −$10. At stock prices above $100, the call has a positive intrinsic value (cells C28:C31). Of course, when the initial premium of $10 is deducted, the profit is decreased as shown in D28:D31.

Exhibit 15.2 shows a graph of the call payoffs from the source data in A17:D31. The horizontal axis is the stock price (A17:A31) as the independent variable in the XY scatter diagram, with the vertical axis showing the payoff associated with each different stock price. This is the typical pattern for the call: the intrinsic value (the upper line) is flat up to the exercise price, and then the payoff increases linearly at a slope of 1.0, indicating that for each $1 increase in the stock price, the intrinsic call value increases by $1. The lower line is the payoff net of the call premium, so it is at −$10 at all stock prices up to $100. Then it increases at the same slope as the intrinsic value line.

Exhibit 15.3 shows a graph of the put payoffs from cells F17:G31 with a pattern just the opposite of the call. At stock prices above the exercise price of $100, the intrinsic value is zero, and the payoff net of the put premium is −$10. But the lower the price of the stock, the greater the profit from

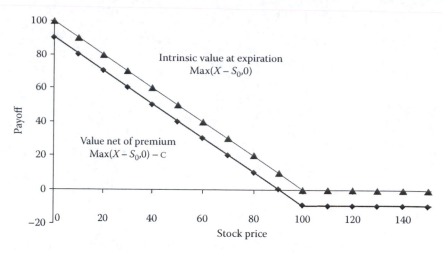

EXHIBIT 15.3   Payoffs at expiration from the put option.

exercising, with the intrinsic value increasing by $1 for each $1 decrease in the price of the stock.

15.2.2 Writing Options

The option writer is the person who sells the option to the buyer and undertakes the obligation to sell the stock in the case of the call, or to buy the stock in the case of the put. Whereas we refer to buying the option as a long position in the option, the seller has a short position. The payoffs for the short position are just the negative of the payoffs from the long positions. The payoffs from the different positions are written as follows:

	Buyer (long)	**Seller (short)**
Call	$\text{Max}(S_T - X, 0)$	$-\text{Max}(S_T - X, 0)$
Put	$\text{Max}(X - S_T, 0)$	$-\text{Max}(X - S_T, 0)$
	or	
	$\text{Max}(-(S_T - X), 0)$	$-\text{Max}(-(S_T - X), 0)$

When you graph the payoff to the writer, it is an upside-down image of the payoff from the option buyer. Exhibit 15.4 shows the payoff to the call writer. Whereas the call buyer loses the call premium if the stock price does not increase, the call writer earns the premium. But if the stock price goes above the exercise price, the call writer loses $S_0 - X$ because she/he has to sell the stock for X when it is worth $S_0 > X$. Therefore, at prices below X, the

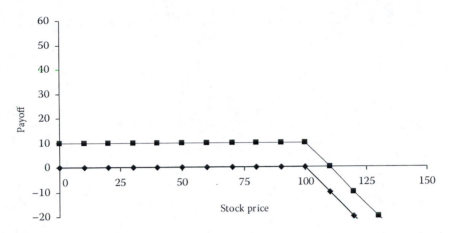

EXHIBIT 15.4 Payoff (top line) and intrinsic value (bottom line) for the call writer.

call writer earns the payoff of the call premium, but at prices above X, the call writer loses an amount equal to what the call buyer earns.

15.2.3 Options as a Leveraged Investment in the Stock

We see from these payoff graphs the profit possibilities from buying options. For a very small investment, just $10 per option per share in this example, you can acquire an opportunity to profit from price moves by a stock. In the case of the call, if the stock price increases, you can earn the same dollar profit as the stock without having to invest nearly as much. So this is like a leveraged, or margined, transaction in the stock. In fact, as we will see in a subsequent section, the payoff pattern from the option is exactly like a leveraged stock position. It is from this fact that we will be able to determine the value of the option. However, prior to discussing the value of the options, we will take a quick look at how we can model combinations of options.

15.2.4 Option Combinations

You need not confine your investment to "plain vanilla" options with just a long position in a call or a put. You can combine them in any number of different mixtures. Many of the combinations have exotic sounding names like strip, spread, and straddle, butterfly spread, bear spread, and even a strangle. Explaining and modeling each of these is beyond the scope of this chapter, so the interested reader is directed to texts devoted to

	A	B	C	D	E	F	G	H
1			Option Payoff			Combinations of Options		
2			at Expiration					
3							Transaction: User Input	
4	Data Input						Indicate 'Buy' or 'Sell'	
5	Interest Rate	5.0%					and Number of Shares	
6	Initial Stock Price	100						
7	Option Features						Call	Put
8	Type of Option		Call	Put		Buy or Sell?	Buy	Buy
9	Exercise Price	X	100	100		Shares	1	1
10	Time to Expiration	T	1	1				
11	Premium to Buy Option		10.00	10.00				
12						Long or Short:	1	1
13								
14							Payoff at Expiration	
15		Stock Price	Call	Put				Option
16		at Expiration	Intrinsic	Intrinsic		Call	Put	Combination
17			Value	Value				
18		0	0	100		0	100	100
19		10				0	90	90
20		20	=IF(G8="Buy",1,IF(G8="Sell",-1,0))*G9			0	80	80
21		30				0	70	70
22		40	0	60		0	60	60
23		50	=IF(H8="Buy",1,IF(H8="Sell",-1,0))*H9			0	50	50
24		60				0	40	40
25		70				0	30	30
26		80	0	20		0	20	20
27		90	0	10	=$G$12*C29	0	10	10
28		100	0	0		0	0	0
29		110	10	0		10	0	10
30		120	20	0		20	0	20
31		130	30	0		30	0	30
32		140	40	0	=$H$12*D29	40	0	40
33		150	50	0		50	=F29+G29	50
34		160	60	0		60		60
35		170	70	0		70	0	70
36		180	80	0		80	0	80
37		190	90	0		90	0	90
38		200	100	0		100	0	100

EXHIBIT 15.5 Payoffs from combinations of calls and puts: a straddle.

derivatives such as Hull (2009) or Dubofsky and Miller (2003). Nevertheless, we can get you started by showing how to adopt a model as simple as Exhibit 15.1 to analyze combinations of options.

Exhibit 15.5 shows a model for analyzing different option combinations. In this case the position is a *straddle*, a combination of buying a call and a put with the same exercise prices and same expiration dates. This combination profits from stock prices moving in either direction away from the exercise price, with payoffs shown in Exhibit 15.6.

In the spreadsheet the input is the same as the example in Exhibit 15.1, and the intrinsic values (payoff at expiration) of the call and put are in columns C and D. Columns F–H model combinations of options. The user enters the transactions in G8:H9 by typing "Buy" or "Sell" in G8 and H8 for the call and the put, respectively. The number of calls or puts bought or sold is entered in G9 and G10. In this example, the transaction is to buy

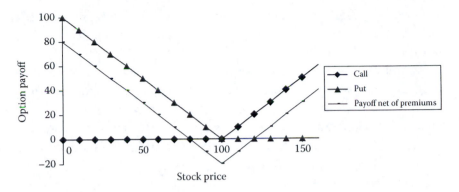

EXHIBIT 15.6   Payoffs from a straddle: Buy one call and one put.

one call and buy one put. The formula in G12 is =IF(G8="Buy",1,IF(G8= "Sell",−1,0))∗G9. If the user has entered "Buy" in G8, this formula returns +1, but if the user entered "Sell," it returns −1; otherwise it returns 0. This gives the correct sign for the transaction, with +1 for buy and −1 for sell, and it is multiplied times the number of shares entered in G9 by the user. The formula in H12 is similar. The results are the payoff amounts for the call and put in F18:G38. Cells H18:H38 add the payoffs from the call and put as the payoff for the combination.

In Exhibit 15.6, the gross payoff (before considering the premium) of the call is zero (following the horizontal axis) up to the exercise price of $100, after which it increases dollar-for-dollar with the stock price. The put payoff is the mirror image going in the other direction, with positive pay-offs for stock prices below $100. The straddle yields a gross payoff shown by the top V-shaped line. When we net out the cost of each of the options at $10 each, the net payoff is shown by the bottom V-shaped line. As you see, net of costs, the straddle profits for any stock price less than $80 or greater than $120. So the straddle is a good bet if you think the stock price will move substantially, but you do not know which way.

If you are not quite sure which way the stock will move, but you are more bullish, thinking a price increase is more likely, you can buy a *strap*, which consists of two calls and one put. This places a bigger bet on the stock price increasing, but still profits if the stock price declines. In the model in Exhibit 15.5, this transaction is input by typing 2 in the call cell G9 and 1 in the put cell H9.

On the other hand, if you are more bearish, but still not sure which way the stock price will move, you can create a *strip*, which is buying one call and two puts. This way if the stock moves down, you have greater profit, but you still have a profit if it goes up. This is input in the model by inserting 1 in G9 and 2 in H9.

As you can see, you can devise most any combination of options by changing the number of each option, and by going long or short by typing in "Buy" or "Sell" in G8 or H8 in the model. There are many different mixes to serve myriad purposes. We leave it to you to explore on your own a range of option combinations and exercise prices with a model such as Exhibit 15.5. For more detail regarding derivative trading strategies consult one of the texts referenced at the start of the chapter. While we do not explain the wide range of option trading strategies, we now build on the model shown in Exhibit 15.5 to demonstrate some of the results of mixing the options with the stock and with bonds.

15.2.5 Mixing Options with the Stock and a Bond

The model shown in Exhibit 15.7 is an extension of the one in Exhibit 15.5 with a section added to allow for buying or selling the stock and a bond. A stock purchase occurs at the initial price originally shown in B6 at $100, with the stock being sold at the option expiration date at the prices shown in E18:E32. A sale of the stock at $t = 0$ is assumed to be a short sale, with the proceeds available. The model user indicates the stock transaction by typing in "Buy" or "Sell" in J8, and the number of shares in J9.

A bond transaction refers to either lending or borrowing money at the interest rate indicated in B5 (in Exhibit 15.5). Lending is indicated by typing in "Buy" in K8, and the amount to be lent in K9. The user indicates borrowing by typing "Sell" in K8, with the amount borrowed in K9. Funds lent or borrowed are repaid at the expiration date of the options.

The option payoffs at expiration are in F18:H32, as before. The payoffs from the stock and bond are in J18:K32. The payoffs from the total mix are in L18:L32. The total cost of the combination of call, put, stock, and bond are in N18:R32.

In the example shown, the transaction consists of buying

- buying one call (G8:G9)
- selling one put (H8:H9)
- selling one share of stock (J8:J9), and
- lending $95.125 at 5% (buying the bond in K8:K9)

	Combinations of Options				Other Securities in Porfolio								
	Transaction: User Input				Transaction: User Input								
	Indicate 'Buy or 'Sell'				Stock	Bond							
	and Number of Shares				(Shares)	($ Amount)							
		Call	Put										
	Buy or Sell?	Buy	Sell		Sell	Buy							
	Shares	1	1		1	95.125							
	Long or Short:	1	-1		-1	95.125							
									Initial Cost				
		Payoff at Expiration			Payoff at Expiration								
Stock Price				Option			Total						
at Expiration	Call	Put	Combination		Stock	Bond	Position		Call	Put	Stock	Bond	Total
0	0	-100	-100		0.00	100.00	0.00		10.00	-10.00	-100	95.125	-4.88
10	0	-90	-90		-10.00	100.00	0.00		10.00	-10.00	-100	95.125	-4.88
20	0	-80	-80		-20.00	100.00	0.00		10.00	-10.00	-100	95.125	-4.88
30	0	-70	-70		-30.00	100.00	0.00		10.00	-10.00	-100	95.125	-4.88
40	0	-60	-60		-40.00	100.00	0.00		10.00	-10.00	-100	95.125	-4.88
50	0	-50	-50		-50.00	100.00	0.00		10.00	-10.00	-100	95.125	-4.88
60	0	-40	-40		-60.00	100.00	0.00		10.00	-10.00	-100	95.125	-4.88
70	0	-30	-30		-70.00	100.00	0.00		10.00	-10.00	-100	95.125	-4.88
80	0	-20	-20		-80.00	100.00	0.00		10.00	-10.00	-100	95.125	-4.88
90	0	-10	-10		-90.00	100.00	0.00		10.00	-10.00	-100	95.125	-4.88
100	0	0	0		-100.00	100.00	0.00		10.00	-10.00	-100	95.125	-4.88
110	10	0	10		-110.00	100.00	0.00		10.00	-10.00	-100	95.125	-4.88
120	20	0	20		-120.00	100.00	0.00		10.00	-10.00	-100	95.125	-4.88
130	30	0	30		-130.00	100.00	0.00		10.00	-10.00	-100	95.125	-4.88
140	40	0	40		-140.00	100.00	0.00		10.00	-10.00	-100	95.125	-4.88

EXHIBIT 15.7 Model for portfolio of options, stock, and bond.

The payoffs at expiration shown in L18:L32 sum the payoffs from each of the individual instruments. For example, at expiration, if the stock price has dropped to $30 (row 21), the call will not be exercised, and the payoff to the call buyer will be zero (F21). The put will be exercised by the put *buyer*, so the put writer will have to buy this stock at the exercise price of $100 when its market value is $30, for a loss to the put writer of $70 (G21). The total payoff from the two options is $0 − $70 = −$70 (H21).

The stock was sold short at $t = 0$ for $100. At expiration, the short position must be covered, so the stock will be purchased at the market price of $30. The −$30 in J21 shows the cost of the stock purchase. A bond was purchased for $95.125 at the start, and at expiration, it matures for $100 (principal + interest), which is shown in K21. The payoffs at expiration from the stock and the bond are −$30 + $100 = $70, and the total payoff from the combination of call, put, stock, and bond is zero summarized as

Call (long)	$0
Put (short)	−$70
Stock (short)	−$30
Bond (long)	±$100
Total payoff at expiration	$0

Note that the payoff from the total position (L18:L32) is zero in every case. That is, no matter what happens, the payoff at expiration nets out to zero. Normally, no one wants an investment that pays zero. But now let us look at what the investment costs at the start of the transaction.

At the start of the transaction ($t = 0$), the cash flows are as follows:

Call (long: buy 1 call)	−$10.00
Put (short: sell 1 put)	+$10.00
Stock (short: sell 1 share)	+$100.00
Bond (long: buy 1 bond)	−$95.125
Total cash flow	+$4.875

The positive cash flow of $4.875 means that the investor collects $4.875 at the start of the transaction from the mix of purchases and sales. This corresponds to the −4.88 shown as the Total cost in R18:R32 of the model.

That is, the cost is negative. Instead of paying out something to make the investment, this investor collects some cash at the start.

This may seem strange and even a little confusing. We have an investment that gives us money at the start, and then gives us nothing at the end. This is an arbitrage opportunity, or what we might call a "money machine." Imagine, if this opportunity persisted, how you could undertake transactions like this for thousands of shares instead of just one. Such an opportunity to make money up front at no cost would seem too good to be true. As they say, "If it seems too good to be true, it probably is." So, what is the catch? The catch is that the prices are in disequilibrium. In the example, the option prices of $10 for both the call and the put are such that the investor has this opportunity to make instant riskless profits. When prices offer such opportunities, investors will rush in to take advantage. The pressures of investors trying to take advantage of the opportunity will result in prices changing to the point where the excess profit opportunity disappears.

The question is, what must the prices of the options be so that there is no longer the opportunity for riskless arbitrage profits? The next section discusses pricing of options and presents option pricing models based on the idea that investors should not be able to engage in transactions that yield arbitrage profits.

15.3 OPTION PRICING MODELS

15.3.1 Binomial Option Pricing

The starting point for modeling option prices is a single period binomial model. This option pricing model derives prices of calls and puts based on the idea just discussed: In equilibrium, the options will be priced so there are no opportunities to make riskless arbitrage profits.

In Chapter 13 we examined a stock price model where the stock price followed a binomial process. The basis of the option model in this section is the binomial stock price model. Assume that at time $t = 0$, a share of stock has a price of $S_0 = \$100$. By the end of the period ($t = 1$), the stock can move up by a multiple (1 + rate of return) of $u = 1.25$ to a price of

$$S_1 = uS_0 = (1.25)100 = \$125,$$

or it can move down by a multiple of $d = 0.80$ to a price of

$$S_1 = dS_0 = (0.80)100 = \$80.$$

	A	B	C	D	E	F	G
1			Binomial Option Model				
2			Single Period				
3							
4	Data Input						
5	Riskless Interest Rate	5.0%					
6	Initial Stock Price	100					
7	Option Features						
8	Type of Option			Call			
9	Exercise Price			X	100		
10	Time to Expiration			T	1		
11							
12			Returns over 1 period				
13		Rate of Return		Wealth Relative Return			
14	Outcome 1	25%		u	1.25	=B20*E14	
15	Outcome 2	-20%		d	0.80		
16							
17		Time	0	*****************	1	**************	
18					Stock Price $_1$	Call Payoff	
19				S u	125	25.0	
20	Stock Price $_0$		100				
21				S d	80	0.0	
22							
23				=B20*E15		=MAX(E21-E9,0)	
24							

EXHIBIT 15.8   Excel model for call option on stock with binomial price moves.

The call option on the stock has an exercise price of $X = \$100$, and the call will expire at time $T = 1$. This is European style call, so it can be exercised only at the expiration date.

Exhibit 15.8 shows a spreadsheet model for this simple option. The input data are shown in lines 1–15, and the payoffs for the stock and the option are shown in lines 19–21. Cell B20 is the beginning price of the stock, and cells E19 and E21 are the stock values at time 1, the date that the call option expires. The stock prices are simply the beginning price multiplied by the up or down multiples, u and d. The payoffs for the call are shown in cells F19 and F21 using the formulas

$$\text{Call Value}_1 = \text{Max}(S_1 - X, 0)$$
$$= \text{Max}(\text{E19} - \text{E9}, 0) = \text{Max}(125 - 100, 0) = \$25,$$
$$\text{or,} \qquad = \text{Max}(\text{E21} - \text{E9}, 0) = \text{Max}(80 - 100, 0) = 0$$

in F19 and F21, respectively.

We see how easy it is to develop a model for the call payoffs for this simple one period binomial case. Modeling the payoffs in more extended cases is not much harder. But we not only want to model the payoffs, we would also like to use the model to explore the value of the call. That is what we do next.

15.3.1.1 Option Pricing by Replicating the Payoffs

To understand option pricing models, it is important to understand the law of one price and arbitrage.* The law of one price is simply the idea that if two investments have the same cash flows, risks, and other features, then they must have the same price. The reason for this is that if they trade at different prices, investors will be able to engage in arbitrage by buying and selling these equivalent securities so that their market prices will adjust until the securities trade at the same price. This principle applies to option pricing: if the market price of an option differs from the prices of other securities that have equivalent cash flows, the prices of the securities should all adjust until they trade at the same price. We use this principle to figure out what the price of the option should be when arbitrage does its job of equilibrating prices.

If we do not know what an option's market price should be, we find other investments whose prices are known and construct a mix of those investments that will replicate the payoffs from the option. If this other investment has the same payoff as the option, then we can conclude that the market value of the option should be the same as this other replicating investment. For example, if we know that a call option has payoffs of either $0 or $25, we try to find other investments, whose prices we know, and mix those known investments in a way so the mix has payoffs that are also $0 or $25. Since we know the prices of these alternative investments, and we can mix them in a way to duplicate the payoffs of the call, the market value of the call adjusts so as to be equal to the cost of constructing this replicating mix.

It turns out that we can duplicate the returns on the call with an appropriate mix of the stock and borrowing. That is, since a call is much like a leveraged investment in the stock, we should be able to engage in a leveraged purchase of the stock that will yield payoffs identical to the call.

Exhibit 15.9 repeats the option payoff model from above, and shows the replicating investment at the bottom in lines 26–34. Line 29 shows the cost of buying one share of stock for $100 at time $t = 0$ (cell B29),

---

* See Berk and DeMarzo (2007, Chapter 3) for a clear discussion of arbitrage and the law of one price.

	A	B	C	D	E	F	G
1			Binomial Option Model				
2			Single Period				
3							
4	Data Input						
5	Riskless Interest Rate	5.0%					
6	Initial Stock Price	100					
7	Option Features						
8	Type of Option			Call			
9	Exercise Price			X	100		
10	Time to Expiration			T	1		
11							
12			Returns over 1 period				
13		Rate of Return		Wealth Relative Return			
14	Outcome 1	25%		u	1.25		
15	Outcome 2	-20%		d	0.80		
16							
17	Time	0		**************** 1		*************	
18				Stock Price 1		Call Payoff	
19				S u	125	25.0	
20	Stock Price 0	100					
21				S d	80	0.0	
22							
23		Replication of Call Payoffs					
24	Time	0	**************** 1	*************			
25							
26	Stock Shares Purchased	1	=-$B$30*EXP($B$5*$E$10)				
27					S d	S u	
28							
29	Stock Purchase	-100.00		Sell Stock	80	125	
30	Borrow	76.10		Repay Debt	-80	-80	
31	Total Cost	-23.90		Total Payoff	0	45	
32							
33			Equivalent Number of Calls to Replicate Payoff			1.80	
34			Cost / Number of Calls			-13.28	
35	=E21*EXP(-B5*E10)						
36			=F31/F19		=B31/F33		
37							

EXHIBIT 15.9 Binomial option model: Replicating portfolio of stock plus borrowing.

and the payoffs from selling the stock at time $t = 1$ for either $80 or $125 (E29 and F29). Line 30 shows the cash flows from borrowing. Assume that at time $t = 0$ we borrow an amount equal to the present value of the "down" payoff from the stock. The "down" payoff when we sell the stock at $t = 1$ is $80. Using continuous discounting at the riskless rate of 5% over one period, this present value is $80\ e^{0.05(1)} = 80(0.9512) = \$76.10$ in cell B30. That is, we borrow $76.10 at the rate of 5% at $t = 0$, and repay $80 (principal plus interest) at $t = 1$. At time $t = 0$, the cost of the stock was $100, but we borrow $76.10 to help finance the stock purchase, so the equity we need to invest is the remainder of $100 − $76.10 = $23.90 (cell B31).

At $t = 1$, when the option expires, the payoff from this leveraged position will be the proceeds from selling the stock less the amount we have to repay on the debt:

Stock Sale$_1$	= $80	$125	(cells E29 and F29)
− Debt Repayment$_1$	= $80	$80	(E30 and F30)
= Net Payoff$_1$	= 0	$45	(E31 and F31)

Note that regardless of whether the stock went up or down, we have to repay the $80 debt. The debt repayment uses up the proceeds of the stock sale in the down state, and leaves $45 in the up state.

The $0 or $45 payoffs do not exactly match the payoffs from the call that are $0 or $25. So we need to adjust the number of calls to duplicate these payoffs. For example, since the $45 payoff in the "up" state is 1.8 times the $25 payoff from the call, we can see that if we bought 1.8 calls, we would have the same payoff as the leveraged stock. So, if the leveraged stock transaction costs $23.90 (cell B31), then the value of one call must be $23.90/1.8 = $13.28 (cell F34). Or, to put it another way, we can scale the leveraged stock transaction down by a factor of 1/1.8 and duplicate the payoffs of the call. That is, buy 1/1.8 = 0.5556 shares of stock, and borrow $42.28 to leverage the investment, and get the same $0 and $25 payoff as the call. The cost of buying 0.5556 shares net of borrowing would be $55.56 − $42.28 = $13.28. Since we can earn the same payoffs of $0 or $25 from this investment as from the call, we certainly should not pay more than $13.28 for the call. If the price of the call was less than this, investors can engage in an arbitrage transaction that will drive up the price of the call and/or drive down the price of the stock until they are in equilibrium.

15.3.1.2 An Arbitrage Transaction

An alternative way to model the option payoffs and value is to model the arbitrage transaction. The basic model is the same, except we construct a transaction so that the payoffs are the same in each state. That way, no matter happens, the investment pays off the same amount, so the investment is riskless.

The transaction we construct is called a covered call. This involves buying the stock and writing a call. Writing a call means that instead of buying call option, we sell the call option (also called taking a short position in the call). When we sell the call option we are undertaking the obligation to sell the stock to the call buyer should the buyer choose to exercise the option to buy the stock. If the call option is in the money at the expiration date $(S_T > X)$, the call buyer will exercise, and we will have to sell the stock at the exercise price, X. When we have to sell the stock at the exercise price of X when it is worth S_T, we lose the difference $S_T - X$. If the call expires out of the money $(S_T < X)$, the call buyer will not exercise, and we do not have to sell the stock. In either case, as the call writer, we earn the call premium, the price that the call buyer paid at the start to buy the call option. As the call writer, the time pattern of cash flows is as follows:

Time: 0 Expiration Date, T
Cash Flow: Call Premium: $+C - \mathrm{Max}(S_T - X, 0)$.

For the binomial case, it is possible to structure the covered call so that no matter what happens, we get the same amount at the expiration date. Exhibit 15.10 shows how this works.

The model in Exhibit 15.10 shows the same binomial case as the previous model. In this case, stock shares are purchased and a call is sold. The results of one particular transaction are shown in the exhibit. The number of shares purchased is in B26, and is input by the model builder. The cost of buying the shares is in B29, and is simply the price per share times the number of shares purchased. B27 is the number of calls written, and B30 is the premium amount the writer receives, which is the number of calls written times the call premium price per call. Assume that we do not yet know the number of shares we will buy, nor the price of the call. The number of shares actually shown in B26 and the call price in B28 are the end product of our computations, but for now let us just use these numbers to understand the model.

The payoffs calculations are shown in columns E and F. The $44.44 and $69.44 in E29 and F29 are the number of shares (0.556 shares in B26) times the price per share of either $80 or $125. The call payoffs are in E30:F30. If the stock price is $80, the call is not exercised, and we get nothing (E30) from the call. If the price is $125, the call will be exercised by the call buyer. We buy the stock for the market price of $125 and sell it to the call buyer for $100, losing the difference of $25 (F30).

The proceeds from the up and down state are the same. In the case where the stock went up, we were able to sell our 0.556 shares of stock for $69.44, and we incurred a loss of $25 from the call. The difference is our net proceeds of $44.44 (F31). In the case where the stock went down, we sold our 0.556 shares for $44.44, and got nothing from the call (E31). Note that in either case, we get $44.44 at $t = 1$. Thus, we have structured our

	A	B	C	D	E	F
1				**Binomial Option Model**		
2				Single Period		
3						
4	**Data Input**					
5	Riskless Interest Rate	5.0%				
6	Initial Stock Price	100				
7	Option Features					
8	Type of Option			Call		
9	Exercise Price			X	100	
10	Time to Expiration			T	1	
11						
12				Returns over 1 period		
13		Rate of Return		Wealth Relative Return		
14	Outcome 1	25%		u	1.25	
15	Outcome 2	-20%		d	0.80	
16						
17	Time	0		*****************	1	*************
18					Stock Price $_1$	Call Value
19				S u	125	25.0
20	Stock Price $_0$	100				
21				S d	80	0.0
22						
23		**Riskless Covered Call**				
24	Time	0	=-B6*B26	***********	1	*************
25					=E19*B26	
26	Stock Shares Purchased	0.556				
27	Call Sold	1	=B28*B27			
28	Call Premium	13.28			S d	S u
29	Stock Purchase	-55.56		Sell Stock	44.44	69.44
30	Call Proceeds	13.28		Call Payoff	0	-25.0
31	Total Cost	-42.28		Total Payoff	44.44	44.44
32					=F31-E31	
33	PV of Payoff	42.28				
34				Payoff Difference		0
35		=E31*EXP(-B5*(E10))				
36						

EXHIBIT 15.10   Binomial option model with riskless covered call.

portfolio so that no matter what happens to the level of stock prices, we get $44.44. This is a riskless transaction.

Had we chosen a different number of shares to purchase, the payoffs would have been different, and it would not necessarily have been a riskless transaction. Before the fact, we do not know precisely the number of shares to buy to construct this riskless transaction. There is formula for this, but for now we will let the spreadsheet model solve this problem.

In the context of the spreadsheet model, we can use Goal Seek to solve for the number of shares to make the payoffs the same in both states. Note in the model, cell F34 is the difference between the payoffs in the two states, F31–E31. We would like to find the number of shares of stock to purchase so that this difference is zero. That is, we want to make the payoffs the same in both states. To do this, on the main menu line use Tools > Goal Seek; on the Goal Seek dialog box, Set Cell F34 > To Value 0 > By Changing Cell B26, click OK. The number of shares will be varied to reach the goal. In this case the number of shares will converge to 0.556 as shown on Exhibit 15.10, and the Payoff Difference in F33 will be zero. Your model has now solved for the risk-free mix of stock and calls. You get $44.44 no matter what happens.

The next step is to solve for the value of the call, which, at this stage, you presumably do not yet know. If you have an investment that has payoff that is certain regardless of the performance of the stock price, then that investment is riskless and should earn the risk-free interest rate. So now we want to find the amount of the call premium that will make the return on the total investment equal to the risk-free interest rate. Once again, we use Goal Seek to find the call premium. Cell B33 is the present value of the riskless payoff. The riskless payoff is $44.44 in F31, and B33 is the present value at the riskless rate. That is,

$$\text{PV of Payoff} = \text{B33} = \text{E31} \times \text{EXP}(-\text{B5} \times (\text{E10})).$$
$$= (44.44)e^{-0.05(1)} = 42.28.$$

We now know that the total cost of the investment mix has to be $42.28 in order for it to earn the riskless rate of 5%. Since we know how much we have to invest in the stock ($55.56 in B29), we need to find the amount we would need to earn as a call premium so the net cost is $42.28. In this case it is obvious that the premium would have to be $55.56 − $42.28 = $13.28. In more complex cases we could use Goal Seek that would find the Call Premium in B28 that would make the Total Cost in B31 equal to the $42.28 in B33.

Note that this method of working out a riskless mix of stock and the call option results in the same value of the call and same number of shares as we obtained with the previous model that used the mix of stock and borrowing in Exhibit 15.9. In this covered call example, it is easier to see how arbitrage would work. If the call premium was more than $13.28, you could sell the call and buy the stock for a total cost lower than $42.28 and thereby earn riskless return greater than the 5% risk-free rate. If the call premium was lower than $13.28, we could devise a mix of buying the call and selling the stock so as to earn a sure rate that exceeded the risk-free rate. Given the pattern of binomial stock prices and the current price of the stock, the only price for the call that will prevent investors from earning a disequilibrium return is the $13.28 that we have solved for. With many investors trading in these instruments, the prices will move to this equilibrium value.

15.3.1.3 The Cox, Ross, and Rubinstein Binomial Option Formula
To demonstrate the basic binomial model of option pricing, we have used a round about method. In fact, in this binomial case, the value of the call that we have solved for can be calculated more directly with a formula that was developed by Cox, Ross, and Rubinstein (1979). The development of their formula follows the same logic that we have presented as a spreadsheet model. In the single-step binomial case, where the length of the step is T years, and r is the risk-free rate per year, the value of the call should be

$$C = [pC_u + (1-p)C_d]e^{-r(T)},$$

where

$$p = \frac{e^{r(T)} - d}{u - d} \quad \text{and} \quad 1 - p = \frac{u - e^{r(T)}}{u - d}. \tag{15.3}$$

Applying this for our example, where $u = 1.25$, $d = 0.80$, $C_u = \$25$, $C_d = 0$, $T = 1$, and at the riskless rate of 5%, the present value of $1 at time $T = 1$ is $e^{-r(T)} = e^{-0.05} = 0.9512$, we have

$$C = [pC_u + (1-p)C_d]e^{-r(T)} = [0.5584(25) + 0.4416(0)](0.9512) = \$13.28$$

where

$$p = \frac{e^{r(T)} - d}{u - d} = \frac{1.0513 - 0.80}{1.25 - 0.80} = \frac{0.2513}{0.45} = 0.5584$$

and

$$1 - p = \frac{u - e^{r(T)}}{u - d} = \frac{1.25 - 1.0513}{1.25 - 0.80} = \frac{0.1987}{0.45} = 0.4416.$$

Note that this is identical to the value generated in our models in Exhibits 15.9 and 15.10. Part of the derivation of the call value is the hedge ratio, H, that is the number of shares of stock purchased for each call sold in the covered call model, which is calculated as

$$H = \frac{C_u - C_d}{uS_0 - dS_0} = \frac{25 - 0}{125 - 80} = 0.556,$$

which matches the calculation in cell B26 of Exhibit 15.10.

602 ■ Introduction to Financial Models for Management and Planning

15.3.1.4 Put Value

Having developed a simple one-step binomial model for a call, we show how to adapt the model to a put. The put option is just the reverse of the call in that it gives the buyer of the put the option to sell the stock for a specified exercise price by the expiration date. Because the payoff pattern of the put is like a reverse image of the pattern for the call, the model of the put is almost identical to the call model, except that we reverse many of the signs. To see how this works, we will use the same single-step binominal data as before, but we modify the model to examine a put option to sell the stock for an exercise price of $X = \$100$ at time $T = 1$. As in the previous case, we assume the current stock price is $S_0 = \$100$ and it will move to either

$$dS_0 = 0.80(100) = \$80, \text{ or } uS_0 = 1.25(100) = \$125$$

by the option expiration date at $T = 1$. In this binomial case, the value of the put at the expiration date will be

$$
\begin{aligned}
P_1 &= \text{Max}(X - S_1, 0) = \text{Max}(-(S_1 - X), 0) \\
&= \text{Max}(100 - 125, 0) = \text{Max}(-25, 0) = 0 \quad \text{when} \quad S_1 = 125 \\
&= \text{Max}(100 - 80, 0) = \text{Max}(20, 0) = 20 \quad \text{when} \quad S_1 = 80.
\end{aligned}
$$

With these payoffs at expiration, we want to know how much the put option is worth at $t = 0$. We can solve for the put value just as we did for the call. We can determine the put value that results when we model either a replicating investment or a riskless portfolio, or, we can also go directly to the solution with the Cox, Ross, and Rubinstein (CRR) formula.

In the case of the call, we found we could replicate the payoffs for the call with a portfolio consisting of a long position in the stock plus borrowing. For the put, we simply reverse the signs of the transaction and duplicate the put payoff with a short position in the stock combined with lending.

15.3.2 Put–Call Parity

We use the same basic model to estimate the equilibrium values of the call and the put. The next step is to show how the values of the put and the call are related. The relationship is called Put–Call Parity. It is based on the idea that because we can replicate the payoffs of a call with a portfolio consisting of the stock, a riskless bond, and a put; and we can replicate the put with a mix of the stock, a bond, and a call. In equilibrium the put and the call prices must be linked.

Assuming that the put and call have the same exercise price and time to expiration, the link between the prices of the put and call can be expressed as

$$C + X e^{-r(T)} = S + P. \qquad (15.4)$$

The left side of the expression is the call price plus the present value of exercise price and the right side is the stock price plus the put price. The expression can be rearranged to express the value of one variable in terms of all the other variables—that is, the call expressed in terms of bond, stock, and put, or the put value expressed in terms of the bond, stock, and call. According to this parity relation, if prices do not conform to this relationship, an arbitrage opportunity will enable investors to make abnormal profits.

For example, suppose prices are such that Equation 15.4 did not hold as an equality. To be specific, suppose

$$C + X e^{-r(T)} < S + P,$$

that is,

$$C + X e^{-r(T)} - S - P > 0. \qquad (15.5)$$

To arbitrage this situation, assume that at time $t = 0$ we sell a call for price C, borrow an amount $(X e^{-r(T)})$ equal to the present value of the exercise price on the call, buy a put for price P, and buy the stock for price S_0. When we add up the cash inflows and outflows from these transactions, we have total flow as shown algebraically in line 6 of Exhibit 15.11, where the inflows are shown with a positive sign and outflows are shown with a negative sign. The algebraic sum of the cash flows in line 6 is identical to Equation 15.5, so it is positive by assumption. This means we have a positive amount of money flowing in at $t = 0$ because the cash inflow from selling the call and borrowing exceeds the outflow from buying a put and buying the stock.

Now look at the cash flows at the expiration date. Each line shows the flow from the purchase or sale, and the algebraic sum of the flows from each of the transactions is shown in line 6. No matter what happens, whether the stock goes up or down, the total payoff at expiration is zero, so there is no uncertainty regarding the payoff. Like the example in Exhibit

		Investment Transaction & Cash Flows at $t=0$		Cash Flows at Expiration of Options	
		Transaction	Flow	Stock Down $S_1 < X$	Stock Up $S_1 > X$
1	Transaction		Flow	Stock Down $S_1 < X$	Stock Up $S_1 > X$
2	Sell Call		$+C$	0	$-(S_1 - X)$
3	Borrow		$+X\,e^{-r(T)}$	$-X$	$-X$
4	Buy Put		$-P$	$+(X - S_1)$	0
5	Buy Stock		$-S_0$	$+S_1$	$+S_1$
6	Sum of Cash Flows		$C + X\,e^{-r(T)} - P - S_0 > 0$ by assumption	$0 - X + (X - S_1) + S_1$ $= 0$	$-(S_1 - X) - X + 0 + S_1$ $= 0$

EXHIBIT 15.11 Cash inflow and outflow from transaction with stock, bond, call, and put.

15.7, the total transaction is one where we have a positive inflow at time $t = 0$, and zero flow at time $t = 1$. That is, we structure the buying and sell-ing so money comes in at $t = 0$, but on a net basis, we do not have to pay anything out at time $t = 1$.

This is the nirvana that every greedy investor dreams about—a situa-tion where money is produced at no cost and no risk. Unfortunately, finan-cial nirvanas seldom exist and if they do exist, it will only be for a fleeting moment. If prices are as assumed in Equation 15.5, investors will quickly notice the disequilibrium, will rush to take advantage of it, and prices will change so that such a profitable arbitrage opportunity quickly disappears. This means that in equilibrium, Equation 15.4 must apply, and Equation 15.5 must hold as an equality. This relationship between prices is what is meant by put–call parity. Put and call prices have to conform to this rela-tionship or there will be arbitrage opportunities that cannot last.

We can build an Excel® model that we can use to check whether prices are conforming to put–call parity and to devise trading strategies if price depart from parity. Exhibit 15.12A is a model for the one-period binomial case allowing the user to input the various buy and sell transactions and see the results in terms of cash flows at times $t = 0$ and 1. Formulas for rows 29–34 are shown in Exhibit 15.12B. The input section, A1:E15, is sim-ilar to the previous binomial models except that current market prices of the call and the put are in D11 and E11 as input data. Lines 19–21 show the binomial stock prices and the call and put payoffs at expiration. The trans-action section is in A27:H34, showing the transactions and the cash flows at times $t = 0$ and 1. As we saw earlier, the put–call parity equation takes account of the prices of the stock, call, and put, and the opportunity to

	A	B	C	D	E	F	G	H	I
1				Binomial Option Model					
2				Put - Call Parity for Binomial Model					
3									
4	Data Input								
5	Riskless Interest Rate	5.0%							
6	Initial Stock Price	100							
7	Option Features								
8	Type of Option			Call	Put				
9	Exercise Price		X	100	100				
10	Time to Expiration		T	1	1				
11	Price of Option			15.00	8.40				
12				Stock Returns over 1 period					
13		Rate of Return		Wealth Relative Return					
14	Outcome 1	25%		u	1.25				
15	Outcome 2	-20%		d	0.80				
16									
17	Time	0				*****************	1	****************	
18					Stock Price ₁		Call Payoff	Put Payoff	
19					S u	125	25.0	0.0	
20	Stock Price ₀	100							
21					S d	80	0.0	20.0	
22									
23									
24	Time		+++++++++	0	++++++++		*********	1	********
25							S d	S u	
26									
27		Price		Transaction					
28				Units	Cash Flow				
29	Stock	100	Buy	1	-100.00		80	125	
30	PV(Exercise)	0.9512	Borrow	100	95.12		-100	-100	
31	Call	15	Sell	1	15.00		0.0	-25.0	
32	Put	8.40	Buy	1	-8.40		20.0	0.0	
33									
34	Portfolio Cost				1.72		0.00	0.00	
35									

EXHIBIT 15.12A   Put and call transactions for put–call parity.

borrow and lend. Stock is in line 29, borrowing or lending in line 30, the call is in line 31 and the put in line 32, with the totals for the portfolio in line 34. For each of these instruments, the current price is carried down from the input section. The user enters the desired transaction in C29:C32

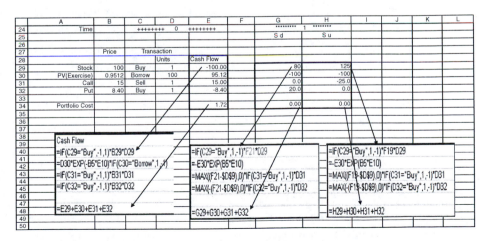

EXHIBIT 15.12B   Put–call parity model with selected formulas.

by entering the words Buy, Borrow, or Sell in the appropriate lines. The number of units of each is in column D, where units refers to shares, and to the dollars borrowed or lent, expressed in terms of maturity value. The time $t = 0$ cash flows from each transaction are in E29:E34. The cash flows at the expiration date from each transaction are shown in G29:H34 for each of the binomial states.

In this example, all the input data are the same as in the previous binomial examples except that the call is priced at a disequilibrium price of $15.00, as compared to the equilibrium price that should be $13.28 as was shown in the previous examples. With the price of the call too high, it should be obvious that we should not buy the call. As the model will show, if we sell the call, combined with the other instruments, we have the opportunity to earn a positive return with no investment and no risk.

Cells C29:D32 show the transactions we have entered. In C29 we enter "Buy" for the stock, and indicate one share in D29; E29 shows the cash outflow at $t = 0$ of buying one share at $100. In C30 we enter "Borrow" to indicate the transaction in the riskless bond. The amount borrowed is $95.12, which is equal to the present value of $100, the exercise price of the options. Line 31 shows the call transaction, where we enter "Sell" in C31, indicating one share in D31. Selling one call brings in a positive cash flow of $15 in E31. Line 32 relates to the put, where we enter "Buy" in C32, and one share in D32, with the amount paid of $8.40 in E32 (negative indicating a cash outflow). In column E we show that we payout $108.40 for the stock and put, we have an inflow of $110.12 from the borrowing and call sale, for a net flow of $1.72 in E34.

The payoffs at the expiration date ($t = 1$) are in columns G and H, based on the binomial stock prices of $80 and $125 for the down and up states, respectively. We buy one share of stock, and then sell it for either $80 or $125 (cell G20 and H29). We borrow $95.12 at 5% at $t = 0$, and repay $100 in both states. The call we sold will pay $0 in the down state, but will cost us $25 in the up state (G31 and H31), and the put we bought will pay $20 in the down state, or 0 in the up state, as shown in G32 and H32. Combining all these $t = 1$ payoffs, we see in G34 and H34 that whatever happens, we have a zero payoff. That is, the payoff is riskless.

If an investment pays zero at $t = 1$ with certainty, it should be worth zero today ($t = 0$). What we see in this case is that this investment that pays zero at $t = 1$ will actually provide us with a positive cash flow at $t = 0$. So it is worth zero, but instead of costing zero, it is as if someone is paying us for the investment. This looks like a great deal. And, in contradiction to the saying that "If it seems too good to be true, it probably is,"—so long as prices are as indicated in the example, it is a virtual money machine. But money machines seldom last long. Investors who are aware of the put–call parity relations, and with the ability to analyze the data with a model such as this, will quickly see and act on the arbitrage opportunity. Prices will change until the arbitrage opportunity disappears. Meanwhile, you, the model builder, will be able to analyze different transaction possibilities by trying them out in your model, and perhaps spot a disequilibrium that can make you rich.

15.3.3 Multi-Step Binomial Option Models

15.3.3.1 Two Periods

In the last section we learned how to model puts and calls when it was assumed that stocks followed a binomial process over one period. In Chapter 13 we noted that even though the binomial model of stock prices can be unrealistic if the length of the single binomial step is too long, we can make the model more realistic if we cut time into shorter and shorter intervals and assume that stock prices follow a binomial process over these very short time intervals. In this section we will continue to use the binomial model for options, but we will use short time intervals to more closely approximate the actual behavior of stock prices.

The first step to understanding the multi-period option models is to understand how we work backward through time, from the future back to the present, in order to derive today's value of the option. Ultimately, we will use many periods, but to get started, assume that we have a two-period model as shown in Exhibit 15.13. Using the same binomial numbers we have used previously, assume that the stock starts out at $t = 0$ at a price of $100 and can go up by $u = 1.25$ or down by $d = 0.80$ by $t = 1$. This yields prices of either $125 or $80 at $t = 1$. Then, from $t = 1$ to $t = 2$, the price can change again by 0.80 or 1.25 to any of the three different prices, $156.3, $100, or $64. These three ending prices are the starting points for our analysis of the options.

As before, assume the exercise price for the call is $X = $100, but assume the expiration date is $t = 2$ instead of $t = 1$. At $t = 2$, the expiration date, the call will be worth $\text{Max}(S_2 - X, 0)$, which will be either $56.3 = \text{Max}(156.3 - 100, 0)$, or $0 = \text{Max}(100 - 100, 0) = \text{Max}(64 - 100, 0)$.

Next, we step back in time to $t = 1$, and calculate the value of the call at that date. It is easy to see that if the current stock price is $80, the call will be worthless because there is no way for the stock to rise above the $100 exercise

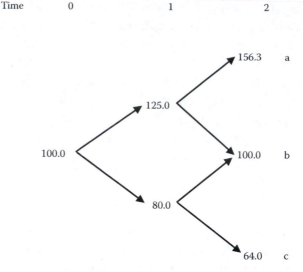

EXHIBIT 15.13 Stock price path over two periods with binomial model.

price by $t = 2$. On the other hand, if the price at $t = 1$ is \$125, it could move to either \$100 or \$156.3 by $t = 2$, so the option will have value.

We use the familiar binomial model at $t = 1$ based on the price of \$125 and the possible moves to \$100 or \$156.3 in the next step. To determine the value of the call at $t = 1$, we can either modify the one-step models shown in Exhibit 15.9 or 15.10 by changing the initial stock price to \$125, or we can go directly to the solution by using the CRR formula (Equation 15.3). Using the CRR formula for the time 1 value, we have

$$C = [pC_u + (1 - p)C_d]e^{-r(T)}$$
$$= [0.5584(56.25) + 0.4416(0)](0.9512) = \$29.88,$$

where the risk neutral probabilities $p = 0.5584$ and $1 - p = 0.4416$ are calculated the same as they were earlier for the single-step problem. Now we know that if the price of the stock at $t = 1$ is \$125, the value of the call will be \$29.88, given that it expires at $t = 2$ with an exercise price of $X = \$100$. Next, we need to take another step back in time from $t = 1$ to $t = 0$.

Now, step back to $t = 0$ when the stock is priced at \$100 and there are two periods to go until the call can be exercised. If the stock moves from

$100 to $125, the call will be worth $29.88 at $t = 1$ as just shown. If the stock moves to $80, the call will be worthless. Once again, we use the CRR formula with the risk neutral probabilities of $p = 0.5584$ and $(1 - p) = 0.4416$, to calculate the value of the call at $t = 0$ as

$$C = [pC_u + (1 - p)C_d]e^{-r(T)}$$
$$= [0.5584(29.88) + 0.4416(0)](0.9512) = \$15.87.$$

For the two period model, we now have the values of the call at each node. The call starts out at $t = 0$ with a value of $15.87, and then either goes to $29.88 or 0 at $t = 1$, depending on what happens to the stock. If $S_1 = \$125$, the call is worth $29.88, but otherwise it remains worthless because there is no chance the stock will rise above $100 by the expiration date at $t = 2$.

Reviewing the solution method for the multi-period problem: We start at the end, when the option will expire, and work backward to the present. Next, we want to apply this method to the case when we use small time increments so we get a better approximation to the continuous movement of stock prices over time.

15.3.3.2 Modeling Option Prices with the Binomial Approximation of Continuous Prices

In Chapter 13 we discussed the model of stock prices where the returns on the stock follow a continuous Weiner process. In that model, over a short interval of time, dt, the return on the stock is generated by the process expressed as

$$\text{Return} = \frac{dP}{P} = \mu dt + \sigma z\sqrt{dt}, \tag{15.6}$$

where P is the price, dP the change in the price, dP/P the rate of return, μ the expected return per period, σ the standard deviation of return per period, and z the standard normal random variable with mean 0 and standard deviation 1. Having introduced the continuous time model, we showed how to use the binomial model to approximate the continuous movement of prices. This approximation chops time into smaller and smaller increments and allows prices to follow a binomial process over these very short intervals. The result is that when the stock price makes small binomial jumps over short time intervals, the pattern of prices over longer periods, made up of many small jumps, closely resembles the pattern that would result from the continuous Weiner process. Now, we use this approximation to model option prices. We use the binomial option model that was introduced in the previous section where we start at the end of the process and work backward, with the time interval for each step being short relative to the time until the option expires.

To approximate the continuous Weiner process of stock returns using the up and down binomial process, the up and down movements are specified as a function of the standard deviation of the continuous return on the stock. These relationships, developed by Cox, Ross, and Rubinstein (1979), are expressed as

$$u = e^{\sigma\sqrt{dt}} \tag{15.7}$$

$$d = e^{-\sigma\sqrt{dt}} = \frac{1}{u} \tag{15.8}$$

and

$$p = \frac{e^{\mu\, dt} - d}{u - d} = \text{probability of an up move, } u, \tag{15.9}$$

where σ is the standard deviation (volatility) of the annual rate of return on the stock, dt the length of the time interval, which in this context is a discrete time interval such as one day or one week, p the risk neutral probability of the stock making the up move, and $(1 - p)$ the risk neutral probability of the stock making the down move in the short, discrete time interval, dt.

To model the options, we will use the same numerical example of stock prices that we used in Chapter 13. The stock is currently priced at $100 and its returns are assumed to follow the process described by Equation 15.6 with the mean return per year on the stock of $\mu = 0.12$, and the standard deviation of annual return is $\sigma = 0.20$. Based on Equations 15.7 and 15.8, using intervals of one week so that $dt = 1/52 = 0.0192$, the stock movement each week will be either $u = 1.0281$ or $d = 0.9726$, with the risk neutral probability of the price moving up calculated as

$$p = \frac{e^{\mu\, dt} - d}{u - d} = \frac{e^{0.12(0.0192)} - 0.9726}{1.0281 - 0.9726} = 0.535, \tag{15.10}$$

and the risk neutral probability of the price moving down by d is $(1 - p) = 0.465$. The call option on the stock expires in 13 weeks, and has an exercise price of $X = \$100$.

Exhibit 15.14A shows the Excel model of the stock price and the option, and Exhibit 15.14B shows the formulas in selected cells of the model. The data input is in A6:B22, and A24:B31 calculate the step length, the weekly stock price step sizes $u = 1.0281$ (B27) and $d = 0.9726$ (B28). The risk neutral probabilities of the up and down moves are in B30 and B31.

Based on this input, the stock will move up by $u = 1.0281$ or down by $d = 0.9726$ each week. Starting from the current price of $100 per share, the price will move to the values shown by the tree of binomial prices in F11:S24. In the first week the price can move to either $102.81 (G11) or $97.26 (G12). If the price at $t = 1$ is $102.81, it can either move up to $105.70 (H11), or down to $100 (H12). If the stock goes down to $97.26 at $t = 1$, it can either move up to $100 (H12), or down to $94.60 (H13). This process continues each week with the price moving up or down to the prices shown in this triangular matrix. The top row (row 11) is the sequence of stock prices moving up each week ($S_t = S_0 u_1 u_2 \ldots u_t$). The bottom diagonal represents the sequence of prices dropping each period ($S_t = S_0 d_1 d_2 \ldots d_t$). The cells in between are the other binomial branches. At each date t, there are $t + 1$ possible prices. So, by week 13 there are 14 possible prices ranging from $143.41 (S11) to $69.73 (S24).

Corresponding to each possible stock price at each date is a call value. These are shown in the F29:S42. For example, the call value of $32.237 in R30 corresponds to the stock value in the same relative location in the stock price matrix; that is, $131.96 in R12. The call value calculations start at the expiration date in week 13 and work backward. Column S corresponds to the expiration date, and the call values in S29:S42 are calculated as the intrinsic value, $Max(S_{13} - X, 0)$. The call values in each earlier period are calculated according to the CRR binomial formula

$$C_t = [pC_{u,t+1} + (1 - p)C_{d,t+1}]e^{-r(T)}, \qquad (15.11)$$

where C_t is the call value at time t, and $C_{u,t+1}$, and $C_{d,t+1}$ are the call value at the next date given that the stock went up or down. For example, suppose the stock price in week 12 is $131.96 (R12), and can move either up to $135.67 (S12) or down to $128.35 (S13). At the expiration date, the value of the call will be either $C_{u,13} = Max(135.67 - 100, 0) = \$35.67$ (S30)

Option Value With Binomial Approximation

$$\frac{dP}{P} = \mu\,dt + \sigma\,z\sqrt{dt}$$

Data Input

Stock Data	
Initial Stock Price	100
Annual Data	
Risk Free Interest Rate	5%
Mean Return	12%
Standard Deviation Per Year	20%
Option Data	
Type	Call
Exercise Price	100
Time to Expiration	13
Length of Sub-Period	1 Week
Sub-Periods Per Year r	52
Step Length	0.0192
(Step Length)$^{1/2}$	0.1387
Up Wealth Relative Return	1.0281
Down Wealth Relative Return	0.9726
Up Probability	0.535
Down Probability	0.465

Tree of Binomial Stock Prices

Step	0	1	2	3	4	5	6	7	8	9	10	11	12	13
Time	0	0.02	0.04	0.06	0.08	0.10	0.12	0.13	0.15	0.17	0.19	0.21	0.23	0.25
Stock Prices	100	102.81	105.70	108.68	111.73	114.88	118.11	121.43	124.84	128.35	131.96	135.67	139.49	143.41
		97.26	100.00	102.81	105.70	108.68	111.73	114.88	118.11	121.43	124.84	128.35	131.96	135.67
			94.60	97.26	100.00	102.81	105.70	108.68	111.73	114.88	118.11	121.43	124.84	128.35
				92.02	94.60	97.26	100.00	102.81	105.70	108.68	111.73	114.88	118.11	121.43
					89.50	92.02	94.60	97.26	100.00	102.81	105.70	108.68	111.73	114.88
						87.05	89.50	92.02	94.60	97.26	100.00	102.81	105.70	108.68
							84.67	87.05	89.50	92.02	94.60	97.26	100.00	102.81
								82.35	84.67	87.05	89.50	92.02	94.60	97.26
									80.10	82.35	84.67	87.05	89.50	92.02
										77.91	80.10	82.35	84.67	87.05
											75.78	77.91	80.10	82.35
												73.71	75.78	77.91
													71.69	73.71
														69.73

=F11*$B$27
=F11*$B$28
=($B$30*G29+$B$31*G30)*EXP(-$B$12*$B$24)

Tree of Binomial Call Values

Step	0	1	2	3	4	5	6	7	8	9	10	11	12	13
Time	0	0.02	0.04	0.06	0.08	0.10	0.12	0.13	0.15	0.17	0.19	0.21	0.23	0.25
Call Values	5.748	7.408	9.375	11.643	14.185	16.957	19.908	22.987	26.165	29.430	32.785	36.282	39.774	43.41
		3.852	5.162	6.788	8.745	11.028	13.602	16.410	19.383	22.467	25.635	28.891	32.237	35.67
			2.354	3.305	4.552	6.140	8.093	10.404	13.026	15.879	18.872	21.947	25.106	28.35
				1.266	1.878	2.737	3.909	5.454	7.412	9.775	12.473	15.377	18.361	21.43
					0.565	0.895	1.396	2.141	3.216	4.711	6.694	9.162	11.979	14.88
						0.187	0.320	0.543	0.910	1.504	2.442	3.873	5.942	8.68
							0.035	0.065	0.122	0.229	0.429	0.803	1.502	2.81
								0.000	0.000	0.000	0.000	0.000	0.000	0.00
									0.000	0.000	0.000	0.000	0.000	0.00
										0.000	0.000	0.000	0.000	0.00
											0.000	0.000	0.000	0.00
												0.000	0.000	0.00
													0.000	0.00
														0.00

=($B$30*$30+$B$31*S31)*EXP(-$B$12*$B$24)
=MAX($12-$B$18,0)

EXHIBIT 15.14A   Call valuation model based on binomial approximation of continuous price changes.

EXHIBIT 15.14B   Selected cell formulas for Exhibit 15.14A.

#	A	B	C	E	F	G	R	S
1	Option Value With Binomial Approximation							
2								
3	$\dfrac{dP}{P} = \mu\,dt + \sigma z \sqrt{dt}$							
4								
5								
6	Data Input				Tree of Binomial Stock Prices			
7								
8	Stock Data			Step	0	1	12	13
9	Initial Stock Price	100		Time	0	=F9+$B$24	=Q9+$B$24	=R9+$B$24
10				Stock				
11	Annual Data			Prices	=B9	=F11*$B$27	=Q11*$B$27	=R11*$B$27
12	Risk Free Interest Rate	0.05				=F11*$B$28	=Q11*$B$28	=R11*$B$28
13	Mean Return	0.12					=Q12*$B$28	=R12*$B$28
14	Standard Deviation Per Year	0.2					=Q13*$B$28	=R13*$B$28
15							=Q14*$B$28	=R14*$B$28
16	Option Data						=Q15*$B$28	=R15*$B$28
17	Type		Call				=Q16*$B$28	=R16*$B$28
18	Exercise Price	100					=Q17*$B$28	=R17*$B$28
19	Time to Expiration	13					=Q18*$B$28	=R18*$B$28
20							=Q19*$B$28	=R19*$B$28
21	Length of Sub-Period		1 Week				=Q20*$B$28	=R20*$B$28
22	Sub-Periods Per Year	52					=Q21*$B$28	=R21*$B$28
23							=Q22*$B$28	=R22*$B$28
24	Step Length	=1/B22						=R23*$B$28
25	(Step Length)$^{1/2}$	=SQRT(B24)						
26					Tree of Binomial Call Values			
27	Up Wealth Relative Return	=EXP(B14+B25)		Step	0	1	12	13
28	Down Wealth Relative Return	=1/B27		Time	0	=F27+$B$24	=Q27+$B$24	=R27+$B$24
29	Up Probability	=((EXP(B13*B24)-B28)/(B27-B28))		Call	=($B$30*G29+$B$31*G30)*EXP(-$B$12*$B$24)	=($B$30*H29+$B$31*H30)*EXP(-$B$12*$B$24)	=($B$30*S29+$B$31*S30)*EXP(-$B$12*$B$24)	=MAX(S11-$B$18,0)
30	Down Probability	=1-B30		Values		=($B$30*H30+$B$31*H31)*EXP(-$B$12*$B$24)	=($B$30*S30+$B$31*S31)*EXP(-$B$12*$B$24)	=MAX(S12-$B$18,0)
31								=MAX(S13-$B$18,0)
32								=MAX(S14-$B$18,0)

or $C_{d,13} = \text{Max}(128.35 - 100,0) = \$28.35$ (S31). Stepping back to week 12 (column R), the call value will be

$$C_{12} = [pC_{u,13} + (1 - p)C_{d,13}] \, e^{-r(T)} = [0.535(35.67) + 0.465(28.35)] = \$32.237,$$

as shown in cell R30. All the call values for week 12 are calculated similarly in column R. Then the process steps back to week 11 in column Q and calculates call value according to Equation 15.11, where the $C_{u,t+1}$ and $C_{d,t+1}$ refer to the call values already calculated for week 12 in column R. We continue to step back week by week until we reach the current call value of $C = \$5.748$ in F29.

In this example, we used step lengths of one week so as to limit the size of the problem in the discussion and exhibits. A more accurate value would result if we cut time into smaller increments. For example, for the call expiring in 13 weeks, we could easily build our model with daily price moves. There would be 65 trading days, and the step length would be $1/260 = 0.0038$ years. The model in Exhibit 15.14A would be expanded to 65 columns instead of 13. The structure would be exactly the same, with the same formulas, we just add columns and rows. By the 65th week there would be 66 possible stock prices, so we would need 66 rows for stock prices and 66 rows for call prices. We will leave it as an exercise for you to calculate the value of the call with the same features as our example, but with daily price changes.

15.4 OPTION PRICING WITH CONTINUOUS STOCK RETURNS: THE BLACK–SCHOLES MODEL

In the previous section, we developed a binomial model of option prices, where we approximated the continuous process with binomial stock price movements. In this section we take the next step and work with the famous Black–Scholes (1973) option model that is based on the assumption that stock returns follow the continuous Weiner process described as

$$\text{Return} = \frac{\mathrm{d}S}{S} = \mu\,\mathrm{d}t + \sigma z\sqrt{\mathrm{d}t}, \tag{15.12}$$

which we introduced in Chapter 12.

The Black–Scholes model of option prices starts with the assumption that the stock returns follow the continuous Weiner process (Equation 15.12) and derives the formula expressing the value of a European call based on the current value of the stock, S_0, the annual rate of volatility of the stock return, σ, the risk-free interest rate, r, the exercise price for the call, X, and the time until expiration, T:

$$C = S_0 N(d_1) - X e^{-r(T)} N(d_2),$$ (15.13)

where

$$d_1 = \frac{\ln(S_0/X) + \left[r + (\sigma^2/2)\right]T}{\sigma\sqrt{T}}$$ (15.14)

$$d_2 = \frac{\ln(S_0/X) + \left[r - (\sigma^2/2)\right]T}{\sigma\sqrt{T}} = d_1 - \sigma\sqrt{T}.$$ (15.15)

The subscripts on d_1 and d_2 do not denote time, they simply distinguish between the two formulas; and the terms $N(d_1)$ and $N(d_2)$ denote the cumulative standard normal probability functions. That is, $N(d_1) = \text{Prob}(x \le d_1)$, where x is a normally distributed random variable with mean 0 and standard deviation 1. In this formula for the cumulative probability, you are asking: "What is the probability that the random variable x will be less than d_1?" The Excel function for the standard normal variable is =NormSDist(d_1), where d_1 is a number or cell reference, or even another formula.

If you have not worked with the Black–Scholes formula before, it may look rather intimidating. However, it is fairly straightforward, and it is very easy to model it in Excel. Exhibit 15.15 shows the Black–Scholes option value calculation. The input is the same example we have been using with a stock priced at $100 with mean annual return of 12% and standard deviation of annual return of 20%. The call has an exercise price of $100, with quarter year to expiration (it is a European call, so it is exercisable only at the expiration date).

In Exhibit 15.15, the input data are in A3:B16, and the Black–Scholes option formulas are shown in the block on the upper right. The computed results are shown in D22:D39, with the value of the option shown as $4.615 in D39. The Excel formulas used at each step of the calculation are shown to the right of the answers. The formulas could be combined for a more concise model, but the step-by-step calculations are shown so you can follow it more easily. For example, "ln(S/X)" is used to calculate d_1 and d_2, and is calculated in D22 as =LN(B6/B15).

	A	B	C	D	E	F	G
1			Black-Scholes Option Value				
2							
3	Data Input						
4							
5	Stock Data			$C = S_0\,N(d_1) - Xe^{-rT}N(d_2)$			
6	Initial Stock Price	100		where			
7							
8	Annual Data			$d_1 = \dfrac{\ln\left(\dfrac{S_0}{X}\right) + (r+\dfrac{\sigma^2}{2})T}{\sigma\sqrt{T}}$			
9	Risk Free Interest Rate	5%					
10	Mean Stock Return	12%					
11	Standard Deviation Per Year	20%					
12							
13	Option Data			$d_2 = \dfrac{\ln\left(\dfrac{S_0}{X}\right) + (r-\dfrac{\sigma^2}{2})T}{\sigma\sqrt{T}} = d_1 - \sigma\sqrt{T}$			
14	Type	Call					
15	Exercise Price	100					
16	Time to Expiration (Years), T	0.25					
17				$\dfrac{dS}{S} = \mu\,dt + \sigma z\sqrt{dt}$			
18							
19							
20							
21							
22		$\ln(S/X)$		0.0000	=LN(B6/B15)		
23		$r + (\sigma^2/2)$		0.0700	=B9+(B11^2)/2		
24		$r + (\sigma^2/2)\,T$		0.0175	=D23*B16		
25		$\sigma(T^{1/2})$		0.1000	=B11*SQRT(B16)		
26							
27		d_1		0.1750	=(D22+D23*B16)/D25		
28		d_2		0.0750	=D27-D25		
29							
30		$N(d_1)$		0.5695	=NORMSDIST(D27)		
31		$N(d_2)$		0.5299	=NORMSDIST(D28)		
32							
33		$S\,N(d_1)$		56.9460	=B6*D30		
34							
35		PV of $1 @ r		0.9876	=EXP(-B9*B16)		
36		$PV(X)$		98.7578	=B15*D35		
37		$PV(X)\,N(d_2)$		52.3310	=D36*D31		
38							
39		Call Value		4.6150	=D33-D37		
40							

EXHIBIT 15.15 Excel model for Black–Scholes option formula.

So now you say to yourself, "Wow! I programmed the famous Black–Scholes option formula in Excel.—So what? What good does that do me? How can I use it?"

First answer: Congratulations. Second answer: If you have the required formula inputs, you are able to easily calculate an approximate* value for the

* The Black–Scholes formula has been shown to have various systematic biases so that the computed value departs systematically from the observed market prices. For more detailed discussions see Cox and Rubinstein (1985, pp. 338–342), Dubofsky and Miller (2003, pp. 532–547), and Hull (2009, Chapter 18).

option. Even if the formula is not exact in real-world applications, it is close enough to give us a good indication of what the option should be worth. If the market price differs greatly from the computed value, it may indicate that there is profitable trading opportunity. On the other hand, it may also indicate that the numbers you plugged into the formula are wrong.

Let us look more closely at the figures you need to plug into the formula. One of the significant things about this and other option formulas is that most of the inputs are readily observable and not subject to guesses or forecasts. There is only one item in the formula that you cannot find out by simply consulting the newspaper: the volatility of the stock. The list of observable inputs are the stock price, S_0, the exercise price of the call, X, the time to expiration, T, and the risk-free interest rate. The item you cannot directly observe is the volatility of the stock return, σ. The items that you do not need to know or estimate include the expected return on the stock, or factors such as beta.

To use the formula, the one item we cannot directly observe and therefore must be estimated is the stock's volatility, σ—that is, the standard deviation of the return. What we would like to have is the volatility for the future period until the option expires. This we can never really know, so past observations of the returns are used to estimate the volatility. As we know, estimates are usually imperfect, and past volatility may not be a good predictor of future volatility. Nevertheless, volatility estimates from past data provide a good starting point and lead to useful valuation estimates.

One of the more interesting applications of the model is its use to estimate "implied volatility." Even though we cannot observe volatility, we can observe the actual option price in the market and the other inputs to the model. By plugging these variables into the model, we can work backward to solve for the volatility that is consistent with, or "implied" by, the observed option price and other variables. This gives us an estimate of the volatility that other investors must be attributing to the stock in the future. If the implied volatility differs from our own prediction, it may signal a trading opportunity—if we think we can make better predictions than the rest of the investors in the market. Implied volatility is also used to estimate volatility for comparable options. For example, we may not have much information about a particular option that is traded infrequently. We turn to other options of similar companies that have similar option features, and that are traded more actively. Using the implied volatility of these comparable companies, we can estimate the volatility for the less actively traded option, and calculate its value.

How do we extract the implied volatility from option prices? Some of the more statistically accurate methods are beyond the scope of this text, but for a good first approximation, we can use our Excel model to solve for the implied volatility. For example, suppose we have the same data as in our previous example, with the stock trading at a price of $100 per share, the exercise price of the call is $100, the call expires in 90 days (quarter year), and the risk-free interest rate is 5%. Unlike the previous example, the call is currently trading at a price of $6.00. What volatility must investors be attributing to the stock in order for the call to trade at a price of $6.00? We solve this easily with our model using Excel's Goal Seek with the following steps: Tools > Goal Seek brings up the Goal Seek dialog box. Fill it in with these entries: Set Cell: F19 (the call value cell), To Value: 6.00, By Changing Cell: B12 (the standard deviation input cell), click OK. The solution value for volatility will be in B12, which is now 27.04%. Assuming the Black–Scholes model is reasonably accurate, the call price of $6.00 implies that investors must be assuming the stock return volatility to be about 27%.

This demonstrates an important aspect of option valuation: the more volatile the stock, the more valuable the option. The reason for this is simply that if the stock price and return are more volatile, there is a higher probability that the call will be in the money when it expires. Exhibit 15.16 shows this link between volatility and call value,[*] where we see the value for the call in our example increasing from $1.73 when $\sigma = 5\%$, to $10.52 when $\sigma = 50\%$.

With this model you can explore many aspects of option valuation. In the exercises at the end of the chapter you will have the opportunity to see the impacts of changing the variables in the pricing model and gaining greater understanding of options.

15.5 SUMMARY

This chapter has introduced you to the basics of options and option pricing. We have presented several different models that allow you to understand options and to analyze their prices. The first section of the chapter explained what options are and showed how to model their payoffs. Once we understand the payoffs at expiration, we can then analyze different combinations of options, stock, and bonds. One of the most useful aspects

---

[*] The sensitivity of the option value to changes in volatility is referred to in the option literature as the "vega."

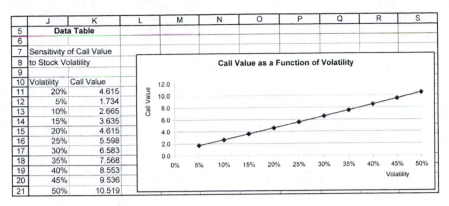

	J	K	L	M	N	O	P	Q	R	S
5	Data Table									
6										
7	Sensitivity of Call Value									
8	to Stock Volatility									
9										
10	Volatility	Call Value								
11	20%	4.615								
12	5%	1.734								
13	10%	2.665								
14	15%	3.635								
15	20%	4.615								
16	25%	5.598								
17	30%	6.583								
18	35%	7.568								
19	40%	8.553								
20	45%	9.536								
21	50%	10.519								

EXHIBIT 15.16   Volatility of stock returns and call value.

of our models is to see and understand how we can mix options with stock and bonds in a way to create a riskless investment. It is from this idea that we get option pricing models.

We looked at two kinds of option models. The first was the binomial model based on the assumption that stock prices will move either up or down. Although this is a very simple model, it not only helps us to understand option payoffs and values, but it can be fine-tuned to approximate the more realistic continuous time model.

The second type of option model is based on the assumption that stock prices move according to a Weiner process in continuous time. The Black–Scholes model is a continuous time model. While the formula looks foreboding at first, we found that it is quite easy to use in a spreadsheet, and we encourage you to construct your own model so you can explore the nuances of this useful formula.

Options, like so many other topics in this text, is a complex topic covered in many specialized articles and books. We cannot hope to cover all the interesting aspects of options in this chapter. But with this introduction to the topic, you should be well prepared to develop and extend your own models of options and option trading.

PROBLEMS

1 Binomial Model of Option Prices

Develop a model to analyze the values of both a call and a put when it is assumed that the stock price follows a binomial process over monthly intervals. Assume the initial stock price is $100 and you can either borrow or lend at the risk-free interest rate of 5% per year. Each month the price of a stock can make an up move of $u = 1.106$, or a down move of $d = 0.904$. The call has an expiration date at the end of 12 months, and the exercise price is $110. The put also expires in one year, and it has an exercise price of $110.

(a) Use your model to structure a mix of the stock and bond (borrowing or lending) so as to replicate the payoffs from the call at expiration. Based on this, how much should the call be worth one month prior to expiration? Is there just one call value at that date?

(b) Use the CRR formula to determine the value of the call at each date at each stock price over the 12-month horizon.

2 Black–Scholes Model

Develop a spreadsheet model of the Black–Scholes option model. Assume that stock prices follow a continuous Weiner process with a mean annual return of 18% and standard deviation of the annual return is 35%. Assume the stock price is currently $100 per share. The call will expire in six months. The exercise price is $110. The risk-free interest rate is 5% annually.

(a) Show the value of the call. Now explore the impact on call value of varying the assumptions.

(b) What happens to the value of the call as time passes? That is, keeping all other factors the same, what happens when you change the time to expiration? Graph the relationship between time to expiration and call value.

(c) What happens when you change the interest rate? Graph the relationship between the interest rate and call value. Briefly explain why you get this relationship.

PART VI

Optimization Models

CHAPTER **16**

Optimization Models for Financial Planning

16.1 INTRODUCTION TO OPTIMIZATION

The modern theory of finance seeks to define the optimal investment and financing decisions for a firm, where optimal is taken to mean maximizing the wealth of the firm's stockholders. The body of financial theory that has been developed is intended to explain the link between decisions and the value of the firm. If the effect of decisions on firm value can be explained, then we should be able to determine which particular policy or set of decisions will maximize the firm's value. The theory of finance, however, is still not so perfected that we have a completely accepted and verified theory of valuation. Consequently, we are not yet in a position to make perfect financial decisions that are the simple result of some calculation. Nevertheless, we do have much of the necessary valuation framework to help us describe the links between decisions and the objective of value maximization. Armed with that framework we should be able to improve financial decision making by constructing decision models that calculate optimal decisions. That is the topic of this chapter—the construction and use of optimization models to aid us in financial planning and decision making by providing the mechanism for calculating the optimal decisions in the context of our financial model.

Chapter 4 explained the construction and use of a simulation model of the firm. Simulation models are being used more widely all the time, and they are useful for exploring the consequences of decisions. However, simulation models have drawbacks: there may be too many decisions to evaluate, too much output for each decision; and, even when the output for each decision alternative is manageable, the output can overwhelm the user.

In addition, when a problem presents many different decision alternatives, the number of different decisions that can be considered may be too large for a person to analyze and evaluate each one. For example, suppose a firm has identified 11 investment projects and wishes to choose the best combination of projects to be included in the firm's capital budget. The portfolio of projects could contain one project, all 11 projects, or any number in between, making the number of different portfolios that could be considered 2047. Most decision makers would rather not be faced with the task of sifting through 2047 alternatives to find the best combination. The optimization approach forces the model user to carefully define the objective of the decision and create a summary statistic. Optimization models then systematically analyze all possible combinations and determine the combination that yields the optimal solution.

The dictionary defines an optimum as "the best or most favorable point," or "the best result obtainable under specific conditions." Optimization is the process of finding the optimum. That is, optimization is the process of searching or solving for a decision variable that will yield the best result obtainable under the conditions faced by the decision maker. For example, Equations 1.1 and 1.2 from Chapter 1 constitute a simple model of the firm's total revenue. They can be combined as one equation of the form

$$\text{Total Revenue} = 500 \times \text{Price per Unit} - 30 \times \text{Price per Unit}^2. \quad (16.1)$$

In this simple model, the decision variable is the Price per Unit. If the objective of the decision is to make Total Revenue as large as possible, the goal is finding the Price per Unit that generates the largest Total Revenue. Optimization is the process or technique of finding this optimal value.

Optimization techniques typically take one of two forms. Some search systematically and efficiently through all the alternatives; others use information about the conditions characterizing an optimum to determine the optimal solution directly. One type of search technique enumerates the various possible values of the decision variable and the value of the objective function that results from that decision, then searches through the list to determine which was best. For example, suppose the prices to be considered by model 16.1 range from $5 to $10 in increments of $1. The Total Revenue associated with each Price per Unit are

Price per Unit ($)	5	6	7	8	9	10
Total Revenue ($)	1750	1920	2030	2080	2070	2000

A search for the highest Total Revenue yields a value of $2080 at a Price per Unit of $8. There is no assurance, however, that there is not another Price per Unit between $7 and $9 that yields a higher Total Revenue. The search could be continued to consider prices in this second interval that might yield even better results.

One drawback of searching for the best alternative is that there may be too many alternatives, making the search process too time-consuming and expensive. In this case, we turn to another method for finding the optimum. One familiar method is to use calculus to determine an extreme point in a mathematical function such as Equation 16.1. By taking the derivative of Equation 16.1 with respect to the variable Price per Unit, equating it to zero and solving for Price per Unit, we obtain the optimal solution of $8.33. At this price, Total Revenue is $2083.33, the largest revenue possible. Any other price produces less revenue.

Exhibit 16.1 is a graph that shows Total Revenue (on the vertical axis) as a function of the Price per Unit (on the horizontal axis). The process of searching for a best price simply calculates values at different points along the curve. Each point represents the Total Revenue earned at a given Price per Unit. Using our knowledge that the maximum is reached when the slope of the curve is zero, calculus solves for that point.

While the techniques of calculus are useful for many problems, the problems of financial modeling are usually too complex to be formulated in a way that can be solved directly with calculus. Fortunately, there are other techniques that are well suited to determining an optimal solution to a financial planning problem. These techniques come under the heading of mathematical programming—LP, non-LP, and dynamic programming. This chapter focuses on the use of LP to find optimal solutions to financial planning problems.

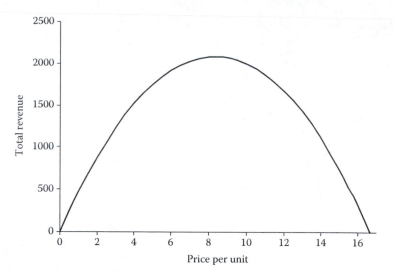

EXHIBIT 16.1   Revenue as function of price.

16.2 CONSTRAINED OPTIMIZATION

Mathematical programming is a technique for solving "constrained optimization" problems. Constrained optimization is the process of finding the optimal solution to a problem in the presence of constraints that limit the choices of the decision variables. For example, consider the problem of determining the quickest (optimal) way to get from New York to San Francisco. Clearly, the quickest way is to travel in a straight line at the speed of light—perhaps being beamed there by the Starship Enterprise. However, technology limits our ability to do that. That is, constraints limit our choices and, ultimately, our ability to reach our goal.

The problem should be restated, saying that we want to find the quickest way to get from New York to San Francisco subject to several constraints. The first constraint is existing technology. Though traveling at the speed of light is preferred, traveling by airplane, train, and automobile are the only modes available. Second, we may want to impose a constraint on our expenditures, limiting our travel expenses to $500. Finally, we want to make a stop in Bishop, California, to visit relatives. Given the transportation constraints, the airplane is clearly the fastest mode of transportation. However, the commercial airlines do not stop in Bishop, so we must consider chartering a plane. Chartering a plane, however, would cause us to exceed our expenditure constraint of $500. So we reject the airplane as a choice. Nor do trains go through Bishop. Therefore, we are constrained to traveling by car, if at all. The car would be a feasible choice if we can make the trip for no more than $500; if not, then the trip is not feasible given our constraints. Given these constraints, the problem becomes finding the quickest route by car that is within our budget.

The pricing problem described by Equation 16.1 can also be formulated as a constrained optimization problem. Assume the objective remains maximizing Total Revenue, but now the Price per Unit must be at least $9 in order to cover our costs. This problem is stated mathematically as

$$\underset{\{Price\ per\ Unit\}}{\text{Maximize Total Revenue}} = 500 \times \text{Price per Unit} - 30 \times \text{Price per Unit}^2 \tag{16.2}$$

subject to

$$\text{Price per Unit} \geq \$9.00 \tag{16.3}$$

The expression to the right of the equal sign in 16.2 is the objective function. It expresses the relationship between the decision variable, Price per Unit, and the objective, Total Revenue. Equation 16.3 is the constraint. It limits our choice of the decision variable to values that are $9 or greater. In Exhibit 16.1 it is obvious that a price of $9 will yield the largest revenue, because Total Revenue declines as the Price per Unit increases for all prices greater than $8.33, and $9 is the lowest price that can be charged without violating the constraint.

16.2.1 Linear Programming

Linear Programming (LP) is a method for formulating and solving certain kinds of constrained optimization problems. It is similar to the constrained maximization pricing example given above, in that the problem has an objective function that is to be either maximized or minimized, has one or more decision variables, and has one or more constraints that limit our choice among decision variables. The difference is that the objective function is linear. There are a number of different algorithms for solving LP problems that are quick and efficient. Quite large and involved problems can be solved easily on a computer. Consequently, if a problem can be formulated as an LP problem, it can usually be solved quite easily.

To understand what problems can be solved with LP, it helps to understand the general structure of an LP problem. An LP problem consists of a linear objective function and linear constraints. What do we mean by linear objective function and constraints? Linear means that the relationships can be drawn as straight lines or planes, rather than curves. Equation 16.4 is an example of a linear objective function.

$$\operatorname*{Max}_{\{X_i\}} Z = 1.17X_1 + 1.22X_2 + 0.47X_3 - 0.02X_4. \tag{16.4}$$

Each X_i represents a decision variable whose value is to be determined so as to maximize the value of the maximand (the value to be maximized), Z; the coefficients 1.17, 1.22, 0.47, and −0.02 are the contribution to the maximand per unit of their corresponding decision variables X_1, X_2, X_3, and X_4. What makes Equation 16.4 linear is that the decision variables are multiplied only by constant coefficients; none are multiplied by other decision variables, including themselves. For example, the expression $1.17X_1 + 1.22X_2$ is linear, but the expressions $1.17X_1X_2$ and $5X_1^2$ are not. The latter two expressions are drawn as curved relationships, whereas the former expressions are drawn as straight lines.

Many problems in economics and finance are non-linear. This was the case in the pricing problem presented earlier, in which the decision variable Price per Unit appeared in the objective function as a squared term, Price per Unit^2. This resulted in the objective function appearing on a graph as a curve (Exhibit 16.1). Another example would be if the objective is to maximize the value of equity, with the decision variables being the amounts of debt and equity financing. This will be a non-linear problem if the discount rate used to determine the value of equity depends on the amount of debt financing, making the coefficients of the debt and equity financing variables depend on each other in a complicated way. Because the problem would be non-linear, it could not be solved with the techniques of LP. It potentially could be solved as a non-LP problem, but the solution process would be more complicated. Fortunately, the increased power of the iterative solution algorithms packaged with today's spreadsheet software, such as Solver in Excel®, render this distinction relatively unimportant when dealing with relatively small-scale problems. Nevertheless, we will focus on linear problems in this chapter.

We now look at the general structure of an LP problem. The standard form for an LP problem is as follows:

$$\operatorname*{Max}_{\{X_i\}} Z = c_1x_1 + c_2x_2 + c_3x_3 + c_4x_4 \tag{16.5}$$

subject to

$$a_{11}x_1 + a_{12}x_2 + a_{13}x_3 + a_{14}x_4 \leq b_1$$
$$a_{21}x_1 + a_{22}x_2 + a_{23}x_3 + a_{24}x_4 \leq b_2 \qquad (16.6)$$
$$a_{31}x_1 + a_{32}x_2 + a_{33}x_3 + a_{34}x_4 \leq b_3,$$

where Equation 16.5 is the objective function, and the Equation 16.6 the constraints; the x_i are the decision variables whose optimal values are to be found by solving the problem; the c_is and a_{ij}s represent the numbers that are the coefficients (the subscripts distinguish among the coefficients so that, for example, a_{32} denotes the coefficient in the third row and second column); and the b_is the constant numbers that are the constraint limitations.

If the problem can be stated in this general form, then it can be solved as a LP problem. Solution algorithms such as the simplex method are usually used to solve the problem, and these algorithms are readily available as computer software. While understanding the solution process can enhance our understanding of the solution, we do not necessarily have to be knowledgeable in the technical details of the solution process to formulate and solve the problem. The reader who is interested in the mathematics of solving the linear program is referred to the operations research books cited in the References at the end of this book, such as Hillier and Hillier (2003) and Ragsdale (2001). We do not emphasize the solution methods here. Rather, we focus on structuring the model with the view toward allowing Excel's Solver to handle the computations.

A short example will help demonstrate how an LP problem is formulated. Assume that your task is to choose how to allocate your investment budget among several alternative investment proposals, and how to finance the investment program. Your decisions are constrained (limited) by limits on your budget, borrowing capacity, and the size of certain investments.

Assume that your objective is to maximize the NPV of your investment portfolio. There are four different investment projects that can be undertaken (projects 1, 2, 3, and 4). Project 1 generates future cash flows with an NPV of $1.17 for each $1 invested; project 2 generates an NPV of $1.22 for each $1 invested; project 3 generates $0.99 for each $1 invested; and project 4 generates $0.47 per dollar invested.

All four projects are attractive, as each has a positive NPV. So you would like to invest as much money as possible. However, there are a number of constraints that limit your choice of investments: you can invest no more than $500 of your own money; you can borrow no more than an additional $200; and you can invest no more than $100 in project 1 and $80 in project 2. These constraints must be taken into account in deciding the best mix of investments and financing.

This decision problem can be set up and solved as a linear program consisting of an objective function (containing five decision variables) and nine constraints. We now develop these expressions. First, we define the decision variables: let Invest 1, Invest 2, Invest 3, and Invest 4 be the amounts to be invested in projects 1, 2, 3, and 4, respectively; and let Borrow represent the amount to be borrowed. The problem is to choose the values of the decision variables Invest 1, Invest 2, Invest 3, Invest 4, and Borrow that will maximize our objective, the NPV of the cash flows from our investment portfolio.

The objective function is the mathematical expression that shows how each decision variable affects the objective. In this case the objective function is written as

$$Z = 1.17 \times \text{Invest } 1 + 1.22 \times \text{Invest } 2 + 0.99 \times \text{Invest } 3$$
$$+ 0.47 \times \text{Invest } 4 - 0.02 \times \text{Borrow}, \tag{16.7}$$

where the coefficient of each decision variable represents the NPV per dollar invested, or in the case of borrowing, the NPV per dollar borrowed.* For example, each dollar invested in project 1 yields $1.17 in NPV, and the decision variable Invest 1 is the number of dollars invested in project 1. Therefore, the product 1.17 × Invest 1 represents the NPV of the cash flows generated by the total investment in project 1. If $300 is invested in project 1, the NPV derived from the investment is

$$1.17 \times \text{Invest } 1 = 1.17 \times 300$$
$$= \$351. \tag{16.8}$$

---

* The borrowing coefficient is the NPV of principal and interest repaid on the debt. It is assumed that the interest rate on the debt is 12% and the discount rate is 10%, which for one period borrowing yields the NPV per dollar borrowed of +$1.00 − ($1.00 × 1.12)/1.10 = −0.02, where the +$1.00 indicates the inflow of $1.00 when the money is borrowed, and −$1.00 × 1.12 represents the outflow of principal and interest when the debt is repaid one period later.

Next, we write the constraint equations, Equation 16.9.

$$
\begin{array}{lr}
\text{Invest 1} + \text{Invest 2} + \text{Invest 3} + \text{Invest 4} - \text{Borrow} & \leq 500 \\
\text{Invest 1} & \leq 100 \\
\text{Invest 2} & \leq 80 \\
\text{Borrow} & \leq 200 \\
\text{Invest 1} & \geq 0 \qquad (16.9) \\
\text{Invest 2} & \geq 0 \\
\text{Invest 3} & \geq 0 \\
\text{Invest 4} & \geq 0 \\
\text{Borrow} & \geq 0
\end{array}
$$

The first expression in Equation 16.9 constrains expenditures, net of borrowing, to be no greater than $500. (When the variable Borrow is transposed to the right side the gross expenditures are $500 plus the amount borrowed.) The next two expressions limit the expenditures for projects 1 and 2 to no more than $100 and $80, respectively. The fourth expression limits the amount borrowed to no more than $200. The last five expressions force the decision variables to be non-negative.

Combining the objective function and constraints yields Equation 16.10.

$$
\underset{\{\text{Invest } i\}}{\text{Max}} \; Z = 1.17 \times \text{Invest 1} + 1.22 \times \text{Invest 2} + 0.99 \times \text{Invest 3}
$$
$$
+ 0.47 \times \text{Invest 4} - 0.02 \times \text{Borrow}
$$

subject to

$$
\begin{array}{lr}
\text{Invest 1} + \text{Invest 2} + \text{Invest 3} + \text{Invest 4} - \text{Borrow} & \leq 500 \\
\text{Invest 1} & \leq 100 \\
\text{Invest 2} & \leq 80 \\
\text{Borrow} & \leq 200 \\
\text{Invest 1} & \geq 0 \qquad (16.10) \\
\text{Invest 2} & \geq 0 \\
\text{Invest 3} & \geq 0 \\
\text{Invest 4} & \geq 0 \\
\text{Borrow} & \geq 0
\end{array}
$$

Using Solver in Excel, the solution is

$$
\begin{aligned}
\text{Invest 1} &= \$100, \\
\text{Invest 2} &= \$80, \\
\text{Invest 3} &= \$520, \\
\text{Invest 4} &= \$0, \text{ and} \\
\text{Borrow} &= \$200.
\end{aligned}
$$

With these decision variables, the optimized objective (the NPV of the investment program) is $Z = \$725.40$. Any other feasible solution to the problem would yield a lower NPV.

This problem is so simple that one does not need a linear program to determine that projects 1 and 2 should be undertaken to their limits because they yield the greatest NPV per dollar invested, that project 3 should be expanded as far as borrowing allows, and that borrowing should be undertaken to its limit. However, the problem demonstrates LP and highlights some of the features of the technique. It is helpful to understand the general structure of an LP problem such as was shown in Equations 16.5 and 16.6. However, in actual practice, we may not have to put the problem in that format. Some computer software oriented to LP can handle messier problems and still solve them. For example, the Excel add-in, Solver, has the capability to treat a spreadsheet problem as a linear program, and solve it even though it is not specifically formulated as a linear programming problem.

16.2.2 Using Solver to Find a Solution

This investment problem can be set up as an Excel spreadsheet and solved for the optimal solution using the Excel add-in "Solver." Exhibit 16.2 is an Excel spreadsheet that shows the data for this problem. Cells E9–I9 represent the decision variables whose value will be determined

	A	B	C	D	E	F	G	H	I	J	K	L
1												
2						Capital Investment Problem						
3												
4												
5			Investment Project		1	2	3		4	Borrowing		
6												
7		Decision Variables										
8			Name of Variable		Invest1	Invest2	Invest3	Invest4	Borrow			
9			Amount of Variable		0	0	0	0	0			
10												
11		NPV per $1 Invested			1.17	1.22	0.99	0.47	-0.02			
12												
13		Objective		0								
14												
15											Constraint	Constraint
16											Usage	Limits
17		Expenditure Constraints										
18		Project 1 Constraint			100	80	520	0	-200		500	500
19		Project 2 Constraint			100						100	100
20		Borrowing Constraint				80					80	80
21		Non-Negativity Constraints						200			200	200
22					100						100	0
23						80					80	0
24							520				520	0
25								0			0	0
26									200		200	0
27												

EXHIBIT 16.2 Spreadsheet for capital investment optimization.

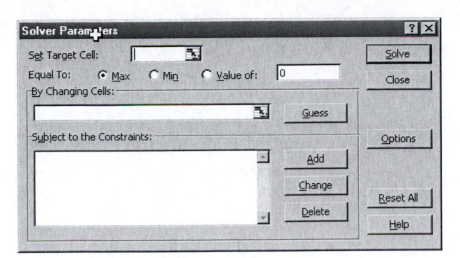

EXHIBIT 16.3   Solver dialog box.

by Solver; cells E11–I11 the coefficients of the decision variables in the objective function, whose values represent the NPVs of $1 invested or borrowed. Cell C13 is the objective function. The formula in C13,

$$=SUMPRODUCT(E9:I9,E11:I11),$$

multiplies each cell in row 9 by its respective cell in row 11, then adds the products, an operation known as vector multiplication. In this case, the result is

$$=E9*E11+F9*F11+G9*G11+H9*H11+I9*I11.$$

Solver provides an optimal solution to the problem by finding the values of the decision variables (E9–I9) that maximize the objective function, C13, subject to the constraints specified in column L, lines 17–25.

Having set up the spreadsheet, we need to tell Solver which cells represent the decision variables, the objective function, and the constraints. We access Solver by choosing Solver from the Tools menu. The Solver dialog box will be displayed as shown in Exhibit 16.3.*

We want Solver to work on the objective function, cell C13 in this example. So C13 is the "Target Cell." To select it as the target cell, click on

---

* If Solver does not appear in Excel's "Tools" menu, select "Add-Ins" and check the "Solver Add-In box."

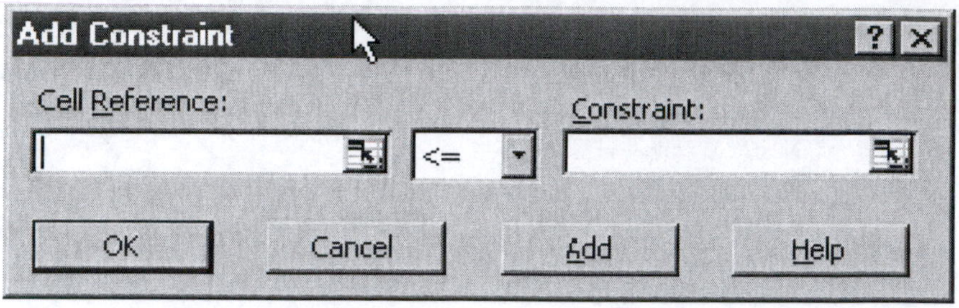

EXHIBIT 16.4   Constraint dialog box.

the square to the right of the "Set Target Cell" box, then click on cell C13, then the square again. C13 should now appear in the box to the right of "Set Target Cell." We select "Max" to indicate that the objective function is to be maximized, and we select the range of cells that are the decision variables for insertion in the "By Changing Cells" blank (cell range E9:I9). To specify the constraints, we click "Add" to obtain the constraint dialog box shown in Exhibit 16.4.

"Cell Reference" is the cell, or range of cells, that contains the formula for the left side of the constraint (cell range J17:J25) and the "Constraint" refers to right side of the constraint equation that is the fixed amount of resource or budget (cell range L17:L25). The middle box is used to specify the direction of the inequality of the constraint. It can also be used to specify an equality or to define the decision variable to be an integer or a binary number (0 or 1). The completed Solver parameter dialog box is shown in Exhibit 16.5.

Click "Solve," and Solver quickly computes the LP solution, with the optimal values of the decision variables inserted in their cells in the spreadsheet as shown in Exhibit 16.6.

The decision variable cells E9–I9 show the amounts to be invested in the four projects as 100, 80, 520, and 0, and the amount to be borrowed as 200. This yields the objective function value of 725.40 (cell C13). All the constrained budget amounts and the borrowing are used up to their limits, as shown in the Constraint Usage column, cells J17–J25.

While this LP model of the capital budgeting problem can be useful, in its simplest form it is limited because it does not take into account the possibility of obtaining additional financing. However, it takes very little elaboration of the model to add various forms of financing, transforming

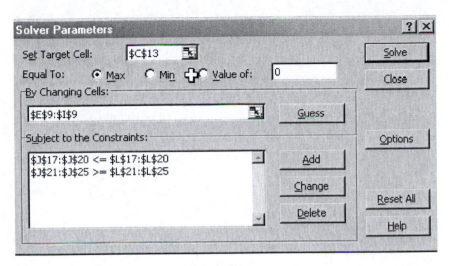

EXHIBIT 16.5   Completed solver dialog box.

it into a more general model for long-term financial planning. We will look at a more detailed example, learn more about using Solver and about interpreting the LP output as we look at a more detailed example in the next section.

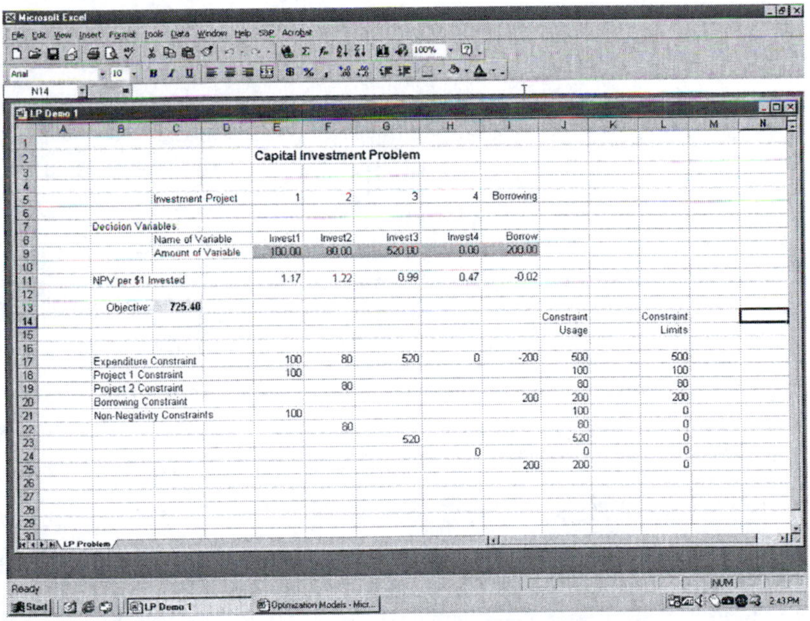

EXHIBIT 16.6   Optimal solution for capital investment problem.

16.2.3 Example: Investment Decisions on a Limited Budget

The capital investment problem solved earlier was a variation of Weingartner's (1967) LP approach to capital budgeting under conditions of capital rationing. In this section, we will look at a more detailed problem. Weingartner pioneered the application of LP to capital budgeting problems, and his formulation is one of the best-known and easiest to understand applications of LP to financial decision problems. Its objective is to choose the best set of investment projects given a limited capital budget.

It is a common mantra that a firm should adopt all the positive NPV projects it can identify. But this assumes a nearly unlimited capital budget. Sometimes, this is not the case. A firm's capital budget may be limited for several reasons: the officers of the firm may set a maximum limit on capital expenditures in each period; the capital expenditures may be limited to the amount of funds generated internally by the operations of the firm; management may assume that there is a limit on the amount of funds that can be raised externally in a given period; management may feel that there is a limit on the rate of expansion of the firm, that is, if the firm tries to expand too rapidly it will overload the firm's managerial resources; or, the officers of the firm may set a limit on the amount of funds that each division can spend for capital projects. Whatever the reason for constraining the capital budget, once the limit is specified the task is choosing the best mix of investment projects while simultaneously ensuring that the budget allocation is not exceeded.

An example will help to demonstrate the problems of capital rationing and the LP approach to solving these problems.

The CFO of the Gartner Corporation asked the managers of its different divisions to submit proposals for major capital expenditures so that he can decide how to allocate the firm's capital budget.

The divisional managers have screened the projects to see that they generally meet the firm's standards for acceptance, which is that each project should have a positive NPV when evaluated at a discount rate that is appropriate for their level of risk. The firm's cost of capital is 10%, and that is the rate used to evaluate the projects. The preliminary screening process has narrowed the field to just four major projects, and now the decision must be made as to which of the various projects should be adopted.

The major project proposals are all quite detailed and involved, but the staff at headquarters has summarized the relevant data relating to the costs and expected cash flows of the projects. Exhibit 16.7 shows this summary data. The four proposals are designated as projects 1–4. Exhibit 16.7 shows the estimated cash flows for each period for each project (expressed in thousands of dollars). Negative numbers indicate cash outlays; positive numbers indicate cash inflows. Cash flows are assumed to occur at the beginning of each period, so that period 1 cash flows are assumed to occur at time $t = 0$ (immediately), and period 2 cash flows are assumed to occur at the beginning of period 2, or one year from now. Project 1 requires an immediate outlay of $120, and then generates positive flows until period 6. Projects 2, 3, and 4 require outlays in periods 1 and 2, but then generate positive cash flows until period 6. All four projects incur costs associated with closing down, hence the negative cash flows in period 6. The NPV and the IRR for each project are shown at the bottom of the exhibit, with NPV evaluated at the 10% cost of capital.

Because all four projects are acceptable with positive NPVs and IRRs greater than the cost of capital, the firm would accept all four if it did not face limits on capital expenditures set by the board of directors. The board has set a maximum of $500 on the first period's capital expenditures and $130 on the second period's capital expenditures. The problem is to decide which portfolio of projects will generate the highest total NPV for the firm without exceeding the capital budget.

The choice of the best portfolio of capital investment projects can be formulated as an LP problem. The objective is to choose the mix of projects

	Gartner Corporation			
	Project Cash Flows			
		Project		
	1	2	3	4
Period				
1	-120.0	-50.0	-90.0	-450.0
2	30.0	-80.0	-90.0	-70.0
3	60.0	200.0	80.0	480.0
4	80.0	100.0	130.0	401.0
5	181.0	50.0	259.0	100.0
6	-1.0	-3.0	1.8	-20.4
NPV	140.0	150.0	170.0	240.0
IRR	43%	68%	41%	29%
Discount Rate		10%		

EXHIBIT 16.7 Cash flows and NPV of investment projects.

that generates the highest NPV for the portfolio as a whole. The main constraints are that the limits on capital expenditures should not be exceeded. The decision variables are which projects to adopt and which to reject.

16.2.3.1 Decision Variables

To set this model up in LP format, the variables must be defined. We number the projects 1–4, letting the subscript i designate a project, so that $i = 3$ refers to project 3. A decision variable is designated as X_i, where $X_i = 1$ if project i is adopted, 0 if it is not. For example, $X_3 = 1$ in the solution means project 3 should be adopted; $X_3 = 0$ means project 3 should be rejected. With four projects ($i = 1, 2, 3, 4$) there will be four decision variables, X_1, X_2, X_3, and X_4, each one of which will be either 0 or 1 in the solution.

16.2.3.2 Objective Function

Let NPV_i denote the NPV of project i. For example, for project 3 in Exhibit 16.7, $NPV_3 = \$170$; for project 4, $NPV_4 = \$240$. Using this notation, the NPV of the portfolio of projects can be expressed in general form as

$$Z = \sum_{i=1}^{4} NPV_i \times X_i \qquad (16.11)$$

or, explicitly as

$$Z = 140X_1 + 150X_2 + 170X_3 + 240X_4. \qquad (16.12)$$

The numbers shown as the coefficients of the X_i are the NPVs of the projects, NPV_i. The product $NPV_i \times X_i$ represents the contribution of project i to the NPV of the portfolio of projects. For example, if in the solution $X_3 = 1$, we have

$$NPV_3 \times X_3 = 170 \times 1$$
$$= \$170. \qquad (16.13)$$

if $X_4 = 0$, we have

$$NPV_4 \times X_4 = 240 \times 0$$
$$= 0. \qquad (16.14)$$

The sum, $\sum_{i=1}^{4} \mathrm{NPV}_i \times X_i$, is the NPV for the whole portfolio. Suppose, for example, the solution values of the decision variables are $X_1 = X_3 = 1$ and $X_2 = X_4 = 0$, then the NPV for the portfolio would be

$$\sum_{i=1}^{4} \mathrm{NPV}_i \times X_i = 140X_1 + 150X_2 + 170X_3 + 240X_4$$
$$= 140 \times 1 + 150 \times 0 + 170 \times 1 + 240 \times 0$$
$$= 310. \tag{16.15}$$

Because all four projects have positive NPV, the portfolio NPV could be made larger by adopting all four projects ($X_i = 1$ for $i = 1, 2, 3, 4$), in which case the portfolio NPV would be $700. However, the essence of the problem is that not all projects can be adopted due to budget limitations.

16.2.3.3 Constraints

Some type of budget constraint is present in virtually all financial LP models. In the mathematical programming context, the limitations imposed by the budget are expressed as constraint equations or inequalities. The budget constraint for a given period t is the mathematical statement that the amount spent for capital investment cannot exceed the budget limit. For period t, the budget constraint is the mathematical expression

$$\text{Total Capital Expenditure}_t \leq \text{Budget Ceiling}_t. \tag{16.16}$$

To express the budget constraint, define Expend_{ti} as the net cash outflow in period t for project i. If there is a net cash *outflow* for the project, Expend_{ti} is *positive*; if there is a net cash *inflow*, Expend_{ti} is *negative*. For example, in Exhibit 16.7 $\mathrm{Expend}_{11} = 120$, meaning that in period 1 project 1 involves a net cash outlay of $120, and $\mathrm{Expend}_{21} = -30$ meaning that in period 2 project 1 produces a net cash inflow (negative outflow) of $30. For project 2, $\mathrm{Expend}_{12} = 50$ and $\mathrm{Expend}_{22} = 80$, meaning that there are outflows of $50 and $80 in periods 1 and 2, respectively.

For the purpose of expressing the budget constraint, Expend_{ti} shows the net incremental cash outflow, regardless of whether or not it is a capital expenditure. For example, project 1's outlay of $120 in period 1 may consist of $100 for plant and equipment and $20 for incremental investment in working capital. Similarly, the net cash inflow in period 2 for project 1 could represent a capital expenditure of $90 offset by a cash inflow from operations for the project of $120.

Using this notation, the cash outflow in period t from project i will be

$$\text{Expend}_{ti} \times X_i,$$

where X_i is the decision variable. If $X_i = 0$, we have

$$\text{Expend}_{ti} \times X_i = \text{Expend}_{ti} \times 0$$
$$= 0,$$

and if the project is accepted, so that $X_i = 1$, we have

$$\text{Expend}_{ti} \times X_i = \text{Expend}_{ti} \times 1$$
$$= \text{Expend}_{ti}.$$

In period t the total expenditures for all four projects will be

$$\sum_{i=1}^{4} \text{Expend}_{ti} \times X_i.$$

The constraint that total project expenditures not exceed the budget ceiling is written for period t as

$$\sum_{i=1}^{4} \text{Expend}_{ti} \times X_i \leq \text{Budget Ceiling}_t. \tag{16.17}$$

There will be a constraint for each period that has a budget ceiling. For example, the immediate (period 1) constraint for Gartner Corporation is that expenditures not exceed \$500, so the constraint is written as

$$120X_1 + 50X_2 + 90X_3 + 450X_4 \leq 500,$$

where the numerical coefficients are the values of Expend_{tj} for $t = 1$ and $j = 1, 2, 3, 4$. Because the firm has set budget limits for periods 1 and 2, the linear program will have two budget constraints, one for each period. The second period budget constraint is

$$-30X_1 + 80X_2 + 90X_3 + 70X_4 \leq 130.$$

In the second period, project 1 generates an inflow of 30. Because out-flows are shown as positive numbers in the budget constraint, this 30 inflow is shown with a negative sign, indicating a negative expenditure. The impact of this is easier to understand if you transpose the $-30X_1$ to the right side of the constraint, so the budget limit becomes $130 + 30X_1$. Now we can see that if project 1 is adopted ($X_1 = 1$), the funds available for the other projects is $160 because the $30 generated by project 1 is available to help cover expenditures for the other projects.

Combining the ingredients developed thus far, our LP can be stated as

$$\underset{\{X_i\}}{\text{Max }} Z = 140X_1 + 150X_2 + 170X_3 + 240X_4 \qquad (16.18)$$

subject to

$$120X_1 + 50X_2 + 90X_3 + 450X_4 \le 500$$
$$-30X_1 + 80X_2 + 90X_3 + 70X_4 \le 130. \qquad (16.19)$$

The solution to this problem, shown in Exhibit 16.8, is that $X_1 = 3.02$, $X_2 = 2.76$, and $X_3 = X_4 = 0$. This means that we should adopt 3.02 units of project 1, 2.76 units of project 2, and none of project 3 or 4. Now we need to consider whether it makes sense to adopt 3.02 units of a project, and if it does not makes sense, what to do about it.

In our discussion to this point we assumed that the decision variables, X_i, would be either 0 or 1, indicating rejection or acceptance of the project. However, the solution above indicates that we should adopt 3.02 units of project 1 and 2.76 units of project 2. That is, we should not only adopt more than 1 unit of project 2, but also adopt it in fractional units. In some instances this may make sense, but in most other instances it is nonsense. For example, if project 2 was a bridge across a river, it may be nonsense to consider building two bridges, and even more nonsense to consider building an additional 76% of a bridge. However, the solution of 2.76 bridges may indicate that we should consider a bridge with greater capacity. This may require that we reformulate the problem and do a new analysis. Nevertheless, for most capital budgeting analyses, it makes sense to consider either accepting or rejecting the project, in which case it does not make sense to recommend a solution of three units of the project, much less 3.02 units. Consequently, we need to limit our analysis to accept–reject decisions.

	A	B	C	D	E	F	G	H	I	J
1						Gartner Corporation				
2					Linear Programming Model for Capital Budgeting					
3										
4										
5	Project *j*			1	2	3	4			
6										
7	Decision Variables, X*j*			3.02	2.76	0.00	0.00			
8										
9	Project NPV*j*			140.0	150.0	170.0	240.0			
10										
11	Portfolio NPV		836.0							
12										
13	Constraints									
14									Capital	Capital
15		Capital Expenditure							Expenditures	Budget
16		Period 1		120	50	90	450		500.0	500
17		Period 2		-30	80	90	70		130.0	130
18										
19				X 1					3.02	1
20					X 2				2.76	1
21						X 3			0.00	1
22							X 4		0.00	1

EXHIBIT 16.8   Gartner Corporation: LP model for capital budgeting.

When mathematical computations are done by computers, we have to tell the system what we want it to do. So, we need a way to tell it that the decision variables should be either 0 or 1. There are two ways we can do this. We can use integer programming so that we only consider decision variables that are integers; or, we can use binary programming so that we only consider decision variables equal to 0 or 1. We discuss the integer approach first.

When only whole projects are to be considered, we confine the decision variables to integer values (0, 1, 2, 3, etc.) by using an integer programming method of solving the problem. Integer programming is a variation of LP that searches for the positive integer values of the decision variables that yield the optimal solution. To move from integer to binary programming, we only need to constrain the solution to be less than or equal to 1, thus

$$X_i \leq 1.$$

Because integer programming considers only positive integer values, this constraint ensures that the only values of the decision variables that will be considered are $X_i = 0$ or $X_i = 1$, where $X_i = 0$ indicates that we reject project *i*, and $X_i = 1$ that we accept it.

Using integer programming, and adding the new constraints, the linear program becomes

$$\text{Max}_{\{X_i\}} Z = 140X_1 + 150X_2 + 170X_3 + 240X_4 \qquad (16.20)$$

subject to

$$120X_1 + 50X_2 + 90X_3 + 450X_4 \leq 500$$
$$-30X_1 + 80X_2 + 90X_3 + 70X_4 \leq 130$$
$$X_1 \qquad\qquad\qquad \leq 1$$
$$X_2 \qquad\qquad \leq 1 \qquad\qquad (16.21)$$
$$X_3 \qquad \leq 1$$
$$X_4 \leq 1$$

Using a solution algorithm that looks only at positive, integer solution values, the solution is $X_1 = 1$, $X_2 = 0$, $X_3 = 1$, and $X_4 = 0$. This means that we adopt projects 1 and 3 and reject projects 2 and 4, resulting in a portfolio NPV of \$310. We only use \$210 of our \$500 budget in period 1, and \$60 of our \$130 budget in period 2.

A second way to derive optimal 0, 1 solutions is to use binary programming. This is similar to the integer approach except that it considers only 0 or 1 as feasible solutions to the problem. In this case, we do not need to specify that $X_i \leq 1$. We need only to tell the system to use a binary solution method. When we do this, we should get the same optimal solution as with the integer approach. For our purposes, we do not need to go into the details of the computation of the solutions for either approach. We will rely on the computer software to reach the solution. The use of Solver for this problem is explained below.

16.2.4 Using Excel and Solver for the Gartner Problem

Exhibit 16.9 shows the Gartner Corporation problem as an Excel spreadsheet.

This spreadsheet is quite simple. All but three of the cells simply contain the numbers shown, as opposed to formulas. The other three cells contain key formulas that are used to complete the LP spreadsheet. These are shown below, with the number in parenthesis on the right indicating the number of the expression in the LP model discussed above.

Key Cell Formulas

Cell	Excel Formula	Purpose	
C11	=SUMPRODUCT(D9:G9,D7:G7)	Objective function	(16.20)
I16	=SUMPRODUCT(D16:G16,D7:G7)	Constraint	(16.21)
I17	=SUMPRODUCT(D17:G17,D7:G7)	Constraint	(16.21)

	A	B	C	D	E	F	G	H	I	J	K
					Gartner Corporation						
1											
2			Integer Linear Programming Model for Capital Budgeting								
3											
4											
5	Project j			1	2	3	4				
6											
7	Decision Variables, Xj			0.00	0.00	0.00	0.00				
8											
9	Project NPV j			140	150	170	240				
10											
11	Portfolio NPV		0.0								
12											
13	Constraints								Capital	Capital	
14									Expenditures	Budget	
15		Capital Expenditure									
16		Period 1		120	50	90	450		0	500	
17		Period 2		-30	80	90	70		0	130	
18											
19				X 1					0.00	1	
20					X 2				0.00	1	
21						X 3			0.00	1	
22							X 4		0.00	1	
23											

EXHIBIT 16.9   Gartner Corporation: Integer LP model for capital budgeting.

The decision variables (the cells changed by Solver) are in cells D7–G7. The coefficients for the objective function are the numbers in cells D9–G9. The coefficients for the constraints are in cells D16:G17. The budget ceilings for periods 1 and 2 are the constants in cells J16 and J17. Cells I16 and I17 perform the multiplication and addition that are shown as the left side of Equation 16.21. Cells I19–I22 simply point back to the decision variables; cells J19–J22 are constraints that these values have to be less than or equal to 1.

To solve this problem with Solver, use Tools/Solver to access the Solver dialog box shown in Exhibit 16.10. This one is already filled in to indicate the objective function (Set Target Cell: C11) and the decision variables (By Changing Cells: D7:G7), and that it is a maximization problem. The constraints are indicated as shown (Subject to the Constraints:). To impose the requirement that decision variables are integers, click "Add" to get the Constraint dialog box (Exhibit 16.11), then click the middle box to show the "int" choice. This constrains the variables indicated to be integers.

Having completed all the necessary choices for Solver, the next step is to tell it to solve the problem. Click the "Solve" button, and the spreadsheet will show the results, including the optimal values for the decision variables, as shown in Exhibit 16.12. The solution spreadsheet shows the effect of the optimally chosen decision variables. With projects 1 and 3 chosen, the NPV of the investment portfolio is $310. Any other feasible combination of projects would yield a smaller portfolio NPV.

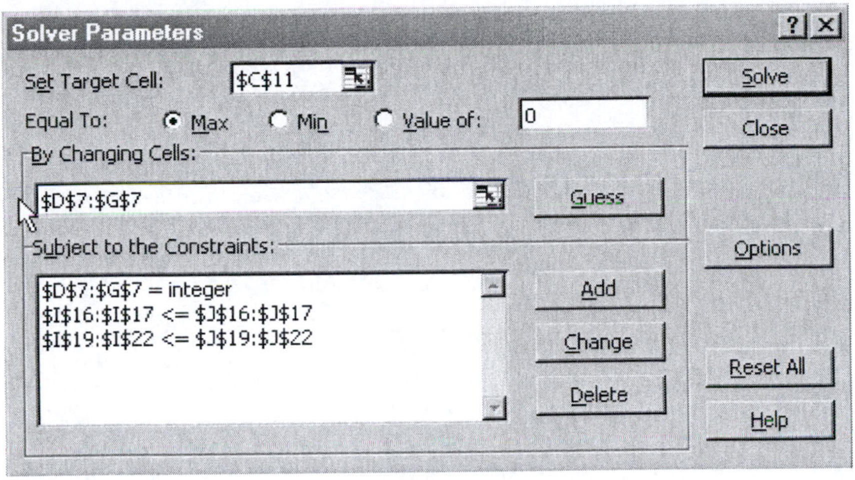

EXHIBIT 16.10   Completed Solver dialog box for base integer LP model.

A feasible program is one that does not violate the constraints. The capital budget allows for $500 to be invested in period 1 and $130 in period 2. However, the optimal solution only uses $210 of the period 1 budget (cell I16) and $60 (cell I17) of the period 2 budget, so it is not using all the budget. With the excess funds in period 1, why not also adopt project 2? It would use only $50 of the unused $290 in period 1. But project 2 requires an additional expenditure of $80 in period 2, making the period 2 expenditures $140, thereby violating the period 2 budget constraint of $130. Adding project 2 is therefore an infeasible solution.

This analysis shows one of the advantages of the LP approach. The LP takes into account all the possible decisions and all the constraints, and then picks the best decision in view of the constraints. In this simple problem, the optimal solution is relatively easy to determine, even without the

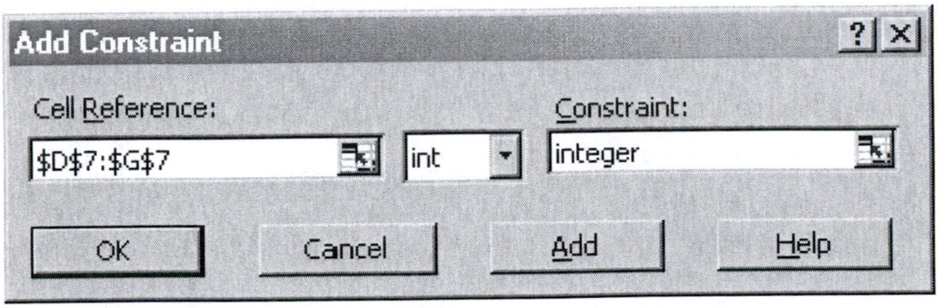

EXHIBIT 16.11   Add Constraint dialog box specifying integer LP program.

	A	B	C	D	E	F	G	H	I	J	K
1					Gartner Corporation						
2		Integer Linear Programming Model for Capital Budgeting									
3											
4											
5	Project j			1	2	3	4				
6											
7	Decision Variables, Xj			1.00	0.00	1.00	0.00				
8											
9	Project NPV j			140	150	170	240				
10											
11	Portfolio NPV		310.0								
12											
13	Constraints									Capital	
14									Capital Expenditures	Budget	
15		Capital Expenditure									
16		Period 1		120	50	90	450		210	500	
17		Period 2		-30	80	90	70		60	130	
18											
19				X 1					1.00	1	
20					X 2				0.00	1	
21						X 3			1.00	1	
22							X 4		0.00	1	
23											

EXHIBIT 16.12   Solved base LP model.

LP. However, in more complicated problems it can be very difficult to check how each possible decision affects the budget without the LP. The LP can handle many different complex constraints and decision combinations, taking into account the costs and benefits of each project relative to all other projects and simultaneously observing all of the constraints.

16.3 ELABORATIONS ON THE BASIC MODEL

This example demonstrates the basic features of the Weingartner LP model for capital budgeting under conditions of capital rationing. But it is too simple to be useful as a general planning model. Fortunately, this base model is easily extended and elaborated on. The elaborations can include

- Mutually exclusive projects
- Contingent projects
- Lending and borrowing
- A general schedule of investment opportunities
- Policy constraints on minimum rates of growth of earnings and dividends
- The issuance of new common stock

With elaborations such as these, the simple base model can be transformed into a more general planning model. We explain just the first three elaborations in the next sections.

16.3.1 Mutually Exclusive Projects

Mutually exclusive projects occur when you can adopt one of two (or more) projects, but not both (or all). That is, you must choose between projects. For example, one choice might be between building a new plant in California or building the same plant in Colorado. Because you can choose one or the other, but not both, the projects are mutually exclusive. Another example would be the starting a project immediately versus delaying the project for a year. The LP format allows us to compare these mutually exclusive choices.

The binary (0, 1) decision variables make it easy to handle mutually exclusive projects in the LP. Suppose that eight projects are being considered, and that projects 3 and 8 are mutually exclusive. A constraint can be added to insure that one (either 3 or 8) but not both projects are adopted. The constraint would be of the form

$$X_3 + X_8 \leq 1. \qquad (16.22)$$

Because the decision variables must be either 0 or 1, Equation 16.22 ensures that we cannot accept both projects. Both projects can be rejected ($X_3 = 0$ and $X_8 = 0$), or one of the projects can be accepted (either $X_3 = 1$ and $X_8 = 0$ or $X_3 = 0$ and $X_8 = 1$). However, both projects cannot be accepted ($X_3 = 1$ and $X_8 = 1$) because in that case $X_3 + X_8 = 2$, violating Equation 16.22. The constraint is easily expanded to include more than two projects. Suppose, for example, there is a set of four projects among which we can choose only one, say projects 2, 3, 7, and 8. Then the constraint would take the form

$$X_2 + X_3 + X_7 + X_8 \leq 1.$$

16.3.2 Contingent Projects

An investment is contingent when it cannot be accepted unless another project is also accepted. For example, suppose project 1 is the purchase of a new computer, and project 2 is the purchase of a software system to be used on the new computer. The computer could be purchased without that particular software, but it would make no sense at all to purchase the software without also purchasing the computer. In this case, project 2 (software) is contingent on the acceptance of project 1 (computer). This type of contingency can be included in the LP with the addition of the constraint

$$X_2 \leq X_1.$$

We usually express the LP constraints with all the decision variables on the left side of the inequality sign and the constant number on the right, thus

$$-X_1 + X_2 \leq 0. \qquad (16.23)$$

If the solution is $X_1 = 0$, Equation 16.23 requires that X_2 also equals 0. When $X_1 = 1$, however, X_2 can be either 0 or 1, depending on its merits. Suppose another project 6 is an alternative software system competing with project 2. Then projects 2 and 6 are mutually exclusive and are contingent on the acceptance of project 1. This would be reflected in the constraint

$$-X_1 + X_2 + X_6 \leq 0.$$

If $X_1 = 1$, either project 2 or 6 could be accepted, but not both.

16.3.3 Lending and Borrowing

As the model is currently structured, nothing can be done to relax a budget constraint that binds, even when extra funds are present in other periods. The ability to transfer excess funds between periods might make it possible to relax such a binding constraint, thereby allowing us to accept additional projects and increase the NPV of the investment portfolio. For example, in the solution to the Gartner Corporation's capital budgeting problem, we had excess funds in period 1 but could not undertake an additional project because it required more funds in period 2 than were available. If we could transfer some of the excess funds from periods 1 to 2, we could undertake additional projects and realize a higher portfolio NPV. We need ways to transfer funds from one period to another. Lending and borrowing are just such mechanisms. Lending transfers funds from earlier periods to later ones; borrowing effectively transfers funds from later periods to earlier ones. When added to the problem, lending and borrowing give us the beginnings of a complete financial planning model. To see how lending and borrowing opportunities can be added to the model, return to the Gartner example.

16.3.3.1 Lending

Lending (e.g., investing in marketable securities such as U.S. Treasury securities) enables the transfer of funds forward from one period to the next. For example, surplus funds in period 1 can be carried forward to period 2. Let $Lend_1$ represent the decision variable that is the amount to be invested in marketable securities in period 1. Assume that there is no limit on the amount that can be lent except for limitations imposed by the funds available for lending, so the decision variable is not required to be an integer variable. Assume that the rate of return earned on funds lent for one period is

$$r_L = 0.10.$$

Assuming a tax rate of $\tau = 0.40$, the after-tax return is

$$r_L(1 - \tau) = 0.10 \times (1 - 0.40)$$
$$= 0.06.$$

Next, the effect of lending must be incorporated into the objective function. Because the objective function is stated in terms of NPV, the coefficients of the lending decision variables in the objective function also must be defined in terms of NPV. The coefficients of the X_is for the investment projects are the NPVs (NPV_i) of the after-tax cash flows generated by these projects. Consequently, the coefficients of the $Lend_1$ decision variables must also be in terms of the NPVs of the after-tax cash flows. Lending $1 in period 1 is a cash outflow of $1 in period 1 and generates a cash inflow after taxes of $1 \times [1 + r_L(1 - \tau)]$ in period 2. Therefore, the NPV of the cash flows from lending $1 in period 1 is

$$-1 + \frac{[1 + r_L(1 - \tau)]}{[1 + k]}, \tag{16.24}$$

where k denotes the discount rate appropriate for discounting the cash inflow from lending.

Suppose the appropriate discount rate is 12%; then the NPV *per dollar lent* in period 1 is

$$-1 + \frac{[1 + 0.10 \times (1 - 0.40)]}{[1.12]} = -0.0536. \qquad (16.25)$$

Given this discount rate,* investing in short-term securities does not add net value to the investment portfolio. However, it allows the transfer of slack or idle funds from one period to a future period in which they may be used in productive capital investments.

Lending also affects two budget constraints. In the objective function, lending $1 reduces the NPV of the portfolio by −$0.0536. Constraints, however, are expressed in terms of cash outflows. In the constraint from which funds are lent (the period during which T-bills are purchased), cash flows out of the budget and into T-bills. Therefore, Lend_1 has a positive coefficient of (+1). In the constraint for the period that receives the funds (the period during which T-bills are sold), cash flows in, relaxing the budget constraint by $1.06 for each dollar lent. Therefore, the coefficient of the lending variable is a negative $1.06 (−1.06).

16.3.3.2 Borrowing

Assume that even though a budget limit has been imposed on capital expenditures, management will allow a limited amount of borrowing to take place. This transfers cash from future periods to the present period. Suppose that the borrowing limit is $200 and that we wish to borrow in period 1 and repay the loan in period 2. Further, assume that the current interest rate on borrowed funds is 12%, denoted by

$$r_d = 0.12.$$

As was the case with lending, borrowing enters the objective function and two budget constraints. Let Borrow_1 represent the decision variable that is the amount to be borrowed in period 1 and repaid in period 2. Borrow_1 is the cash inflow to the firm in period 1 that is available to help

---

* The discount rate is a crucial variable in determining the value of lending or borrowing. Whether the NPV of lending or borrowing is positive or negative depends on whether the discount rate is higher or lower than the lending or borrowing rate. The borrowing rate would be appropriate as the discount rate because it represents the highest return that can be earned on an extra $1 at this level of risk. That is, an extra $1 can be used to repay debt and save (i.e., earn) the borrowing rate.

finance other investment projects. Retiring the debt in period 2 (principal and interest net of taxes) results in a cash outflow of

$$[1 + r_d(1 - \tau)] \times \text{Borrow}_1 = [1 + 0.12 \times (1 - 0.40)] \times \text{Borrow}_1$$
$$= 1.072 \times \text{Borrow}_1.$$

Similar to lending's coefficient in the objective function, the coefficient of the borrowing decision variable in the objective function is the NPV of one dollar borrowed. Borrowing $1 in period 1 generates a cash inflow of $1 at the date of the loan and a cash outflow after taxes of $[1 + r_d(1 - \tau)]$ in period 2. Therefore, the NPV of the cash flows from period 1 borrowing is

$$\left\{ 1 - \frac{[1 + r_d(1 - \tau)]}{(1 + k)} \right\},$$

where k is the discount rate. We assume that the borrowing rate, $r_d = 0.12$, equals the discount rate, $k = 0.12$, because the borrowing rate is the opportunity cost of funds. Then the coefficient (the NPV per dollar borrowed) of Borrow_1 in the objective function is

$$\left\{ 1 - \frac{[1 + 0.12(1 - 0.40)]}{(1 + 0.12)} \right\} = 0.0429,$$

and the borrowing variable will appear in the objective function as

$$+0.0429 \times \text{Borrow}_1.$$

To add the lending and borrowing opportunities to the Gartner LP (shown earlier as Equations 16.20 and 16.21), the objective function and constraints are expanded with the new terms shown in Exhibit 16.13, with the complete expressions shown in Exhibit 16.14. In the exhibit the negative coefficient of the borrowed amount, $-1 \times \text{Borrow}_1$ in Equation 16.27a indicates a cash inflow from borrowing in period 1. To understand this, transpose $-1 \times \text{Borrow}_1$ to the right side of Equation 16.27a. With the change of sign we see that Borrow_1 adds to the funds available for investment. The positive coefficient $1.072 \times \text{Borrow}_1$ in Equation 16.27b represents the after-tax cash outflow

Equation in Gartner Model	Lending Terms Added	Borrowing Terms Added
16.26a	−0.0536 Lend$_1$	+0.0429 Borrow$_1$
16.27a	+1.0 Lend$_1$	−1 Borrow$_1$
16.27b	−1.06 Lend$_1$	+1.07 Borrow$_1$
New: Borrowing Constrait 16.27g		Borrow$_1$ ≤ 200

EXHIBIT 16.13 Terms added to LP for lending and borrowing.

resulting from repayment of the debt. The last Equation 16.27g limits the total amount that can be borrowed in period 1 to no more than $200. The fact that all borrowing is repaid in period 2, coupled with the fact that there is no further borrowing in period 2, assures that all borrowing is repaid from the cash flows generated by the investment projects.

After adding borrowing and lending, the new solution is $X_1 = 1$, $X_2 = 1$, $X_3 = 1$, and Lend$_1$ = 38.94, as shown in Exhibit 16.15. The addition of the opportunity to lend and borrow increases the portfolio NPV (the objective function) from $310 in the original solution to a value of $457.91. This is because we can now accept project 2, which previously was rejected due to insufficient funds in period 2. Transferring unused funds from periods 1 to 2 (lending) relaxed the constraint.

The program still does not accept project 4 even though it has the largest NPV. Why not? Because project 4 requires a net cash outflow of $450 in period 1. If we were to drop project 2 and borrow $190, we could meet project 4's cash needs in period 1. The problem is that there will not be enough cash in period 2 to repay the debt. If you borrow $190 in period 1 to finance project 4, you would not have enough funds in period 2 to cover the $70 outflow required by project 4 and repay the $190 plus interest. Had

Max Z	$140X_1$ $+150X_2$ $+170X_3$ $+240X_4$ -0.0536 Lend$_1$ $+0.0429$ Borrow$_1$	16.26a
subject to:		
	$120X_1$ $+50X_2$ $+90X_3$ $+450X_4$ $+1 \times$ Lend$_1$ $-1 \times$ Borrow$_1$ ≤ 500	16.27a
	$30X_1$ $+80X_2$ $+90X_3$ $+70X_4$ -1.06 Lend$_1$ $+1.072$ Borrow$_1$ ≤ 130	16.27b
	X_1 ≤ 1	16.27c
	X_2 ≤ 1	16.27d
	X_3 ≤ 1	16.27e
	X_4 ≤ 1	16.27f
	Borrow$_1$ ≤ 200	16.27g

EXHIBIT 16.14 Constrained capital budgeting problem with borrowing and lending between periods.

Gartner Corporation
Linear Programming Model for Capital Budgeting

Project j	1	2	3	4	Lending	Borrowing		
Decision Variables, X_j	1.00	1.00	1.00	0.00	38.94	0.00		
Project NPV_j	140.0	150.0	170.0	240.0	-0.0536	0.0429		
Portfolio NPV	457.91							

Constraints							Capital Expenditures	Capital Budget
Capital Expenditure								
Period 1	120	50	90	450	1	-1	298.9	500
Period 2	-30	80	90	70	-1.06	1.072	98.7	130
X 1							1.00	1
X 2							1.00	1
X 3							1.00	1
X 4							0.00	1

EXHIBIT 16.15 Solved LP model for capital budgeting with borrowing and lending.

sufficient funds been available in period 2 to repay the borrowing in period 1, the LP would have recommended borrowing in period 1 to undertake project 4. To see this, let us revise the problem by increasing our period 2 budget constraint to $250 (from $130). The problem is now stated in Exhibit 16.16.

The only difference from Exhibit 16.14 is the change in the right side of Equation 16.29b from 130 to 250. Solving this problem yields the optimal solution of $X_1 = 1$, $X_2 = 1$, $X_3 = 0$, $X_4 = 1$, $Lend_1 = 0$, and $Borrow_1 = 120.66$. The value of the objective function is $Z = 535.18$. The new solution is shown in Exhibit 16.17. Increasing the period 2 budget limit increased the firm's borrowing capacity in period 1 by making more funds available for debt service in period 2. This permitted the adoption of project 4, which is

Max Z $140X_1 +150X_2 +170X_3 +240X_4 -0.0536\ Lend_1 +0.0429\ Borrow_1$		16.26a
subject to:		
$120X_1 +50X_2 +90X_3 +450X_4$ $+1 \times Lend_1$ $-1 \times Borrow_1$	≤ 500	16.29a
$30X_1 +80X_2 +90X_3 +70X_4$ $-1.06\ Lend_1$ $+1.072\ Borrow_1$	≤ 250	16.29b
X_1	≤ 1	16.29c
X_2	≤ 1	16.29d
X_3	≤ 1	16.29e
X_4	≤ 1	16.29f
$Borrow_1$	≤ 200	16.29g

EXHIBIT 16.16 Constrained capital budgeting problem with increased budget limit for period 2.

Gartner Corporation
Linear Programming Model for Capital Budgeting

Project j	1	2	3	4	Lending	Borrowing		
Decision Variables, X_j	1.00	1.00	0.00	1.00	0.00	120.66		
Project NPV_j	140.0	150.0	170.0	240.0	-0.0536	0.0429		
Portfolio NPV	535.18							
Constraints								
							Capital Expenditures	Capital Budget
Capital Expenditure								
Period 1	120	50	90	450	1	-1	499.3	500
Period 2	-30	80	90	70	-1.06	1.072	249.3	250
X 1							1.00	1
X 2							1.00	1
X 3							0.00	1
X 4							1.00	1

EXHIBIT 16.17   LP model for capital budgeting with increased budget limit for period 2.

expensive but has a very high NPV. Without the additional $120 in period 2's budget, there was no way to borrow the funds needed to finance project 4's large initial investment in period 1 and repay the loan in period 2.

The period 1 cost of projects 1, 2, and 4 is $120 + $50 + $450 = $620. Given a budget constraint of $500, we need to borrow $120 to be able to finance them. Yet, the LP solution calls for borrowing $120.66 instead of $120, so it is borrowing $0.66 more than we seem to need. The reason for this is that the LP will try to increase the value of any decision variable that increases the value of the objective function, and there is a positive coefficient on the borrowing variable in the objective function. This points out one of the idiosyncrasies of including borrowing in an LP financial planning model: Unless we formulate the problem carefully, the model will typically push borrowing to its limit. In many cases this may seem irrational. We will discuss borrowing in greater detail in the next section.

With this simple problem, we have seen how the LP takes account of the inter-relationships between investment, financing, and budgeting. Problems larger than this that would send an analyst to the mental ward are easily solved with LP.

16.4 BORROWING IN THE LP PLANNING MODEL

One of the most difficult problems in developing an LP model for financial planning is the treatment of debt financing. We need a way to capture the advantages of debt financing and at the same time properly take account of the risks imposed by debt financing. The theory of finance says that a firm financed entirely with equity can increase its value by adding moderate and prudent amounts of debt, primarily due to the tax advantages of debt financing. There is a point, however, where the risks and potential costs of financial distress due to borrowing begin to outweigh the advantages of debt. Debt financing should not be expanded beyond that point. The ideal planning model would assess these trade-offs and arrive at an optimal mix of equity and debt financing. While it is difficult to construct this ideal model as a linear program because of the non-linearities inherent in this capital structure problem, we can handle some of this complexity in an LP model.

It is not difficult to show the tax advantage of debt in the LP model.* Following the adjusted present value approach by Myers and Pogue (1974), we calculate the value of an investment project as if it is financed entirely with equity, and then add the value of the tax savings from interest. The pure equity value is obtained by discounting the project cash flows at the cost of unleveraged equity; the tax savings from interest are calculated separately at the cost of debt.

Consider the following example. Suppose an investment is financed entirely with equity and that the investment has a level of risk that would require that its cash flows be discounted at a rate k_u (where subscript u denotes that it is "unleveraged," i.e., has no debt financing). The unleveraged value of the cash flows would be

$$\sum_{t=1}^{n} \frac{\text{Cash Flow}_t}{(1+k_u)^t}. \tag{16.30}$$

To account for the effects of debt financing, we add a term that is the present value of the tax savings resulting from the tax deductibility of interest rather than adjusting the discount rate to show the effects of financial leverage. Therefore, the value of the leveraged cash flows would be

$$\sum_{t=1}^{n} \frac{\text{Cash Flow}_t}{(1+k_u)^t} + \tau \sum_{t=1}^{n} \frac{\text{Interest}_t}{(1+k_d)^t}. \tag{16.31}$$

* The standard method for evaluating investments and financing alternatives is to discount expected future cash flows at the firm's WACC, denoted by k_a. To the extend that k_a depends on the proportional mix of debt and equity financing, the present value of a cash flow, discounted at k_a, is a non-linear function of the financing mix. That is, if the proportion of debt is changed, the present value of a future cash flow changes in a non-linear way, violating the requirement that the model be linear with respect to the decision variables, one of which is the proportion of debt financing.

The term on the right is the present value of these tax savings, because the tax shield from interest in a given year is $\tau \times \text{Interest}_t$, where τ is the tax rate and k_d the interest rate on the debt. This value is linearly related to the amount of debt financing because $\tau \times \text{Interest}_t$ is a linear function of the amount of debt financing. This approach allows us to avoid the complications of non-linearity that would occur if we used the WACC as the discount rate.*

To see how these ideas can be applied in an LP model, return to the four-project investment decision problem represented by Exhibit 16.14. As before, X_1, X_2, X_3, and X_4 are the investment decision variables and Borrow_1 is the decision variable representing the amount borrowed in the first period. Assume that the NPVs for the investment projects were based on the cost of equity capital as though the firms were financed entirely with equity. Even with the addition of borrowing to the problem we still use the same project NPVs in the objective function of the LP. We account for the effect of borrowing through the coefficient of the decision variable Borrow_1 in the objective function, rather than using the WACC as the discount rate. This coefficient is the NPV of $1 borrowed. In general terms, for debt with an n-period life, this coefficient is

$$\tau \sum_{t=1}^{n} \frac{k_d}{(1+k_d)^t},$$

where k_d is the borrowing rate on the debt. With the borrowing rate at 12% per year, and the tax rate is 40%, the borrowing coefficient is

$$\tau \sum_{t=1}^{1} \frac{k_d}{(1+k_d)^t} = \tau \frac{k_d}{(1+k_d)}$$

$$= 0.40 \times \frac{0.12}{1.12}$$

$$= 0.0429$$

---

* This approach has the same effect as if the cash flows were discounted at the firm's WACC, where the cost of equity changes with the debt-to-equity ratio according to the equation

$$k_s = k_u + \frac{D}{E}(k_u - r_f)(1 - \tau),$$

where k_u is the discount rate for a pure equity (unleveraged) cash flow and r_f the risk-free rate of interest. This expression, developed by Modigliani and Miller (1958), leads to the result that the WACC declines as the proportional amount of debt financing increases.

per dollar borrowed at the beginning of the first period and repaid in period 2. The borrowing decision variable appears in the objective function as

$$0.0429 \times \text{Borrow}_1.$$

The shortcoming of this approach is that the positive coefficient on Borrow_t in the objective function encourages the use of debt beyond the point most practitioners would consider prudent. This means that without constraints, an LP planning model tends to generate solutions that recommend excessive expansion of debt financing. Debt appears an especially attractive source of funds in planning models for two reasons: first, borrowing relaxes the budget constraint and thereby provides funds for expanding investments; second, borrowing generates tax savings because the interest payments are tax deductible. However, as everyone but the linear program knows, excessive borrowing can be hazardous to a firm's financial health. So, like cigarettes, debt should carry warnings of impending doom that limit its use by the linear program.

There are numerous ways to limit the use of debt financing in the LP model, but none is entirely satisfactory. One way that has already been introduced is to limit the amount to be borrowed by adding a constraint to the LP in the form

$$\text{Borrow}_t \leq \text{Borrowing Limit}_t. \tag{16.32}$$

This approach is appropriate when either the lender or the borrower sets a limit (e.g., a revolving credit agreement with a maximum cap). The problem with this approach is that, because of the apparent advantage of borrowing, the LP will usually push borrowing right to this limit without telling the user when it is best to borrow less than the maximum.

Financial ratios also can be used to set borrowing limits. For example, it is common to set a target for the ratio of total debt to total assets. This is easily done as an LP constraint. Suppose you regard the prudent limit for debt to be 25% of total assets. The overall constraint would be

$$\text{Total Debt}_t < 0.25 \times \text{Total Assets}_t. \tag{16.33}$$

Here, Total Debt$_t$ consists of debt that is already outstanding, denoted by Debt$_0$, plus any new borrowed funds as specified by the decision variable Borrow$_t$. Similarly, Total Assets$_t$ consists of initial assets, Assets$_0$, plus additions specified by the investment decisions recommended by the model. For example, in period 1, we have

$$\text{Total Assets}_1 = \text{Assets}_0 + \sum_{i=1}^{4} \text{Asset}_i\, X_i,$$

where Asset_i is the amount of assets purchased in period 1 for investment project i. With these definitions of assets and debt, the constraint for the first period would be written as

$$\text{Debt}_0 + \text{Borrow}_1 = 0.25 \times \left(\text{Assets}_0 + \sum_{i=1}^{4} \text{Asset}_i X_i \right). \qquad (16.34)$$

Following the LP convention of putting all decision variables on the left side of the inequality and all constants (non-decision variables) on the right, the constraint for the first period becomes

$$\text{Borrow}_1 - 0.25\sum_{i=1}^{4} \text{Asset}_i X_i < 0.25 \times \text{Assets}_0 - \text{Debt}_0. \qquad (16.35)$$

Similar constraints would be used for each period in the planning horizon.

16.4.1 Risk Constraint

Another way to constrain debt financing is to use a risk constraint. With this approach, management subjectively determines the maximum acceptable probability that the firm will be unable to service its debt. A constraint is imposed that prohibits debt from expanding past the point at which the probability of insolvency is unacceptable.

Define the risk of insolvency as the probability that the cash flow will be insufficient to service the debt. Suppose we are willing to live with the risk that there is a 10% probability that our firm will be unable to service its debt from its operating cash flow. Using this information about the probability distribution of the cash flows, we can impose a constraint on the amount of debt so that we do not exceed the acceptable level of risk of insolvency. For example, suppose as a first approximation that the firm's free cash flow is normally distributed with a mean (expected value) of

$$E(\text{Cash Flow}) = \$100{,}000$$

and a standard deviation of

$$\sigma_{\text{Cash Flow}} = \$60{,}000.$$

Using a standard normal probability table, we find the number of standard deviations below the expected value below that we expect a draw less than 10% of the time (1.28 standard deviations). Translating this into the expected value and standard deviation of our firm's cash flow yields $23,200, as shown in Equation 16.36.

$$E(\text{Cash Flow}) - 1.28\sigma_{\text{Cash Flow}} = 100,000 - 1.28 \times 60,000$$
$$= 100,000 - 76,800$$
$$= \$23,200. \tag{16.36}$$

Thus, there is a 10% probability that the cash flow will be less than $23,200 and a 90% probability that the cash flow will be greater than $23,200. Constraining the debt level such that debt service is no more than $23,200 limits the chance that we will be unable to service the debt to 10%. If the annual debt service is equal to 20% of the outstanding debt, the period 1 constraint on debt would be

$$0.20(\text{Debt}_0 + \text{Borrow}_1) \leq 23,200 \tag{16.37}$$

or, putting the constant terms on the right-hand side,

$$0.20(\text{Borrow}_1) \leq 23,200 - 0.20(\text{Debt}_0). \tag{16.38}$$

The difficulty with this approach is that the probability distribution of cash flows may be dependent on the investment projects accepted, thereby making it necessary to incorporate the cash flow distribution assumptions in our constraints. Myers and Pogue (1974, p. 593) explain how to treat this type of complication.

When we develop our LP on a spreadsheet, we can use the spreadsheet's statistical function NORMINV(•) to find the critical level of income as shown in Exhibit 16.18.

In this model, the user specifies the risk of default in cell F4 and the parameters of the probability distribution in cells F6 and F7. The $23,107 in cell F9 is the output of the Excel function for the inverse of the normal probability function shown in Equation 16.39

$$= \text{NORMINV}(\text{probability, mean, standard deviation})$$
$$= \text{NORMINV}(\text{F4, F6, F7}) \tag{16.39}$$
$$= \text{NORMINV}(0.10, 100,000, 60,000).$$

	A	B	C	D	E	F	G	H
1								
2								
3								
4		Maximum Acceptable Risk of Default				10%		
5		Probability Distribution for Cash Flows						
6			Mean			100,000		
7			Standard Deviation			60,000		
8								
9			Critical Point			23,107		
10								
11								
12			Formula in Cell F9:		=NORMINV(F4,F6,F7)			
13								
14								
15								

EXHIBIT 16.18   Identifying the critical income level for a given acceptable limit on default risk.

This function takes the level of acceptable risk specified in F4 (10%) and the mean and standard deviation parameters specified in F6 and F7, and returns the critical point of the distribution where, in this example, 10% of the distribution is below and 90% is above. In the debt Equation 16.34 of the linear program, the =NORMINV(•) function would be inserted in place of the $23,200 now shown. In this way, the user has the option of easily varying the acceptable risk to explore different policies or scenarios.

16.5 SUMMARY

This chapter introduced you to the ideas and techniques of using LP for solving financial planning problems. The advantages of using a constrained optimization technique like LP are that it forces the planner to specify the objective, the constraints, and the linkages between the decision variables and the objective; and most important, it provides a systematic way to identify the best set of decisions.

The essential features of an LP model for financial planning include the decision variables that represent the amounts or levels of the different decisions that the planner must make, the objective function that expresses mathematically the links between the decisions and the objective, and the constraints that express the various limitations on the planner's decisions.

The constraints in a financial planning model usually include expressions that balance the system, that link one period to the next, that limit access to financing, that represent limitations set by the firm's policies, and that express accounting definitions or technical relations between variables. The crucial constraints are those that balance the system by requiring assets to equal liabilities or sources of funds to equal uses of funds. The limits on access to financing usually refer to limits on borrowing when it is one of the decision variables, limits that prevent the model from pushing borrowing beyond what would be considered prudent. Policy constraints take many forms, including dividend payout ratios, target growth rates in earnings and dividends, and cash balance requirements.

If too many constraints are imposed on the model, the problem may not have a feasible solution. That is, there may be no solution that simultaneously satisfies all constraints. Consequently, the user should not add constraints unless they are absolutely necessary to the problem being modeled. In fact, the modeler should make every effort to keep the model as simple as possible, using only the minimum number of constraints and decision variables necessary. The more complex the model, the greater the likelihood that it will confuse the user and contain undetected errors.

Models are merely guides to decision making; their results should not be enshrined. The user must constantly be alert for errors and illogical results. Finally, the results should be recalculated as new information is received, in case the new information leads to a revision of the results, or even a restructuring of the model itself.

Naturally, not all problems of financial planning require LP—or any kind of optimization technique, for that matter—for their solution. But there are many financial planning problems that are handled best with an optimization technique, and they often can easily be fitted into the LP format. These include problems of working capital management, capital budgeting under conditions of capital rationing, and more general long-range financial planning. This chapter has focused on problems of planning the firm's capital budgeting program. The next chapter will discuss models that deal with managing working capital.

PROBLEMS

1 Optimal Capital Investment Portfolio: Prager Corporation

The CEO of the Prager Corporation has asked the managers of its various divisions to submit proposals for major capital expenditure programs. The proposals have been submitted and now the decision must be made as to which of the proposed projects should be adopted.

Summary data regarding the costs and expected cash flows of the proposed projects are shown in Table 16.1. There are 12 proposed

TABLE 16.1   Data for Problem 1

Optimal Capital Investment Portfolio

CASH FLOWESTIMATES FOR PROPOSED INVESTMENT PROJECTS

PROJECT:		A	B	C	D	E	F	G	H	I	J	K	L
TIME													
	0	-23	-200	-600	-100	-150	-420	-200	-10	0	-400	-80	-50
	1	4.78	-150	0	-475	30	80	-600	-4	-100	-200	0	400
	2	4.78	20	200	-250	35	90	280	5	-475	-300	150	50
	3	4.78	80	350	290	40	180	280	8	-250	400	0	-510
	4	4.78	120	350	290	40	180	280	8	290	350	0	0
	5	4.78	150	100	290	40	130	280	8	290	300		
	6	4.78	180	0	290	40	0	280	0	290	300		
	7	4.78	90		290	40		280		290	250		
	8	4.78	70		0	40		100		290	0		
	9	4.78	50			35		100		0			
	10	12.78				10		-950					
	11	0				0		0					
CAPITAL INVESTMENT	0	23	200	600	100	150	420	200	10	0	400	80	50
	1	-4.78	150	0	475	-30	-80	600	4	100	200	0	-400
	2	-4.78	-20	-200	250	-35	-90	-280	-5	475	300	-150	-50
DISCOUNT RATE		0.18	0.15	0.15	0.15	0.15	0.15	0.15	0.15	0.15	0.15	0.15	0.15
PRESENT VALUE of													
InitialCosts													
Cash Inflows													
NET PRESENT VALUE													
INTERNAL RATE OF RETURN													
PRESENT VALUE INDEX													

projects, designated as projects A–L. The numbers in the body of the table are the estimated cash flows (in thousands of dollars), with negative numbers indicating cash outflows, and positive numbers indicating inflows. All project cash flows (with the exception of project A) should be discounted at the firm's cost of capital of 15%. Project A has a risk-adjusted cost of capital of 18%.

All 12 projects meet Prager's acceptance criterion: they have positive NPV at the cost of capital of 15%, and their IRRs equal or exceed 15%. If there were no limitations on the firm's ability to accept projects, all 12 would be adopted. However, the board of directors has set limits on capital expenditures. The maximum that can be spent right now (time 0) is set at $950; for the end of the coming year (time 1), the limit is $400; and for year 2 (time 2) the limit is $200. Given the limits set by the directors, it must now be decided which of these 12 projects should be undertaken. The portfolio of projects that is chosen should be the one which maximizes the NPV of the portfolio as a whole, while not exceeding the limited capital budget.

Assignment

(a) For each of the projects, calculate in a spreadsheet the present value of the costs and the cash inflows, calculate the NPV, the IRR, and the present value index (profitability index).

(b) Formulate this investment problem to be solved as a linear program (with continuous decision variables), and solve it.

(c) Now, set the problem up to solve with integer (0,1) variables.

(d) Assume that the company can borrow up to $100 at time 0. These funds can be used to finance investment beyond the expenditure limits set by the directors. Assume that for each $1 borrowed at time 0, $1.20 must be repaid (including interest) at time 2, and the NPV of $1 borrowed is −$0.02. Add borrowing as a decision variable to your problem and determine the new solution.

(e) Assume you can lend money at times 0 and 1. You can lend any positive amount—that is, there is no constraint on the amount you can lend. For each $1 lent there is an outflow of $1 at the time of the lending, and an inflow of $1.10 one period later. The NPV of lending is −0.015 for each $1 lent. Add lending as a decision variable to the model used in part (c) and solve it to see if the lending opportunity changes any of your choices.

For each of the different cases (b)–(e), explain your solution and explain how and why the solution changes as you proceed down the list.

2 AOC Corporation

The CEO of AOC Corporation has asked the managers of its various divisions to submit proposals for major capital expenditure programs. The proposals have been submitted and now the decision must be made as to which of the proposed projects should be adopted.

Summary data regarding the costs and expected cash flows of the proposed projects are shown in Table 16.2 (all values are in thousands). Eleven projects have been proposed, designated as projects A–K. The numbers in the body of the table are the estimated cash flows (in thousands of dollars), with negative numbers indicating cash outflows and positive numbers indicating inflows. Project K is contingent on the adoption of project H. Also, project E is simply project D delayed by one year, project I is project H delayed by one year, and project J is project C delayed by one year (treat each pair as two mutually exclusive projects). Discount all project cash flows at the firm's cost of capital of 15%.

All 11 projects meet AOC's acceptance criterion: they have positive NPVs when evaluated using the firm's cost of capital and their IRRs equal or exceed 15%. If there were no limitations on the firm's ability to accept projects, all 11 would be adopted. However, the board of directors has set limits on capital expenditures. They wish to preserve cash should more attractive opportunities present themselves in the future. The maximum amounts that can be invested are $500 for the current year (year 0), $300 for year 1, and $150 in year 2. Given the limits set by the directors, it must now be decided which of these 11 projects to undertake.

The following instructions walk you through how to set up both of these tasks. Begin by finding the portfolio that maximizes firm value.

Your Assignment

(a) In a spreadsheet, calculate each project's NPV and IRR.

(b) Model 1. Formulate this investment problem to be solved as a linear program, and then solve it. Be sure to constrain the portfolio decision variables to be binary and select Solver's "Nonnegative solutions" option.

(c) Model 2. Copy the Model 1 worksheet and make the following changes. Assume that the company can borrow up to $200 in years 0 and 1 (from years 1 and 2, respectively). These funds can be used to finance investment beyond the expenditure limits set by the directors. Assume that for each $1 borrowed, $1.046 must be repaid (after taxes) the following year. The NPV of $1 borrowed in year 0 and repaid in year 1 is $0.022; the NPV of

TABLE 16.2   Data for Problem 2

CASH FLOW ESTIMATES FOR PROPOSED INVESTMENT PROJECTS

TIME	PROJECT: A	B	C	D	E	F	G	H	I	J	K
0	-150	0	-135	-275	0	-285	-100	-420	0	0	-150
1	100	-250	-200	120	-275	165	-250	250	-420	-135	100
2	50	-185	-100	95	120	155	135	250	250	-200	75
3	50	150	200	75	95	75	140	100	250	-100	75
4	50	150	200	75	75	40	245	100	100	200	75
5	50	150	200	75	75	40	245	0	100	200	0
6	50	150	200	75	75	40	245	0	0	200	0
7	50	150	200	75	75	40	245	0	0	200	0
8	50	150	200	75	75	40	50	0	0	200	0
9	50	150	0	75	75	0	50	0	0	200	0
10	0	150	0	75	75	0	-900	0	0	0	0
11	0	0	0	0	75	0	0	0	0	0	0

$1 borrowed in year 1 and repaid in year 2 is $0.021.* Add these borrowing decision variables and parameters to your problem and have Solver find the new solution.

(d) Model 3. Copy the Model 2 worksheet and make the following changes. You may observe slack cash in the capital constraints for years 0 and 1. Assume that you can lend this slack cash to years 1 and 2, respectively. Each $1 lent constrains the capital budget in the year lent by $1 and relaxes the capital budget in the year received by $1.02. The NPV of lending from year 0 to year 1 is –$0.047 for each $1 lent; from year 1 to year 2 it is –0.044 for each $1 lent. Add these lending decision variables and parameters to your model and solve it.

- Does the addition lending change your portfolio choices for this model?

(e) Tabulate the NPVs for these three scenarios and report them.

- Why did the addition of borrowing and/or lending increase the NPV of your optimal portfolio of projects? (HINT: track the budget slack in each year.)

---

* The assumption underlying these values, and those for lending are a tax rate of 34%, a borrowing (and discount rate of 7%), and a T-bill rate (earned on money lent) of 3%.

CHAPTER 17

Planning and Managing Working Capital with LP

17.1 OPTIMIZING WORKING CAPITAL DECISIONS

One of the tasks of a financial manager is to develop and execute plans for managing working capital. The management of working capital involves issues such as deciding what minimum amount of cash should be maintained, how to invest excess cash, which marketable securities should be purchased, which should be sold, how receivables and payables should be managed, how much funds should be borrowed and in what maturity, and when outstanding debt should be repaid. The planning and management of the current accounts involves a plethora of problems with a myriad of potential solutions. Various approaches have been offered for the handling of individual problems such as cash management. The current accounts are interrelated, however, and the separate management of each account as an independent problem may be suboptimal. What is necessary is a method for considering the numerous interrelationships and making decisions that would be consistent with the objective of maximizing the value of the firm.

LP is well suited to handling some of the problems and interdependencies of current assets and liabilities. Numerous LP models of the current accounts have been presented; two of the earliest and best known are those by Robichek, Teichroew, and Jones (1965), and Orgler (1970).

Robichek, Teichroew, and Jones (1965) developed an LP model to analyze short-term financing decisions. The problem they addressed was how to fund the seasonal demands for funds with different short- and intermediate-term funding sources over a short-term planning horizon. The sources of funds they considered included (1) a line of credit; (2) the pledging of account receivable (AR); (3) delays in payments on trade account payable (AP); and (4) a term loan. In addition, excess cash could be invested in short-term marketable securities. The decision variables were the amounts of each of these funding sources to be used in each of the periods of the planning horizon, and the objective was the minimization of the interest cost over the planning horizon. The planning horizon for their example was 12 months. Their model would generate recommended financing decisions for each of the 12 months in the planning horizon. The general structure of their model can be summarized as follows:

Objective:	**Minimize: Total Net Interest Cost over the Horizon**
Constraints:	
Line of Credit:	
Total borrowing on line of credit	≤ Credit limit
cash balance	≥ Required compensating balance
Payments on line	≤ Balance on line of credit
Pledging of AR:	
Receivables pledged	≤ Lender limit on receivables
Repayment on receivable loans	≤ Amount borrowed
Stretching (Delayed Payment) of AP:	
Payables stretched	≤ (X%) Total payables
Term Loan:	
Amount borrowed	≤ Lender limit on loan
Total borrowing from all sources	≤ Policy limit on borrowing
Cash Balance:	
Cash balance	≥ Desired cash balance

Cash Flow:

Cash balance
$$= \text{Beginning cash} +$$
$$\text{Cash inflow} - \text{Cash outflow}$$

Robichek, Teichroew, and Jones (1965) demonstrated their model with a 12-month example. It had 73 decision variables and 134 constraints. It is interesting to note that with the technology of the time (1964), using a Fortran program on an IBM 7090 mainframe computer, it took about 15 minutes to compile the program and 5 minutes to solve it. Solving it now on a PC with Solver would seem almost instantaneous.

Yair Orgler (1970) developed a LP model for working capital decisions that considered cash balance targets, marketable securities, AP, and a line of revolving credit. His model of working capital decisions had the objective of maximizing the net earnings over a finite horizon. The maximization of net earnings (revenue minus cost) over the planning horizon has the effect of maximizing the future value of the NWC accounts, evaluated at the end of the planning horizon. This has the same decision result as the maximization of present value, but it is an easier formulation in the working capital context. The decision variables include the amounts to be paid on trade AP, the amounts to be borrowed and repaid on a line of credit, purchases and sales of marketable securities, and the amount of cash to keep on hand. The constraints are similar to Robichek, Teichroew, and Jones:

Payments on AP	≤ Payable balances
Borrowing on line of credit	≤ Credit limit
Securities sold	≤ Securities held
Cash balance	≥ Desired cash balance
Cash balance	= Beginning cash +
	Cash inflow − Cash outflow

One of the unique features of Orgler's approach is the use of unequal time periods. In the short-term planning context, the size of the problem can quickly get out of hand because of the number of periods and the number of different decision variables. For example, we may need to plan for daily or weekly transactions in securities and other short-term accounts, and if we have a six-month planning horizon, we can end up with an overwhelming number of decision variables. To make the problem more manageable, Orgler divides the planning horizon into periods of unequal length. For example, a six-month horizon is divided into six periods, with the first two periods of one day each, the third period lasting 10 days, followed by periods of 20, 60, and 90 days. The advantage of this unequal period planning horizon is that it is unnecessary to make decisions regarding daily actions, say, four months hence, but it is necessary to worry about tomorrow and the next day. The decision problem is solved to generate decision recommendations for the next two days, and tentative decisions for the longer periods after that. Then the problem is solved again as time passes and market conditions change. In this way it is generating recommendations every couple of days for current decisions. The recommendations for periods further in the future are tentative and are revised as those periods approach the present. This unequal period approach decreases the unnecessary detail of the plan for periods further in the future, while allowing more detailed consideration of the near term.

Even with the efficiency of unequal periods, the problem can easily be quite large. In the small six-month example presented by Orgler, there were 69 decision variables and 25 constraints. As he notes, "a cash management model for a large corporation will include, approximately, 300 to 500 constraints and 2,000 to 4,000 variables" (Orgler, 1970, p. 101). Although Orgler's model is considered smaller than more realistic applications, it is still sufficiently large that the reader quickly gets lost in the detail. We look at a considerably smaller example so that we can understand the principles and structure without being overwhelmed by detail.

17.2 WORKING CAPITAL DECISIONS FOR THE STILIKO COMPANY

The manager of the working capital accounts of the Stiliko Corporation faces the task of planning the firm's short-term security investments, short-term borrowing, and managing the cash account. Over the next four weeks the firm's operating cash flows will alternate between inflows and outflows. The cash flows expected in each of the next four weeks are as follows:

Stiliko Corporation
Schedule of Cash Flows for the Next Four Weeks

Week	1	2	3	4
Cash flow	−3000	+6000	−4000	+10,000

The decisions available to help manage the working capital accounts include purchasing and selling marketable securities and borrowing on a line of credit.

17.2.1 Marketable Securities

At the start of the planning period, the firm has a portfolio of marketable securities that will mature according to the following schedule:

Schedule of Maturing Marketable Securities				
Week	1	2	3	4
Maturing securities	1000	0	500	2000

These securities were purchased at a discount, and the maturing amounts are the face values of the securities, representing the amount of cash that will be received. New securities can be purchased at a discount from face value, with the discount based on an annual yield to maturity of 10%, or a weekly rate of 0.192%. For purchasing any new securities, the policy will be to purchase only securities that will mature in two weeks. None of the securities will be sold prior to maturity.

17.2.2 Borrowing

The company has a bank line of credit that will allow it to increase its short-term borrowing up to a balance of $5000 at an annual interest rate of 12%, or 0.231% per week. Interest will accrue on the debt at 0.231% per week, with principal and accrued interest repaid after the end of the planning horizon.

17.2.3 Cash Balance

The current cash balance is $2000, and the firm would like to carry a minimum cash balance of $3000.

17.2.4 Time

Assume that all cash flows, security transactions, and borrowing occur at the end of each week. Thus, the first exogenous cash flow (an outflow of $3000) will occur on Friday of the first week, and securities scheduled to mature will provide funds the same day.

17.2.5 Objective

The objective of the decisions is to maximize the value of NWC at the end of the four-week planning horizon. Value of NWC is defined as assets minus liabilities. In this model the assets consist of cash and marketable securities, and the sole liability is the debt balance (line of credit). Thus, the objective function is

$$\text{Max } W = \text{Cash}_4 + \text{Marketable Securities}_4 - \text{Debt Balance}_4.$$

At the horizon date, Marketable Securities$_4$ is the value of securities held at the end of the fourth week that mature in a subsequent period. Debt balance$_4$ is the balance, including accrued interest, at the end of the fourth week. The terminal value of NWC will include all earnings on investments and all costs of borrowing, so it takes account of the time value of money. Consequently, the maximization of terminal value is identical with the maximization of present value.

17.2.6 Decision Variables

The model will have 12 decision variables: four security purchase variables: SP_{13}, SP_{24}, SP_{35}, and SP_{46}, four borrowing variables: B_1, B_2, B_3, and B_4, and four cash balance variables: C_1, C_2, C_3, and C_4. These decision variables are defined as follows.

17.2.6.1 Security Purchases

SP_{tj} denotes purchases at the end of week t that mature in week j, expressed as the face value of the securities. With the restriction that we will consider purchasing only securities that mature in two weeks, the variables are

$$SP_{13}, SP_{24}, SP_{35}, \text{ and } SP_{46}.$$

Let SP_{0j} denote the securities held at the beginning of the planning period. With the initial endowment of securities, we have

$$SP_{01} = 1000, SP_{02} = 0, SP_{03} = 500, \text{ and } SP_{04} = 2000.$$

17.2.6.2 Borrowing

B_t denotes the principal amount borrowed at the end of week t. The borrowing variables are

$$B_1, B_2, B_3, \text{ and } B_4.$$

17.2.6.3 Cash Balance

The cash balance is the result of our other decisions, and we want to maintain a minimum cash balance of at least $3000. Since the balance itself is a result, it is not really a decision variable. Nevertheless, for expressing our model let C_t represent the cash balance at the end of week t, with $C_0 = \$2000$ being the initial cash balance with which we are endowed.

17.2.7 Constraints

We will have the following types of constraints that limit our decisions:

Borrowing:	Debt Balance$_t$ ≤ Borrowing Limit$_t$
Security purchases:	Total Security Purchases$_t$ ≤ Funds Available$_t$
Cash balance:	Cash Balance$_t$ ≥ Minimum cash
Cash balance and flow:	Cash Balance$_t$ = Cash Balance$_{t-1}$ + Cash Inflow$_t$ − Cash Outflow$_t$

Now let us fill in the complete model so we can find a solution. We will explain the various coefficients as we proceed.

Exhibit 17.1 shows the completed model set up as a LP tableau. The decision variables are listed across the top. The objective function is shown as row 0, and the constraints are shown as rows 1–12.

17.2.8 Objective Function

The objective is to maximize the value of NWC at the end of the fourth week. NWC consists of cash and securities, less the debt balance. Cash at the end of the fourth week is C_4. At the end of the fourth week, the security portfolio consists of the securities that mature in later periods, which would be the securities purchased in week 3 to mature in week 5, SP_{35}, and the securities purchased in week 4 to mature in week 6, SP_{46}. There are no other securities at that time because we are only considering securities that mature in two weeks. As of the end of week 4, the value of these securities is the present value of the maturity payment (face value) to be received in one and two weeks, respectively:

$$0.99808SP_{35} + 0.99617SP_{46}.$$

These coefficients are simply the present value factors $1/(1 + r)^t$ for $t = 1, 2$, where the discount rate is the weekly rate $r = 0.192\%$, based on the annual rate of 10%. At the end of week 4, securities purchased in week 3 to mature in week 5, SP_{35}, now have just one week to maturity, so their value is $1/1.00192 = 0.99808$ per dollar of face value. Similarly, the securities purchased in week 4 to mature in week 6 have two weeks to mature and their value is $1/(1.00192)^2 = 0.99617$ per dollar of face value.

The debt balance at the end of the fourth week includes principal and accrued interest. With no debt repaid during the planning horizon, the principal and interest is

$$1.00694B_1 + 1.00463B_2 + 1.00231B_3 + B_4.$$

	C_1	C_2	C_3	C_4	SP_{13}	SP_{24}	SP_{35}	SP_{46}	B_1	B_2	B_3	B_4		
0	C_1	C_2	C_3	C_4			$.99808SP_{35}$	$.99617SP_{46}$	$-1.00694B_1$	$-1.00463B_2$	$-1.00231B_3$	$-B_4$		
1	C_1												$\geq$	3000
2		C_2											$\geq$	3000
3			C_3										$\geq$	3000
4				C_4									$\geq$	3000
5									B_1				$\leq$	5000
6									B_1	$+B_2$			$\leq$	5000
7									B_1	$+B_2$	$+B_3$		$\leq$	5000
8									B_1	$+B_2$	$+B_3$	$+B_4$	$\leq$	5000
9	C_1				$+.99617SP_{13}$				$-B_1$				$=$	0
10	$-C_1$	C_2				$+.99617SP_{24}$				$-B_2$			$=$	6000
11		$-C_2$	C_3				$+.99617SP_{35}$				$-B_3$		$=$	−3500
12			$-C_3$	C_4				$+.99617SP_{46}$				$-B_4$	$=$	12000

EXHIBIT 17.1 Stiliko Corporation. Linear programming model for working capital management. Table of objective function, decision variables and constraints

At a nominal annual interest rate of 12%, the weekly rate is $r_d = 0.231\%$. For each dollar borrowed in week 1, the amount owed at the end of week 4 is $(1 + r_d)^3 = 1.00231^3 = 1.00694$. The coefficients for borrowing in weeks 2 and 3 are calculated similarly with two and one weeks of compounding, respectively. The coefficient for B_4 is 1 because no interest has yet accrued.

Combining terms, the value of the NWC at the end of the fourth week is

$$C_4 + 0.99808SP_{35} + 0.99617SP_{46} - 1.00694B_1 - 1.00463B_2 - 1.00231B_3 - B_4.$$

This is the objective function that is to be maximized. Note that the other decision variables (cash and security purchases) for weeks 1, 2, and 3 do not appear in the objective function because their effects are reflected in the week 4 values of cash, securities, and borrowing. For example, the cash balance in week 4 derives from the cash in week 3 plus the cash flows resulting from security transactions and borrowing during the week. Similarly, the security portfolio held at the end of week 4 is the result of purchases in earlier weeks.

17.2.9 Constraints

Lines 1–4 are the cash balance constraints, $C_t \geq 3000$, that require that the cash balance at the end of each week be at least $3000. Lines 5–8 are the constraints on borrowing that limit the total principal amount borrowed to be less than $5000 in each week. In week 1 the total borrowed is B_1, but with none of the debt repaid, each week's new borrowing adds to the principal balance so that the balance at the end of week 4 will be $B_1 + B_2 + B_3 + B_4$. Lines 9–12 are the cash flow equations that balance the model. At the end of a week, the cash balance is calculated as

$$\text{Cash}_t = \text{Cash}_{t-1} + \text{Securities Matured}_t - \text{Securities Purchased}_t$$
$$+ \text{Borrowing}_t + \text{Exogenous Cash Flow}_t.$$

Expressed in terms of our notation, at the end of the first week the cash balance is

$$C_1 = C_0 + SP_{01} - 0.99617SP_{13} + B_1 + (-3000),$$

where the -3000 is the exogenous cash outflow of week 1. With the beginning cash balance, $C_0 = 2000$, and the initial endowment of securities that mature in week 1, $SP_{01} = 1000$, this equation becomes

$$C_1 = 2000 + 1000 - 0.99617SP_{13} + B_1 - 3000.$$

Putting the decision variables to the left and the constant terms on the right side, we have

$$C_1 + 0.99617SP_{13} - B_1 = 2000 + 1000 - 3000,$$

or

$$C_1 + 0.99617SP_{13} - B_1 = 0,$$

which is the way it is shown in line 9 of the exhibit. In week 2 the cash balance is

$$C_2 = C_1 + SP_{02} - 0.99617SP_{24} + B_2 + 6000.$$

With the initial security endowment of $SP_{02} = 0$, this is shown in line 10 as

$$-C_1 + C_2 + 0.99617SP_{24} - B_2 = 6000.$$

The remaining cash flow equations are similarly constructed.

Exhibit 17.2 displays an Excel® spreadsheet set up in LP format so that it can be solved with Solver. The lines are numbered in column B to correspond to the equation numbers in Exhibit 17.1. The decision variables are labeled in spreadsheet row 21, with the actual decision variables in cells D22:O22. The numbers in the LP tableau are the coefficients of the decision variables. The numbers in cells D26:O26 are the coefficients of the decision variables that appear in the objective function, and the value of the objective function is calculated in cell C26 as =SUMPRODUCT (D22:O22,D26:O26). The constraint section shows just the coefficients. For example, the "1" in cell D31 is the coefficient of decision variable C_1 in constraint #1, which is $1 \cdot C_1 \geq 3000$, with the minimum cash balance of $3000 shown in cell Q31. The product of the coefficient (1 in cell D31) and the decision variable C_1 (cell D22) is calculated in cell P31 as =D31 × D22, in the column labeled "Resource Used."

To solve the problem as a LP, we use Tools/Solver to call up the Solver dialog box as is shown as the superimposed image in Exhibit 17.3. In the Solver dialog box, we have designated cell C26 as the objective function (Target Cell) to be maximized, and cells D22:O22 as the decision variables (Changing Cells). The constraints are shown in the "Subject to the Constraints" section. Having specified the objective function, decision variables, and constraints in the dialog box, we click "Options" and specify

Stilko Corporation
Working Capital Decision Model
Linear Programming

Input Data: Assumptions

	C0				
Initial Cash Balance	2000				

Exogenous Cash Flows

Week	1	2	3	4	
ECF	-3000	6000	-4000	10000	

Security Portfolio: Maturing Securities

Week	1	2	3	4
SP 0 t	1000	0	500	2000

Annual Borrow Rate	12.00%
Weekly	0.2308%
Annual Lending Rate	10.00%
Weekly	0.1923%

Policy Constraints

Target Cash	3000
Maximum Debt	5000

Decision Variables

Cash Balances

	C1	C2	C3	C4	Security Purchases					Borrowing			
					SP13	SP24	SP35	SP46	B1	B2	B3	B4
	0.00	0.80	0.00	0.00	0.00	0.00	0.00	0.00	0.00	0.00	0.00	0.00

Objective Function

Maximize													
0	0.00				1.00000		0.99808	0.99616	-1.00694	-1.00462	-1.00231	-1.00000

Constraints

													Resource Used	Resource Available	
													0	3000	Minimum
													0	3000	Cash
													0	3000	
													0	3000	

Cash Balance

1		1											0	5000	Borrowing
2			1						1				0	5000	
3				1					1	1			0	5000	
4					1				1	1	1		1		

Borrowing

5										1			
6											1		
7												1	
8													1

Cash Flow

9	CF1	1				0.996165		0.996165	0.996165	-1				0	0	Cash
10	CF2	-1	1				0.996165				-1			0	6000	Flow
11	CF3		-1	1				0.996165				-1		0	-3500	
12	CF4			-1	1								-1	0	12000	

EXHIBIT 17.2 Excel spreadsheet in linear programming format.

Stilko Corporation
Working Capital Decision Model
Linear Programming

Input Data: Assumptions

	C0
Initial Cash Balance	2000

Exogenous Cash Flows

Week	1	2	3	4
ECF	-3000	6000	-4000	10000

Security Portfolio: Maturing Securities

Week	1	2	3	4
SP 0.1	1000	500	2000	

Annual Borrow Rate	12.00%
Weekly	0.2308%
Annual Lending Rate	10.00%
Weekly	0.1923%
Policy Constraints	
Target Cash	3000
Maximum Debt	5000

Decision Variables

Cash Balances

C1	C2	C3	C4
3,000.00	9,000.00	3,000.00	15,000.00

Security Purchases

SP13	SP 24	SP35	SP 46
0.00	0.00	2,509.82	0.00

Borrowing

B1	B2	B3	B4
3,000.00	0.00	0.00	0.00

Objective Function
Maximize
14,930.30

Constraints

Cash Balance

	Resource Used	Resource Available	
0	100000	100000	
1	3000	3000	Minimum
2	9000	5000	Cash
3	3000	3000	
4	15000	3000	

Borrowing

5	3000	6000	Borrowing
6	3000	5000	
7	3000	5000	
8	3000	5000	

Cash Flow

9 (CF1)	0	0	Cash
10 (CF2)	6000	6000	Flow
11 (CF3)	-3500	-3500	
12 (CF4)	12000	12000	

Solver Parameters

Set Target Cell: $C$26
Equal To: ⦿ Max ○ Min ○ Value of: 0
By Changing Cells:
$D$22:$Q$22

Subject to the Constraints:
$P$31:$P$34 >= $Q$31:$Q$34
$P$36:$P$39 <= $Q$36:$Q$39
$P$41:$P$44 = $Q$41:$Q$44

[Guess]
[Add] [Change] [Delete]
[Solve] [Close] [Options] [Reset All] [Help]

EXHIBIT 17.3   Solution to Exhibit 17.2 with superimposed Solver dialog box.

that we want to "Assume Non-Negative" in order to require the decision variables to be non-negative. When we click the Solve button, we obtain the solution shown in the body of the spreadsheet of Exhibit 17.3.

The solution calls for borrowing $3000 in period 1 so that we can fund the exogenous outflow and bring our cash balance up to the target. In period 3, we purchase $2509 of securities that will mature in period 5. This optimal plan will generate a net asset value at the end of period 4 of $14,484, which is the largest balance that is feasible given our resources, obligations, and policy constraints. This solution results in a cash balance that meets or exceeds the target of $3000 in each period and borrowing that never exceeds the limit.

It is important to make sure that the solution makes sense. As the size of a problem expands, it is increasingly difficult to sort through the results to see that everything balances and we are not getting nonsense results. Financial statements that summarize the results in a familiar format help us to check the results. Exhibit 17.4 is part of the spreadsheet showing cash flow, working capital balances, and security transactions, so we can see

Stiliko Corporation
Financial Statements

Cash Flow

	Week	1	2	3	4
Exogenous Cash Flow		-3,000	6,000	-4,000	10,000
Maturing Securities		1,000	0	500	2,000
Security Purchases		0	0	2,500	0
Amount Borrowed		3,000	0	0	0
Net Cash Inflow		1,000	6,000	-6,000	12,000
Beginning Cash		2,000	3,000	9,000	3,000
Ending Cash		3,000	9,000	3,000	15,000

Working Capital Balance
Net Asset Balance

		1	2	3	4
Cash		3,000	9,000	3,000	15,000
Marketable Securities @ market value		2,487	2,491	4,496	2,505
Debt Balance		3,000	3,007	3,014	3,021
Net Assets		2,487	8,484	4,482	14,484

Security Purchases

		To mature in week:						Securities Purchased @ Face	@ Market
	Week Purchased	1	2	3	4	5	6		
@ Face Value	0	1,000	0	500	2,000			3,500	3,480
	1			0				0	0
	2				0			0	0
	3					2,510		2,510	2,500
	4						0	0	0
Total Maturing Securities		1,000	0	500	2,000	2,510	0		

	Week	1	2	3	4
Security Portfolio @market value					
Maturing in 1 Period		0	499	1,996	2,505
Maturing in 2 Periods		498	1,992	2,500	0
Maturing in 3 Periods		1,989	0		
Market Value of Securities		2,487	2,491	4,496	2,505

EXHIBIT 17.4 Summary of solution for Stiliko problem.

the results of the recommended program. The top section is the cash flow showing each of the transactions that contributes to the cash balance at the end of the week. The NWC balance is shown in the second section, representing the working capital assets net of borrowing. The balance at the end of week 4 was the value that was maximized. Security transactions are shown as security purchases and the portfolio balance at the end of each week. The last line is the balance in the security portfolio, stated in terms of market value. These statements help us to see how the various transactions feed into the cash flow and the account balances, so we can be sure that the model is doing what we want it to do. As we will see in the larger model in the next section, without this kind of check, we could easily develop a model that gives us nonsense results without our knowing they are nonsense.

17.3 ELABORATIONS AND EXTENSIONS

The Stiliko example was simple so that we could see how to construct a working capital model without getting too bogged down with the detail of too many decision variables and too many planning periods. A more realistic model would usually be far more complex with more decision variables and planning periods. Now we look at a few elaborations that we might make that could help us handle some more complicated problems.

17.3.1 Debt Repayment

Our Stiliko example allowed us to borrow, but it did not provide for the possibility of repaying any of the debt during the planning period. We can easily add decision variables for repaying debt. If we add the option to repay debt prior to maturity, we have to modify the model by adding decision variables, modifying the objective function, and adding constraints. Assume that Stiliko can repay debt in weeks 2, 3, or 4 (we assume there is no debt outstanding at the beginning, so no repayment is possible in week 1). Let R_t be the amount of debt repaid in period t.

17.3.2 Objective

Given that our objective is to maximize the value of NWC (working capital assets—debt) at the end of the horizon, the repayment variables affect the objective directly by way of the debt balance at the end, and indirectly by affecting the asset balances. If we repay debt, we do not have as much debt outstanding at the end of the horizon, and by reducing principal during the planning period, we save interest expense we would otherwise have to pay. The interest saved will show up as increased asset balances (cash or marketable securities) at the end of the horizon. The general form of our objective function is the same as before:

$$\text{Max } W = \text{Cash}_4 + \text{Marketable Securities}_4 - \text{Debt Balance}_4.$$

The debt balance at the end of week 4 will now show debt repayments:

$$\text{Debt Balance}_4 = 1.00694B_1 + 1.00463B_2 + 1.00231B_3 + B_4$$
$$- 1.00463R_2 - 1.00231R_3 - R_4,$$

where the negative coefficients represent the principal and interest saved from the repayment date to the horizon. For example, debt repaid in week 2 saves two weeks' interest by the end of the fourth week. At a weekly interest rate of 0.231%, the principal and interest per dollar repaid is 1.00463. Adding these terms to the objective function, the expanded objective function is

$$\text{Max } W = C_4 + 0.99808SP_{35} + 0.99617SP_{46} - 1.00694B_1 - 1.00463B_2$$
$$- 1.00231B_3 - B_4 + 1.00463R_2 + 1.00231R_3 + R_4.$$

17.3.3 Constraints

Debt repayment will require new constraints so that the amount repaid does not exceed the outstanding debt, and the repayment variables will appear in the borrowing limit constraints as well as the cash flow constraints.

17.3.3.1 Repayment

The amount repaid in any period cannot exceed the outstanding debt:

$$R_2 \qquad\qquad \le B_1$$
$$R_2 + R_3 \qquad \le B_1 + B_2$$
$$R_2 + R_3 + R_4 \quad \le B_1 + B_2 + B_3.$$

Putting the decision variables on the left side of the inequality yields

$$-B_1 + R_2 \qquad\qquad\qquad \le 0$$
$$-B_1 - B_2 + R_2 + R_3 \qquad\qquad \le 0$$
$$-B_1 - B_2 - B_3 + R_2 + R_3 + R_4 \quad \le 0.$$

17.3.3.2 Borrowing Limit
The total debt outstanding cannot exceed $5000:

$$B_1 \leq 5000$$
$$B_1 + B_2 - R_2 \leq 5000$$
$$B_1 + B_2 + B_3 - R_2 - R_3 \leq 5000$$
$$B_1 + B_2 + B_3 - R_2 - R_3 - R_4 \leq 5000.$$

17.3.3.3 Cash Flow Constraint
The cash flow constraint requires that the ending cash equals the beginning cash plus net cash inflows. The repayment of debt needs to be added to these constraints by showing the repayment as an outflow. The general form of the cash flow constraint is

$$\text{Cash}_t = \text{Cash}_{t-1} + \text{Securities Matured}_t - \text{Securities Purchased}_t$$
$$+ \text{Borrowing}_t - \text{Repayment}_t + \text{Exogenous Cash Flow}_t.$$

For example, the period 2 constraint would now be

$$-C_1 + C_2 + 0.99617SP_{24} - B_2 + R_2 = 6000,$$

where we have transposed the decision variables to the left side, so R_2 appears with a positive sign. The other cash flow expressions would be similarly modified by adding the repayment variable. Once these additional repayment variables are added to the model, we can solve it to find the optimal pattern of borrowing and repayment over the planning horizon.

17.3.4 Another Source of Debt Financing: Accounts Payable (AP)
In the simple Stiliko problem, we considered only a single type of bank debt. We could easily expand the model to consider several types of loans with different terms and constraints. Most borrowing sources would be handled similarly to the borrowing in this model. However, one source of funds that is treated somewhat differently is trade credit. Trade credit is included in both the working capital models by Robichek, Teichroew, and Jones, and Orgler. In their models, they assume that trade credit is a spontaneous source of financing so that AP is related to the firm's production operation and is taken as exogenously given in the context of the working capital LP. Thus, in their models AP financing is not a decision variable. However, the pattern of payments is considered a decision variable to be included in the LP model. Let us see how we might handle AP in our model.

Assume that Stiliko purchases materials from its suppliers on credit terms that grant a 2% discount for payment in the next week after the purchase, and require that the account be paid in three weeks. (Our terms are more stringent than normal practice, so we can demonstrate the idea and still fit within the framework of our simple four-week model.) The schedule of purchases on credit is given below, where the weeks with negative numbers indicate weeks prior to the start of our planning period. The amount purchased is the credit purchase in that week, and the last two columns are the weeks when the bill must be paid to receive the discount, and the week when the total balance is due, respectively. For example, the 900 purchased in week 0 (the week prior to the start of the planning period) must be paid in full by the end of week 3, but if it is paid by the end of week 1, the 2% discount is taken.

Week of Purchase	Amount Purchased	Week to Receive Discount	Week to Pay Total
−2	200	−1	1
−1	800	0	2
0	900	1	3
1	1000	2	4
2	700	3	5
3	600	4	6
4	400	5	7

17.3.5 Decision Variables

Let P_{jt} represent the payment made in week t on a credit purchase made in week j. For example, $P_{-1\,2}$ is the payment made in week 2 on a purchase made in week −1, the week prior to the start of our planning period. Assume that none of the purchases prior to week 1 have been paid, nor have the discounts been taken. Our decisions start in week 1, so we cannot take discounts on purchases in week −2 or −1, but if we pay this week on purchases made in week 0, we can take the discount. There are 12 decision variables in our model, designated as

$$P_{-2\,1},$$
$$P_{-1\,1}, \quad P_{-1\,2},$$
$$P_{0\,1}, \quad P_{0\,2}, \quad P_{0\,3},$$
$$P_{1\,2}, \quad P_{1\,3}, \quad P_{1\,4},$$
$$P_{2\,3}, \quad P_{2\,4},$$
$$P_{3\,4}.$$

Within our four-week model, we are not dealing with payment decisions after period 4.

17.3.6 Objective

Our objective function is the NWC balance at the end of period 4, with the AP balance appearing with a negative sign:

$$\text{Max W} = \text{Cash}_4 + \text{Marketable Securities}_4 - \text{Debt Balance}_4 - \text{AP}_4.$$

By the end of week 4, we will have paid for all purchases made up through week 1. We may still owe on purchases made in weeks 2, 3, and 4. The amount still owed on purchases made in week 2 is

$$\text{Purchase}_2 - 1.0204P_{23} - P_{24}.$$

The amount owed on purchases in week 3 is

$$\text{Purchase}_3 - 1.0204P_{34}.$$

Of course, we have not yet made any payments on week 4 purchases, so we owe Purchase$_4$. The AP balance at the end of period 4 would be

$$\text{AP}_4 = \text{Purchase}_2 - 1.0204P_{23} - P_{24} + \text{Purchase}_3 - 1.0204P_{34} + \text{Purchase}_4.$$

With the purchases specified in the table above, this equation is

$$\begin{aligned} \text{AP}_4 &= 700 - 1.0204P_{23} - P_{24} + 600 - 1.0204P_{34} + 400 \\ &= 1700 - 1.0204P_{23} - P_{24} - 1.0204P_{34}. \end{aligned}$$

This equation is subtracted from the rest of the objective function to reflect the AP balance owed at the end of the fourth week. In the objective function, it is unnecessary to show the constant, $1700, so the terms subtracted from the rest of the objective function are just the decision variables with their coefficients:

$$-1.0204P_{23} - P_{24} - 1.0204P_{34}.$$

The coefficient 1.0204 accounts for the 2% discount for early payment to convert the amount paid to the face amount owed on the account. That is, suppose we owe $100 face amount. With the 2% discount if we pay within one week, we can pay off the $100 liability by paying $98. The face amount that we have paid off is

$$[1/(1 - R_2 0.02)] \times 98 = 1.0204 \times 98 = 100.$$

Since the decision variable, P_{jt}, is the dollar amount paid, for the early payments, we multiply it by the 1.0204 coefficient to specify the face amount of the account that is paid.

17.3.7 Constraints

We have two constraints specific to AP. First, we do not want to pay more than we owe; and second, we must pay the total amount due on each purchase by the end of the third week. In addition to these constraints, we need to include the payment variables in the cash flow constraints, with the payments shown as outflows. The A/P specific constraints would be as follows:

P_{-21}	$= 200$	(17.1a)
P_{-11}	≤ 800	(17.1b)
$P_{-11} + P_{-12}$	$= 800$	(17.1c)
$1.0204P_{01}$	≤ 900	(17.1d)
$1.0204P_{01} + P_{02}$	≤ 900	(17.1e)
$1.0204P_{01} + P_{02} + P_{03}$	$= 900$	(17.1f)
$1.0204P_{12}$	≤ 1000	(17.1g)
$1.0204P_{12} + P_{13}$	≤ 1000	(17.1h)
$1.0204P_{12} + P_{13} + P_{14}$	$= 1000$	(17.1i)
$1.0204P_{23}$	≤ 700	(17.1j)
$1.0204P_{23} + P_{24}$	≤ 700	(17.1k)
$1.0204P_{34}$	≤ 600	(17.1l)

Equation 17.1a requires that the payables balance of purchases made in week −2 be paid in full in week 1. Equation 17.1d says that payments in week 1 on purchases in week 0 cannot exceed the $900 owed. The coefficient 1.0204 converts the payment with the discount taken to the face amount owed. Equation 17.1f says that the total of payments in weeks 1, 2, and 3 on week 0 purchases must completely repay the amount owed. Similar interpretations apply to the other expressions. This formulation requires that the accounts be paid in full by the end of the normal payment period. A further extension suggested by Robichek, Teichroew, and Jones would allow stretching of AP beyond the normal period.

17.3.8 Selling Securities

Our four-week example allowed us to purchase securities and we held the securities until they matured. We did not consider the possibility of selling securities from our portfolio prior to maturity. When we add the opportunity to sell securities prior to maturity, we expand the range of methods for obtaining funds when we need them. To consider selling securities, we add decision variables representing the sale of securities, and we add constraints. The primary constraint that we have to add is a restriction that says that we cannot sell more securities than we own. Of course, we have to include security sales in the cash flow constraints, with the proceeds from the sale shown as inflows at the date of the sale. The security sales affect the objective function by decreasing the value of securities held at the horizon, and by either adding cash or decreasing the debt balances at the end if we borrow less or use the proceeds to repay debt.

17.3.9 Decision Variables

Let $SS_{t\,j}$ denote the sale of securities in week t that mature in week j, with the values expressed in terms of face value securities sold. In our original Stiliko problem we had an initial endowment of securities, $SP_{0\,1} = 1000$, $SP_{0\,3} = 500$, and $SP_{0\,4} = 2000$. We allowed the purchase of securities that matured in two weeks, so we had the four purchase variables $SP_{1\,3}$, SP_{24}, SP_{35}, and SP_{46}. Given the beginning portfolio and allowed purchases, prior to maturity we could sell securities that mature in periods 1, 3, 4, 5, and 6. Presumably, we could sell securities in any week, which would require 13 additional decision variables. So we do not get too complicated, we will decrease the number of new variables by considering only the possibility of selling securities in weeks 1 and 2. With this simplification, we have the following four decision variables: SS_{13}, SS_{23}, SS_{14}, and SS_{24}. In week 1 or 2 we cannot sell securities maturing in week 5 or 6 because we will not have acquired them yet.

17.3.10 Constraints

Assuming that we are not considering short sales, we cannot sell securities we do not own, so we need constraints that say

$$\text{Security Sales} \leq \text{Securities Held}.$$

These constraints would be formulated as follows:

Sales in week 1:

$$SS_{13} \leq 500$$
$$SS_{14} \leq 2000,$$

where the constant amounts on the right side are the initial endowments, $SP_{03} = 500$ and $SP_{04} = 2000$, respectively.

Sales in week 2:

$$SS_{23} \leq SP_{03} + SP_{13} - SS_{13},$$
$$SS_{24} \leq SP_{04} + SP_{14} - SS_{14}.$$

With the initial endowments shown as constants, and putting the decision variables on the left, the selling constraints for both periods would be written as

$$SS_{13} \qquad\qquad\qquad \leq 500$$
$$SS_{14} \qquad\qquad\qquad \leq 2000$$
$$SS_{13} + SS_{23} - SP_{13} \quad \leq 500$$
$$SS_{14} + SS_{24} - SP_{14} \quad \leq 2000.$$

We would show similar constraints for each period as we expanded the problem to consider more decision variables.

In the cash flow constraints, we would show the inflow from security sales. Since the variables are defined in terms of face value of the securities, the cash derived from sales would be the discounted value at the date of the sale. The weekly rate earned on securities is 0.192%, so the proceeds per dollar of face value of securities sold would be $0.99808, $0.99617, and $0.99426 for securities sold 1, 2, and 3 weeks prior to maturity. (These figures would be adjusted downward to account for transactions costs of selling.) The cash flow constraints for weeks 1 and 2 would be modified to show the cash inflow from security sales as

Week 1: $\quad C_1 + 0.99617SP_{13} - 0.99617SS_{13} - 0.99426SS_{14} - B_1 = 0,$

Week 2: $\quad -C_1 + C_2 + 0.99617SP_{24} - 0.99808SS_{23} - 0.00617SS_{24} - B_2 = 6000.$

Each subsequent cash flow constraint would also have to be modified to take account of the fact that if securities are sold in week 1 or 2, the securities that would otherwise have matured in week 3 or 4 would not be available to generate an inflow. For example, in week 3 the modified constraint would be

$$\text{Week 3:} \quad -C_2 + C_3 + 0.99617 SP_{35} + SS_{13} + SS_{23} - B_3 = -3500,$$

where the terms SS_{13} and SS_{23} reflect the fact the sales in periods 1 and 2 of securities that mature in week 3 will reduce the cash flow in period 3. Their coefficients are 1 because their value and cash flow are at face value in the week the securities mature.

17.3.11 Objective Function

The objective is

$$\text{Max } W = \text{Cash}_4 + \text{Marketable Securities}_4 - \text{Debt Balance}_4.$$

If securities are sold during the planning period, the direct effect is to reduce the securities held at the end of the horizon. Taking account of possible sales in week 1 or 2, the ending security balance is expressed as

$$\text{Marketable Securities}_4 = \sum_{t=0}^{4} 0.9981\, SP_{t,5} + \sum_{t=0}^{4} 0.9962\, SP_{t,6} - 0.9981\, SS_{1,5}$$
$$- 0.9981\, SS_{2,5} - 0.9962\, SS_{1,6} - 0.9962\, SS_{2,6}.$$

The summation terms reflect all purchases of securities maturing after period 4, including the initial endowments. The last four terms represent the sales in periods 1 and 2 of securities maturing beyond the horizon. However, in our particular problem, the only securities potentially held at the horizon would be those purchased in week 3 or 4, and the only sales we are considering are those in week 1 or 2. Consequently, the security sales would not appear directly in the objective function. The indirect effects could be to increase the cash balance or reduce the debt balance when sales proceeds are used to repay debt, or to decrease the need for borrowing. Furthermore, if the funds are reinvested, sales in week 1 or 2 could increase the ending security balance by way of purchases in week 3 or 4.

17.4 SUMMARY

This chapter is an introduction to the use of LP for solving problems of working capital management. We looked at a simple working capital problem to learn the principles. The problem we examined included decisions about cash balances, security purchases, and borrowing. We discussed how the model could be extended to handle debt repayment, management of AP, and security sales. The model would have to be expanded greatly to handle more realistic problems. But the problem we examined helps us to understand how to structure the larger problems.

PROBLEM

1 Orgler Corporation: Optimization of Working Capital Decisions

At the end of December 2010 the CEO of the Orgler Corporation has asked you to develop the best plan for meeting the firm's working capital needs for the next two months. You are to develop an optimization model that will enable you to manage the cash account and manage the firm's short-term investments and financing over the next four weeks.

The objective of this set of decisions is to maximize the net value of the working capital accounts at the end of the four-week planning horizon. Net value means: Short-term Assets – Short-term Liabilities, where accrued interest is included in the values of assets and liabilities.

You have gathered the following information that relates to this problem:

Over the next four weeks the firm's operating cash flows will alternate between inflows and outflows as the firm carries on its business. The following table shows these flows. Assume that the flows for each week occur all on one day at the end of the week. Positive numbers mean inflows, and negative numbers mean outflows.

Week	1	2	3	4
Cash flow	−8000	4000	−4500	+7000

The firm has a portfolio of short-term securities that will mature during the planning horizon. The face value of the securities that will mature at the end (Friday) of each of the four weeks is shown as follows:

Week	1	2	3	4	5
Securities maturing	1000	500	2000	3500	0

Period 5 means "after the end of the planning horizon." At the end of each of these weeks, these securities will mature for their face value. These are discount securities that have an annual yield of 8%. Their market values are calculated as the present value of the face value, discounted at the 8% annual rate. Assume each week has seven days.

If the firm has excess cash, it will be used either to repay debt or to invest in marketable securities. If marketable securities are purchased, they are purchased at the discounted price based on the 8% annual yield. You can purchase securities that will mature either one or two weeks after the purchase date. But no securities will be purchased to mature later than one week after the end of the four-week planning horizon. No securities can be sold prior to maturity.

The resources that are available to help the firm meet its objectives include its cash balance, its marketable securities, and its ability to borrow. The current cash balance is $5000. The marketable securities have already been explained. Its borrowing is based on a line of credit. The company can borrow up to $5000 at the end of any week. It can repay part or all of the debt at the end of any subsequent week. When debt is repaid, the repayment includes interest accrued on the debt. Over the four-week horizon, the firm does not have to repay any debt, nor does it have to pay interest. Of course, if it does not pay interest, the interest accrues. The interest rate on the debt is 9% per year. Ignore the effects of income taxes. The firm has a target desired cash balance of $3000.

Assignment

Set this problem up as an LP problem that can be solved with Solver. Solve the problem and explain the solution.

References

Ackoff, R. L. 1970. *A Concept of Corporate Planning.* New York: Wiley.

Albright, S. C. 2003. *Learning Statistics with StatTools: A Guide to Statistics Using Excel an Palisade's StatTools Software.* Newfield, NY: Palisade Corporation.

Altman, E. 1968. Financial Ratios, Discriminant Analysis and the Prediction of Corporate Bankruptcy. *Journal of Finance,* 23:4 (September), 589–609.

Altman, E. 1993. *Corporate Financial Distress and Bankruptcy,* 2nd Edition. New York, NY: Wiley.

Andersen, A. and A. Weiss. 1984. Forecasting: The Box–Jenkins Approach. In: *The Forecasting Accuracy of Major Time Series Methods,* S. Makridakis (ed.). New York: Wiley.

Arditti, F. 1973. The Weighted Average Cost of Capital: Some Questions on Its Definition, Interpretation and Use. *Journal of Finance,* 28:4 (September), 1001–1007.

Bell, F. W. and N. B. Murphy. 1968. Costs in Commercial Banking: A Quantitative Analysis of Bank Behavior and Its Relationships. Federal Reserve Bank of Boston Research Report No. 41, Boston, April.

Benninga, S. 2008. *Financial Modeling,* 3rd Edition. Cambridge, MA: MIT Press.

Berk, J. and P. DeMarzo. 2007. *Corporate Finance.* Boston, MA: Pearson/Addison-Wesley.

Bernhard, R. H. 1969. Mathematical Programming Models for Capital Budgeting—A Survey, Generalization, and Critique. *Journal of Financial and Quantitative Analysis,* 4:2 (June), 111–158.

Bierwag, G. 1987. *Duration Analysis: Managing Interest Rate Risk*. Cambridge, MA: Ballinger Publishing.

Black, F. and M. Scholes. 1973. The Pricing of Options and Corporate Liabilities. *Journal of Political Economy*, 81 (May/June), 637–659.

Blumenthal, P. 1983. *Financial Model Preparation*. New York: American Institute of Certified Public Accountants.

Boquist, J., G. Racette, and G. Schlarbaum. 1975. Duration and Risk Assessment for Bonds and Common Stocks. *Journal of Finance*, 30(December), 1360–1365.

Boudoukh, J., M. Richardson, and R. Whitelaw. 1995. Expect the Worst. *Risk*, 8, 100–101.

Box, G. and G. Jenkins. 1976. *Time Series Analysis: Forecasting and Control*. San Francisco: Holden-Day.

Bradley, M., G. Jarrell, and E. Kim. 1984. On the Existence of an Optimal Capital Structure: Theory and Evidence. *Journal of Finance*, 37, 857–878.

Brealey, R., S. Myers, and F. Allen. 2008. *Principles of Corporate Finance*, 9th Edition. New York: McGraw-Hill.

Brigham, E. and M. Ehrhardt. 2008. *Financial Management: Theory and Practice*, 12th Edition. Cincinnati, OH: South-Western Publishing.

Bruner, R. 2004. *Applied Mergers and Acquisitions*. Hoboken, NJ: Wiley.

Bryant, J. W., ed. 1982. *Financial Modeling in Corporate Management*. New York: Wiley.

Carleton, W. 1970. An Analytical Model for Long Range Financial Planning. *Journal of Finance*, 25:2 (May), 291–315.

Carleton, W. T., C. L. Dick, and D. Downes. 1973. Financial Policy Models: Theory and Practice. *Journal of Financial and Quantitative Analysis*, 8:5 (December), 691–709.

Carleton, W., G. Kendall, and S. Tandon. 1974. An Application of the Decomposition Principle to the Capital Budgeting Problem in a Decentralized Firm. *Journal of Finance*, 29:3 (June), 815–827.

Charnes, A. and W. Cooper. 1959. Chance Constrained Programming. *Management Science*, 6:1 (October), 73–79.

Charnes, A. A., W. W. Cooper, and M. H. Miller. 1959. Application of Linear Programming to Financial Budgeting and the Costing of Funds. *Journal of Business*, 32:1 (January), 20–46.

Charnes, A. and S. Thore. 1966. Planning for Liquidity in Financial Institutions: The Chance Constrained Method. *Journal of Finance*, 21(December), 649–674.

Chen, H. Y. 1967. Valuation under Uncertainty. *Journal of Financial and Quantitative Analysis*, 2(September), 313–325.

Clark, J., T. Hindelang, and R. Prichard. 1989. *Capital Budgeting: Planning and Control of Capital Expenditures*, 3rd Edition. Englewood Cliffs, NJ: Prentice Hall.

Cohen, K. and F. Hammer. 1966. *Analytical Methods in Banking*. Homewood, IL: Richard D. Irwin.

Cohen, K. and F. Hammer. 1967. Linear Programming and Optimal Bank Asset Management Decisions. *Journal of Finance*, 22(May), 147–165.

Cooper, I. and K. Nyborg. 2008. Tax Adjusted Discount Rates With Investor Taxes and Risky Debt. *Financial Management* 37(Summer), 365–379.

Copeland, T., T. Koller, and J. Murrin. 2000. *Valuation: Measuring and Managing the Value of Company*. New York: Wiley.

Copeland, T. E., J. F. Weston, and K. Shastri. 2005. *Financial Theory and Corporate Policy*, 4th Edition. Reading, MA: Addison-Wesley.

Cox, J., J. Ingersoll, and S. Ross. 1985. A Theory of the Term Structure of Interest Rates. *Econometrica*, 53, 385–407.

Cox, J. and S. Ross. 1976. The Valuation of Options for Alternative Stochastic Processes. *Journal of Financial Economics*, 3 (January–March), 145–166.

Cox, J., S. Ross, and M. Rubinstein. 1979. Option Pricing: A Simplified Approach. *Journal of Financial Economics*, 7 (October), 229–264.

Cox, J. and M. Rubinstein. 1985. *Options Markets*. Englewood Cliffs, NJ: Prentice Hall.

Cragg, J. and B. Malkiel. 1968. The Consensus and Accuracy of Some Predictions of the Growth of Corporate Earnings. *Journal of Finance*, 23(1), 19–27.

Damodaran, A. 2002. *Investment Valuation: Tools and Techniques for Determining the Value of Any Asset*, 2nd Edition. New York: Wiley.

Dantzig, G. 1963. *Linear Programming and Extensions*. Princeton, NJ: Princeton University Press.

Davis, B. E., G. J. Caccappolo, and M. A. Chaudry. 1973. An Econometric Planning Model for American Telephone and Telegraph Company. *Bell Journal of Economics and Management Science*, 4:1 (Spring), 29–56.

DeLurgio, S. A. 1998. *Forecasting Principles and Applications*. Boston, MA: Irwin/McGraw-Hill.

DePamphilis, D. 2005. *Mergers, Acquisitions, and Other Restructuring Activities: An Integrated Approach to Process, Tools, Cases and Solutions*, 3rd Edition. Burlington, MA: Elsevier Academic Press.

Doane, D. and L. Seward. 2006. *Applied Statistics in Business and Economics*. Irwin Professional Publications.

Donaldson, G. 1961. *Corporate Debt Capacity: A Study of Corporate Debt Policy and the Determination of Corporate Debt Capacity*. Cambridge, MA: Division of Research, Harvard School of Business Administration.

Donaldson, G. 1962. New Framework for Corporate Debt Policy. *Harvard Business Review*, 40(March–April), 117–131.

Dorfman, R., P. Samuelson, and R. Solow. 1985. *Linear Programming and Economic Analysis*. New York: McGraw-Hill.

Drucker, P. F. 1959. Long Range Planning: Challenge to Management Science. *Management Science*, 5, 238–249.

Dubofsky, D. and T. Miller. 2003. *Derivatives: Valuation and Risk Management*. New York: Oxford University Press.

Duesenberry, J., G. Fromm, L. Klein, and E. Kuh, eds. 1965. *The Brookings Quarterly Econometric Model of the United States*. Chicago: Rand McNally.

Duffie, D. and J. Pan. 1997. An overview of value at risk. *The Journal of Derivatives*, 4, 7–49.

Elton, E. J., M. J. Gruber, S. Brown, and W. Goetzman. 2006. *Modern Portfolio Theory and Investment Analysis*, 6th Edition. New York: Wiley.

Emery, G. 1984. Measuring Short Term Liquidity. *Journal of Cash Management*, 4(July/August), 25–32.

Eppen, G. and E. Fama. 1968. Solutions for Cash-Balance and Simple Dynamic-Portfolio Problems. *Journal of Business*, 41(January), 94–112.

Evans, M. 1969. *Macroeconomic Activity: Theory, Forecasting, and Control*. New York: Harper & Row.

Evans, J. and D. Olson. 1998. *Introduction to Simulation and Risk Analysis*. Upper Saddle River, NJ: Prentice Hall.

Fabozzi, F. J. 2004. *Bond Markets, Analysis, and Strategies*, 5th Edition. NJ: Pearson/Prentice Hall.

Fama, E. and M. Miller. 1972. *The Theory of Finance*. New York: Holt, Rinehart & Winston.

Fielitz, B. and T. Loeffler. 1979. A Linear Programming Model for Commercial Bank Liquidity Management. *Financial Management*, 8:3 (Autumn), 41–49.

Fogarty, D., J. Blackstone, and T. Hoffman. 1991. *Production and Inventory Management*, 2nd Edition. Cincinnati, OH: South-Western.

Foster, G. 1986. *Financial Statement Analysis*. Englewood Cliffs, NJ: Prentice Hall.

Frame, J. D. 2003. *Managing Risk in Organizations*. San Francisco, CA: Jossey-Bass.

Francis, J. C. 1983. Financial Planning and Forecasting Models: An Overview. *Journal of Economics and Business*, 35, 185–300.

Francis, J. C. and D. Rowell. 1978. A Simultaneous Equation Model of the Firm for Financial Analysis and Planning. *Financial Management*, 7:1 (Spring), 29–44.

Gentry, J., R. Vaidyanathan, and H. W. Lee. 1990. A Weighted Cash Conversion Cycle. *Financial Management*, 19(Spring), 90–99.

Gershefski, G. 1968. *The Development and Application of a Corporate Financial Model*. Oxford, OH: The Planning Executives Institute.

Gershefski, G. 1969. Building a Corporate Financial Model. *Harvard Business Review*, 47:4 (July/August), 61–72.

Gershefski, G. 1970. Corporate Models: The State of the Art. *Management Science* 16:6 (February), B303–B312.

Gitman, L. T. 2003. *Principles of Managerial Finance*, 10th Edition. Reading, MA: Addison-Wesley.

Goldberg, L. R., G. Miller, and J. Weinstein. 2007. Beyond Value at Risk: Forecasting Portfolio Loss at Multiple Horizons. MSCI Barra Working Paper.

Graham, J. 2000. How Big Are the Tax Benefits of Debt? *Journal of Finance*, 55, 1901–1941.

Granito, M. 1984. *Bond Portfolio Immunization*. Lexington, MA: Lexington Books.

Greene, W. H. 2000. *Econometric Analysis*. Upper Saddle River, NJ: Prentice Hall.

Grove, M. 1966. A Model of the Maturity Profile of the Balance Sheet. *Metroeconomica*, 18(April), 40–55.

Grove, M. 1974. On Duration and the Optimal Maturity Structure of the Balance Sheet. *Bell Journal of Economics and Management Science*, 5:2 (Autumn), 696–709.

Harke, J. E., D. Wichern, and A. G. Reitsch. 2004. *Business Forecasting*, 8th Edition, Upper Saddle River, NJ: Prentice Hall.

Hadley, G. 1962. *Linear Programming*. Reading, MA: Addison-Wesley.

Hadley, G. and T. Whiten. 1963. *Analysis of Inventory Systems*. Englewood Cliffs, NJ: Prentice Hall.

Haley, C. and L. Schall. 1979. *The Theory of Financial Decisions*, 2nd Edition. New York: McGraw-Hill.

Hamada, R. 1969. Portfolio Analysis, Market Equilibrium and Corporation Finance. *Journal of Finance*, 24(March), 13–31.

Hamada, R. 1972. The Effect of the Firm's Capital Structure on the Systematic Risk of Common Stocks. *Journal of Finance*, 27(May), 435–452.

Hanke, J. E. and A. G. Reitsch. 1998. *Business Forecasting*. Upper Saddle River, NJ: Prentice Hall.

Hanke, J. E., D. Wichern, and A. G. Reitsch. 2004. *Business Forecasting*, 8th Edition. Upper Saddle River, NJ: Prentice Hall.

Haugen, R. A. and N. L. Baker. 1996. Commonality in the Determinants of Expected Stock Returns. *Journal of Financial Economics*, 41, 401–439.

Haugen, R. and D. Wichern. 1974. The Elasticity of Financial Assets. *Journal of Finance*, (September), 1229–1240.

Haugen, R. and D. Wichern. 1975. The Intricate Relationship Between Financial Leverage and the Stability of Stock Prices. *Journal of Finance*, 30(December), 1283–1292.

Helfert, E. 1991. *Techniques of Financial Analysis*, 7th Edition. Homewood, IL: Richard D. Irwin.

Hendricks, D. 1996. Evaluation of Value-at-Risk Models Using Historical Data. *FRBNY Economic Policy Review*, 2 (April), 39–69.

Hertz, D. B. 1964. Risk Analysis in Capital Investment. *Harvard Business Review*, 42:1 (January–February), 95–106.

Higgins, R. C. 1981. Sustainable Growth Under Inflation. *Financial Management*, 10:4 (Autumn), 36–40.

Higgins, R. C. 1996. *Analysis for Financial Management*. Homewood, IL: Business One Irwin.

Higgins, R. C. 2001. *Analysis for Financial Management*, 6th Edition. Irwin/ McGraw-Hill.

Hill, G. and R. Fildes. 1984. The Accuracy of Extrapolation Methods: An Automatic Box–Jenkins Package SIFT. *Journal of Forecasting*, 3, 319–323.

Hillier, F. 1963. The Derivation of Probabilistic Information for the Evaluation of Risky Investments. *Management Science*, 9:3 (April), 443–457.

Hillier, F. and M. Hillier. 2003. *Introduction to Management Science: A Modeling and Case Studies Approach with Spreadsheets*, 2nd Edition. Boston, MA: McGraw-Hill.

Hitchner, J. R. 2003. *Financial Valuation: Applications and Models*. Hoboken, NJ: Wiley.

Ho, T. and S. B. Lee. 2004. *The Oxford Guide to Financial Modeling*. New York, NY: Oxford University Press.

Ho, T. and S. B. Lee. 2005. *Securities Valuation: Applications of Financial Modeling*. New York, NY: Oxford University Press.

Hooke, J. C. 1997. *M&A: A Practical Guide to Doing the Deal*. New York: Wiley.

Hull, J. C. 2009. *Options, Futures, & Other Derivatives*, 7th Edition. Upper Saddle River, NJ: Pearson/Prentice Hall.

Iriji, Y., F. K. Levy, and R. C. Lyon. 1963. A Linear Programming Model for Budgeting and Financial Planning. *Journal of Accounting Research*, 1:2 (Autumn), 198–212.

Jacob, N. L. and R. R. Pettit. 1988. *Investments*. Homewood, IL: Richard D. Irwin.

Kallberg, J. G. and K. Parkinson. 1996. *Corporate Liquidity: Management and Measurement*. Burr Ridge, IL: Irwin/McGraw-Hill.

Kaplan, P. D. 1998. Asset Allocation Models Using the Markowitz Approach. Ibbotson Associates Working Paper.

Keeley, R. H. and R. Westerfield. 1972. A Problem in Probability Distribution Techniques for Capital Budgeting, *Journal of Finance*, 27 (June), 703–709.

Kester, G. 1987. A Note on Solving the Balancing Problem. *Financial Management*, 16:1 (Spring), 52–54.

Law, A. and W. Kelton. 2000. *Simulation Modeling and Analysis*, 3rd Edition. Burr Ridge, IL: Irwin/McGraw-Hill.

Leibowitz, M. L. and S. Kogelman. 1994. *Franchise Value and the Price/Earnings Ratio*. Charlottesville, VA: The Research Foundation of the Institute of Chartered Financial Analysts.

Levine, D. M., M. Berenson, and D. Stephan. 1999. *Statistics for Managers Using Microsoft Excel*, 2nd Edition. Upper Saddle River, NJ: Prentice Hall.

Libert, G. 1984. The M-Competition with a Fully Automatic Box–Jenkins Procedure. *Journal of Forecasting*, 3, 325–328.

Linter, J. 1965. Security Prices, Risk, and Maximal Gains from Diversification. *Journal of Finance*, 20:4 (December), 587–616.

Levine, D. M., M. Berenson, and D. Stephan. 2007. *Statistics for Managers Using Microsoft Excel*, 5th Edition. Upper Saddle River, NJ: Prentice-Hall.

Lummer, S. L., M. W. Riepe, and L. Siegel. 1994. Taming Your Optimizer: A Guide Through the Pitfalls of Mean-Variance Optimization. In: *Global Asset Allocation: Techniques for Optimizing Portfolio Management*, J. Lederman and R. A. Klein (eds). New York: Wiley.

Makridakis, S. 1986. The Art and Science of Forecasting. *International Journal of Forecasting*, 2, 15–39.

Makridakis, S. and M. Hibon. 1997. ARMA Models and the Box–Jenkins Methodology. *Journal of Forecasting*, 16, 147–163.

Makridakis, S., S. Wheelwright, and V. McGee. 1983. *Forecasting: Methods and Applications*. New York: Wiley.

Maness, T. and J. Zietlow. 2002. *Short-Term Financial Managment*, 2nd Edition. Cincinnati, OH: South-Western.

Mao, J. C. T. 1968. Application of Linear Programming to Short Term Financing Decisions. *The Engineering Economist*, 13:4, 221–241.

Markowitz, H. 1952. Portfolio Selection. *Journal of Finance*, 7:1 (March), 77–91.

Markowitz, H. 1959. *Portfolio Selection: Efficient Diversification of Investment*. New York: Wiley.

Masson, D. and D. Wikoff. 1995. *Essentials of Cash Management*, 5th Edition. Bethesda, MD: Treasury Management Association.

Mathur, K. and D. Solow. 1994. *Management Science: The Art of Decision Making.* Englewood Cliffs, NJ: Prentice Hall.

Mayes, T. and T. Shank. 2007. *Financial Analysis with Microsoft Excel.* 4th ed. Mason, OH: Thomson, South-Western.

Merton, R. 1973. Theory of Rational Option Pricing. *Bell Journal of Economics and Management Science,* 4 (Spring), 141–183.

Merville, L. and L. Tavis. 1974. A Total Real Asset Planning System. *Journal of Financial and Quantitative Analysis,* 9 (January), 107–115.

Meyer, H. I. 1977. *Corporate Financial Planning Models.* New York: Wiley.

Michaud, R. O. 1989. The Markowitz Optimization Enigma: Is "Optimized" Optimal? *Financial Analysts Journal,* 45:1 (January–February), 31–42.

Michaud, R. O. and R. O. Michaud. 2008. *Efficient Asset Management: A Practical Guide to Stock Portfolio Optimization and Asset Allocation.* New York: Oxford University Press.

Miles, J. A. and J. R. Ezzell. 1980. The Weighted Average Cost of Capital, Perfect Capital Markets, and Project Life: A Clarification. *Journal of Financial and Quantitative Analysis,* XV(3)(September), 719–730.

Miller, M. 1977. Debt and Taxes. *Journal of Finance,* 32, 261–276.

Miller, M. H. and F. Modigliani. 1961. Dividend Policy, Growth, and the Valuation of Shares. *Journal of Business,* XXXIV(4)(October), 411–433.

Miller, M. and D. Orr. 1966. A Model of Demand for Money by Firms. *Quarterly Journal of Economics,* 80:3 (August), 413–435.

Modigliani, F. and M. Miller. 1958. The Cost of Capital, Corporation Finance, and the Theory of Investment. *American Economic Review,* 48(June), 261–297.

Modigliani, F. and M. Miller. 1963. Corporate Income Tax and the Cost of Capital. *American Economic Review,* 53 (June), 433–443.

Morris, J. R. 1975. An Application of the Decomposition Principle to Financial Decision Models. *Journal of Financial and Quantitative Analysis,* 10:1 (March), 37–65.

Morris, J. R. 1976a. On Corporate Debt Maturity Strategies. *Journal of Finance,* 31:1 (September), 29–37.

Morris, J. R. 1976b. A Model for Corporate Debt Maturity Decisions. *Journal of Financial and Quantitative Analysis,* 11:3 (September), 339–357.

Morris, J. 1982. Taxes, Bankruptcy Costs and the Existence of an Optimal Capital Structure. *Journal of Financial Research,* 5:3, 285–299.

Morris, J. R. 2004. Reconciling the Equity and Invested Capital Method of Valuation When the Capital Structure is Changing. *Business Valuation Review,* 23:1 (March), 36–46.

Morris, J. R. 2006. Growth in the Constant Growth Model. *Business Valuation Review,* 25:4 (Winter), 153–162.

Mossin, J. 1966. Equilibrium in a Capital Asset Market. *Econometrica,* 4 (October), 768–783.

Mun, J. 2006. *Real Option Analysis: Tools and Techniques for Valuing Stategic Investments and Decisions,* 2nd Edition. Hoboken, NJ: Wiley.

Myers, S. 1977. Determinants of Corporate Borrowing. *Journal of Financial Economics,* 5, 147–175.

Myers, S. 1984. The Capital Structure Puzzle. *Journal of Finance*, 39, 575–592.

Myers, S. C. 1974. Interactions of Corporate Financing and Investment Decisions— Implications for Capital Budgeting. *Journal of Finance*, 29:1 (March), 1–25.

Myers, S. and G. Pogue. 1974. A Programming Approach to Corporate Financial Management. *Journal of Finance*, 29 (May), 579–599.

Nantell, T. J. and C. R. Carlson. 1975. The Cost of Capital as a Weighted Average. *Journal of Finance*, 30:5 (December), 1343–1355.

Narayanan, M. P. and V. K. Nanda. 2004. *Finance for Strategic Decision Making: What Non-Financial Managers Need to Know*. San Francisco, CA: Jossey-Bass.

Naslund, B. 1966. A Model of Capital Budgeting under Risk. *Journal of Business*, 39:2 (April), 257–271.

Naylor, T. 1971. *Computer Simulation Experiments with Models of Economic Systems*. New York: Wiley.

Naylor, T. 1973. Corporate Simulation Models. *Simulation Today*, (August), 16.

Naylor, T. 1979. *Corporate Planning Models*. Reading, MA: Addison-Wesley.

Naylor, T. and H. Schauland. 1976. A Survey of Users of Corporate Planning Models. *Management Science*, 22:9 (May), 927–957.

Neter, J. 1996. *Applied Linear Regression Models*. Burr Ridge, IL: Richard D. Irwin.

Oliva, M.-A. and L. A. Rivera-Batiz. 1999. Mergers and Acquisitions, Bargaining and Synergy Traps. In: *Current Trends in Economics: Third International Meeting of the Society for the Advancement of Economic Theory*, A. Alkan, C. Aliprant, and N. Yannelis, eds, pp. 325–346. Berlin: Springer.

Orgler, Y. 1969. An Unequal Period Model for Cash Management Decisions. *Management Science*, 15 (October), B77–92.

Orgler, Y. 1970. *Cash Management: Methods and Models*. Belmont, CA: Wadsworth Publishing Co.

O'Donovan, T. 1983. *Short-Term Forecasting: An Introduction to the Box–Jenkins Approach*. Chichester, England: Wiley.

Palisade Corporation. 2007. *Guide to Using @Risk: Analysis and Simulation Add-In for Microsoft Excel*, Version 5. Newfield, NY: Palisade Corporation (December).

Pindyck, R. S. and D. L. Rubinfeld. 1998. *Econometric Models and Economic Forecasts*, 4th Edition. New York: McGraw-Hill.

Pogue, G. and R. Bussard. 1972. A Linear Programming Model for Short Term Financial Planning under Uncertainty. *Sloan Management Review*, 13 (Spring), 69–98.

Pratt, S. and R. Grabowski. 2008. *Cost of Capital: Applications and Example*, 3rd Edition. Hoboken, NJ: Wiley.

Pratt, S., R. Reilly, and R. Schweihs. 2000. *Valuing a Business: The Analysis and Appraisal of Closely Held Companies*, 4th Edition. New York: McGraw-Hill.

Ragsdale, C. T. 2001. *Spreadsheet Modeling and Decision Analysis: A Practical Introduction to Management Science*, 3rd Edition. Cincinnati, OH: South-Western College Publishing.

Reilly, F. and K. Brown. 2006. *Investment Analysis and Portfolio Management*, 8th Edition. Mason, OH: Thomson/South-Western.

Richards, V. and E. Laughlin. 1980. A Cash Conversion Cycle Approach to Liquidity Analysis. *Financial Management*, 9(Spring), 32–38.

Robechek, A. and S. C. Myers. 1966. Conceptual Problems in the Use of Risk-Adjusted Discount Rates. *Journal of Finance*, 21 (December), 727–730.

Robichek, A. A., D. Teichroew, and J. M. Jones. 1965. Optimal Short Term Financing Decision. *Management Science*, 12(September), 1–36.

Ross, S., R. Westerfield, and J. Jaffe. 2008. *Corporate Finance*, 8th Edition. New York, NY: McGraw-Hill.

Rubinstein, M. 2002. Markowitz's "Portfolio Selection": A Fifty-Year Retrospective. *Journal of Finance*, 57, 1041–1045.

Sartoris, W. and M. Spruill. 1974. Goal Programming and Working Capital Management. *Financial Management*, 3:1 (Spring), 67–74.

Saunders, A. 1999. *Credit Risk Management: New Approaches to Value at Risk and Other Paradigms*. New York: Wiley.

Schrieber, A. 1970. *Corporate Simulation Models*. Seattle: University of Washington Press.

Sengupta, C. 2004. *Financial Modeling Using Excel and VBA*. New York, NY: Wiley.

Sharpe, W. F. 1964. Capital Asset Prices: A Theory of Market Equilibrium under Conditions of Risk. *Journal of Finance*, 19:3 (September), 425–442.

Shefrin, H. 2007. *Behavioral Corporate Finance*. New York: McGraw-Hill.

Smith, K. 1979. *Guide to Working Capital Management*. New York: McGraw-Hill.

Standard and Poor's 2002. Research Insight (Compustat Database). Englewood, CO: Standard and Poor's.

Stowe, J., T. Robinson, J. Pinto, and D. McLeavey. 2007. *Equity Asset Valuation*. Hoboken, NJ: Wiley.

Strang, G. 2003. *Introduction to Linear Algebra*. Wellesley, MA: Wellesley-Cambridge Press.

Texter, P. A. and J. K. Ord. 1989. Forecasting Using Automatic Identification Procedures: A Comparative Analysis. *International Journal of Forecasting*, 5, 209–215.

Traenkle, J., E. Cox, and J. Bullard. 1975. *The Use of Financial Models in Business*. New York: Financial Executives Research Foundation.

Tyran, M. W. 1980. *Computerized Financial Forecasting and Performance Reporting*. Englewood Cliffs, NJ: Prentice Hall.

Vanderweide, J. and S. Maier. 1985. *Managing Corporate Liquidity: An Introduction to Working Capital Management*. New York: Wiley.

Vasicek, O. 1977. An Equilibrium Characterization of the Term Structure. *Journal of Financial Economics*, 5, 177–188.

Wagner, H. M. 1969. *Principles of Operations Research: With Applications to Managerial Decisions*. Englewood Cliffs, NJ: Prentice Hall.

Warren, J. and J. Shelton. 1971. A Simultaneous Equation Approach to Financial Planning. *Journal of Finance*, 26:5 (December), 1123–1142.

Weingartner, H. M. 1966a. Capital Budgeting of Interrelated Projects: Survey and Synthesis. *Management Science*, 12:7 (March), 485–516.

Weingartner, H. M. 1966b. Criteria for Programming Investment Project Selection. *Journal of Industrial Economics*, 15:1 (November), 65–76.

Weingartner, H. M. 1967. *Mathematical Programming and the Analysis of Capital Budgeting Problems*. Chicago: Markham Publishing Co.

Weingartner, H. M. 1977. Capital Rationing: Authors in Search of a Plot. *Journal of Finance*, 32:5 (December), 1403–1431.

Whitbeck, V. S. and M. Kisor. 1963. A New Tool in Investment Decision Making. *Financial Analysts Journal*, 19:3 (May–June), 55–62.

White, G., A. Sondhi, and D. Fried. 2002. *The Analysis and Use of Financial Statements*, 3rd Edition. Upper Saddle River, NJ: Prentice Hall.

Winston, W. 1996. *Simulation Modeling Using @Risk*. Belmont, CA: Duxbury Press.

Winston, W. 2000. *Financial Models Using Simulation and Optimization*. Newfield, NY: Palisade Corporation.

Winston, W. 2001. *Financial Models Using Simulation and Optimization, II*. Newfield, NY: Palisade Corporation.

Zangari, P. 1996. An Improved Methodology for Measuring VaR. JP Morgan RiskMetrics Monitor. Second Quarter, 7–25.

Index